AF510708

LINEAR COLLIDER PHYSICS IN THE NEW MILLENNIUM

ADVANCED SERIES ON DIRECTIONS IN HIGH ENERGY PHYSICS

Published

Vol. 1 – High Energy Electron–Positron Physics (*eds. A. Ali and P. Soding*)
Vol. 2 – Hadronic Multiparticle Production (*ed. P. Carruthers*)
Vol. 3 – CP Violation (*ed. C. Jarlskog*)
Vol. 4 – Proton–Antiproton Collider Physics (*eds. G. Altarelli and L. Di Lella*)
Vol. 5 – Perturbative QCD (*ed. A. Mueller*)
Vol. 6 – Quark–Gluon Plasma (*ed. R. C. Hwa*)
Vol. 7 – Quantum Electrodynamics (*ed. T. Kinoshita*)
Vol. 9 – Instrumentation in High Energy Physics (*ed. F. Sauli*)
Vol. 10 – Heavy Flavours (*eds. A. J. Buras and M. Lindner*)
Vol. 11 – Quantum Fields on the Computer (*ed. M. Creutz*)
Vol. 12 – Advances of Accelerator Physics and Technologies (*ed. H. Schopper*)
Vol. 13 – Perspectives on Higgs Physics (*ed. G. L. Kane*)
Vol. 14 – Precision Tests of the Standard Electroweak Model (*ed. P. Langacker*)
Vol. 15 – Heavy Flavours II (*eds. A. J. Buras and M. Lindner*)
Vol. 16 – Electroweak Symmetry Breaking and New Physics at the TeV Scale (*eds. T. L. Barklow, S. Dawson, H. E. Haber and J. L. Siegrist*)
Vol. 17 – Perspectives on Higgs Physics II (*ed. G. L. Kane*)
Vol. 18 – Perspectives on Supersymmetry (*ed. G. L. Kane*)

Forthcoming

Vol. 8 – Standard Model, Hadron Phenomenology and Weak Decays on the Lattice (*ed. G. Martinelli*)

Advanced Series on
Directions in High Energy Physics — Vol. 19

LINEAR COLLIDER PHYSICS IN THE NEW MILLENNIUM

Editors

Keisuke Fujii
KEK, Japan

David J Miller
UCL, London, UK

Amarjit Soni
Brookhaven National Lab, USA

World Scientific

NEW JERSEY · LONDON · SINGAPORE · BEIJING · SHANGHAI · HONG KONG · TAIPEI · CHENNAI

Published by

World Scientific Publishing Co. Pte. Ltd.
5 Toh Tuck Link, Singapore 596224
USA office: 27 Warren Street, Suite 401-402, Hackensack, NJ 07601
UK office: 57 Shelton Street, Covent Garden, London WC2H 9HE

British Library Cataloguing-in-Publication Data
A catalogue record for this book is available from the British Library.

LINEAR COLLIDER PHYSICS IN THE NEW MILLENNIUM

Copyright © 2005 by World Scientific Publishing Co. Pte. Ltd.

All rights reserved. This book, or parts thereof, may not be reproduced in any form or by any means, electronic or mechanical, including photocopying, recording or any information storage and retrieval system now known or to be invented, without written permission from the Publisher.

For photocopying of material in this volume, please pay a copying fee through the Copyright Clearance Center, Inc., 222 Rosewood Drive, Danvers, MA 01923, USA. In this case permission to photocopy is not required from the publisher.

ISBN 981-238-908-3

Printed by FuIsland Offset Printing (S) Pte Ltd, Singapore

Preface

The results from a linear electron positron collider, working in the centre of mass energy range from about 200 to 500 GeV and onwards to the region of 1 TeV, will be both strategically important and rich in detail. An international consensus [1] continues to grow — that this machine is needed, alongside the large Hadron Collider at CERN (the LHC), to address the fundamental questions on the origin of electroweak symmetry breaking: Is there a Higgs boson? Is Supersymmetry there as well? If neither the Higgs nor Supersymmetry exists, then what is the new physics that makes the standard model work so well, giving mass to particles and interconnecting the forces? Could it involve extra, hidden, spatial dimensions?

As was seen with the LEP and SLC colliders, an e^+e^- machine with the luminosity and precision required to address strategic questions will also shed light (and generate Ph.D. theses) on a rich variety of physics. Examples include the detailed properties of W and Z bosons and of the top quark, QCD, beauty, charm and tau-lepton physics; perhaps also the origins of CP violation. And if Higgs bosons are found, or Supersymmetry, their detailed properties will also be explored in ways which will be inaccessible to LHC.

This book is aimed at graduate students and others who want to understand the physics to be done at a next generation linear collider. It presents the whole range of topics which can be studied there, explaining the strategic situation in some theoretical depth and reporting the extensive simulations which have shown that clean samples of many different final states can be separated and measured, and how the important parameters can be extracted from them.

In August 2004 a panel of "wise persons" recommended [2] that the machine should be built using the superconducting technology that had been developed in the TESLA project. This choice was endorsed by ICFA (the International Committee for Future Accelerators) and is supported by all participating laboratories from around the world. At the time of going to press the accelerator designers who have worked on both the warm

(X−band) and the superconducting technologies [3] have got together in a Global Design Effort to produce plans for the International Linear Collider [4, 5]. A Conceptual Design Report for the machine is planned for the end of 2006, with commitment to construct in 2009 and completion in 2015. Government funding agencies from around the world are already meeting informally to review progress and they have welcomed the technology choice as a critical step in moving forward towards the design of a linear collider.

The Editors
Keisuke Fujii
David J. Miller
Amarjit Soni

References

1. International consensus document http://sbhep1.physics.sunysb.edu/ grannis/lc_consensus.html
2. Executive summary of the International Technology Recommendation Panel decision http://www.fnal.gov/directorate/icfa/execsum_ITRP.pdf
3. International Linear Collider Technical Review Committee; Second report, 2003. SLAC-R-606. This reviews all available linac technologies in great detail. http://www.slac.stanford.edu/pubs/slacreports/slac-r-606.html
4. The webpage of the International Committee for Future Accelerators reports progress on this process - see: http://www.fnal.gov/directorate/icfa/icfa_home.html
5. There is a new public webpage for the GDE at http://www.interactions.org/linearcollider/gde/

Contents

List of Contributors

D. Atwood
Iowa State Univ., Ames, Iowa; USA

G. A. Blair
RHUL, Egham, Surrey; UK

P. N. Burrows
Univ. of London, London; UK

J. L. Feng
Univ. of California, Irvine; USA

K. Fujii
KEK, Tsukuba, Ibaraki; Japan

J. F. Gunion
Univ. of California, Davis; USA

H. E. Haber
Univ. of California, Santa Cruz; USA

J. L. Hewett
Stanford Linear Acc. Center, Menlo Park; USA

W. Kilian
Univ. of Karlsruhe, Karlsruhe; Germany

R. V. Kooten
Indiana Univ., Bloomington, Indiana; USA

D. J. Miller
UCL, London; UK

K. Mönig
DESY, Zeuthen; Germany

M. M. Nojiri
Kyoto Univ, Kyoto; Japan

M. E. Peskin
Stanford Linear Acc. Center, Menlo Park; USA

A. Soni
Brookhaven National Lab., Upton, NY; USA

Y. Sumino
Tohoku Univ., Sendai; Japan

T. Takahashi
Hiroshima Univ., Hiroshima; Japan

CHAPTER 1

The Machine and Detector

G.A. Blair[a] and D.J. Miller[b]
[a]*Department of Physics, RHUL, Egham, Surrey, UK*
[b]*Department of Physics and Astronomy, UCL, Gower St., London, UK*

1. Introduction

Most of the physics topics covered in this book require higher e^+e^- Centre of Mass (CM) energies than the $\simeq$ 208 GeV which has been reached at LEP (and the $\simeq$ 92 GeV reached at the Stanford SLC). This chapter explains why going to higher energies at the required luminosity poses major new problems in accelerator technology, and describes the schemes being developed to solve those problems. We then discuss the demands on the detector and show how they can be met by extensions of the techniques already used at LEP and SLC, backed up by new ideas which have been developed for the Large Hadron Collider at CERN and for the BaBar and Belle beauty factories.

Considerable R&D has been done on the collider designs and they are now almost ready for a firm decision to go ahead, though continued R&D will be needed to finalise all of the subsystems. The detectors will also need substantial amounts of R&D to ensure that they can do exactly what is needed for a reasonable cost.

2. The Machine

2.1. *Required Energy and Luminosity*

The cross sections for various e^+e^- processes are shown in Fig. 1. Most of them fall like $1/s$, where s is the square of the CM energy. The scale is set by the muon pair production cross section which falls from $\simeq$ 3.5 pb at 200 GeV to $\simeq$ 0.14 pb at 1 TeV. The instantaneous luminosity of a

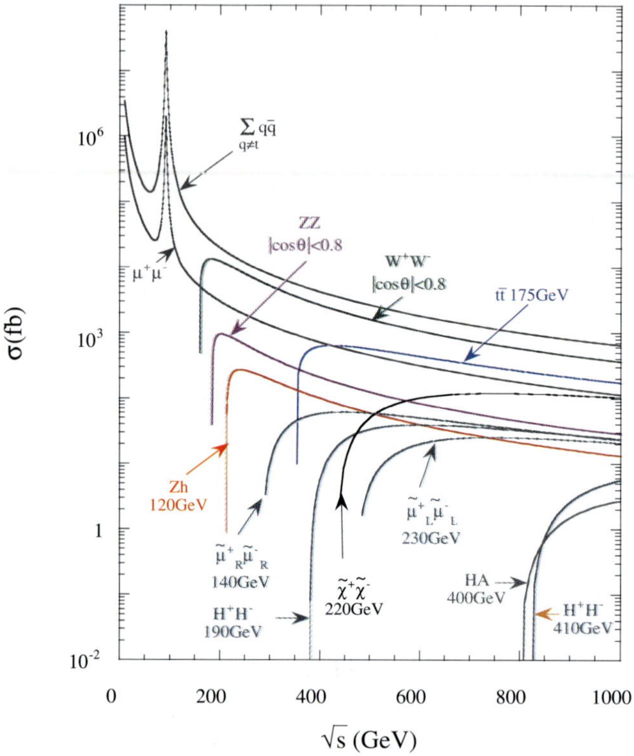

Fig. 1. Cross sections for some real and possible processes in the Linear Collider energy range.

collider is measured in events per cm^2 of cross section per second. LEP was achieving peak luminosities of about 10^{32} $cm^{-2}s^{-1}$ at 200 GeV in 1999 and had an integrated luminosity over the year of $\simeq 250$ pb^{-1} per experiment. This gave a useful sample of $\simeq 20000$ $q\bar{q}$ pairs ($\sigma_{q\bar{q}} \simeq 20$ pb) but only a few dozen Z^0Z^0 events ($\sigma_{ZZ} \simeq 0.1$ pb).

The Linear Collider will start its programme after the start of the CERN LHC, which is primarily built to search for new particles and effects. If there is a light Higgs boson, for instance, as may have been glimpsed first by the LEP experiments [1] it should be seen at the LHC (or the Fermilab Tevatron) before the Linear Collider starts up. But the Linear Collider is complimentary to the LHC and will be particularly important for making precision studies of both established and new physics processes. High precision generally requires large statistics, though different physics studies have different luminosity needs and may be best done at different energies.

Extensive results are available from feasibility studies of the physics which can be done at a linear collider [2–11]. Much of the rest of this book is based upon those sources. The result of these studies and workshops in Europe, North America and Asia is a consensus on what the major features of the programme should be. If there is a light Higgs boson ($m_h \leq 140$ GeV) then the e^+e^- collider will be where its properties are measured with high precision (see Chapter 2, below). To measure its branching ratio to $b\bar{b}$ with a precision of 4%, for instance, will require at least 200 fb^{-1} at $\sqrt{s}$ in the range 350 to 500 GeV. To make this measurement in a year the average luminosity would have to be 2×10^{34} cm^{-1}s^{-1} (assuming 10^7 useful seconds per year). At 350 GeV the study of the excitation curve for $t\bar{t}$ production would need 50 fb^{-1} distributed over 10 closely spaced points in $\sqrt{s}$ to measure the top quark mass to a precision of about 100 MeV (see Chapter 3, below). To determine the $t\bar{t}h$ Yukawa coupling by measuring the rate for radiation of a 120 GeV Higgs from $t\bar{t}$ final states would require 1000 fb^{-1} at $\sqrt{s} = 800$ GeV. If there is no light Higgs boson then the first signs of new physics may come in precision studies of the W^+W^- final state. A sample of 500 fb^{-1} at $\sqrt{s} \geq 500$ GeV would give a good start.

On the basis of these physics needs, proposals are now being prepared for machine designs which could have a luminosity of at least $\sim 5 \times 10^{33}$ cm^{-1}s^{-1} at 350 GeV startup, rising within a couple of years to $\simeq 3 \times 10^{34}$ cm^{-1}s^{-1} at 500 GeV, followed by a clear upgrade path to reach $\sim 5 \times 10^{34}$ at 800 or 1000 GeV.

2.2. *Why a Linear Collider?*

The reader may ask why we should go back to a Linear Collider for increased energies and luminosities after the circular LEP machine achieved significantly higher energy and luminosity than the original SLAC Linear Collider (SLC). The answer has two parts:
i) There is no alternative. Synchrotron radiation in a circular machine like LEP rises as (Energy)^4/Radius. LEP at 208 GeV was already radiating 3% of the beam energy per turn so a 500 GeV circular machine would be prohibitively expensive to build and to run.
ii) Accelerator engineers have learned a lot from SLC [12], and from other studies, which makes them confident that a future single-pass machine can achieve the required luminosities at high energy.

A simplified form of the equation for the luminosity is

$$L = \nu N_1 N_2 / (4\pi\sigma_x^* \sigma_y^*) \tag{1}$$

where ν is the number of collisions per second, N_1 and N_2 are the numbers of particles in the two colliding bunches and σ_x^*, σ_y^* are the RMS widths of the colliding spots in the horizontal and the vertical directions. Liouville's theorem tells us [13, 14] that quantities called the emittances ϵ_x and ϵ_y, the horizontal and vertical phase-space areas of a beam, are invariants of its motion through a perfect focussing lattice. If the lattice also accelerates the beam then the invariant quantity is actually the emittance times the momentum of the particle, so the emittance must shrink like $1/\gamma$ as the momentum rises, an effect called "adiabatic cooling" (simply put, in an accelerating cavity the longitudinal momentum increases but tranverse momenta are unchanged, so the range of divergence angles decreases). The size of the spot at the collision point, for a given final focus system, is given by $\sigma_{x,y}^* = \sqrt{\epsilon_{x,y}\beta_{x,y}^*}$ (neglecting dispersion) where $\beta_{x,y}^*$ are measures of the horizontal or vertical focussing strengths of the lens systems which squeeze the beam on either side of the collision point. Since both ϵ_x and ϵ_y shrink like $1/\gamma$ the effect of adiabatic cooling in a perfect machine is to make the luminosity from equation 1 grow proportionally to the beam energy, *i.e.* proportionally to $\sqrt{s}$. This is highly desirable but very hard to achieve.

High luminosity requires large beam currents, proportional to N_1 and N_2, and/or small beam cross sectional area $\sigma_x\sigma_y$, also called spot size. High current in a single pass collider implies high power consumption since all the bunch energy is dumped after each collision. To avoid impossibly high running costs, the realistic way of going to higher luminosity is to concentrate upon reducing the spot size. SLC already achieved considerable improvements in luminosity by reducing the vertical spot size from $(3.5\,\mu\text{m})^2$ in 1989 to $(0.9\,\mu\text{m})^2$ in 1998 [12]. The goal for the next generation of 500 GeV linear colliders is to reduce it further to ~ 5 nm $\times$ 400 nm. Linear colliders have an advantage over circular machines like LEP in that they do not suffer from the accumulating effects of beam–beam interactions which gives a "tune-shift" [13, 14] that blows up the spot size.

Dealing with tiny beam spots, each containing $\sim 10^{10}$ charged particles, brings a list of new problems [15, 16]:

(1) Before making a useful collision either or both particles may radiate energy through beamstrahlung [17], a process like synchrotron radiation in which an individual beam particle reacts with the collective electromagnetic field of the whole opposing bunch. Beamstrahlung is a

serious problem for physics analyses which require precise knowledge of the CM energy; scans across a particle production threshold for instance, or constrained fits to determine particle masses. Fortunately, a strategy has been developed to greatly reduce the beamstrahlung at a given luminosity. All current 500 GeV e^+e^- collider designs specify a much larger value ($\simeq$ 400 nm) for σ_x, the horizontal beam size, than for σ_y ($\simeq$ 5 nm) because the charge is spread out horizontally, for a given beamspot area [$\sigma_x\sigma_y$ on the bottom line of eq. (1)] such flat ribbons of beam have much smaller associated electromagnetic fields, which keeps the mean beamstrahlung losses down to a few percent. After beamstrahlung the distribution of collision energies is very non-gaussian, with a spike of some tens of percent of events having less than 1% loss, and a long tail running out to losses of more than 50%, the exact shape varying from bunch crossing to bunch crossing.

(2) Each bunch is disrupted by the electromagnetic field of the other. The effect is different for e^+e^- and e^-e^- collisions. For e^+e^- the orbits of the colliding particles are deflected inward towards the centre of gravity of the opposing bunch, producing an enhancement of effective luminosity by a factor which can easily exceed $\times 2$. Conversely, the orbits in e^-e^- collisions are pushed away from the centre of the opposing bunch and the effective luminosity drops. For "round beams" with $\sigma_x \simeq \sigma_y$ the disruption could be so severe that the outgoing beams would blow up rapidly and collide with the beam-pipe and magnets soon after the collision point, causing background and radiation problems. But, as with beamstrahlung, by using flat ribbon beams the effective electromagnetic fields from the two bunches are greatly reduced and the resulting disruption effects can be accommodated.

(3) It is difficult to monitor the size and position of bunches with a vertical spot size of less than 10 nm; less than $\sim 1/40$ of the wavelength of visible light. No mechanical system could do it. A scanned wire would be much too fat, and would be physically cut-through by the localised intensity of ionisation. A "laser wire" [18] can be scanned through a bunch-train — with a downstream detector monitoring the intensity of Compton-scattered photons — but the smallest conceivable waist-size from a visible light laser is about 1μm, far too large to resolve the spot. Nevertheless, an ingenious monitor has been devised and demonstrated [19] in tests with the Final Focus Test Beam (FFTB) experiment at the SLC. It set up a standing-wave pattern of laser interference fringes across the plane where the beam was focussed. When the lateral posi-

tion of the bunch train was swept slowly across the fringe pattern the Compton scattering rate was modulated by the intensity variation in the fringes. By measuring the depth of the minima of the Compton rate it was possible to deduce the transverse size of the spot. With low backgrounds and UV laser light the method could work for spot sizes down to about 10 nm, though it is hard to see how to fit it into the crowded detectors discussed below.

(4) Even if the spot size can be determined, before collisions can be produced it is also necessary to bring the two spots together to within a few nm. Here the experience from SLC is crucial. They observed that once the spots are within a few μm they begin to deflect one another — a longer-range precursor of the disruption effect. The displacement of the leading bunches of the outgoing train can be sensed by the off-centre signal in a beam-pickup monitor a few metres downstream and the result used to generate a correction signal which drives a fast kicker magnet a few metres before the intersection point. This feedback system brings successive bunches closer and closer together until collision is achieved.

(5) To get the beam-based correction to work at all it will be necessary to stop the components of the final focus from moving around between bunch trains due to ground vibration — seismic or man made. Mechanical micromovers on the magnets, linked to precision survey devices, should be able to do this for the machines at present planned. A substantial R&D programme is needed.

In the following sections we discuss the details of the main machine proposals which are now on offer to overcome the problems listed above. They are, inevitably, presented as rival projects by consortia of laboratories, but much of the fundamental R&D on the technologies needed has been done in inter-lab collaborations which cross the boundaries of the particular proposals. Taking two examples — there are many others — the laboratories collaborating in the vital FFTB experiment at SLAC included KEK, Berkeley, Saclay, DESY, Novosibirsk, Max Planck (Munich) and many more; and the vital work on electropolishing the superconductor surfaces to reach higher field gradients with the TESLA cavities was done by a KEK, CERN, DESY, Saclay collaboration. This makes it very likely that, when agreement has been reached on one linac technology and one site, a new consortium of participating laboratories will emerge to build and operate the machine. Already ICFA (the International Committee for

Future Accelerators) is looking at ways of running such a consortium as a Global Accelerator Network [20].

2.3. *The Proposed Machines*

Table 1. Design parameters for NLC/JLC X-band, TESLA and CLIC.

Linear Collider Design Parameters, October 2000						
	NLC/JLC		TESLA		CLIC	
CMS Energy (GeV)	490	888	500	800	3000	5000
Luminosity ($10^{34}\text{cm}^{-2}\text{s}^{-1}$)	2.2	3.4	3.4	5.8	10.0	10.0
RF Frequency (GHz)	11.4	11.4	1.3	1.3	30	30
Gradient (MV/m)	50.2	50.2	22	34	150	172
Two Linac Length (km)	5.4	9.9	33	33	27.5	40
Trains per second	120	120	5	4	100	50
Bunches per train	190	190	2820	4886	154	154
Bunch spacing (ns)	1.4	1.4	337	176	0.7	0.7
Particles per bunch (10^9)	7.5	7.5	20	14	4	4
Spot size σ_x/σ_y (nm)	245/2.7	200/2.2	553/5	391/2.8	43/1	31/ 0.78
AC power (MW)	(134)	(243)	100	169	300	290
	(estimated)					

Table 1 shows design parameters for the three main Linear Collider designs which are being developed: TESLA by an international consortium based on DESY in Hamburg (see Volume I of Ref.[8]); JLC/NLC by a collaboration including SLAC in California ("Next Linear Collider" [21]), KEK in Japan ("Japanese Linear Collider" [22]), and other major American laboratories; CLIC by CERN in Geneva, Switzerland. All the machines have the same major constituent parts, though the technologies used are different.

We have left out of the table the interesting room temperature C-band design [23] which is also being developed in Japan. This could be an economical choice for a 350 GeV collider.

TESLA (see Fig. 2) uses superconducting radio frequency (RF) cavities for the main linac, excited in a simple standing-wave mode by L-band (3 GHz) RF power. Their advantage is their very low power dissipation, measured as the Q-value. This means that they can store RF power for the relatively long period of about 1 millisecond, during which time thousands of bunches can be accelerated, with intervals of hundreds of nanoseconds between bunches. There is a critical field strength above which the niobium

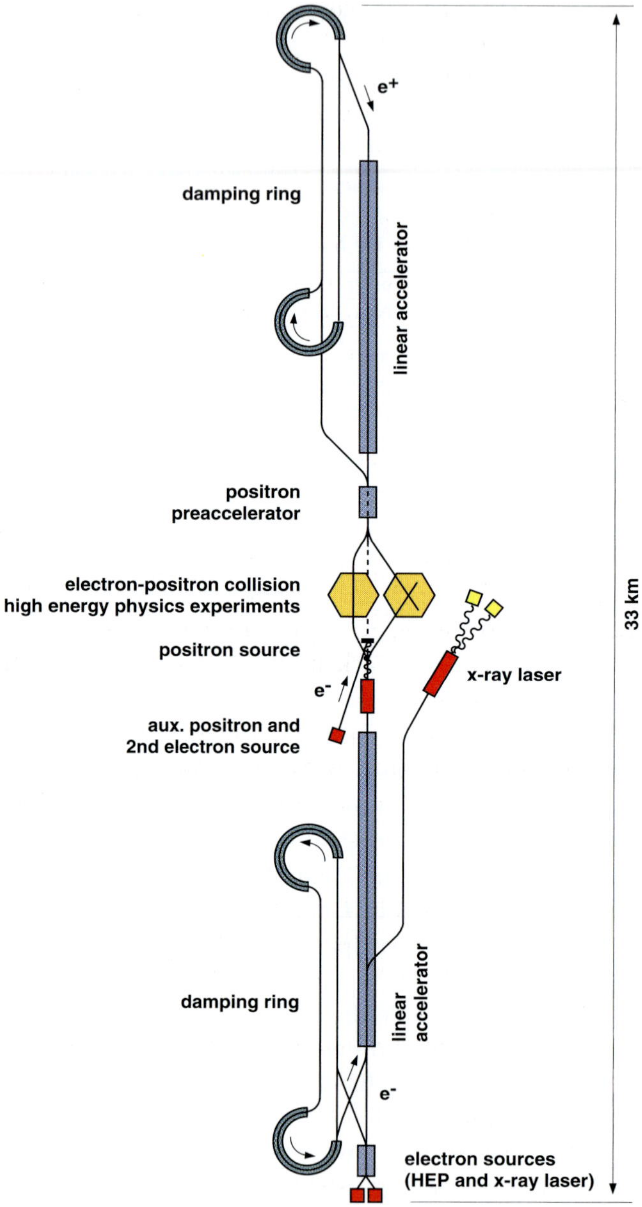

Fig. 2. Layout of the TESLA collider.

superconductor will quench. This corresponds to a maximum effective accelerating potential gradient of 50 MV per metre. Imperfections in the fabrication of the cavities and their assembly into machine modules have so far limited the actual accelerating gradient in the TESLA Test Facility to ≤ 25 MV/m, sufficient for the proposed 500 GeV CM energy initial stage of TESLA[8]. Each new batch of cavities tends to reach a higher gradient, and there are strong grounds for planning an eventual upgrade of the machine to 800 GeV, based on the achievement already of 40 MV/m before quenching in single cavities which have been electropolished.

The NLC/JLC design uses room temperature copper cavities excited in the X-band (11 GHz); see Fig. 3 for the latest version of the NLC concept. Because the dissipation is much greater in normally conducting cavities they can be loaded with power for a much shorter time than can superconducting cavities. Only $\simeq 100$ bunches can be accelerated for each RF pulse, with an inter bunch gap of 1.4 ns. There are 120 bunch-trains per second. New high power klystrons are being developed to drive these cavities. Figure 4 shows a JLC concept for twin tunnels with the klystrons and their modulators (the large boxes in the larger tunnel) separately accessible for maintenance. Power is transferred to the linac tunnel through waveguides in the cross tunnels. There has been a setback in the last two years to the X-band development when test cavities run at full power have shown signs of plasma erosion, close to the input coupler. This is thought to be due to breakdown in high fields from localised higher excitation modes [24], or close to sharp edges inside the cavities [25]. An intense programme of tests is now under way with modified cavities to establish that the peak accelerating field can indeed be taken to higher values at higher frequencies, as had always been supposed. Current cavity designs are already good enough for the 500 GeV version of the NLC/JLC with $\simeq 70$ MV/m unloaded acceleration gradient, but the 1 TeV design will require improvements if the length of the linac is not going to scale with energy.

CLIC [26] also uses room temperature copper cavities, but at a frequency of 30 GHz where high powered klystrons do not exist. Instead, the power to drive the main beam is derived from a second parallel linac running at lower energy. Although the drive frequency in the second linac is much less than in the main linac, the electron bunches in the drive beam are, as always, very short compared with the wavelength of the drive RF or with the dimensions of the cavities, so the wakefields which they excite in the cavities are at high frequencies. Suitably designed pickups are used to extract 30 GHz power from specially designed cavities and deliver it

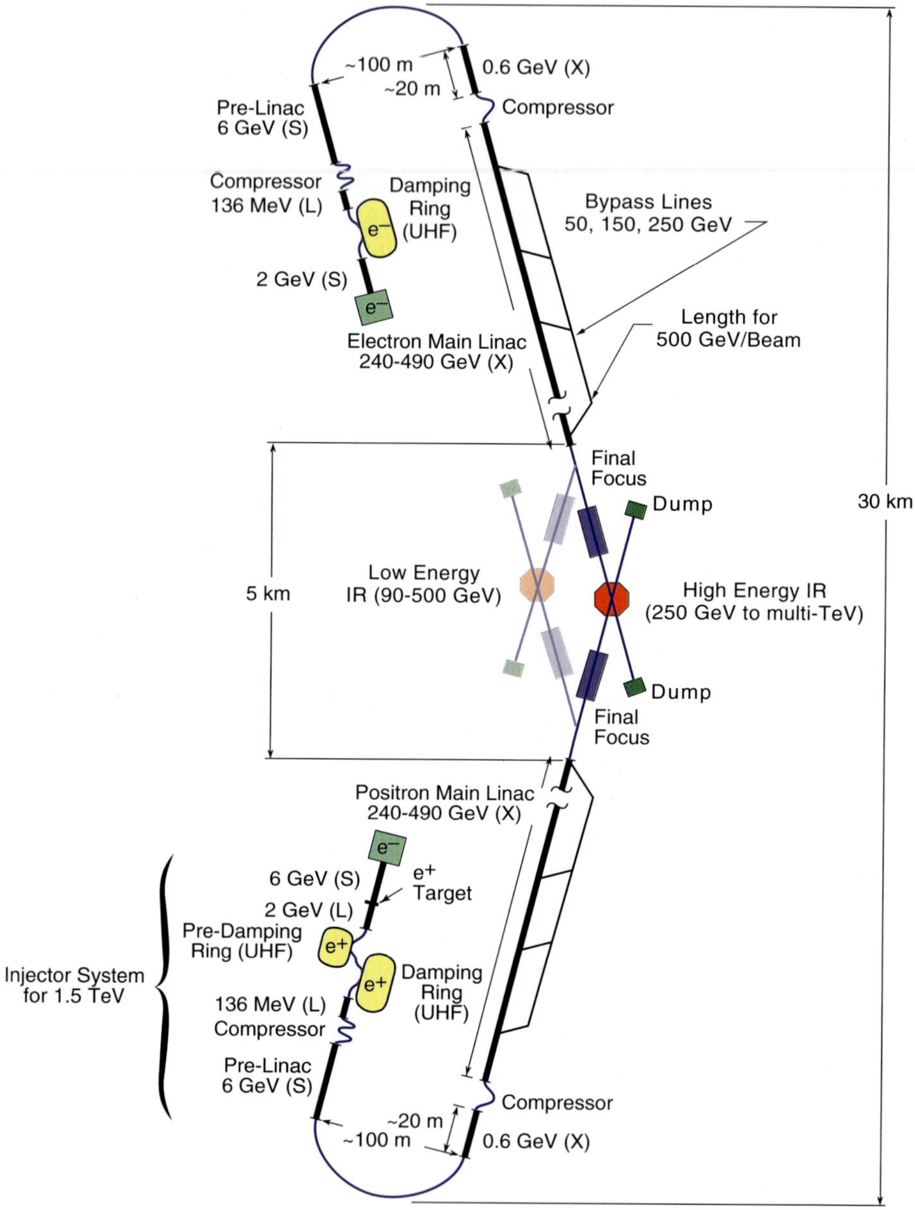

Fig. 3. Conceptual layout of the NLC.

Fig. 4. Artists concept of the JLC tunnels. The linac is in the small tunnel on the right, linked by waveguides to the klystrons on the left.

via short waveguides to the accelerating cavities of the main linac where gradients of $\simeq 120$ MV/m will be produced.

2.4. *X-ray Free Electron Laser*

An important by-product is very likely to follow from the achievement of high-intensity high quality linac beams with energies above ~ 30 GeV. These beams have exactly the properties required to feed an X-ray free

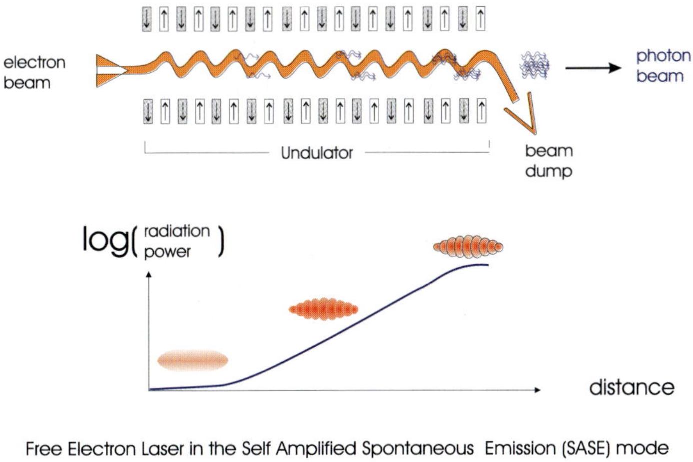

Fig. 5. How a free electron laser works. The undulator must be some tens of metres long to achieve saturation of the radiation power in the output photon pulse.

electron laser (FEL), working at $\simeq 1$ Angstrom wavelength — the most important region for studies of solid and liquid materials, biological molecules, chemical processes etc. Conventional laser cavities cannot be made at these wavelengths because there are no suitable mirrors. Instead, a low-emittance high energy electron bunch is made to oscillate transversely through many cycles in a long undulator magnet (see Fig. 5). Synchrotron radiation photons from successive undulations will be in phase with one another and a process called Self Amplification of Spontaneous Emission occurs (SASE), building up over tens of metres to produce a beam which is laterally coherent, monochromatic and less than one picosecond in length. This has already been demonstrated at UV wavelengths at the TESLA test facility. The intensity of X-ray FEL beams will be many orders of magnitude greater than those available from synchrotrons (see Fig. 6). A first picture

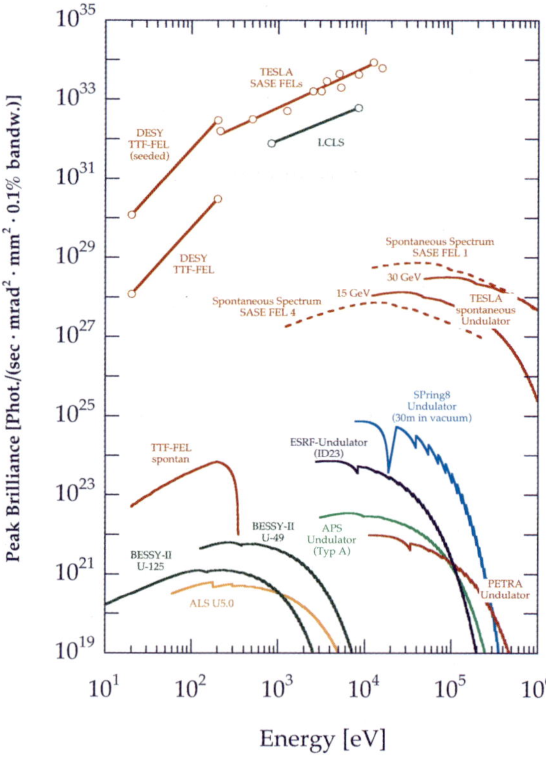

Fig. 6. Comparison of the brilliance of X-ray sources, including FELs.

of some of the potential applications is given in the TESLA TDR [8]. The laboratories DESY, SLAC and KEK are all actively working to develop these applications, using respectively their superconducting, X-band or C-band accelerator technologies, and to build up communities of scientists who could become co-users of the collider laboratory.

2.5. *Electron and Positron Sources*

"High brightness" (low emittance) electron guns are needed by all kinds of electron accelerators, from TV tubes to synchrotron light sources. The physics requirements for the Linear Collider add the extra demand that the beam should be polarised. The SLC has demonstrated that 70% polarisation can be produced robustly by a gun in which a strained GaAs photocathode is stimulated by an intense spot of polarised laser light. These techniques are being developed further [27] and many physics studies assume that 80% electron polarisation will be available.

Existing positron sources — and the designs for NLC/JLC and CLIC — use an electron beam at relatively low energy to produce bremsstrahlung gamma rays from a high Z foil. These are allowed to interact in a second foil (labelled "e^+ target" on Fig. 3) to make unpolarised electron-positron pairs. For TESLA, however, it is intended to make the gamma rays by introducing an undulator magnet into the high energy electron beam in the main linac (undulator is a shaded box on Fig. 2 just below the collision region). This was first suggested to ensure the correct time interval between the positron bunches. But it gives an added possibility. Since the gamma rays from an undulator can be polarised it should be possible to produce polarised positrons at the conversion target. R&D is still needed on this, but first estimates suggest that 50 to 60% positron polarisation can be achieved.

2.6. *Damping Rings*

The initial emittance of the positron beam from a pair-production target is limited only by the aperture and focussing power of the capture optics (0.03 m-rad for NLC/JLC). It is very much larger than the emittance required at the entrance to the main linac ($1.5\ 10^{-4}$ m-rad). And the emittance of useful electron guns is also bigger than required. The emittances can be reduced by storing the positrons or electrons in a damping ring where they are continually refocussed and reaccelerated while they produce synchrotron radiation which carries away entropy and cools the beam. The natural synchrotron radiation due to bending around the ring is enhanced by adding strong "wiggler magnets" with much larger $\int \mathbf{B}.\mathbf{dl}$. The Accelerator Test Facility at KEK [28] is a prototype wiggler-damped storage ring of 139 m circumference specifically designed as a protoype for a high energy Linear Collider. Special instrumentation has been developed to monitor the beam size as the emittance shrinks. Initial results show that it works well and that eventual emittances might be even smaller than those specified in the current NLC/JLC design. Both electrons and positrons in NLC/JLC have a main damping ring, with a special predamping ring for the positrons (see Fig. 3).

TESLA also needs damping rings for both positrons and electrons. These are unconventional in shape (see Fig. 2) because they are also used to set up the very long TESLA bunch train. If all 2820 bunches were stored end to end at the linac spacing of 337 ns the rings would need to be nearly 300 km in circumference. But that is not necessary. The bunches will actually be at 20 ns spacing in the ring and will be kicked out separately when needed. This still requires a circumference of 17 km, compared with only 300 m for the main damping rings of NLC where the bunch train is so much shorter. To economise in tunneling the major parts of the circumference of the TESLA damping rings are actually straight (but filled with wigglers) and are housed alongside the linacs in the main tunnel. There are arcs at each end which carry the beam from the "up" straight section to the "down" section. Due to the swelling at either end, with a slim parallel centre section, this shape is known as a "dogbone" damping ring. It has not yet been prototyped but extensive simulations show no serious difficulties.

2.7. *Final Focus*

As mentioned in Sec. 2.2 above, a large part of the increased luminosity at a future Linear Collider, compared with SLC, comes from a making a beam spot which is only $\simeq$ 5 nm high at collision. To achieve such a small spot, the beam from the linac has to be delivered to the collision point by a special final focus optical system which has the appropriate vertical and horizontal demagnifications and has very low chromaticity so that the finite spread of beam momentum does not degrade the quality of the spot. The FFTB experiment at the SLAC linac demonstrated that the required demagnification could be achieved. It had special quadrupoles with precision movers, all mounted on pillars which went down to bedrock and with a very precise survey system. The laser interference method developed by the KEK group (see above, Sec. 2.2, item 3) demonstrated that effective spot sizes of about 60 nm rms had been produced, corresponding to the demagnification required for a future Linear Collider. (Smaller spots were not possible at the FFTB; the input emittance from SLC was bigger than in NLC.) There has been a recent breakthrough [29] in the design of the final focus for the NLC which has reduced the tunnel length required from 2×1.8 km (for beams up to 750 GeV) to less than 0.4 km. These ideas are now being considered for TESLA too.

2.8. *Crossing Angle and Number of Collision Points*

The long time between individual bunches at TESLA will allow each bunch to be separately deflected out of the linac into a beam dump after it has got clear of the final focus quadrupole magnets close to the collision point. There is no danger of colliding with the next opposing bunch on its way in to collision. But the bunches at NLC/JLC or CLIC are so close together that outgoing bunches would collide with the incoming bunches even before they reached the nearest quadrupoles. To avoid this in these machines the particles collide at a finite crossing angle, typically $\sim$20 mrad. In the published design The main collision point at TESLA is designed to have a zero crossing angle.

There is considerable debate about whether there should only be one collision point at a future Linear Collider, or whether two separate final focus lines should be built, feeding two separate experiments placed side by side. In TESLA a second collision point would have a finite crossing angle to allow the $\gamma\gamma$ and γe options [see Eq. (2.10) below] to be installed. In the Snowmass 2001 version of the NLC [21] design the main linac tunnels are not quite collinear, but are aligned to give a crossing angle of about 20 mrad for beams above 1 TeV (much exaggerated in Fig. 3) which might eventually be produced as the machine is upgraded — or replaced by a CLIC type of collider — and which could only be bent by a very small amount in the final focus without serious emittance dilution due to synchrotron radiation at the bend. A second collision region is envisaged at NLC for lower energy beams which would tolerate larger bends in the final focus. As well as catering for $\gamma\gamma$ and γe this could also allow a GigaZ option [see Eq. (2.11) below].

Experience at the SPS, the Tevatron, at LEP and at HERA has shown that competition between users on a circular storage ring leads to the very effective utilisation of the shared beam, with rapid development of new techniques — often going far beyond what had been envisaged when the experiments were designed (e.g. the achievement of errors of 5 parts in 10^4 on the luminosity at the Z^0 at LEP [30]). At any machine an extra experiment adds of the order of 10% or so to the cost of the whole project, with a few more % at a Linear Collider for a second pair of final focus beamlines. It is not often realised that many storage rings operate with their luminosity limited by the beam-beam tuneshift. This can be shared equally between intersection points or could all be concentrated at one. But the luminosity at any collision point is strongly related to the tuneshift there, so the situation at a circular collider is not totally different from what it would be at a linear collider; the management can decide the share of the available luminosity seen by each user. At a storage ring the total luminosity does not simply increase proportionally to the number of interaction points, and the distribution of luminosity to the experiments is controlled by adjusting the beam optics. At a linear collider the distribution of the fixed total luminosity would be controlled by diverting bunch trains to one or other experiment. Since second (even third and fourth) detectors have been the rule at circular colliders — with great physics benefit — there is indeed a strong case that a second collision region and detector will be needed at the Linear Collider.

2.9. *Measuring Beam Energy and Luminosity Spectrum*

An experimental run on a collider produces files of event-by-event data which are reconstructed and analysed to extract the physics measurements. During any period of running the machine will fluctuate in its performance for a long list of possible reasons: the final focus being shaken by a truck driving down a nearby highway, the polarised laser on the electron target failing and having to be replaced, a wiggler magnet tripping off for ten minutes in the positron damping ring etc. Each such event will disturb the beams and alter vital properties such as energy, luminosity and polarisation for some part of the running period. In order to reconstruct cross sections and other physically interesting quantities it is necessary to monitor the beams continually and keep a record of the history of fluctuations.

The energy of the beams before collision can change slightly from its intended value for many reasons, the most obvious being the tripping off of an accelerating klystron. Some physics analyses need very precise knowledge of the energy. For instance, measuring the top quark mass to $\simeq 50$ MeV in a threshold scan (see Chapter 3 below) needs a precision of 1 part in $\simeq 3,500$ on the energy. Measuring the W mass to 8 MeV in the GigaZ option would need 1 part in $\simeq 10^5$. With a finite crossing angle it is possible to put spectrometer magnets in the "used beam" lines before the dump: such a system was used at SLC[31]. Single bunches of uncollided positrons and electrons would be monitored to give the beam energy and its spread, perhaps once per second or once per bunchtrain. Such a downstream spectrometer would not work at TESLA where the bunches have to be kicked out to the dump by a pulsed magnet with a time-varying field that will vary slightly from bunch to bunch. It will be necessary instead to introduce a carefully surveyed set of spectrometer magnets into both beams before the collision point, with ultra precise beam position monitors to sense the orbits of the particles before and after the bend. This method has been prototyped in LEP2 [32] with some success. With sufficient R&D there is no reason why techniques cannot be developed to measure the mean beam energy with the precision required, and to monitor the (probably non-gaussian) spread about the mean. This spread of energies imposes an inescapable spread of Centre of Mass values $\sqrt{s}$.

The effective Centre of Mass energies of all physics processes are further spread out by beamstrahlung [see Sec. 2.2, bullet (1) above]. By the time an electron and positron meet and collide to produce a high-energy event, they have both been deflected by the intense electromagnetic field of the opposite bunch, and they may already have lost some of their energy through beamstrahlung. Instead of simply measuring the luminosity of the machine for a run at a given energy it will be necessary to determine the integrated luminosity spectrum $\partial L/\partial\sqrt{s}$ in fb^{-1} GeV^{-1} for any period of data taking. We would like to be able to use a well defined physics process to monitor the luminosity since the events from that process will be stored with all of the physics data and will share the same luminosity and energy history as the physics channels to be studied. At LEP we used e^+e^- Bhabha scattering events [30, 33], detected by special precision electromagnetic calorimeters at small angles where the matrix element is dominated by t-channel photon exchange, giving larger statistics than any annihilation channel, even at the peak of the Z^0. The same method may well be used to monitor the total luminosity at the Linear Collider, but it does not have sufficient sensitivity to the exact energy of each event to give the required precision on $\partial L/\partial\sqrt{s}$ (calorimeter energy resolution is only $\simeq 1\%$; better than 0.1% is needed). One promising approach has been suggested [34] which measures the distribution of acollinearity angles (see Fig. 7) of the outgoing e^+ and e^- in Bhabha scattering events in the endcap region of the detector ($80 < \theta_{\rm scatter} < 300$ mrad) where the statistics are higher

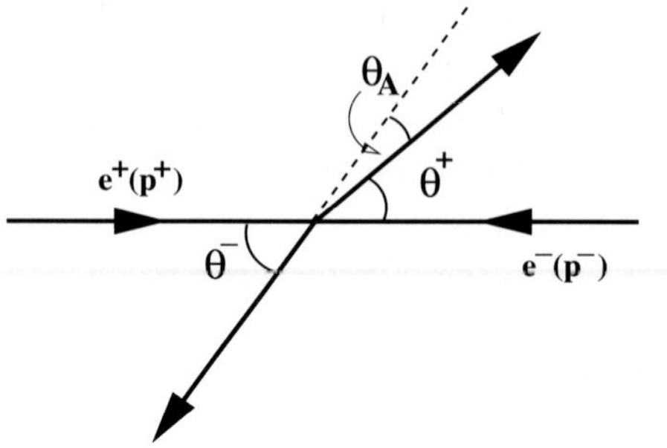

Fig. 7. Definition of the acollinearity angle θ_A in Bhabha scattering.

than in any major physics channel (except QCD $\gamma\gamma$ events). The acollinearity angle is actually sensitive to $(p^+ - p^-)$, the difference between the two beam momenta, whereas $\sqrt{s} \simeq (p^+ + p^-)$, but the distribution of θ_A can be unfolded to give the separate distributions of p^+ and p^- which can then be combined to give the distribution of $\sqrt{s}$. Simulations [35, 36] of the technique suggest that it will achieve the precision required, though more work is needed to determine how precise it can be. When designing the detector it is particularly important that well-surveyed precision tracking is provided for electrons in the 80 to 300 mrad region.

2.10. *The $\gamma\gamma$, γe and e^-e^- Options*

Compton scattering of laser light by an electron beam (see Fig. 8) can transfer a large fraction of the beam into scattered photons whose direction is close to that of the initial electrons. If this is done a few millimetres before the collision point in a Linear Collider it is possible to produce such intense

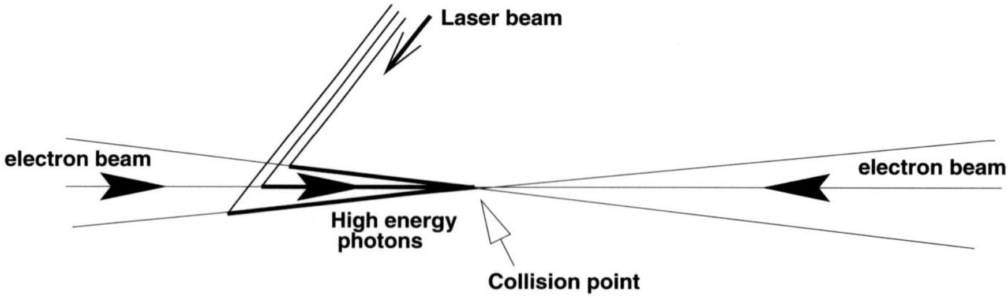

Fig. 8. Compton scattering of laser light to give high energy $e^-\gamma$ collisions. For $\gamma\gamma$ collisions laser light is scattered from both incoming e^- beams.

photon beams that the effective luminosity for high energy $e^-\gamma$ or $\gamma\gamma$ collisions is up to $\simeq 0.5$ of the luminosity for e^-e^+ collisions. The spectrum may be peaked at $\simeq 0.8$ of the lepton beam energy, or spread out to lower energies, depending on the polarisations of the incoming electrons and of the laser. This technique was first suggested by a group from Novosibirsk [37]. The technology for this "Compton collider" is not as well developed as that for the basic e^+e^- Linear Collider, but consultation with laser technologists and accelerator designers has not exposed any showstoppers [38]. Sufficiently powerful lasers exist, delivering between 1 and 10 Joules per pulse in the near infra-red, and schemes are being devised to match them to the bunch structures of TESLA and of the NLC/JLC project.

The Compton collider option requires a finite beam crossing angle in order to allow enough space around the outgoing beams, after collision, to accommodate the low energy, large angle, scattered beam electrons and the large numbers of e^+e^- pairs produced. These particles should not be allowed to hit the closest quadrupole, for its sake and for the sake of the backscattered backgrounds they would produce in the detector. Extra space will also be required to get the laser beams into the intersection region. Because of these requirements the detector for $e^-\gamma$ and $\gamma\gamma$ physics is likely to have worse acceptance for tracks close to the beam direction (≤ 150 mrad) than would a detector dedicated only to e^+e^-.

Some physics studies, especially in Supersymmetry (see Chapter 4 below), will benefit from the use of e^-e^- collisions as an alternative to e^+e^-. This will require provision of an extra polarised electron gun at the "positron end" of the collider; something which is also needed for the Compton collider.

2.11. *The GigaZ Option*

Despite the impressive precision already achieved on Electroweak and other processes with LEP1 and SLD at the Z^0 peak and with LEP2 above the W^+W^- threshold, measurements of many interesting quantities are still statistics-limited. It is argued in Chapter 6 below that a return to the LEP/SLC energy region with $\simeq 100$ times the statistics would greatly improve the constraints on the Standard Model — and might be crucial in choosing between its extensions. The aims would be to produce a sample of 10^9 Z^0 and take a further $\simeq 100$ fb^{-1} at $\simeq 161$ GeV. Because of their tiny beamspots and large beam currents, both the TESLA and the NLC/JLC designs could achieve the luminosity required to collect this quantity of data in about a year. But provision has to be made for this option from the beginning. (The name "GigaZ" is usually taken to include running both at $\simeq 92$ GeV and $\simeq 161$ GeV C. of M.). The linac will only operate efficiently when working close to its maximum load; it is not efficient to produce 46 GeV beams by accelerating gently over the full length of a 250 GeV linac. Instead a bypass beamline has to be provided, extracting the beam from the linac at the point where it has reached 46 GeV and carrying it passively to the collision point. A number of bypass lines are shown for the NLC in Fig. 3. It is straightforward to add more bypass lines to TESLA also (one is shown on Fig. 2 leading to the X-ray laser). Accelerating to a lower energy means there is less adiabatic cooling (see Sec. 2.2 above), so the expected luminosity is likely to be $\simeq 5 \times 10^{33} \text{cm}^{-2}\text{s}^{-1}$ at the Z^0.

3. Machine Detector Interface

The Linear Collider project, by its very nature, does not have a well-defined dividing line between where the accelerator ends and where the detector begins. The interplay between machine and detector issues is very strong at the Linear Collider and influences directly both the layout of the detector at low angles and the final stages of the machine.

As discussed in Sec. 2.2, the need for high luminosity requires the beams to be tightly focused and this leads to final quadrupoles situated close to the interaction point, together with beam diagnostics extending well into the detector. In addition, the need to tag particles down to very small angles and the requirement of a small beam-pipe radius for the vertex detector both have direct implications for the machine design, as discussed in Sec 4.2.2.

3.1. *Backgrounds*

In this section the main backgrounds that influence the design of the detector and machine are discussed. Some of these backgrounds are "machine related" in that they are due to off-momentum particles in the beam colliding with beam-line elements, or due to synchrotron radiation arising from bending or focusing of the beam upstream of the interaction point. Other background sources, such as from pair creation or beamstrahlung, are related to the luminosity because they are a direct result of the interaction between the colliding bunches. While the first type of background can be reduced by optimising the layout and operation of the machine, the second type implies a compromise between luminosity and level of background.

The obvious fact that the Linear Collider lies essentially in a straight line from electron source to interaction point, means that any beam backgrounds produced tangential to the beam, such as muons produced from the beam halo scraping the beam-line collimators, may have a clear path to the detector sitting downstream. It has been estimated [39] that approximately one high energy muon is produced for every 10^4 electrons lost in the beam-line. These muons will travel long distances before being stopped in material and many will make it downstream to the detector. The resulting tracks could lead to a high occupancy in the central trackers, though present indications are that the rate is tolerable. Estimates of the machine halo in the various Linear Collider options are being constantly refined.

Although the backgrounds at the Linear Collider are mostly electromagnetic in origin, significant numbers of neutrons are produced via giant dipole resonances from $\sim$10 MeV photons interacting with beam-line elements. In addition, large numbers of neutrons are produced in the beam-dumps downstream of the interaction region and care must be taken to avoid back-shine migrating to the detector. Neutrons in the detector can be problematic because, in addition to causing radiation damage, they can lead to spurious hits in gaseous central trackers. While this effect can be reduced by an appropriate choice of gas, this is an additional constraint on the detector design. Present indications [40] are that the neutron flux levels are acceptable, with a few thousand neutrons entering the detector volume per bunch crossing, however if the fluxes close to the vertex detector are higher than expected, then this may have implications for the choice of technology, as discussed in Sec. 4.2.2.

At the Linear Collider the need for high luminosity requires the electron and positron beams to be very tightly focused, which in turn requires very high powered doublets of quadrupoles situated close to the interaction point. As a result, intense swathes of synchrotron radiation will be created approximately tangential to the electron tracks in the focusing beam and care must be taken that none of this synchrotron radiation intersects directly any material near sensitive detectors.

In addition an intense forward flux of beamstrahlung photons, produced by the colliding bunches, together with the highly-disrupted spent beams themselves must all be extracted cleanly. This can be achieved by a careful layout of the detector and the limiting apertures of the beam-line elements, however higher order effects such as back-scattering of photons will occur and needs full simulation to quantify. The implications of such effects to the design of the mask region are discussed below.

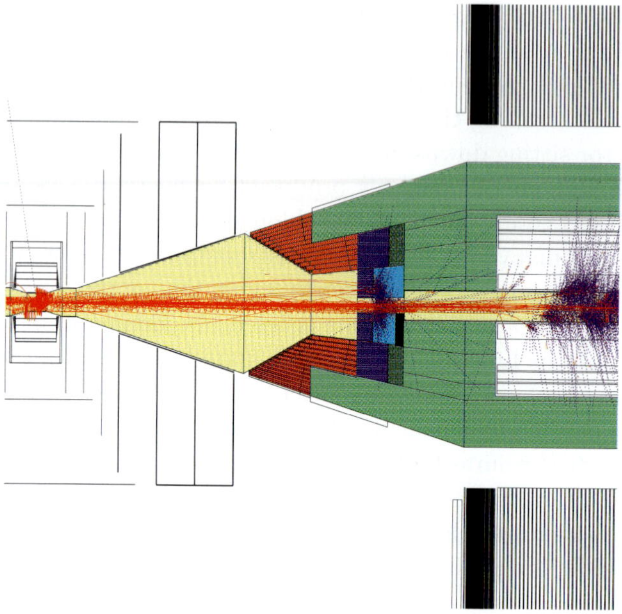

Fig. 9. Simulation of e^+e^- pair production via beam-beam interactions in the TESLA detector, with a magnetic field of 4 T. The scale is relatively expanded by a factor two in the vertical direction. Only about 0.1% of a single bunch crossing is shown.

Other backgrounds arise from the interaction of the bunches themselves. A major bunch–bunch related background is the production of e^+e^- pairs as a result of the enormous electric fields created when the bunches collide [41]. These pairs are produced in great numbers, typically $\sim 10^5$ per bunch crossing, with average energies of approximately 2 GeV per pair [40]. The electrons and positrons will then follow helical paths whose radial extent is inversely proportional to the magnetic field of the detector, discussed in more detail in Sec. 4 (see Fig. 9) and the envelope containing these pairs must not intersect any of the sensitive detectors such as the vertex detectors or low-angle calorimeters. This constraint is one of the main motivations for a large magnetic field and magnitudes of order 3–4 T are currently being considered.

Another source of background proportional to the luminosity is the production of hadrons in so-called minijets. These have transverse momentum typically a few GeV/c and are produced in collisions between the virtual photons associated with the colliding electrons and positrons ($\gamma^*\gamma^*$ processes). The total rate depends on the hadronic structure of the photon, as discussed in Chapter 9. For the 0.5–1 TeV machines under study, simulations predict that these processes will result in about one charged track per bunch crossing ending up in the central tracker [40]. Consequently this background is not a problem for detector design, although these processes will be a significant physics background and will also result in additional tracks superposed on unrelated events in the same bunch crossing. These effects will be much more serious at a multi-TeV machine, where about 40 charged tracks per crossing can be expected [42].

In summary, machine-related backgrounds have been the subject of much study and have influenced directly the design of the detectors, especially at low angles. While present indications from detailed simulations are that the various backgrounds can be handled, it should be remembered that there are a lot of uncertainties and assumptions going into these calculations and that at least an order of magnitude safety factor should be included in the sub-detector specifications.

3.2. *Mask Region*

New physics processes, such as decays of supersymmetric particles (Chapter 4) with small energy release, could lead to signals at very small angles from the beam direction. Some established processes also need to be measured down to low angles, especially $\gamma^*\gamma^*$ scattering and Bhabha scattering for luminosity monitoring. It will therefore be important to have an hermetic detector that is sensitive to particles at very small polar angles. However, as described in the previous section, the backgrounds at very low angles from the bunch–bunch interactions are high and ultimately limit how close a detector can be placed to the beam-line. This is illustrated in Fig. 9, which shows a simulation of the background e^+e^- pair-production mentioned above, and subsequent tracking of the products through the detector. While most of the particles produced by this interaction escape the detector, some then hit and form showers in the final quadrupole. As a result, many secondary electrons are back-scattered towards the detector and dedicated shielding is needed. This shielding is provided by the so-called "mask", a cone of dense material such as tungsten, pointing in towards the interaction point as shown in Fig. 9 with an expanded vertical scale. The mask also helps reduce the number of neutrons entering the detector.

The detailed design of the mask must balance the desire to detect particles at low angles with the primary function of shielding the main detector volume from the back-scattered electrons. Additional back-scattering from the inner mask itself must be avoided and careful positioning of low-density material such as graphite in front of the tungsten at low angles can help [41].

The limit in accessible polar angle is set by the envelope of the pairs, which in turn depends on the beam parameters and the detector magnetic field. For the TESLA design [8], the proposal is to install a calorimeter (the LAT, shown red Fig. 9) to measure high energy electrons at angles down to $\sim$ 28 mrad. The ultimate aim is to be hermetic down to $\sim$ 20 mrad, for instance to be able to tag $\gamma\gamma$ events from the scattered electron. While such events will provide very useful information on the structure of the photon, if untagged they are also a significant background to supersymmetric processes, especially when mass differences between the supersymmetric particles are small.

At lower angles, down to approximately 5 mrad, a luminosity monitor (the LCAL, shown full of shower particles in Fig. 9) is needed simply to count the rate of Bhabha events and thereby provide a fast measurement of the luminosity variation within a single bunch-train. The backgrounds in this region will be too high to attempt any reasonable energy measurement. These low-angle detectors will have to be radiation hard and survive electromagnetic doses of order 1 MGy per year. Several detector technologies such as diamond, low-temperature silicon or lead tungstate [43] are presently being studied [8].

In summary, the mask region is very much the realm of machine–detector interface and is of central importance to the Linear Collider physics programme. Both luminosity monitors and electromagnetic calorimeters sit adjacent to each other, in a region of very high machine-related backgrounds. In addition, sensitive devices such as beam position monitors and laser-based beam diagnostics will also occupy this region, making it a very challenging area to design.

4. Requirements for the Detectors

Lepton colliders have the unique advantage that the CM energy of the fundamental interaction is in many channels known exactly from the beam energy (modulo beamstrahlung effects, which can be measured, see Sec.2.9 above). This allows very high precision measurements of cross-sections and particle masses via threshold scans and the use of kinematic constraints in the reconstruction of the final state. However not all channels can use this constraint, for example WW scattering or fusion processes, where two neutrinos are also intrinsic to the production. In such cases the event kinematics must be reconstructed from the final state particles alone. Indeed, the quality of the jet reconstruction may determine whether an hadronic W^+W^- final state can be differentiated from a Z^0Z^0 one, or whether key fundamental measurements, such as the higgs self trilinear coupling, can be measured at all. The need for excellent jet resolution is central to the detector design and will require the combination of both tracking and calorimetry in the jet reconstruction. This procedure is called "energy flow" and is discussed further in Sec. 4.1.

As will be discussed in Sec. 4.2, the detector must possess excellent tracking capability in order to be able to reconstruct jets accurately and also to measure single high-momentum tracks well. This requires the presence of a uniform and well-measured magnetic field throughout the tracking volume. The solenoid that provides such a field will be a large and very significant component of the detector. It must be superconducting. The ability to place much of the calorimetry (Sec. 4.3) inside the solenoid has significant impact on the detector energy flow performance and both "large" detectors with magnetic fields of 3–4 T with solenoid inner radius of order 3 m (Ref. [8]) and "small" detectors with fields of order 6 T have been studied to date [44].

In addition to very high performance jet reconstruction, the detector must provide excellent flavour identification to determine the charges of the heavy quarks as well as to distinguish between b, c and uds quarks within multi-jet final states. This capability will provide a rich programme of physics ranging from the measurement of the couplings of the higgs bosons (Chapter 2) and the top-quark (Chapter 3) to the reconstruction of complicated cascade decays of supersymmetric particles (Chapter 4). Flavour identification is discussed further in Sec. 4.2.2.

A considerable amount of work is presently going into design and performance studies for a detector at the Linear Collider and only a few key points can be touched on in this chapter. It is not possible to do justice here to many interesting ideas such as triggerless event readout or to include any detailed discussion of the outer sub-detectors, so the reader is referred to the reports from the international study groups for more details [11]. While the basic concept and general layout of the detectors presently being considered are similar to those at LEP and SLC, the quantitative improvements in global performance and sub-detector specifications should not be taken for granted and they will require significant R&D to achieve.

4.1. *Energy Flow*

Energy flow is the procedure whereby individual objects (tracks, hit clusters etc.) are reconstructed in the tracking detectors and in the calorimeter and combined optimally into jets or complete events. Both calorimetry and tracking will need to be optimised over a wide range of energies, multiplicities and solid angle to achieve the required energy flow performance.

Taking as an example the production of h^0Z^0 at $\sqrt{s} = 500$ GeV and $m_H = 115$ GeV, the hadronic final state yields on average approximately 470 GeV of visible energy divided into four jets made up of charged particles, photons and neutral hadrons. The event by event fluctuations are large, but typically there are about 40 charged particles, 45 photons and 5 neutral hadrons, contributing 60%, 25% and 15% respectively to the total visible energy. On average about 5% (with large fluctuations) of the energy is carried by V^{0}s (K^0_S and Λ decays) so these must also be taken into account. This demonstrates that, although $\sqrt{s}$ is greater than that at LEP/SLC, the key to excellent jet performance is still the accurate reconstruction of large numbers of lower-energy particles, with energies typically less than a few GeV.

At lower energies, tracking provides the best momentum measurement for charged particles because the error on the transverse momentum p_T scales as:

$$\delta\left(\frac{1}{p_T}\right) = C_{\text{track}} \tag{2}$$

where C_{track} is a constant, with target values of order a few$\times 10^{-5}$ GeV^{-1}c. At higher energies, the calorimetry wins because the error of the measurement of the energy E scales as:

$$\frac{\Delta E}{E} = \frac{C_{\text{cal}}}{\sqrt{E}} \tag{3}$$

where C_{cal} is a constant which differs for the electromagnetic and hadronic calorimeters; values of 10–15% and 35–50% (with E in GeV) respectively are typical goals for the detectors currently being considered. The energy flow procedure involves the identification of clusters in the calorimeter and, where possible, the association of these clusters to charged tracks. For each particle, the calorimetric and tracking measurements are combined to provide the best estimate of the particle momentum. For lower energy ($\lesssim 150$ GeV) electrons and for charged hadrons the tracking will dominate the precision, whereas for very high energy electrons (and of course for all photons) the electromagnetic calorimeter will provide the accuracy. Neutral hadrons will require good performance from the calorimeters at all energies. The implications for the tracking detectors and calorimeters are discussed further in Secs. 4.2 and 4.3 respectively.

When discussing energy flow performance for jet reconstruction, account should be taken of intrinsic limiting factors such as visible energy loss due to escaping neutrinos (originating from heavy flavour decays) or migration of particles between jets due to fragmentation effects. While the former can be compensated to a limited extent by identifying any leptons in jets and making approximate corrections, the latter cannot be corrected and will be dependent on the jet-finding algorithm used. The problem of migration of particles between jets and the overlapping of jets themselves are effects that can be compounded by the detector field and geometry. Charged particles from one jet will have their tracks bent by the magnetic field and as a result their calorimeter clusters may overlap with those from a neighbouring jet. This effect can be reduced by increasing the granularity of the calorimeter or by decreasing the magnetic field or the tracking volume radius. The first solution would increase the cost of the detector whereas the latter two would degrade the momentum resolution of the tracker (as discussed in Sec. 4.2). The final decision on geometry and field will require detailed Monte-Carlo studies.

In summary, the energy flow performance will dominate the measurements of many physics channels at the Linear Collider. The method has worked well at LEP where overall energy flow performance of approximately $60\%/\sqrt{E}$ has been achieved [45]. A significant advance on this number will require a lot of effort, however $30\%/\sqrt{E}$ appears to be a feasible goal [8].

4.2. *Tracking*

The detector tracking system needs to identify tracks and to measure their momentum vectors and impact parameters with high precision. In this section these two main rôles are discussed and key points in detector design are outlined.

4.2.1. *Momentum Vector Measurement*

The importance of tracking to the energy flow measurement was emphasised in Sec. 4.1. The large number of intermediate-momentum tracks within a typical jet means that the momentum resolution, given by Eq. (2), must be very good.

Excellent tracking is also important to measure very high momentum tracks, for example the muon tracks from the recoiling $Z^0 \to \mu^+\mu^-$ in $h^0 Z^0$ "higgsstrahlung" production. A high precision momentum measurement of the muon tracks, combined with the knowledge of $\sqrt{s}$ from the machine, yields a very good measurement of the higgs mass, a method which works even if the higgs itself decays invisibly. Given a global tracking performance of $C_{\text{track}} = 5\times10^{-5}$ a precision of 70 MeV on a higgs mass of 120 GeV could be achieved [46] from this channel alone, with 40 MeV precision achievable by combining all channels for the standard model higgs (see discussion in Chapter 2).

The momentum precision provided by a large number N of position measurements along a track length is given by the well-known result [47]

$$C_{\text{track}} = \frac{\sigma_{\text{point}}}{0.3B(R_O - R_I)^2}\sqrt{\frac{720}{N+4}} \tag{4}$$

where σ_{point} is the precision of the $r\phi$ measurement of the individual track hits, B is the magnetic field and R_O, R_I are respectively the outer and inner radii of the tracking detector. This equation illustrates that for a given tracking technology, the largest gains can be made by maximising R_O, due to the larger lever-arm for measuring the track curvature.

The global tracking performance depends upon several measurements of the same track from different tracking sub-detectors. Closest to the interaction point lies the silicon vertex detector whose primary goal, discussed below, is to detect and measure the displaced vertices of long-lived heavy flavoured quarks inside jets (and any other long-lived exotica that may be discovered). This detector will also provide a small number N_{vxd} of high-precision points ($\sigma_{\text{point}} \approx$ few μm) at small radii, effectively decreasing the value of R_I in the global track fit. If $N_{\text{vxd}} > 3$ then the vertex detector can also be used as an independent tracking device with its own pattern recognition and momentum measurement. This capability is especially interesting when adjacent tracks are very close together, requiring exceptional two-track resolution to detect. Details of such a design, utilising five layers of CCD detectors, can be found in Ref. [8].

The choice of the central tracking system, outside the vertex detector, depends on the global detector philosophy. The "small" detector concept aims for the necessary tracking performance by employing a high magnetic field together with a lower number of very high precision point measurements, which implies a purely silicon-based technology. This approach has the additional advantage that the two-hit resolution is excellent throughout

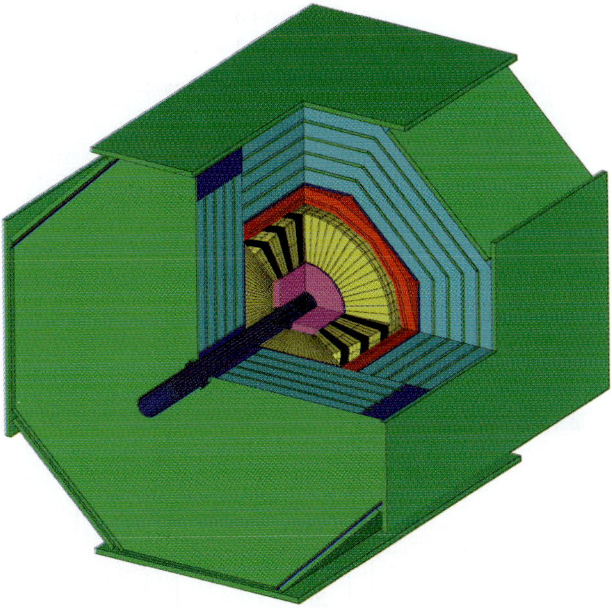

Fig. 10. The detector concept proposed for JLC.

the tracking volume and that the occupancy of the tracker is manageable, even in the high background environment to be expected at multi-TeV linear colliders. However the studies for physics at a 0.5–1 TeV collider have so far tended to favour a "large" detector because of the natural advantage that a long lever-arm gives to the tracking performance. One such detector is that proposed for the JLC [48], whose overall layout is shown in Fig. 10. A complementary approach for TESLA is shown in Fig. 11.

A "large" detector will need a sizeable gaseous central tracking volume such as a Time Projection Chamber (TPC) [8] or a jet chamber [49]. These detectors have the advantage that they consist of very little material with correspondingly low multiple scattering, so that the momentum resolution remains good for lower-momentum tracks. They provide, cost effectively, a large number of measurement points N in Eq. (4), together with some additional particle identification from dE/dx. The large N also facilitates pattern recognition, especially for V^0's decaying well within the tracking volume without any hits at all in the vertex detector. Good pattern recognition is also important to identify and follow the trajectories of lower-momentum spiralling tracks, that is those with $p_T <\sim 0.3BR_O$.

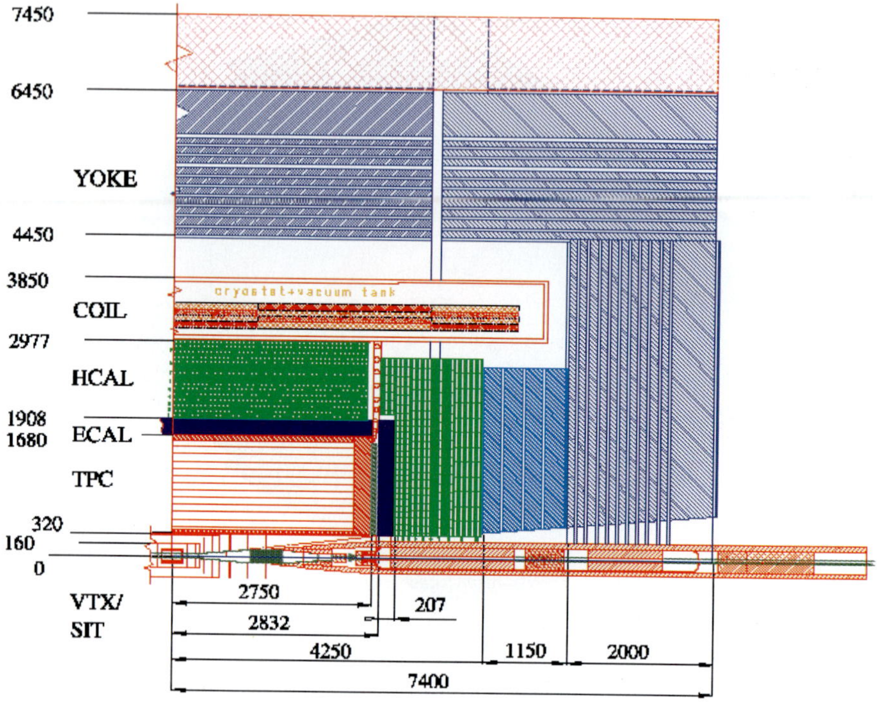

Fig. 11. The detector concept proposed for TESLA.

These will leave a large number of hits throughout the tracking volume and have associated calorimeter clusters positioned and pointing far away from their original trajectory. Such issues are important because, as discussed in Sec. 4.1, lower-momentum tracks and V^0's contribute a significant amount to the overall energy-flow performance of the detector.

In the TESLA TDR [8] a TPC is proposed as a central tracker with $\sigma_{\text{point}} \approx 140\ \mu\text{m}$, $B = 4$ T, $N \sim 200$, $R_I = 0.4$ m and $R_O = 1.6$ m. This results in $C_{\text{track}}^{\text{TPC}} = 1.5 \times 10^{-4}\ \text{GeV}^{-1}\text{c}$. This performance, combined with information from the vertex detector and from an additional layer of silicon detector at a radius just less than that of the inner TPC, can provide a $C_{\text{track}}^{\text{global}}$ of $5 \times 10^{-5}\ \text{GeV}^{-1}\text{c}$. It should be remembered that to achieve such an excellent goal also requires the magnetic field to be uniform and well-mapped to an accuracy of better than 10^{-3} and the systematics to be kept below 10μm.

At low angles, cylindrical detectors such the vertex detector and central tracker lose their power and need to be supplemented by additional devices such as silicon discs in the very forward region, as shown in Fig. 9. High tracking accuracy is required down at these low angles in order to measure the luminosity spectrum from the acollinearity of Bhabha events, as discussed in Sec. 2.9.

4.2.2. *Vertexing Systems*

Much of the pioneering physics at the Linear Collider will centre on the detection of particles with finite lifetime. This will apply to channels ranging from standard model topquark physics and higgs branching ratios to supersymmetric processes and exotic cascade decays. For example, one of the major tasks of the Linear Collider is to map out the mass and couplings of the higgs boson(s) (Chapter 2). The dominance of the branching ratio $h^0 \to b\bar{b}$ for higgs masses less than $2m_W$ means that the b jet tagging efficiency will be one of the most important aspects of detector performance for measurements in the higgs sector. If the origin of mass via the higgs mechanism is to be tested, then the Yukawa couplings of a given particle must be measured and shown to be proportional to the mass of that particle. This will involve measuring accurately the branching ratios of the higgs decays to b and c quarks, τ leptons and gluons, all of which will require vertex reconstruction to (anti-)tag the underlying flavour.

Physics beyond the standard model may also provide signatures of particles with measurable lifetimes. For instance supersymmetric scenarios (such as GMSB [50] or scenarios where R-parity is violated) can lead to particles with measurable lifetimes (Chapter 4). In addition, supersymmetric cascade decays can produce complicated final states, where flavour identification will help in understanding the specific decay-chains involved.

Tracking-based flavour identification relies on the detection and measurement of tracks with significant impact parameters, arising from the decay of the heavy b- and c-quarks. As is now standard, this is done by measuring very accurately the full helix parameters of the tracks in the vicinity of the interaction point using high precision silicon detectors. Where possible the tracks are then formed into vertices, which will be displaced from the primary vertex if the tracks originate from a particle with a finite lifetime, a B or D meson or a τ lepton. Although $\sqrt{s}$ is higher than for LEP/SLC, the final state energy will in general be shared among a correspondingly larger number of particles so the Lorentz boost factors for individual particles will not be significantly greater. As a result, the demands on the vertexing will remain stringent both in the need to measure very short lifetimes as well as to deal with greater complexity. It is amusing to note that a 250 GeV τ will have a mean lifetime $\gamma ct \approx 1$ cm and so will often decay well within the vertex detector volume — and there will be about 400,000 $\tau\tau$ events per 500 fb^{-1}.

The topological vertex structure within jets provides much valuable information about the charge and flavour of the quarks involved in the fundamental process. The fact that a B meson usually decays first into a D meson means that two vertices will be present in a single b jet and this, together with information such as the invariant mass of the daughter tracks and the charged multiplicity, will allow b quark jets to be distinguished from c quark jets. Detailed reconstruction of the vertex charges will allow in some instances b quarks to be distinguished from $\bar{b}$ and similarly c from $\bar{c}$. The efficiency and purity of flavour tagging that can be achieved with such a detector is shown in Fig. 12, where information from the event topology and kinematics has also been included via a neural network [8].

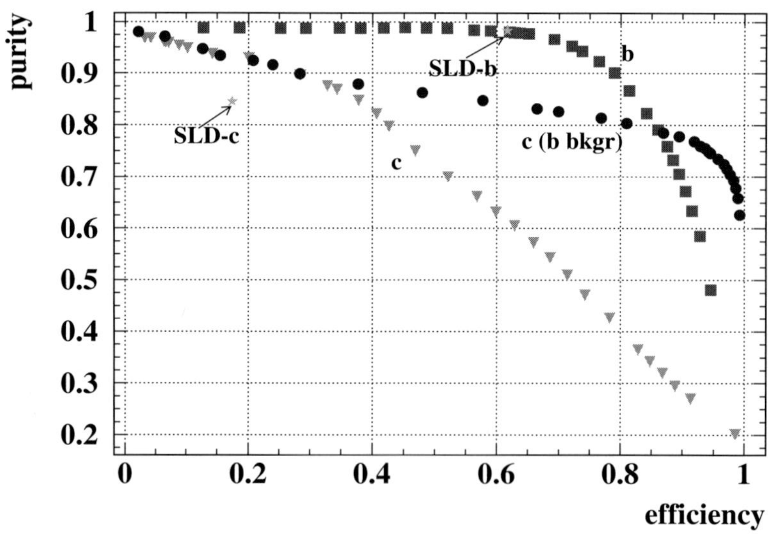

Fig. 12. Efficiency and purity of flavour tagging using a five-layer CCD vertex detector, including information from the event topology and jet mass in a neutral network. The circles show the c-tag performance with b-background only (*i.e.* no uds quarks). The performance at the SLD is indicated for comparison.

Such high performance vertexing will require a pixel-based approach with the first layer as close as feasible to the interaction point, requiring a beam-pipe of small radius in this region. This point is highly non-trivial because reducing the beam-pipe radius by a few mm has major consequences for the layout of the machine beam delivery system, where the positions of beam-line elements such as limiting apertures against synchrotron radiation depend crucially on the limiting aperture at the interaction point. The issue becomes one of balancing backgrounds and integrated luminosity against vertexing performance. For TESLA, a value of 15 mm for the inner radius of the vertex detector was chosen [8] with the maximum length along the beam constrained by the envelope of tracks from the beam-related pair background, as illustrated in Fig. 9.

The impact parameter precision $\sigma_{\rm ip}$ can be expressed as:

$$\sigma_{\rm ip} = \sqrt{a^2 + \left(\frac{b^2}{p\sin^{\frac{3}{2}}\theta}\right)^2} \tag{5}$$

where a is determined by the point resolution of the detector (typically a few μm) and b is the multiple scattering term, which is largest for small momentum p and low polar angle θ (the performance degrades at smaller θ because the particles must traverse a greater distance through material). Current detector designs [8] quote values of a and b of order 4 μm. As illustrated in Sec. 4.1 a typical event has charged tracks with mean energy $\sim$2 GeV and so the multiple scattering term is roughly equal to that from the intrinsic spatial resolution, resulting in $\sigma_{\rm ip} \sim 5$ μm at $\theta = \pi/2$. This example emphasises the importance of keeping the material budget low and motivates R&D into producing ever thinner detectors as well as ever more precise ones.

The enormous potential physics payoff justifies a significant effort in pushing the technology of the vertex detector as far as possible. Promising options currently include CCD, CMOS pixels or active hybrid pixels. The first two can be made very thin, to about 0.06% radiation length per layer with measurement precision of a few μm whereas the active hybrid pixels are more radiation hard, which may be a crucial factor if neutron backgrounds turn out to be higher than presently expected.

4.3. *Calorimetry*

The task of the calorimeter is to measure the energy of individual particles especially those that are invisible to the tracking such as photons and neutral hadrons. The calorimeter should be as hermetic as possible and not have any cracks projecting towards the interaction point. This is important to avoid false signatures of missing energy. Particles escaping through cracks in the calorimeter also reduce the sensitivity of the detector to new missing-mass signals such as expected in theories like supersymmetry or those with extra spatial dimensions. For good hermeticity, the detector must be sensitive down to very low polar angles. The end-cap calorimeters are typically sensitive down to $\theta \simeq 85$ mrad and below this, specialised calorimeters must be installed in the mask region, as discussed in Sec. 3.2. The calorimeter is divided into an inner electromagnetic calorimeter ECAL and an outer hadronic one HCAL. These are now discussed in turn.

4.3.1. *Electromagnetic Calorimeter*

The ECAL is designed primarily to identify and measure the energies, positions and directions of photons and high energy electrons and its performance is critical to many important physics channels.

A high performance ECAL can only work well if the electromagnetic particles do not shower before they enter it and so the material between the interaction point and the ECAL must be kept to a minimum. This has major implications for the global detector design. In particular, we may choose to put the solenoid that supplies the magnetic field for the detector either inside, within or outside the calorimeter. From the point of view of cost, the smaller the diameter of the solenoid the better, however even with extensive R&D it would not be possible to reduce its thickness to less than about half a radiation length. This would be disastrous for the ECAL, where good resolution requires as much of the original particle energy to be detected as possible. As a result the solenoid must lie outside the ECAL and the inner and central tracking detectors must be made of light, thin materials. This constraint applies especially to the forward direction where the end-plates of tracking volumes can contain a lot of material, including electronics, cables and water cooling and where, as discussed in Sec. 2.9, electrons need to be measured very well, in order to determine the luminosity spectrum. To some extent early showering can be detected and compensated for by pre-shower detectors, placed directly in front of the inner face of the ECAL. Such detectors can also play a valuable rôle in supplying additional track hits to strengthen the tracking performance in the forward direction [8].

Referring again to the discussion of energy flow in Sec. 4.1, about a quarter of the energy in hadronic events is made up of photons with mean energy less than about 3 GeV. Measuring these photons well will require both a small value of $C_{\rm cal}^{\rm em}$ in Eq. (3) and also a very fine granularity to distinguish adjacent photons and thereby optimise the jet mass resolution. Fine granularity would also enable single prong decays $\tau \to \nu_\tau \pi^+(n\pi^0)$ to be tagged by resolving and associating any nearby electromagnetic clusters from the π^0s to the single charged track. To meet all of these requirements, the ECAL must be made of a dense material with a small Molière radius, with a readout granularity matched to this radius. Such an approach leads to a compact detector that can fully contain the electromagnetic showers, requiring typically about 27 radiation lengths of calorimeter depth. Designs currently under active study [51] include a very high granularity (1 cm cell size) silicon-tungsten option or a projective "shashlik" option (3 cm cell size) consisting of lead-scintillator.

The longitudinal segmentation of the ECAL must also be sufficient to be sensitive to the shower shape of photons that do not come from the interaction point. Such non-pointing photons can arise from neutralino decays in the GMSB scenario (Chapter 4). The highest granularity detectors are sufficient to this task and provide transverse position accuracy of order $(2/\sqrt{E})$ mm, angular pointing precision of order $(70/\sqrt{E})$ mrad and energy resolution of order $(11–14)/\sqrt{E}$ % (where E is in GeV).

4.3.2. *Hadronic Calorimeter*

The HCAL is designed primarily to detect and measure the energy of neutral hadrons, as well as to aid in muon identification. Just as for the ECAL, the performance of the HCAL will be degraded by any insensitive, or "dead", material in front of it and this also has implications for the location of the solenoid, as well as for other dead material such as cables and supports.

A "large" superconducting solenoid such as the 4 T design proposed in Ref. [52] has a central coil within a cryostat that is about 85 cm thick, consisting of 1.6 interaction lengths consisting mostly of steel, which is a considerable amount of dead material. A compromise must be found between the number of hadronic interaction lengths of HCAL inside the solenoid and cost. The favoured approach is to pack as much sensitive HCAL as possible inside a large solenoid and then to catch any hadronic tails that make it through the solenoid into the instrumented iron outside. The design proposed in Ref. [8] has ~4.5 interaction lengths of HCAL in the barrel region inside the solenoid and ~8 in the forward direction, where there is no constraint from the solenoid. This leads to hadronic leakage of a few percent in the barrel and an order of magnitude less in the forward direction. The iron must be there in any case to act as a return yoke for the magnetic field and, when instrumented, it forms the muon detection system as well as the hadronic tail catcher. Hadronic energy from the tail catcher can be used either to correct the energy in the primary hadronic shower or, in some channels, to veto the event. The overall layout of this approach can be seen in the detector concepts illustrated in Figs. 10 and 11.

Several options also exist for the HCAL technology. One approach uses a sandwich of steel plates as absorbers interleaved with scintillator tiles as the sensitive medium [8, 53]. Compensation for the difference in response of electromagnetic and hadronic particles within the showers will need to be performed offline via software, an approach that needs high granularity to be successful. High granularity is also necessary for efficient muon identification and for the separation of neutral hadrons within jets. Also under study [51] is a digital approach to the HCAL, where hits from the minimum ionising tracks in the shower are simply counted, the number of hits being proportional to the total hadronic energy. Such an approach avoids the need to measure the amplitude of energy deposits, while still providing an accurate measurement.

In summary, many interesting possibilities for calorimeters are under active study and it seems highly feasible that the required detector performance will be attained. However, much R&D remains to be done before a detailed design can be finalised and the task of translating this design into a practical engineering structure should not be underestimated. Financial constraints will be particularly strong, with calorimeters taking up a significant fraction of the total detector cost.

References

1. The 4 LEP collaborations, "Search for the standard model Higgs Boson at LEP" (preliminary results) ALEPH 2001-066, DELPHI 2001-113, L3 Note 2699, OPAL PN 479; will be in proceedings of EPS HEP Conference 2001, Budapest July 12–18 2001; and of LP '01 symposium Rome 23–28 July 2001. `http://opal.web.cern.ch/Opal/pubs/physnote/html/pn483.html`
2. DESY Laboratory reports "123 series":– ed P.M.Zerwas. "e^+e^- Collisions at 500 GeV: the Physics Potential" proceedings of 1991 workshops, DESY 92-123A and DESY 92-123B; proceedings of the 1992/93 worskhops, DESY 93-123C; "e^+e^- Collisions at TeV Energies: the Physics Potential", proceedings of 1994/95 workshops, DESY 96-123D; ed. R.Settles "e^+e^-

Linear Colliders: Physics and Detector Studies", proceedings of the first ECFA/DESY Study, DESY 97-123E.

3. LCWS Fermilab October 24–28 2000; Proceedings edited by A.Para and H.E.Fisk, AIP Conference Proceedings volume 578, American Institute of Physics, Melville NY (2001) ISBN 0-7354-0017-2.
4. LCWS Sitges April 28–May 5 1999; Proceedings edited by E.Fernandez and A.Pacheco, Universita Autonoma de Barcelona Servei de Publicacions, Belaterra (2000) ISBN 84-490 1977-X.
5. Proceedings of the Workshop on Physics and Experiments with Linear Colliders, Morioka-Appi, Japan, September 1995, eds A. Miyamoto, Y. Fujii, T. Matsui, S. Iwata, World Scientific, Singapore (1996).
6. Proceedings of the Workshop on Physics and Experiments with Linear Colliders, Wiakoloa, Hawaii, April 1993, eds. F.A. Harris, S.L. Olsen, S. Pakvasa, X. Tata, World Scientific, Singapore (1993).
7. Proceedings of the Workshop on Physics and Experiments with Linear Colliders, Saariselka, Finland, September 1991, eds. R. Orava, P. Eerola, M. Nordberg, World Scientific, Singapore (1992).
8. "TESLA The Superconducting Electron-Positron Linear Collider with an Integrated X-Ray Laser laboratory", Technical Design Report DESY 2001-0011, ECFA 2001-209, in 5 volumes. `http://tesla.desy.de/tdr/`.
9. "Linear Collider Physics Resource Book for Snowmass '01", American linear collider working group; BNL-52627, CLNS 01/1729, FERMILAB-Pub-01/058-E, LBNL-47813, SLAC-R-570, UCRL-ID-143810-DR, LC-REV-2001-074-US, `http://www.desy.de/~lcnotes/2001/074/LCBook/index.html`.
10. K. Abe *et al.*,"Particle Physics at JLC" ACFA Linear Collider Working Group Report, Autumn 2001. `http://acfahep.kek.jp/acfareport/` Updated at ACFA LC Symposium 12 February 2003 `http://conference.kek.jp/acfalc`.
11. Web index of Linear Collider Notes `http://www.desy.de/~lcnotes/LCnotes_welcome.html`.
12. P. Raimondi *et al.*, "Luminosity Upgrades for the SLC", Proceedings of 1999 Particle Accelerator Conference, ibid, p. 3384.
13. "An Introduction to Particle Accelerators", E.J.N. Wilson, Oxford University Press, (2001).
14. H.Wiedemann, "Particle Accelerator Physics; Basic Principles and Linear Beam Dynamics", Springer-Verlag (Berlin 1993) ISBN 3-540-56550-7.
15. R. Brinkmann. "Technology and Challenges of Linear Colliders", Proceedings of 1999 Particle Accelerator Conference, ibid, p. 16.
16. T.Raubenheimer. "Accelerator Physics Challenges in Future Linear Colliders", Proceedings of 1999 Particle Accelerator Conference, New York, 29 Mar–2 April 1999, eds. A.Luccio, W.MacKay; IEEE, Piscataway, NJ, USA (1999)IEEE99CH36366, p. 240.
17. O. Napoly and D. Schulte, "Beam-beam effects at the TESLA Linear Collider", DESY TESLA-2001-15.
18. P. Tenenbaum, T. Shintake SLAC-PUB-8057 (1999)

19. T.Shintake *et al.*, “Experiments of nanometer spot size monitor at FFTB using laser interferometer”; Proc. of Particle Accelerator Conference Dallas, Texas May 1–5 1995, (IEEE 1996) 0-7803-3053-6/96, p. 2444.
T.Shintake, Proposal of a nanometer beam size monitor for e^+e^- linear colliders”, Nucl. Instr. and Meth. **A311**, 453 (1992).
20. A.Wagner “Accelerators to span the globe”, CERN Courier **40**, 5, June 2000, p. 19. `http://www.cerncourier.com/main/article/40/5/15`.
21. “2001 report on the Next Linear Collider”; Report submitted to Snowmass ’01, ed. N.Phinney, SLAC. FERMILAB-Conf-01-075-E, LBNL-47935, SLAC-R-571, UCRL-ID-144077.
22. Latest full account is KEK Report 92-16, “JLC-I” December, 1992. `http://www-jlc.kek.jp/JLC.proposal-e.html`.
23. H Matsumoto, talk given at the 5th ACFA Workshop on Physics and Detector at a Linear Collider, July 2002 `http://www.icepp.s.u-tokyo.ac.jp/acfa/docs/H.Matsumoto.pdf`.
24. J. Frisch *et al.* “Studies of Breakdown in High Gradient X-Band Accelerator Structures using Acoustic Emission” SLAC-PUB-9649.
25. F. le Pimpec *et al.* “Autopsy on an RF processed X-band travelling wave structure”, SLAC LCC Technical Note-0097 Rev 1, August 2002. `http://www-project.slac.stanford.edu/lc/ilc/TechNotes/LCCNotes/lcc_notes_index.htm`.
26. R.W.Assman *et al.* “A 3 teV e^+e^- Linear Collider based on CLIC Technology”, CERN report 2000-008, July 2000. `http://ps-div.web.cern.ch/ps-div/CLIC/General.html`.
27. A.V. Subashiev *et al.* ”Spin Polarised Electrons, Generation and Applications”, Phys. Low Dim. Struct. 1/2, 1 (1999), also SLAC-PUB8035(1998).
28. KEK Accelerator Test Facility `http://atfweb.kek.jp/`.
29. P. Raimondi and A.Seryi, “A novel final focus design for future linear colliders”, Phys. Rev. Lett. **86**, 3889 (2001).
30. e.g. The OPAL Collaboration, G. Abbiendi *et al.*, Eur. Phys. J **C19**, 587 (2001).
31. J.Kent *et al.* “Precision measurement of the SLC beam energy” SLAC PUB 4922, LBL-26977. Presented at IEEE Particle Accelerator Conference, Chicago, March 20–23, 1989.
32. P.B.Renton for the LEP Energy Working Group, “Measurement of beam energy at LEP2”, Proceedings of ICHEP2000, Osaka 27 July–2 August 2000, eds C.S.Lim and T.Yamanaka. World Scientific (Singapore) 2001, p. 687.
33. e.g. The OPAL collaboration, G.Abbiendi *et al.*, Eur. Phys. J. **C14**, 373 (2000); The ALEPH collaboration, D. Bederede *et al.*, Nucl. Instr. and Meth. **A365**, 117 (1995).
34. M.N Frary and D.J.Miller, “Monitoring the Luminosity Spectrum” DESY 92-123A, 379 (1992), `http://www.hep.ucl.ac.uk/lc/documents/frarymiller.pdf`.
35. K.Moenig “Measurement of the Differential Luminosity using Bhabha events in the Forward-Tracking region at TESLA” LC Notes LC-PHSM 2000-60. `http://www-flc.desy.de/lcnotes/notes/LC-PHSM-2000-060.ps.gz`.

36. S.T.Boogert and D.J.Miller, "Questions about the measurement of the e^+e^- luminosity spectrum at a future linear collider" Proceedings of LCWS, Jeju Island, Korea, 2002, eds. J.S. Kang and H. Park, to be published by World Scientific. `http://www.hep.ucl.ac.uk/lc/documents/sb_dm.ps`.
37. I.F.Ginzburg *et al.* JETP Lett. **34**, 491 (1982).
38. see e.g. V.Telnov in Appendix 1 of the TESLA TDR, Ibid, DESY 2001-0011, ECFA 2001-209. `http://tesla.desy.de/tdr/`.
39. Markiewicz, T (2000) "Summary of the Machine Detector Interface Sessions", (2000) Proc. LCWS2000 (Fermilab) [3].
40. Büßer, K. (2000) "Mask Design and Background Studies for TESLA", (2000) Proc. LCWS2000 (Fermilab) [3].
41. D. Schulte (1996) "Study of Electromagnetic and Hadronic Background in the Interaction Region of the TESLA Collider", Thesis, Univ. Hamburg. TESLA 97-08.
42. Schulte, D. (2000) "Backgrounds at CLIC", (2000) Proc. LCWS2000 (Fermilab) [3].
43. H. M. Araujo, A. Stahl and D. J. Miller "A Lead Tungstate Detector with Integrated Light Readout for the TESLA Luminosity Calorimeter". LC-DET-2001-80.
44. Brau,J.E. (1999) "Overview of the American Detector Models", Proc. Worldwide Study on Physics and Experiments with Future Linear e^+e^- Colliders, Sitges (1999), p. 881.
45. ALEPH Collaboration, D.Buskulic *et al.*, Nucl. Inst. Meth. **A360**, 481 (1995).
46. Battaglia, M. (2001) "Precision Measurements of the light Neutral Higgs at the Linear Collider",Proc. Worldwide Study on Physics and Experiments with Future Linear e^+e^- Colliders, Fermilab (2000).
47. Gluckstern, R.L., Nucl. Instrum. Methods **24**, 284 (1963).
48. Matsui,T. (1999), "Overview of the JLC Detector Model" Proc. Worldwide Study on Physics and Experiments with Future Linear e^+e^- Colliders, Sitges (1999), p. 728.
49. Fujii,K. (2000), "JLC Central Tracker R&D", Proc. Worldwide Study on Physics and Experiments with Future Linear e^+e^- Colliders, Fermilab (2000).
50. For a review, see Giudice,G.F. and Ratazzi,R., Phys. Rep. **322**, 419 (1999). Experimental prospects are covered in: Ambrosanio, S. and Blair, G.A. "Measuring gauge mediated supersymmetry breaking parameters at a 500-GeV e^+e^- linear collider. Eur. Phys. J. **C12**, 287 (2000).
51. Brient, J.C. (2000) "The Calorimetery for the e^+e^- FLC", (2000) Proc. LCWS2000 (Fermilab) [3].
52. CMS Collaboration (1997) "The Magnet Project", Technical Design Report, CERN/LHCC 97-10.
53. Takeshita,T.(2000)"Tile/Fibre Hadron Calorimeter Performance", Proc. Worldwide Study on Physics and Experiments with Future Linear e^+e^- Colliders, Fermilab (2000).

CHAPTER 2

Higgs Physics at the Linear Collider

John F. Gunion
Davis Institute for High Energy Physics
University of California, Davis, CA 95616, U.S.A.
E-mail: jfgucd@higgs.ucdavis.edu

Howard E. Haber
Santa Cruz Institute for Particle Physics
University of California, Santa Cruz, CA 95064, U.S.A.
E-mail: haber@scipp.ucsc.edu

Rick Van Kooten
Department of Physics
Indiana University, Bloomington, IN 47405, U.S.A.
E-mail: rickv@paoli.physics.indiana.edu

We review the theory of Higgs bosons, with emphasis on the Higgs scalars of the Standard Model and its non-supersymmetric and supersymmetric extensions. After surveying the expected knowledge of Higgs boson physics after the Tevatron and LHC experimental programs, we examine in detail expectations for precision Higgs measurements at a future e^+e^- linear collider (LC). A comprehensive phenomenological profile can be assembled from LC Higgs studies (both in e^+e^- and $\gamma\gamma$ collisions). The Giga-Z option can provide important constraints and consistency checks for the theory of electroweak symmetry breaking.

1. Introduction

This chapter shows how an e^+e^- linear collider (LC) can contribute to our understanding of the Higgs sector through detailed studies of the physical Higgs boson state(s). Although this subject has been reviewed several times in the past [1, 2, 3, 4, 5, 6], there are at least two reasons to revisit the subject. First, the completion of the LEP Higgs search, together with precise measurements from SLC, LEP, and the Tevatron, provides a clearer idea of what to expect. The simplest interpretation of these results point to a light Higgs boson with (nearly) standard couplings to W and Z. The key properties of such a particle can be investigated with a LC with a center-of-mass energy of $\sqrt{s} = 500$ GeV. Second, the luminosity expected from the LC is now higher: 200–300 $\mathrm{fb}^{-1}\mathrm{yr}^{-1}$ at $\sqrt{s} = 500$ GeV, and 300–500 $\mathrm{fb}^{-1}\mathrm{yr}^{-1}$ at $\sqrt{s} = 800$ GeV. In particular, detailed designs have been developed for the LC by the American/Asian collaboration (the NLC/JLC design [7, 8]) and by the DESY-based European collaboration (the TESLA design [9]). At $\sqrt{s} = 500$ GeV, the nominal 10^7 sec year integrated luminosities are 220 $\mathrm{fb}^{-1}\mathrm{yr}^{-1}$ and 340 $\mathrm{fb}^{-1}\mathrm{yr}^{-1}$ for the NLC/JLC and TESLA designs, respectively [10]. Consequently, several tens of thousands of Higgs bosons should be produced in each year of operation. With such samples, numerous precision Higgs measurements become feasible, and will provide fundamental insights into the properties of the Higgs boson(s) and the underlying dynamics of electroweak symmetry breaking.

The e^+e^- LC with center-of-mass energy $\sqrt{s}$ can also be designed to operate as a photon-photon collider. This is achieved by using Compton backscattered photons in the scattering of intense laser photons on the initial polarized e^+e^- beams [11, 12]. The resulting $\gamma\gamma$ center of mass energy is peaked at about $0.8\sqrt{s}$ for the appropriate choices of machine parameters. The luminosity achievable as a function of the photon beam energy depends strongly on the machine parameters (in particular, the choice of laser polarizations). The photon collider provides additional opportunities for Higgs physics [13]–[19].

Finally, we note that substantial improvements are possible for precision measurements of m_W, m_t and electroweak mixing angle measurements at the LC [20]. But, the most significant improvements can be achieved at the Giga-Z [21], where the LC operates at $\sqrt{s} = m_Z$ and $\sqrt{s} \simeq 2m_W$. With an integrated luminosity of 50 fb^{-1}, one can collect 1.5×10^9 Z events and about 10^6 W^+W^- pairs in the threshold region. Employing a global fit to the precision electroweak data in the Standard Model, the anticipated fractional Higgs mass uncertainty achievable would be about 8%. This would provide a stringent test for the theory of the Higgs boson, as well as very strong constraints on any new physics beyond the Standard Model that couples to the W and Z gauge bosons.

There is an enormous literature on the Higgs boson and, more generally, on possible mechanisms of electroweak symmetry breaking. It is impossible to review all theoretical approaches here. To provide a manageable, but nevertheless illustrative, survey of the LC capabilities, we focus mostly on the Higgs boson of the Standard Model (SM), and on the Higgs bosons of the minimal supersymmetric extension of the Standard Model (MSSM). Although this choice is partly motivated by simplicity, a stronger impetus comes from the precision data collected over the past few years, and some other related considerations.

The SM, which adds to the observed particles a single complex doublet of scalar fields, is economical. It provides a good fit to the precision electroweak data. Many extended models of electroweak symmetry breaking possess a limit, called the decoupling limit [22], that is experimentally almost indistinguishable from the SM at low energies. These models agree with the data equally well, and even away from the decoupling limit they usually predict a weakly coupled Higgs boson whose mass is at most several hundred GeV. Thus, the SM serves as a basis for discussing the Higgs phenomenology of a wide range of models, all of which are compatible with present experimental constraints.

However, the SM suffers from several theoretical problems, which are either absent or less severe with weak-scale supersymmetry. The Higgs sector of the minimal supersymmetric extension of the Standard Model (MSSM) is a constrained two-Higgs-doublet model (2HDM), consisting of two CP-even Higgs bosons, h^0 and H^0, a CP-odd Higgs boson, A^0, and a charged Higgs pair, $H^\pm$. The MSSM is especially attractive because the superpartners modify the running of the strong, weak, and electromagnetic gauge couplings in just the right way as to yield unification at about 10^{16} GeV [23]. For this reason, the MSSM is arguably the most compelling extension of the SM. This is directly relevant to Higgs phenomenology, because a theoretical bound requires the mass of the lightest CP-even Higgs boson h^0 of the MSSM to be less than about 135 GeV (in most non-minimal supersymmetric models, $m_{h^0} \lesssim 200$ GeV), as discussed in Section 5.2. Furthermore, the MSSM offers, in some regions of parameter space, very non-standard Higgs phenomenology, so the full range of possibilities in the MSSM can be used to indicate how well the LC performs in non-standard scenarios. Thus, we use the SM to show how the LC fares when there is only one observable Higgs boson, and the MSSM to illustrate how extra fields can complicate the phenomenology. We also use various other models to illustrate important exceptions to conclusions that would be drawn from the SM and MSSM alone.

The rest of this chapter is organized as follows. Section 2 presents, in some detail, the argument that one should expect a weakly coupled Higgs boson with a mass that is probably below about 200 GeV. In Section 3, we summarize the theory of the SM Higgs boson. In Section 4, we review the expectations for Higgs discovery and the determination of Higgs boson properties at the Tevatron and LHC. In Section 5, we introduce the Higgs sector of the MSSM and discuss its theoretical properties. The present direct search limits are reviewed, and expectations for discovery at the Tevatron and LHC are described in Section 6. In Section 7, we treat the theory of the non-minimal Higgs sector more generally. In particular, we focus on the decoupling limit, in which the properties of the lightest Higgs scalar are nearly identical to those of the Standard Model Higgs boson, and discuss how to distinguish the two. We also discuss some non-decoupling exceptions to the usual decoupling scenario. In Section 8, we briefly discuss the case of a Higgs sector containing triplet Higgs representations and also consider the Higgs-like particles that can arise if the underlying assumption of a weakly coupled elementary Higgs sector is *not* realized in Nature.

Finally, we turn to the program of Higgs measurements that can be carried out at the LC, focusing on e^+e^- collisions at higher energy, but also addressing the impact of the Giga-Z operation and $\gamma\gamma$ collisions. This material summarizes and updates the results of the American Linear Collider Working Group that were presented in [10]. The measurement of Higgs boson properties in e^+e^- collisions is outlined in Section 9, and includes a survey of the measurements that can be made for a SM-like Higgs boson for all masses up to 500 GeV. We also discuss measurements of the extra Higgs bosons that appear in the MSSM. Because the phenomenology of the decoupling limit mimics that of the SM Higgs boson, we emphasize how the precision that stems from high luminosity helps to diagnose the underlying dynamics. In Section 10, we outline the impact of the Giga-Z operation on constraining and exploring various scenarios. In Section 11, the most important gains from a photon-photon collider are reviewed. We end this chapter with some brief conclusions in Section 12.

2. Expectations for Electroweak Symmetry Breaking

With the recent completion of experimentation at the LEP collider, the Standard Model of particle physics appears close to final experimental verification. After more than ten years of precision measurements of electroweak observables at LEP, SLC and the Tevatron, no definitive departures from

Standard Model predictions have been found [24]. In some cases, theoretical predictions have been checked with an accuracy of one part in a thousand or better. However, the dynamics responsible for electroweak symmetry breaking have not yet been directly identified. Nevertheless, these dynamics affect predictions for currently observed electroweak processes at the one-loop quantum level. Consequently, the analysis of precision electroweak data can already provide some useful constraints on the nature of electroweak symmetry breaking dynamics.

In the Standard Model, electroweak symmetry breaking dynamics arise via a self-interacting complex doublet of scalar fields, which consists of four real degrees of freedom. Renormalizable interactions are arranged in such a way that the neutral component of the scalar doublet acquires a vacuum expectation value, $v = 246$ GeV, which sets the scale of electroweak symmetry breaking. Hence, three massless Goldstone bosons are generated that are absorbed by the $W^{\pm}$ and Z, thereby providing the resulting massive gauge bosons with longitudinal components. The fourth scalar degree of freedom remains in the physical spectrum, and it is the CP-even neutral Higgs boson of the Standard Model. It is further assumed in the Standard Model that the scalar doublet also couples to fermions through Yukawa interactions. After electroweak symmetry breaking, these interactions are responsible for the generation of quark and charged lepton masses.

The global analysis of electroweak observables provides a good fit to the Standard Model predictions. Such analyses take the Higgs mass as a free parameter. The electroweak observables depend logarithmically on the Higgs mass through its one-loop effects. The accuracy of the current data (and the reliability of the corresponding theoretical computations) already provides a significant constraint on the value of the Higgs mass. In [24], the constraints of the global precision electroweak analysis yield a one-sided 95% CL upper limit of $m_{h_{\rm SM}} \leq 193$ GeV at 95% CL. Meanwhile, direct searches for the Higgs mass at LEP achieved a 95% CL limit of $m_{h_{\rm SM}} > 114.4$ GeV [25].

One can question the significance of these results. After all, the self-interacting scalar field is only one model of electroweak symmetry breaking; other approaches, based on different dynamics, are also possible. For example, one can introduce new fermions and new forces, in which the Goldstone bosons are a consequence of the strong binding of the new fermion fields [26]. Present experimental data does not sufficiently constrain the nature of the dynamics responsible for electroweak symmetry breaking. Nevertheless, one can attempt to classify alternative scenarios and study the constraints of the global precision electroweak fits and the implications for phenomenology at future colliders. Since electroweak symmetry dynamics must affect the one-loop corrections to electroweak observables, constraints on alternative approaches can be obtained by generalizing the global precision electroweak fits to allow for new contributions at one-loop. These enter primarily through corrections to the self-energies of the gauge bosons (the so-called "oblique" corrections). Under the assumption that new physics is characterized by a mass scale $M \gg m_Z$, one can parameterize the leading oblique corrections by three constants, S, T, and U, first introduced by Peskin and Takeuchi [27]. In almost all theories of electroweak symmetry breaking dynamics, $U \ll S$, T, so it is sufficient to consider a global electroweak fit in which $m_{h_{\rm SM}}$, S and T are free parameters. (The zero of the S–T plane must be defined relative to some fixed value of the Higgs mass, usually taken to be 100 GeV.) New electroweak symmetry breaking dynamics could generate non-zero values of S and T, while allowing for a much heavier Higgs mass (or equivalent). Various possibilities have been classified by Peskin and Wells [28], who found that all models in the literature allowing a significantly heavier Higgs boson also generate experimental signatures of new physics at the TeV scale. At the LC, such new physics could be studied either directly by producing new particles or indirectly by improving precision measurements of electroweak observables.

In this chapter, we mainly consider the case of a light weakly coupled Higgs boson, corresponding to the simplest interpretation of the precision electroweak data. Nevertheless, this still does not fix the theory of electroweak symmetry breaking. It is easy to construct extensions of the scalar boson dynamics and generate non-minimal Higgs sectors. Such theories can contain charged Higgs bosons and neutral Higgs bosons of opposite (or indefinite) CP-quantum numbers. Although some theoretical constraints exist, there is still considerable freedom in constructing models which satisfy all known experimental constraints. Moreover, in most extensions of the Standard Model, there exists a large range of parameter space in which the properties of the lightest Higgs scalar are virtually indistinguishable from those of the Standard Model Higgs boson. Once the Higgs boson is discovered, one of the challenges for experiments at future colliders is to detect deviations from the properties expected for the Standard Model Higgs boson, in order to better constrain the underlying scalar dynamics.

Although the Standard Model provides a remarkably successful description of the properties of the quarks, leptons and spin-1 gauge bosons at energy scales of $\mathcal{O}(100)$ GeV and below, the Standard Model is not the ultimate theory of the fundamental particles and their interactions. At an energy scale above the Planck scale, $M_{\rm PL} \simeq 10^{19}$ GeV, quantum gravitational effects become significant and the Standard Model must be replaced by a more fundamental theory that incorporates gravity. Furthermore, on theoretical grounds it is very likely that the Standard Model breaks down at some energy scale, Λ. In this case, the Standard Model degrees of freedom are no longer adequate for describing the physics above Λ and new physics must enter. Thus, the Standard Model is not a *fundamental* theory; at best, it is an *effective field theory* [29]. At energies below the scale Λ, the Standard Model (with higher-dimension operators to parameterize the new physics at the scale Λ) provides an extremely good description of all observable phenomena.

To assess the potential of future experiments, it is essential to ask how large Λ can be. After all, if Λ is as large as $M_{\rm PL}$, it does not matter much whether the Standard Model is a fundamental or an effective field theory. The energy scale Λ arises when analyzing the scale dependence of the Higgs self-coupling and the Higgs-top quark Yukawa coupling, so the value of Λ depends on the Higgs mass $m_{h_{\rm SM}}$ and the top quark mass m_t. (Recall that $m^2_{h_{\rm SM}} = \frac{1}{2}\lambda v^2$ and $m_t = h_t v/\sqrt{2}$, where λ is the Higgs quartic self-coupling and h_t is the Higgs-top quark Yukawa coupling.) If $m_{h_{\rm SM}}$ is too large, then the perturbatively evolved Higgs self-coupling λ blows up at some scale Λ [30, 31]. If $m_{h_{\rm SM}}$ is too small (compared to m_t), then the (perturbative) Higgs potential develops a second (global) minimum at a large value of the scalar field of order Λ [32]. Thus, new physics must enter at or below Λ, in order that the true minimum of the theory correspond to the observed SU(2)×U(1) broken vacuum with $v = 246$ GeV for scales above Λ.

The arguments just given are based on perturbation theory so, taken on their own, are not completely persuasive. Once the couplings become strong, it is *a priori* conceivable that a novel, non-perturbative scaling behavior would rescue the theory. This possibility has been checked, mostly in an approximation neglecting the gauge couplings [30]. Let us consider first the case of large Higgs mass (large λ), so the Yukawa coupling can be neglected. By a number of methods, particularly numerical lattice methods [33, 34] and strong coupling expansions [35], one finds that non-perturbative renormalization effects rapidly bring the self-coupling back into the perturbative regime. This is the so-called triviality phenomenon: if one starts with finite Λ and tries to take the limit $\Lambda \to \infty$, one ends up with a renormalized self-coupling $\lambda \to 0$. A similar phenomenon occurs for large Yukawa couplings. In these studies [36], the vacuum is never unstable (non-perturbatively). Instead, strong renormalization effects drive the Yukawa coupling back into the perturbative regime.

The non-perturbative studies very much favor a finite Λ, despite exhaustive attempts [34] to find non-trivial behavior for $\Lambda \to \infty$. They also justify the use of perturbation theory, because the renormalized self- and Yukawa couplings end up being small enough at scales below Λ. Thus, with the perturbative analysis and a given value of Λ, one can compute the minimum and maximum Higgs mass allowed. The results of such an analysis (with shaded bands indicating the theoretical uncertainty of the result) are illustrated in Fig. 1 [31, 37]. From the upper bound one sees that if $\Lambda \sim 10^3$ GeV $= 1$ TeV, then $m_{h_{\rm SM}} \lesssim 800$ GeV. If Λ is more remote, say $\Lambda \gtrsim 10^6$ GeV, then $m_{h_{\rm SM}} \lesssim 300$ GeV.

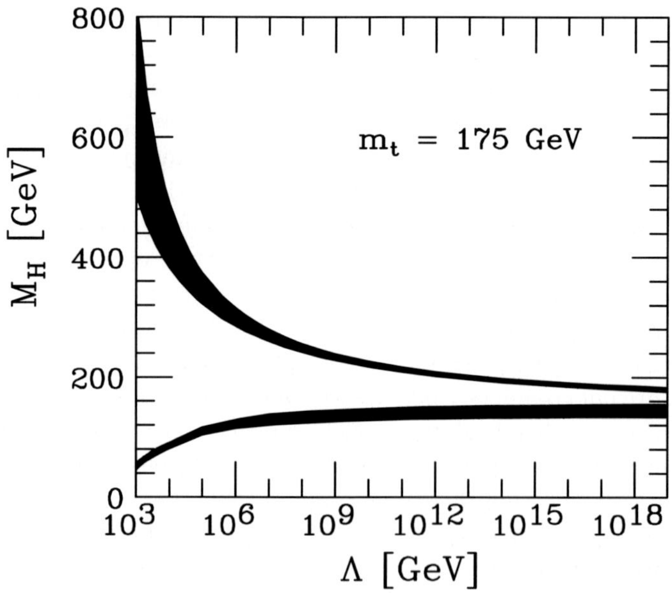

Fig. 1. The upper [30, 31] and the lower [32] Higgs mass bounds as a function of the energy scale Λ at which the Standard Model breaks down, assuming $m_t =$ 175 GeV and $\alpha_s(m_Z) = 0.118$. The shaded areas above reflect the theoretical uncertainties in the calculations of the Higgs mass bounds from [37].

In the Higgs mass range 130 GeV $\lesssim m_{h_{\rm SM}} \lesssim$ 180 GeV, it seems that an effective Standard Model could survive all the way to the Planck scale.[a] However, such a possibility seems unlikely, based on the following "naturalness" argument [39]. In an effective field theory, masses and dimensionless couplings are calculable in terms of parameters of a more fundamental theory that describes physics at the energy scale Λ. All low-energy couplings and fermion masses are logarithmically sensitive to Λ. In contrast, scalar squared-masses are *quadratically* sensitive to Λ. The Higgs mass (at one-loop) has the following heuristic form:

$$m_h^2 = (m_h^2)_0 + \frac{cg^2}{16\pi^2}\Lambda^2 \,, \tag{1}$$

where $(m_h^2)_0$ is a parameter of the fundamental theory and c is a constant, presumably of $\mathcal{O}(1)$, that depends on the physics near the scale Λ. The "natural" value for the scalar squared-mass is $g^2\Lambda^2/16\pi^2$. Thus, the expectation for Λ is

$$\Lambda \simeq \frac{4\pi m_h}{g} \sim \mathcal{O}(1~\text{TeV})\,. \tag{2}$$

If Λ is significantly larger than 1 TeV then the only way for the Higgs mass to be of order the scale of electroweak symmetry breaking is to have an "unnatural" cancellation between the two terms of Eq. (1). This seems highly unlikely given that the two terms of Eq. (1) have completely different origins.

An attractive theoretical framework that incorporates weakly coupled Higgs bosons and satisfies the constraint of Eq. (2) is that of "low-energy" or "weak-scale" supersymmetry [40, 41]. In this framework, supersymmetry is used to relate fermion and boson masses and interaction strengths. Since fermion masses are only logarithmically sensitive to Λ, boson masses will exhibit the same logarithmic sensitivity if supersymmetry is exact. Since no supersymmetric partners of Standard Model particles have yet been found, supersymmetry cannot be an exact symmetry of nature. Thus, Λ should be identified with the supersymmetry breaking scale. The naturalness constraint of Eq. (2) is still relevant. It implies that the scale of supersymmetry breaking should not be much larger than 1 TeV, to preserve the naturalness of scalar masses. The supersymmetric extension of the Standard Model

[a] The constraint on Λ due to vacuum stability in Fig. 1 can be relaxed somewhat if one allows for the electroweak vacuum to be metastable, with a lifetime greater than the age of the universe. The analysis of [38] finds that for a sufficiently long-lived electroweak vacuum, the Higgs mass lower limit of 130 GeV just quoted is reduced to about 115 GeV.

would then replace the Standard Model as the effective field theory of the TeV scale.

One advantage of the supersymmetric approach is that the effective low-energy supersymmetric theory *can* be valid all the way up to the Planck scale, while still being natural. The unification of the three gauge couplings at an energy scale close to the Planck scale, which does not occur in the Standard Model, is seen to occur in the minimal supersymmetric extension of the Standard Model [23], and provides an additional motivation for seriously considering the low-energy supersymmetric framework. However, the fundamental origin of supersymmetry breaking is not known at present. Without a fundamental theory of supersymmetry breaking, one ends up with an effective low-energy theory characterized by over 100 unknown parameters that in principle would have to be measured by experiment. This remains one of the main stumbling blocks for creating a truly predictive supersymmetric model of fundamental particles and their interactions. Nevertheless, the Higgs sectors of the simplest supersymmetric models are quite strongly constrained and exhibit very specific phenomenological profiles.

3. The Standard Model Higgs Boson-Theory

In the Standard Model, the Higgs mass, $m^2_{h_{\rm SM}} = \frac{1}{2}\lambda v^2$, depends on the Higgs self-coupling λ. Since λ is unknown at present, the value of the SM Higgs mass is not predicted, although other theoretical considerations (discussed in Section 2) place constraints on the Higgs mass, as exhibited in Fig. 1. The Higgs couplings to fermions [gauge bosons] are proportional to the corresponding particle masses [squared-masses]. As a result, Higgs phenomenology is governed primarily by the couplings of the Higgs boson to the $W^\pm$ and Z and the third-generation quarks and leptons. Two-loop-induced Higgs couplings are also phenomenologically important: the $h_{\rm SM}gg$ coupling (g is the gluon), which is induced by the one-loop graph in which the Higgs boson couples to a virtual $t\bar{t}$ pair, and the $h_{\rm SM}\gamma\gamma$ coupling, for which the W^+W^- loop contribution is dominant. Further details of the SM Higgs boson properties are given in [1].

3.1. *Standard Model Higgs Boson Decay Modes*

The Higgs boson mass is the only unknown parameter in the Standard Model. Thus, one can compute Higgs boson branching ratios and production cross sections as a function of $m_{h_{\rm SM}}$. The branching ratios for the dominant decay modes of a Standard Model Higgs boson are shown as a function of

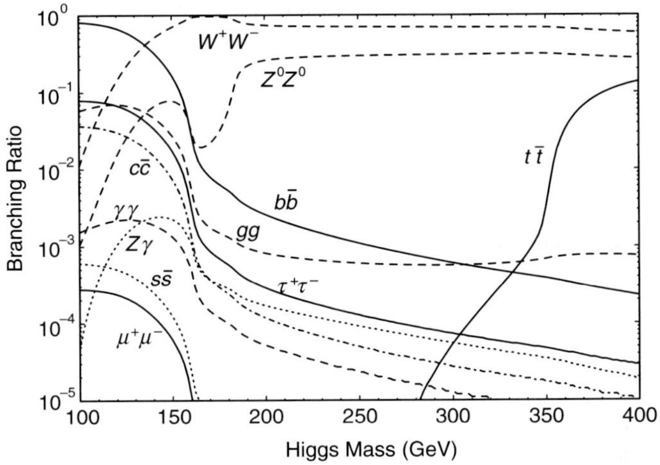

Fig. 2. Branching ratios of the dominant decay modes of the Standard Model Higgs boson. These results have been obtained with the program HDECAY [42], and include QCD corrections beyond the leading order.

Higgs boson mass in Fig. 2. Note that subdominant channels are important to establish a complete phenomenological profile of the Higgs boson, and to check consistency (or look for departures from) Standard Model predictions. For 115 GeV $\sim m_{h_{\rm SM}} \lesssim 2m_W$ many decays modes are large enough to measure, as discussed in Section 9.

For $m_{h_{\rm SM}} \lesssim 135$ GeV, the main Higgs decay mode is $h_{\rm SM} \to b\bar{b}$, while the decays $h_{\rm SM} \to \tau^+\tau^-$ and $c\bar{c}$ can also be phenomenologically relevant. In addition, although one–loop suppressed, the decay $h_{\rm SM} \to gg$ is competitive with other decays for $m_{h_{\rm SM}} \lesssim 2m_W$ because of the large top Yukawa coupling and the color factor. As the Higgs mass increases above 135 GeV, the branching ratio to vector boson pairs becomes dominant. In particular, the main Higgs decay mode is $h_{\rm SM} \to WW^{(*)}$, where one of the W's must be off-shell (indicated by the star superscript) if $m_{h_{\rm SM}} < 2m_W$. For Higgs bosons with $m_{h_{\rm SM}} \gtrsim 2m_t$, the decay $h_{\rm SM} \to t\bar{t}$ begins to increase until it reaches its maximal value of about 20%.

Rare Higgs decay modes can also play an important role. The one-loop decay $h_{\rm SM} \to \gamma\gamma$ is a suppressed mode. For $m_W \lesssim m_{h_{\rm SM}} \lesssim 2m_W$, BR($h_{\rm SM} \to \gamma\gamma$) is above 10^{-3}. This decay channel provides an important Higgs discovery mode at the LHC for 100 GeV $\lesssim m_{h_{\rm SM}} \lesssim 150$ GeV. At the LC, the direct observation of $h_{\rm SM} \to \gamma\gamma$ is also possible in $Zh_{\rm SM}$ production despite its suppressed branching ratio. The partial width $\Gamma(h^0 \to \gamma\gamma)$ is also significant in that it governs the Higgs production rate at a $\gamma\gamma$ collider.

3.2. *Standard Model Higgs Boson Production at the LC*

In the Standard Model there are two main processes to produce the Higgs boson in e^+e^- annihilation. These processes are also relevant in many extensions of the Standard Model, particularly near the decoupling limit, in which the lightest CP-even Higgs boson possesses properties nearly identical to those of the SM Higgs boson. In the "Higgsstrahlung" process, a virtual Z boson decays to an on-shell Z and the $h_{\rm SM}$, as depicted in Fig. 3(a).

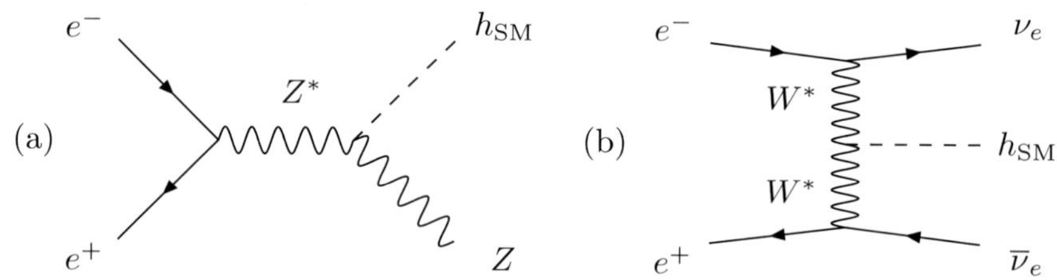

Fig. 3. Main production processes for Higgs production in e^+e^- annihilation. (a) Higgsstrahlung. (b) WW fusion.

The cross section for Higgsstrahlung rises sharply at threshold to a maximum a few tens of GeV above $m_h + m_Z$, and then falls off as s^{-1}, as shown in Fig. 4. The associated production of the Z provides an important trigger for Higgsstrahlung events. In particular, in some theories beyond the Standard Model, the Higgs boson decays into invisible modes, in which case the ability to reconstruct the Higgs boson mass peak in the spectrum of the missing mass recoiling against the Z will be crucial. The other production process is called "vector boson fusion", where the incoming e^+ and e^- each emit a virtual vector boson, followed by vector boson fusion to the $h_{\rm SM}$. Figure 3(b) depicts the W^+W^- fusion process. Similarly, the ZZ fusion process corresponds to $e^+e^- \to e^+e^-h_{\rm SM}$. In contrast to Higgsstrahlung, the vector boson fusion cross section grows as $\ln s$, and thus is the dominant Higgs production mechanism for $\sqrt{s} \gg m_{h_{\rm SM}}$. The cross section for WW fusion is about ten times larger than that for ZZ fusion. Nevertheless, the latter provides complementary information on the $ZZh_{\rm SM}$ vertex. Note that

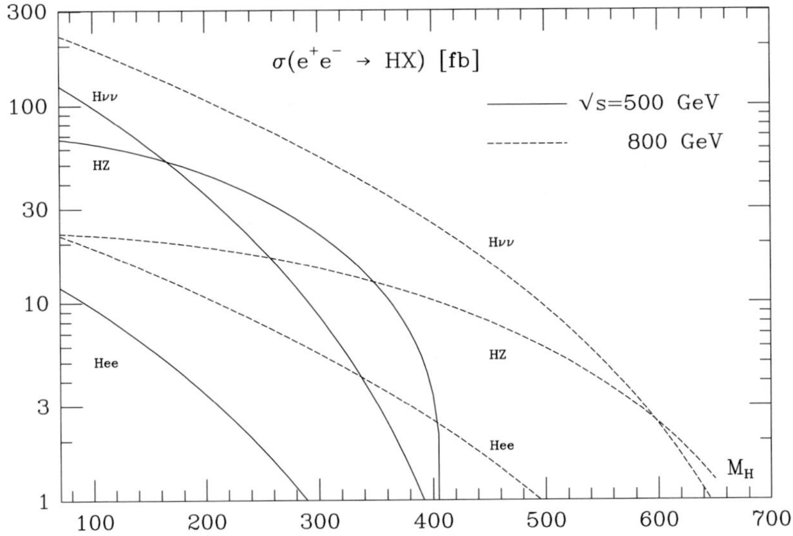

Fig. 4. Cross sections for Higgsstrahlung ($e^+e^- \to Zh_{\rm SM}$) and Higgs production via W^+W^- fusion ($e^+e^- \to \nu\bar{\nu}h_{\rm SM}$) and ZZ fusion ($e^+e^- \to e^+e^-h_{\rm SM}$) as a function of $m_{h_{\rm SM}}$ for two center-of-mass energies, $\sqrt{s} = 500$ and 800 GeV [5].

at an e^-e^- collider, the Higgsstrahlung and W^+W^- fusion processes are absent, so that ZZ fusion is the dominant Higgs production process [43].

Other relevant processes for producing Higgs bosons are associated production with a fermion-antifermion pair, and multi-Higgs production. Among production mechanisms of the former class, only $e^+e^- \to t\bar{t}h_{\rm SM}$ has a significant cross section, around the femtobarn level in the Standard Model, as depicted in Fig. 5. As a result, if $m_{h_{\rm SM}}$ is small enough (or $\sqrt{s}$ is large enough), this process can be used for determining the Higgs–top quark Yukawa coupling. The cross section for double Higgs production ($e^+e^- \to Zh_{\rm SM}h_{\rm SM}$) is even smaller, of order 0.1 fb for 100 GeV $\lesssim m_{h_{\rm SM}} \lesssim$ 150 GeV and $\sqrt{s}$ ranging between 500 GeV and 1 TeV. With sufficient luminosity, the latter can be used for extracting the triple Higgs self-coupling.

At the $\gamma\gamma$ collider, a Higgs boson is produced as an s-channel resonance via the one-loop triangle diagram. Every charged particle whose mass is generated by the Higgs boson contributes to this process. In the Standard Model, the main contributors are the $W^\pm$ and the t-quark loops. See Section 11 for further discussion.

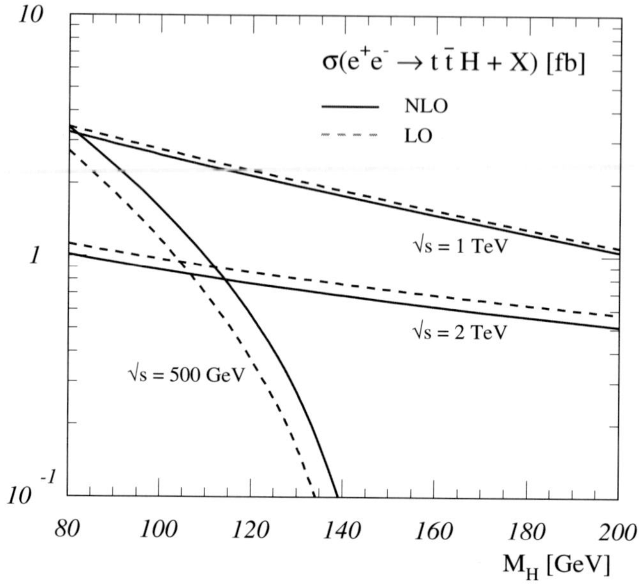

Fig. 5. Cross-sections for $e^+e^- \to t\bar{t}h_{\rm SM}$ in fb for three choices of center-of-mass energy. The dashed lines correspond to the tree-level result [44], and the solid lines include the next-to-leading order QCD corrections [45].

4. SM Higgs Searches before the Linear Collider

4.1. *Direct Search Limits from LEP*

The LEP collider completed its final run in 2000, and presented tantalizing hints for the possible observation of the Higgs boson. The combined data from all four LEP collaborations exhibits a slight preference for the signal plus background hypothesis for a Standard Model Higgs boson mass in the vicinity of 116 GeV, as compared to the background only hypothesis [25]. However, the excess over the expected background was not sufficiently significant to support a claim of discovery or even an "observation" of evidence for the Higgs boson. A more conservative interpretation of the data yields a 95% CL lower limit of $m_{h_{\rm SM}} > 114.4$ GeV.

4.2. *Implications of Precision Electroweak Measurements*

Indirect constraints on the Higgs boson mass within the SM can be obtained from confronting the SM predictions with results of electroweak precision

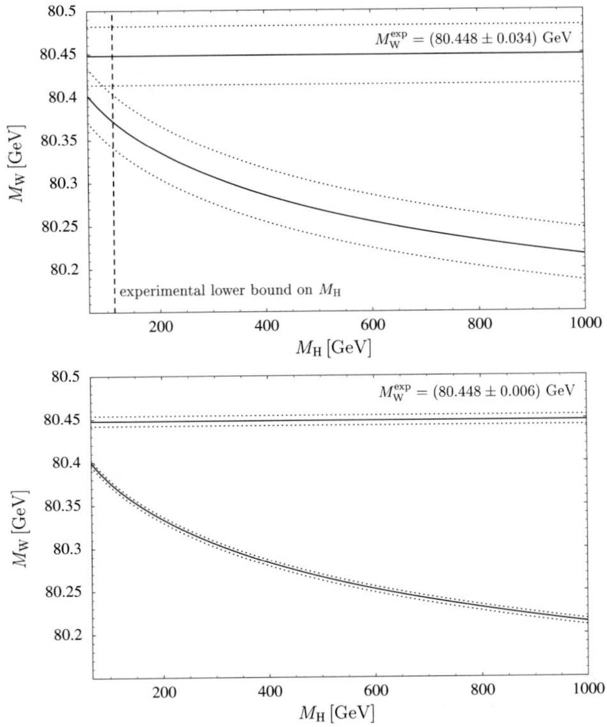

Fig. 6. The prediction for m_W as a function of $m_{h_{SM}}$ is compared with the experimental value of m_W (a) for the current experimental accuracies of m_W and m_t, and (b) for the prospective future accuracies at the LC with the Giga-Z option and a precise W threshold scan, assuming the present experimental central values [46]. The present experimental 95% CL lower bound on the Higgs-boson mass, $m_{h_{SM}} = 114.4$ GeV, is indicated in (a) by the vertical dashed line.

measurements. In the case of the top quark mass, the indirect determination turned out to be in remarkable agreement with the actual experimental value [24]. In comparison, to obtain constraints on $m_{h_{SM}}$ of similar precision, much higher accuracy is required for both the experimental results and the theory predictions. This is due to the fact that the leading dependence of the precision observables on $m_{h_{SM}}$ is only logarithmic, while the dominant effects of the top-quark mass enter quadratically.

Figure 6(a) shows the currently most precise result for m_W as function of $m_{h_{SM}}$ in the SM, and compares it with the present experimental value of m_W. The calculation incorporates the complete electroweak fermion-loop contributions at $\mathcal{O}(\alpha^2)$ [46]. Based on this result, the remaining theoretical uncertainty from unknown higher-order corrections has been estimated to be about 6 MeV [46]. This is about a factor five smaller than the uncertainty induced by the current experimental error on the top-quark mass, $\Delta m_t^{\rm exp} = \pm 5.1$ GeV, which presently dominates the theoretical uncertainty. Fig. 6(b) shows the prospective situation at a future e^+e^- linear collider after the Giga-Z operation and a threshold measurement of the W mass (keeping the present experimental central values for simplicity), which are expected to reduce the experimental errors to $\Delta m_W^{\rm exp} = 6$ MeV and $\Delta m_t^{\rm exp} = 200$ MeV. The plot clearly shows a considerable improvement in the sensitivity to $m_{h_{\rm SM}}$ achievable at the LC via very precise measurements of m_W and m_t. Since furthermore the experimental error of $\sin^2\theta_{\rm eff}$ is expected to be reduced by almost a factor of 20 at Giga-Z, the accuracy in the indirect determination of the Higgs-boson mass from all data will improve by about a factor of 10 compared to the present situation [21].

4.3. *Expectations for Tevatron Searches*

The upgraded Tevatron began taking data in the spring of 2001. This is the only collider at which the Higgs boson can be produced until the LHC begins operation in 2007. The Tevatron Higgs Working Group [47] presented a detailed analysis of the Higgs discovery reach at the upgraded Tevatron. Here, we summarize the main results. Two Higgs mass ranges were considered separately: (i) 100 GeV$\lesssim m_{h_{\rm SM}} \lesssim$ 135 GeV and (ii) 135 GeV$\lesssim m_{h_{\rm SM}} \lesssim$ 190 GeV, corresponding to the two different dominant Higgs decay modes: $h_{\rm SM} \to b\bar{b}$ for the lighter mass range and $h_{\rm SM} \to WW^{(*)}$ for the heavier mass range.

In mass range (i), the relevant production mechanisms are $q_i\bar{q}_j \to Vh_{\rm SM}$, where $V = W$ or Z. In all cases, the dominant $h_{\rm SM} \to b\bar{b}$ decay was employed. The most relevant final-state signatures correspond to events in which the vector boson decays leptonically ($W \to \ell\nu$, $Z \to \ell^+\ell^-$ and $Z \to \nu\bar{\nu}$, where $\ell = e$ or μ), resulting in $\ell\nu b\bar{b}$, $\nu\bar{\nu}b\bar{b}$ and $\ell^+\ell^- b\bar{b}$ final states. In mass range (ii), the relevant production mechanisms include $gg \to h_{\rm SM}$, $V^*V^* \to h_{\rm SM}$ and $q_i\bar{q}_j \to Vh_{\rm SM}$, with decays $h_{\rm SM} \to WW^{(*)}$, $ZZ^{(*)}$. The most relevant phenomenological signals are those in which two of the final-state vector bosons decay leptonically, resulting in $\ell^+\ell^-\nu\bar{\nu}$ or $\ell^\pm\ell^\pm jjX$, where j is a hadronic jet and X consists of two additional leptons (either charged or neutral). For example, the latter can arise from $Wh_{\rm SM}$ production followed by $h_{\rm SM} \to WW^{(*)}$, where the two like-sign W bosons decay

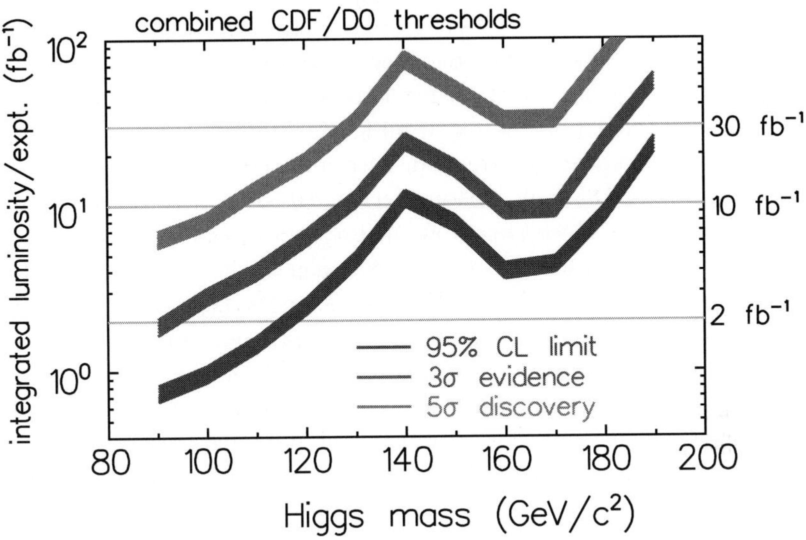

Fig. 7. The integrated luminosity required per experiment, to either exclude a SM Higgs boson at 95% CL or discover it at the 3σ or 5σ level, as a function of the Higgs mass. These results are based on the combined statistical power of both experiments. The curves shown are obtained by combining the $\ell\nu b\bar{b}$, $\nu\bar{\nu}b\bar{b}$ and $\ell^+\ell^-b\bar{b}$ channels using the neural network selection in the low-mass Higgs region (90 GeV $\lesssim m_{h_{\rm SM}} \lesssim$ 135 GeV), and the $\ell^\pm\ell^\pm jjX$ and $\ell^+\ell^-\nu\bar{\nu}$ channels in the high-mass Higgs region (135 GeV $\lesssim m_{h_{\rm SM}} \lesssim$ 190 GeV). The lower edge of the bands is the calculated threshold; the bands extend upward from these nominal thresholds by 30% as an indication of the uncertainties in b-tagging efficiency, background rate, mass resolution, and other effects. Taken from [47].

leptonically, and the third W decays into hadronic jets. In this case X is a pair of neutrinos.

Figure 7 summarizes the Higgs discovery reach versus the total integrated luminosity delivered to the Tevatron (and by assumption, delivered to each detector). As the plot shows, the required integrated luminosity increases rapidly with Higgs mass to 140 GeV, beyond which the high-mass channels play the dominant role. With 2 fb^{-1} per detector, the 95% CL limits will barely extend the expected LEP limits, but with 10 fb^{-1}, the SM Higgs boson can be excluded up to 180 GeV if the Higgs boson does not exist in that mass range.

With further machine improvements, the Tevatron may provide a total integrated luminosity as high as 15 fb^{-1} before the LHC starts to yield significant results. If such an integrated luminosity could be reached and if $m_{h_{\rm SM}} \simeq 115$ GeV, then the Tevatron experiments will be able to achieve a 5σ discovery of the Higgs boson. If no Higgs events are detected, the LEP limits will be significantly extended, with a 95% CL exclusion possible up to about $m_{h_{\rm SM}} \simeq 185$ GeV.[b] Moreover, evidence for a Higgs boson at the 3σ level could be achieved up to about $m_{h_{\rm SM}} \simeq 175$ GeV.

Evidence for or discovery of a Higgs boson at the Tevatron would be a landmark in high energy physics. However, even if a Higgs boson is seen, the Tevatron data would only provide a very rough phenomenological profile of the Higgs boson. In contrast, the LHC and the LC (the latter with greater precision) could measure enough of its properties to verify that the interactions of the Higgs boson provide the dynamics responsible for the generation of mass for the vector bosons, quarks and charged leptons.

4.4. *Expectations for LHC Searches*

At the LHC, the ATLAS and CMS detectors have been specifically designed to guarantee observation of a SM Higgs boson, regardless of its mass (below about 1 TeV). The most important production processes for the $h_{\rm SM}$ are the gluon fusion process, $gg \to h_{\rm SM}$, and the vector boson fusion process, $WW \to h_{\rm SM}$. In particular, for $m_{h_{\rm SM}} \lesssim 130$ GeV the important discovery modes are $gg, WW \to h_{\rm SM} \to \gamma\gamma$, $\tau^+\tau^-$. At high luminosity, $q_i\bar{q}_j \to W^\pm h_{\rm SM}$ and $gg \to t\bar{t}h_{\rm SM}$ with $h_{\rm SM} \to \gamma\gamma$ and $h_{\rm SM} \to b\bar{b}$ should also be visible. For $m_{h_{\rm SM}} > 130$ GeV, $gg \to h_{\rm SM} \to ZZ^{(*)} \to 4\ell$ is extremely robust except for the small mass region with $m_{h_{\rm SM}}$ just above $2m_W$ in which $h_{\rm SM} \to WW$ is allowed and $\text{BR}(h_{\rm SM} \to ZZ^*)$ drops sharply. In this region, $gg, WW \to h_{\rm SM} \to WW \to \ell\nu\ell\nu$ provides a strong Higgs signal. For $m_{h_{\rm SM}} > 300$–400 GeV, the final states $h_{\rm SM} \to WW \to \ell\nu jj$ and $h_{\rm SM} \to ZZ \to \ell\ell\nu\nu$, where the $h_{\rm SM}$ is produced by a combination of gg and WW fusion, provide excellent discovery channels for Higgs masses up to about 1 TeV (*i.e.*, well beyond the $m_{h_{\rm SM}} \sim 800$ GeV limit of viability for the $h_{\rm SM} \to 4\ell$ mode). These results are summarized in Fig. 8, from which we observe that the net statistical significance for the $h_{\rm SM}$, after combining channels, exceeds 10σ for all $m_{h_{\rm SM}} > 80$ GeV, assuming accumulated

[b]The Higgs mass region around 140 GeV might require more luminosity, depending on the magnitude of systematic errors due to uncertainties in b-tagging efficiency, background rate, the $b\bar{b}$ mass resolution, *etc.*

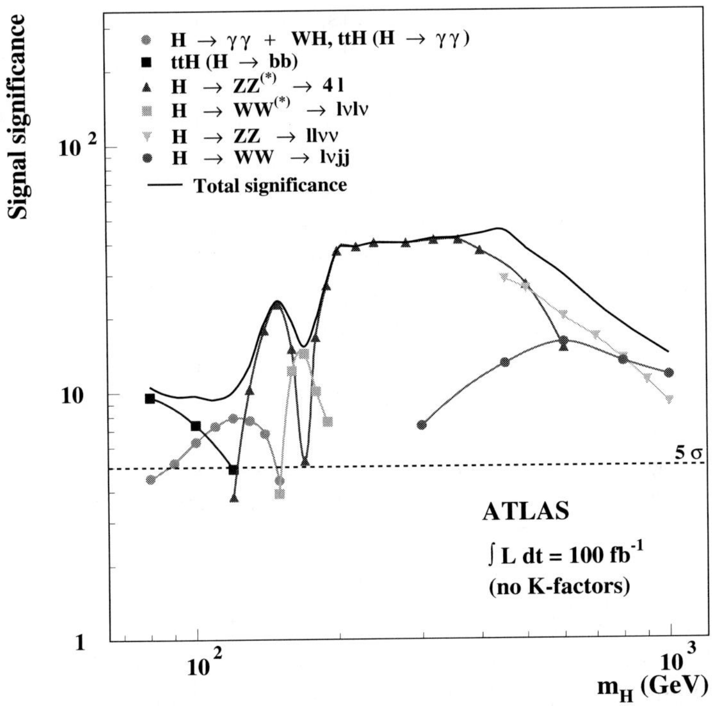

Fig. 8. Statistical significance levels as a function of the Higgs mass for the ATLAS experiment at the LHC, assuming an integrated luminosity of 100 fb^{-1}. Taken from [48].

luminosity of $L = 100\ \text{fb}^{-1}$ at the ATLAS detector [48]. Similar results are obtained by the CMS collaboration [49, 50], with the $h_{\rm SM}\gamma\gamma$ discovery mode even more prominent for $m_{h_{\rm SM}} \lesssim 150$ GeV.

Precision measurements for a certain number of quantities will be possible, depending upon the exact value of $m_{h_{\rm SM}}$. For example, in [51] it is estimated that $m_{h_{\rm SM}}$ can be measured to $< 0.1\%$ for $m_{h_{\rm SM}} < 400$ GeV and to 0.1–1% for $400 < m_{h_{\rm SM}} < 700$ GeV. Using the 4ℓ final state, $\Gamma^T_{h_{\rm SM}}$ can be determined for $m_{h_{\rm SM}} \gtrsim 250$ GeV from the shape of the 4ℓ mass peak. Various ratios of branching ratios and a selection of cross sections times branching ratios can be measured in any given mass region. Some early estimates of possibilities and achievable accuracies appear in [2]. A more recent, but rather optimistic parton-level theoretical study [52, 53] (see Fig. 9) finds that if $m_{h_{\rm SM}} \lesssim 200$ GeV then good accuracies can be achieved

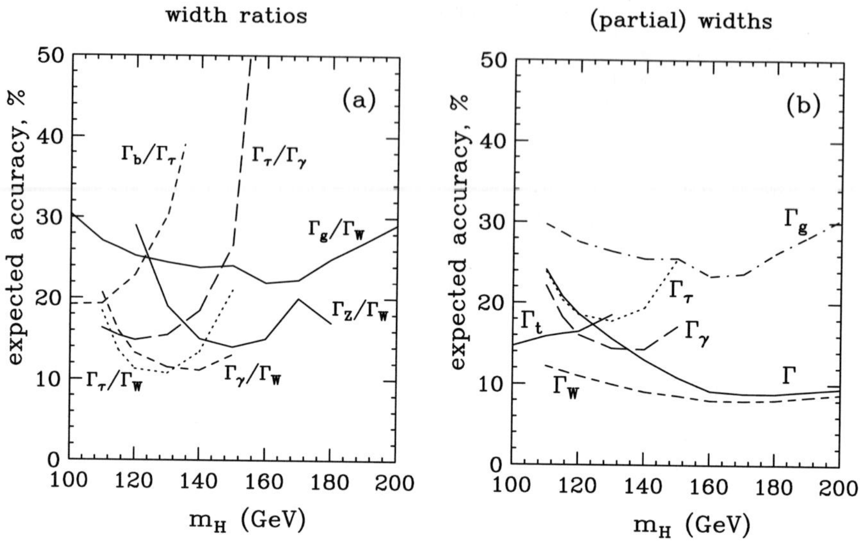

Fig. 9. Relative accuracy expected at the LHC with 200 fb^{-1} of data for (a) various ratios of Higgs boson partial widths and (b) the indirect determination of partial and total widths. Expectations for width ratios assume W, Z universality; indirect width measurements also assume b, τ universality and a small branching ratio for unobserved modes. Taken from the parton-level analysis of [53].

for many absolute partial widths and for the total width provided: (a) WW fusion production can be reliably separated from gg fusion; (b) the WW/ZZ coupling ratio is as expected in the SM from the SU(2)×U(1) symmetry; (c) the WW^* final state can be observed in both gg and WW fusion; and (d) there are no unexpected decays of the h_{SM}. Errors estimated in this way for $L = 200\ \text{fb}^{-1}$ of accumulated data are given in Fig. 9. However, errors found in recent ATLAS studies of a few of these channels are substantially larger [54]. For example, for $L = 300\ \text{fb}^{-1}$ ATLAS finds that the smallest error on Γ_τ/Γ_W is achieved for $m_{h_{\text{SM}}} \sim 130$ GeV and is of order $\pm 27\%$ (as compared to the $\pm 11\%$ of Fig. 9) and that the error rises rapidly for $m_{h_{\text{SM}}}$ away from 130 GeV, reaching $\pm 60\%$ for $m_{h_{\text{SM}}} = 150$ GeV. Invisible Higgs decays have also been addressed in the theoretical work [55]; CMS simulations show some promise for this channel.

5. Higgs Bosons in Low-energy Supersymmetry

The Higgs sector of the MSSM, the simplest realistic model of low-energy supersymmetry, consists of the two-Higgs-doublet extension of the Standard

Model plus the corresponding superpartners. Two Higgs doublets, one with $Y = +1$ and one with $Y = -1$, are needed in order that gauge anomalies due to the higgsino superpartners are exactly canceled. In particular, the Higgs sector contains eight scalar degrees of freedom: one complex $Y = -1$ doublet, $\boldsymbol{\Phi_d} = (\Phi_d^0, \Phi_d^-)$ and one complex $Y = +1$ doublet, $\boldsymbol{\Phi_u} = (\Phi_u^+, \Phi_u^0)$. This notation reflects the fact that in the MSSM, the interaction Lagrangian that describes the Higgs couplings to fermions is constrained by supersymmetry and obeys the following property: Φ_d^0 couples exclusively to down-type fermion pairs and Φ_u^0 couples exclusively to up-type fermion pairs. This pattern of Higgs–fermion couplings defines the Type-II two-Higgs-doublet model [56, 57].

When the Higgs potential is minimized, the neutral components of the Higgs fields acquire vacuum expectation values[c] $\langle\Phi_d^0\rangle = v_d/\sqrt{2}$ and $\langle\Phi_u^0\rangle = v_u/\sqrt{2}$, where the normalization has been chosen such that $v^2 \equiv v_d^2 + v_u^2 = (246\text{ GeV})^2$. The ratio of vacuum expectation values is denoted by

$$\tan\beta \equiv v_u/v_d\,. \tag{3}$$

The physical Higgs spectrum consists of a charged Higgs pair

$$H^\pm = \Phi_d^\pm \sin\beta + \Phi_u^\pm \cos\beta\,, \tag{4}$$

one CP-odd scalar

$$A^0 = \sqrt{2}\left(\mathrm{Im}\,\Phi_d^0 \sin\beta + \mathrm{Im}\,\Phi_u^0 \cos\beta\right)\,, \tag{5}$$

and two CP-even scalars:

$$\begin{aligned} h^0 &= -(\sqrt{2}\,\mathrm{Re}\,\Phi_d^0 - v_d)\sin\alpha + (\sqrt{2}\,\mathrm{Re}\,\Phi_u^0 - v_u)\cos\alpha\,, \\ H^0 &= (\sqrt{2}\,\mathrm{Re}\,\Phi_d^0 - v_d)\cos\alpha + (\sqrt{2}\,\mathrm{Re}\,\Phi_u^0 - v_u)\sin\alpha\,, \end{aligned} \tag{6}$$

(with $m_{h^0} \le m_{H^0}$). The angle α arises when the CP-even Higgs squared-mass matrix (in the Φ_d^0—Φ_u^0 basis) is diagonalized to obtain the physical CP-even Higgs states. The Goldstone bosons, $G^\pm$ and G^0, which are orthogonal to $H^\pm$ and A^0 respectively, provide the longitudinal components of the massive $W^\pm$ and Z via the Higgs mechanism.

In two-Higgs-doublet models with Type-II Higgs–fermion couplings, the $Y = -1$ Higgs doublet generates mass for "up"-type quarks and the $Y = +1$ Higgs doublet generates mass for "down"-type quarks (and charged leptons)

[c]The phases of the Higgs fields can be chosen such that the vacuum expectation values are real and positive. That is, the tree-level MSSM Higgs sector conserves CP, which implies that the neutral Higgs mass eigenstates possess definite CP quantum numbers.

[58, 59]. The tree-level relations between the quark masses and the Higgs–fermion Yukawa couplings (using 3rd family notation) are given by:

$$h_b = \frac{\sqrt{2}\,m_b}{v_d} = \frac{\sqrt{2}\,m_b}{v\cos\beta}\,, \tag{7}$$

$$h_t = \frac{\sqrt{2}\,m_t}{v_u} = \frac{\sqrt{2}\,m_t}{v\sin\beta}\,. \tag{8}$$

Radiative corrections to these relations will be examined in Section 5.2.

At this stage, $\tan\beta$ is a free parameter of the model. However, theoretical considerations suggest that $\tan\beta$ cannot be too small or too large. The crudest bounds arise from unitarity constraints. If $\tan\beta$ becomes too small, then the Higgs coupling to top quarks becomes strong and the tree-unitarity of processes involving the Higgs-top quark Yukawa coupling is violated. A rough lower bound advocated by [60], $\tan\beta \gtrsim 0.3$, corresponds to a value of h_t in the perturbative region. A similar argument involving h_b would yield $\tan\beta \lesssim 120$. A more restrictive theoretical constraint is based on the requirement that Higgs–fermion Yukawa couplings, h_t and h_b, remain finite when running from the electroweak scale to some large energy scale Λ, above which new physics enters. The limits on $\tan\beta$ depend on the choice of Λ. Integrating the Yukawa coupling renormalization group equations from the electroweak scale to Λ (allowing for the possible existence of a supersymmetry-breaking scale, $m_Z \leq M_{\rm SUSY} \leq \Lambda$), one can determine the range of $\tan\beta$ for which h_t and h_b remain finite. This exercise has been carried out at two-loops in [61]. Suppose that the low-energy theory at the electroweak scale is the MSSM, and that there is no additional new physics below the grand unification scale of $\Lambda = 2\times 10^{16}$ GeV. Then, for $m_t = 175$ GeV, the Higgs–fermion Yukawa couplings remain finite at all energy scales below Λ if $1.5 \lesssim \tan\beta \lesssim 65$. Note that this result is consistent with the scenario of radiative electroweak symmetry breaking in low-energy supersymmetry based on supergravity, which requires that $1 \lesssim \tan\beta \lesssim m_t/m_b$.[d] Thus, we expect $\tan\beta$ to lie in the range $1 \lesssim \tan\beta \lesssim 55$.[e]

---

[d]Here, the quark masses are evaluated at m_Z. We take $m_t(m_Z) \simeq 165$ GeV and $m_b(m_Z) \simeq 3$ GeV.

[e]The lower bound on $\tan\beta$ can be taken to be $\tan\beta \gtrsim 2.4$, based on the LEP MSSM Higgs search discussed in Section 6.1.

5.1. *MSSM Higgs Sector at Tree Level*

The supersymmetric structure of the theory imposes constraints on the Higgs sector of the model. As a result, all Higgs sector parameters at tree level are determined by two free parameters: $\tan\beta$ and one Higgs mass, conveniently chosen to be m_{A^0}. In particular,

$$m_{H^\pm}^2 = m_{A^0}^2 + m_W^2\,, \tag{9}$$

and the CP-even Higgs bosons h^0 and H^0 are eigenstates of the following squared-mass matrix (with respect to the Re Φ_d^0–Re Φ_u^0 basis):

$$\mathcal{M}_0^2 = \begin{pmatrix} m_{A^0}^2 \sin^2\beta + m_Z^2\cos^2\beta & -(m_{A^0}^2 + m_Z^2)\sin\beta\cos\beta \\ -(m_{A^0}^2 + m_Z^2)\sin\beta\cos\beta & m_{A^0}^2\cos^2\beta + m_Z^2\sin^2\beta \end{pmatrix}. \tag{10}$$

The eigenvalues of $\mathcal{M}_0^2$ are the tree-level squared-masses of the two CP-even Higgs scalars

$$m_{H^0,h^0}^2 = \frac{1}{2}\left(m_{A^0}^2 + m_Z^2 \pm \sqrt{(m_{A^0}^2 + m_Z^2)^2 - 4m_Z^2 m_{A^0}^2\cos^2 2\beta}\right), \tag{11}$$

Note that Eq. (11) yields an upper bound to the tree-level mass of the light CP-even Higgs boson: $m_{h^0}^2 \leq m_Z^2|\cos 2\beta| \leq m_Z^2$. Radiative corrections can significantly increase this upper bound by as much as 50% as described in Section 5.2. Nevertheless, it is already apparent that the MSSM favors a CP-even Higgs boson whose mass is not much larger than m_Z, a result that is consistent with the inferred Higgs mass limit from precision electroweak measurements discussed in Section 2.

From the above results, one also obtains:

$$\cos^2(\beta-\alpha) = \frac{m_{h^0}^2(m_Z^2 - m_{h^0}^2)}{m_{A^0}^2(m_{H^0}^2 - m_{h^0}^2)}\,, \tag{12}$$

where α is the angle that diagonalizes the CP-even Higgs squared-mass matrix [see Eq. (6)]. In the convention where $\tan\beta$ is positive (*i.e.*, $0 \leq \beta \leq \pi/2$), the angle α lies in the range $-\pi/2 \leq \alpha \leq 0$.

The limit of $m_{A^0} \gg m_Z$ is of particular interest. The expressions for the Higgs masses and mixing angle simplify in this limit and one finds

$$m_{h^0}^2 \simeq m_Z^2\cos^2 2\beta\,, \tag{13}$$

$$m_{H^0}^2 \simeq m_{A^0}^2 + m_Z^2\sin^2 2\beta\,, \tag{14}$$

$$m_{H^\pm}^2 = m_{A^0}^2 + m_W^2\,, \tag{15}$$

$$\cos^2(\beta-\alpha) \simeq \frac{m_Z^4\sin^2 4\beta}{4m_{A^0}^4}\,. \tag{16}$$

Two consequences are immediately apparent. First, $m_{A^0} \simeq m_{H^0} \simeq m_{H^\pm}$, up to corrections of $\mathcal{O}(m_Z^2/m_{A^0})$. Second, $\cos(\beta-\alpha)=0$ up to corrections of $\mathcal{O}(m_Z^2/m_{A^0}^2)$. This is the *decoupling* limit [22, 62] because when m_{A^0} is large, the effective low-energy theory below the scale of m_{A^0} contains a single CP-even Higgs boson, h^0, whose properties are nearly identical to those of the Standard Model Higgs boson, $h_{\rm SM}$.

The phenomenology of the Higgs sector is determined by the various couplings of the Higgs bosons to gauge bosons, Higgs bosons and fermions. The couplings of the two CP-even Higgs bosons to W and Z pairs are given in terms of the angles α and β by

$$g_{h^0VV} = g_V m_V \sin(\beta-\alpha)\,, \qquad g_{H^0VV} = g_V m_V \cos(\beta-\alpha)\,, \tag{17}$$

where $g_V \equiv 2m_V/v$ for $V=W$ or Z. There are no tree-level couplings of A^0 or $H^\pm$ to VV. The couplings of one gauge boson to two neutral Higgs bosons are given by:

$$g_{h^0A^0Z} = \frac{g\cos(\beta-\alpha)}{2\cos\theta_W}\,, \qquad g_{H^0A^0Z} = \frac{-g\sin(\beta-\alpha)}{2\cos\theta_W}\,. \tag{18}$$

As noted earlier, the pattern of Higgs–fermion couplings in the MSSM are those of the Type-II two-Higgs-doublet model [56, 57]. The couplings of the neutral Higgs bosons to $f\bar{f}$ relative to the Standard Model value, $gm_f/2m_W$, are given by

$$\begin{aligned}
h^0b\bar{b}\ \ (\text{or } h^0\tau^+\tau^-):&\quad -\frac{\sin\alpha}{\cos\beta} = \sin(\beta-\alpha) - \tan\beta\cos(\beta-\alpha)\,,\\
h^0t\bar{t}:&\quad \frac{\cos\alpha}{\sin\beta} = \sin(\beta-\alpha) + \cot\beta\cos(\beta-\alpha)\,,\\
H^0b\bar{b}\ \ (\text{or } H^0\tau^+\tau^-):&\quad \frac{\cos\alpha}{\cos\beta} = \cos(\beta-\alpha) + \tan\beta\sin(\beta-\alpha)\,,\\
H^0t\bar{t}:&\quad \frac{\sin\alpha}{\sin\beta} = \cos(\beta-\alpha) - \cot\beta\sin(\beta-\alpha)\,,\\
A^0b\bar{b}\ \ (\text{or } A^0\tau^+\tau^-):&\quad \gamma_5\tan\beta\,,\\
A^0t\bar{t}:&\quad \gamma_5\cot\beta\,,
\end{aligned} \tag{19}$$

(the γ_5 indicates a pseudoscalar coupling), and the charged Higgs boson couplings to fermion pairs (with all particles pointing into the vertex) are given by

$$g_{H^-t\bar{b}} = \frac{g}{\sqrt{2}m_W}\left[m_t \cot\beta\, P_R + m_b \tan\beta\, P_L\right],$$
$$g_{H^-\tau^+\nu} = \frac{g}{\sqrt{2}m_W}\left[m_\tau \tan\beta\, P_L\right], \tag{20}$$

where $P_{R,L} \equiv \frac{1}{2}(1 \pm \gamma_5)$ are the right and left-handed projection operators, respectively.

The neutral Higgs couplings to fermion pairs [Eq. (19)] have been written in such a way that their behavior can be immediately ascertained in the decoupling limit ($m_{A^0} \gg m_Z$) by setting $\cos(\beta - \alpha) = 0$. In particular, in the decoupling limit, the couplings of h^0 to vector bosons and fermion pairs are equal to the corresponding couplings of the SM Higgs boson.

The region of MSSM Higgs sector parameter space in which the decoupling limit applies is large, because $\cos(\beta-\alpha)$ approaches zero quite rapidly once m_{A^0} is larger than about 200 GeV, as shown in Fig. 10. As a result, over a significant region of the MSSM parameter space, the search for the lightest CP-even Higgs boson of the MSSM is equivalent to the search for the Standard Model Higgs boson.

5.2. *The Radiatively Corrected MSSM Higgs Sector*

When one-loop radiative corrections are incorporated, the Higgs masses and couplings depend on additional parameters of the supersymmetric model that enter via the virtual loops. One of the most striking effects of the radiative corrections to the MSSM Higgs sector is the modification of the upper bound of the light CP-even Higgs mass, as first noted in [63]. When $\tan\beta \gg 1$ and $m_{A^0} \gg m_Z$, the *tree-level* prediction of $m_{h^0} = m_Z$ corresponds to the theoretical upper bound for m_{h^0}. Including radiative corrections, this theoretical upper bound is increased, primarily because of an incomplete cancellation of the top-quark and top-squark (stop) loops. (These contributions would cancel if supersymmetry were exact.) The relevant parameters that govern the stop sector are the average of the two stop squared-masses: $M^2_{\rm SUSY} \equiv \frac{1}{2}(M^2_{\tilde{t}_1} + M^2_{\tilde{t}_2})$, and the off-diagonal element of the stop squared-mass matrix: $m_t X_t \equiv m_t(A_t - \mu\cot\beta)$, where A_t is a soft supersymmetry-breaking trilinear scalar interaction term, and μ is the supersymmetric Higgs mass parameter. The qualitative behavior of the radiative corrections can be most easily seen in the large top squark mass limit, where, in addition, the splitting of the two diagonal entries and the

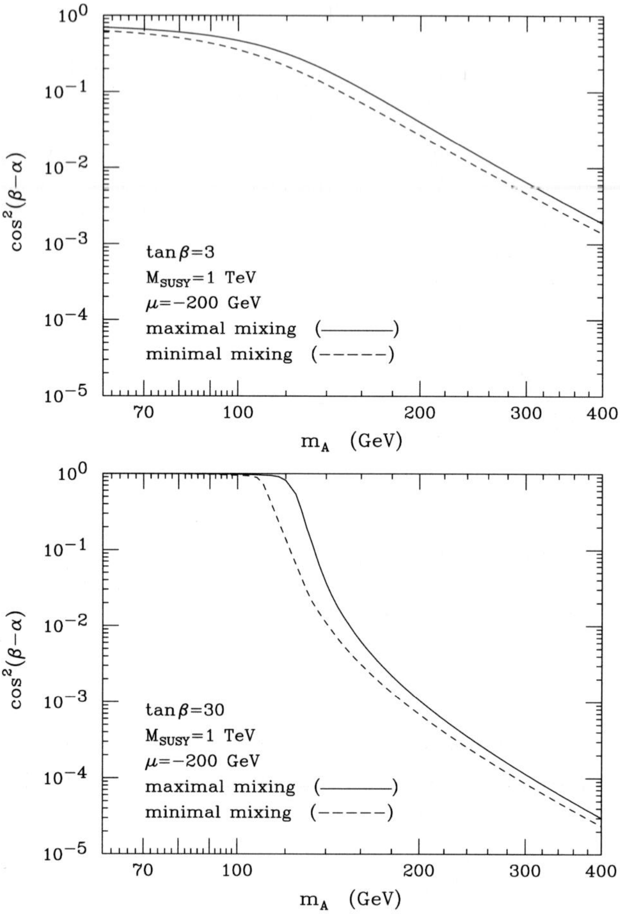

Fig. 10. The value of $\cos^2(\beta - \alpha)$ is shown as a function of m_{A^0} for two choices of $\tan\beta = 3$ and $\tan\beta = 30$. When radiative-corrections are included, one can define an approximate loop-corrected angle α as a function of m_{A^0}, $\tan\beta$ and the MSSM parameters. In the figures above, we have incorporated radiative corrections, assuming that the top-squark masses are given by $M_{\rm SUSY} = 1$ TeV. In addition, two extreme cases for the squark mixing parameters are shown (see Section 5.2 for further discussion of the radiative corrections and their dependence on the supersymmetric parameters). The decoupling effect expected from Eq. (16), in which $\cos^2(\beta - \alpha) \propto m_Z^4/m_{A^0}^4$ for $m_{A^0} \gg m_Z$, continues to hold even when radiative corrections are included.

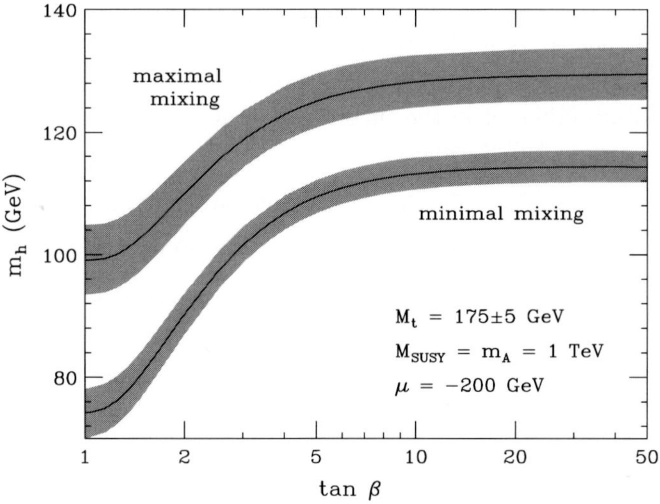

Fig. 11. The radiatively corrected light CP-even Higgs mass is plotted as a function of $\tan\beta$, for the maximal mixing [upper band] and minimal mixing cases. The impact of the top quark mass is exhibited by the shaded bands; the central value corresponds to $m_t = 175$ GeV, while the upper [lower] edge of the bands correspond to increasing [decreasing] m_t by 5 GeV.

off-diagonal entry of the stop squared-mass matrix are both small in comparison to $M^2_{\rm SUSY}$. In this case, the upper bound on the lightest CP-even Higgs mass is approximately given by

$$m^2_{h^0} \lesssim m^2_Z + \frac{3g^2m_t^4}{8\pi^2 m_W^2}\left[\ln\left(\frac{M^2_{\rm SUSY}}{m_t^2}\right) + \frac{X_t^2}{M^2_{\rm SUSY}}\left(1 - \frac{X_t^2}{12M^2_{\rm SUSY}}\right)\right]. \quad (21)$$

More complete treatments of the radiative corrections include the effects of stop mixing, renormalization group improvement, and the leading two-loop contributions, and imply that these corrections somewhat overestimate the true upper bound of m_{h^0} (see [64] for the most recent results). Nevertheless, Eq. (21) correctly illustrates some noteworthy features of the more precise result. First, the increase of the light CP-even Higgs mass bound beyond m_Z can be significant due to the m_t^4 enhancement of the one-loop radiative correction. Second, the dependence of the light Higgs mass on the stop mixing parameter X_t implies that (for a given value of $M_{\rm SUSY}$) the upper bound of the light Higgs mass initially increases with X_t and reaches its *maximal* value at $X_t \simeq \sqrt{6}M_{\rm SUSY}$. This point is referred to as the *maximal mixing* case (whereas $X_t = 0$ is called the *minimal mixing* case).

Taking m_{A^0} large, Fig. 11 illustrates that the maximal value of the lightest CP-even Higgs mass bound is realized in the case of maximal mixing at large $\tan\beta$. Allowing for the uncertainty in the measured value of m_t and the uncertainty inherent in the theoretical analysis, one finds for $M_{\rm SUSY} \lesssim 2$ TeV that $m_{h^0} \lesssim m_h^{\rm max}$, where

$$\begin{aligned} m_h^{\rm max} &\simeq 122~\text{GeV}, \quad \text{minimal stop mixing,} \\ m_h^{\rm max} &\simeq 135~\text{GeV}, \quad \text{maximal stop mixing.} \end{aligned} \tag{22}$$

The h^0 mass bound in the MSSM quoted above does not apply to non-minimal supersymmetric extensions of the Standard Model. If additional Higgs singlet and/or triplet fields are introduced, then new Higgs self-coupling parameters appear, which are not significantly constrained by present data. For example, in the simplest non-minimal supersymmetric extension of the Standard Model (NMSSM), the addition of a complex Higgs singlet field S adds a new Higgs self-coupling parameter, λ_S [65]. The mass of the lightest neutral Higgs boson can be raised arbitrarily by increasing the value of λ_S, analogous to the behavior of the Higgs mass in the Standard Model. Under the assumption that all couplings stay perturbative up to the Planck scale, one finds in nearly all cases that $m_{h^0} \lesssim 200$ GeV, independent of the details of the low-energy supersymmetric model [66]. Moreover, if the perturbative unification scale is significantly lower than the Planck scale, then the lower bound on the Higgs mass can be substantially larger [67].

Radiative corrections also modify the tree-level expressions for m_{H^0} and $m_{H^\pm}$ [Eqs. (11) and (9)]. However, since $m_{H^0} \simeq m_{H^\pm} \simeq m_{A^0}$ when $m_{A^0} \gg m_Z$, the impact of the radiative corrections on the heavier Higgs masses is less pronounced that those described above. In Fig. 12, we exhibit the masses of the CP-even neutral and the charged Higgs masses as a function of m_{A^0}. Note that $m_{H^0} \geq m_h^{\rm max}$ for all values of m_{A^0} and $\tan\beta$, where $m_h^{\rm max}$ is to be evaluated depending on the top-squark mixing, as indicated in Eq. (22).

Radiative corrections also significantly modify the tree-level values of the Higgs boson couplings to fermion pairs and to vector boson pairs. As discussed above, the tree-level Higgs couplings depend crucially on the value of $\cos(\beta - \alpha)$. In the first approximation, when radiative corrections of the Higgs squared-mass matrix are computed, the diagonalizing angle α is modified. This provides one important source of the radiative corrections of the Higgs couplings. In Fig. 10, we show the effect of radiative corrections on the value of $\cos(\beta-\alpha)$ as a function of m_{A^0} for different values of the squark mixing parameters and $\tan\beta$. One can then insert the radiatively corrected

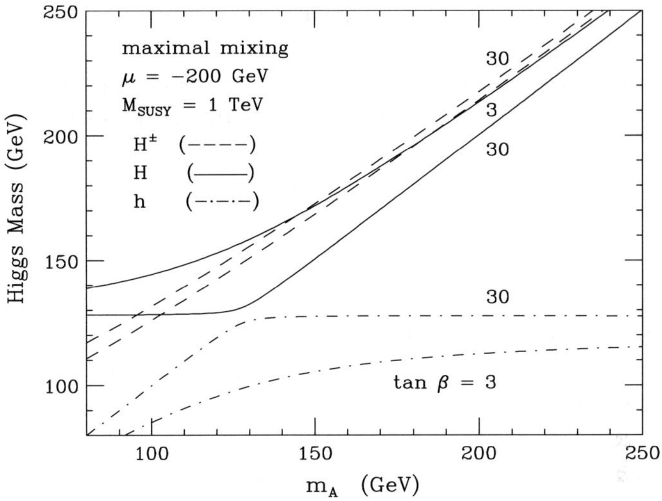

Fig. 12. Lightest CP-even Higgs mass (m_{h^0}), heaviest CP-even Higgs mass (m_{H^0}) and charged Higgs mass ($m_{H^\pm}$) as a function of m_{A^0} for two choices of $\tan\beta = 3$ and $\tan\beta = 30$. The slight increase in the charged Higgs mass as $\tan\beta$ is increased from 3 to 30 is a consequence of the radiative corrections.

value of α into eqs. (17)–(19) to obtain radiatively improved couplings of Higgs bosons to vector bosons and to fermions.

At large $\tan\beta$, there is another potentially important class of radiative corrections in addition to those that enter through the modified value of α. These corrections arise in the relation between m_b and $\tan\beta$ and depend on the details of the MSSM spectrum (which enter via loop-effects). At tree level, the Higgs couplings to $b\bar{b}$ are proportional to the Higgs–bottom-quark Yukawa coupling, h_b. Deviations from the tree-level relation between h_b and m_b [Eq. (7)] due to radiative corrections are calculable and finite [68, 69, 70, 71, 72]. One of the fascinating properties of such corrections is that in certain cases the corrections do *not* vanish in the limit of large supersymmetric mass parameters. These corrections grow with $\tan\beta$ and therefore can be significant in the large $\tan\beta$ regime. In the supersymmetric limit, bottom quarks only couple to Φ_d^0. However, supersymmetry is broken and the bottom quark will receive a small coupling to Φ_u^0 from radiative corrections,

$$-\mathcal{L}_{\text{Yukawa}} \simeq h_b \Phi_d^0 b\bar{b} + (\Delta h_b)\Phi_u^0 b\bar{b}\,. \tag{23}$$

Because the Higgs doublet acquires a vacuum expectation value, the bottom quark mass receives an extra contribution equal to $(\Delta h_b)v_u$. Although Δh_b is one–loop suppressed with respect to h_b, for sufficiently large values of $\tan\beta$ ($v_u \gg v_d$) the contribution to the bottom quark mass of both terms in Eq. (23) may be comparable in size. This induces a large modification in the tree level relation,

$$m_b = \frac{h_b v_d}{\sqrt{2}}(1+\Delta_b)\,, \tag{24}$$

where $\Delta_b \equiv (\Delta h_b)\tan\beta/h_b$. The function Δ_b contains two main contributions: one from a bottom squark–gluino loop (depending on the two bottom squark masses $M_{\tilde{b}_1}$ and $M_{\tilde{b}_2}$ and the gluino mass $M_{\tilde{g}}$) and another one from a top squark–higgsino loop (depending on the two top squark masses $M_{\tilde{t}_1}$ and $M_{\tilde{t}_2}$ and the higgsino mass parameter μ). The explicit form of Δ_b at one–loop in the limit of $M_S \gg m_b$ is given by [69, 70, 71]:

$$\Delta_b \simeq \frac{2\alpha_s}{3\pi}M_{\tilde{g}}\mu\tan\beta\, I(M_{\tilde{b}_1}, M_{\tilde{b}_2}, M_{\tilde{g}}) + \frac{Y_t}{4\pi}A_t\mu\tan\beta\, I(M_{\tilde{t}_1}, M_{\tilde{t}_2}, \mu), \tag{25}$$

where $\alpha_s = g_s^2/4\pi$, $Y_t \equiv h_t^2/4\pi$, and contributions proportional to the electroweak gauge couplings have been neglected. In addition, the function I is defined by

$$I(a,b,c) = \frac{a^2b^2\ln(a^2/b^2) + b^2c^2\ln(b^2/c^2) + c^2a^2\ln(c^2/a^2)}{(a^2-b^2)(b^2-c^2)(a^2-c^2)}\,, \tag{26}$$

and is manifestly positive. Note that the Higgs coupling proportional to Δh_b is a manifestation of the broken supersymmetry in the low energy theory; hence, Δ_b does not decouple in the limit of large values of the supersymmetry breaking masses. Indeed, if all supersymmetry breaking mass parameters (and μ) are scaled by a common factor, the correction Δ_b remains constant.

Similarly to the case of the bottom quark, the relation between m_τ and the Higgs–tau-lepton Yukawa coupling h_τ is modified:

$$m_\tau = \frac{h_\tau v_d}{\sqrt{2}}(1+\Delta_\tau). \tag{27}$$

The correction Δ_τ contains a contribution from a tau slepton–neutralino loop (depending on the two stau masses $M_{\tilde{\tau}_1}$ and $M_{\tilde{\tau}_2}$ and the mass parameter of the $\widetilde{B}$ ("bino") component of the neutralino, M_1) and a tau sneutrino–chargino loop (depending on the tau sneutrino mass $M_{\tilde{\nu}_\tau}$, the mass parameter of the $\widetilde{W}^\pm$ component of the chargino, M_2, and μ). It is given by [70, 71]:

$$\Delta_\tau = \frac{\alpha_1}{4\pi}M_1\mu\tan\beta\, I(M_{\tilde{\tau}_1}, M_{\tilde{\tau}_2}, M_1) - \frac{\alpha_2}{4\pi}M_2\mu\tan\beta\, I(M_{\tilde{\nu}_\tau}, M_2, \mu), \tag{28}$$

where $\alpha_2 \equiv g^2/4\pi$ and $\alpha_1 \equiv g'^2/4\pi$ are the electroweak gauge couplings. Since corrections to h_τ are proportional to α_1 and α_2, they are typically smaller than the corrections to h_b.

From Eqs. (23) and (24) we can obtain the couplings of the physical neutral Higgs bosons to $b\bar{b}$. At large $\tan\beta$, the dominant corrections to Eq. (19) are displayed below:[f]

$$
\begin{aligned}
h^0 b\bar{b}: &\qquad -\frac{\sin\alpha}{\cos\beta}\frac{1}{1+\Delta_b}\left[1-\frac{\Delta_b\cot\alpha}{\tan\beta}\right],\\
H^0 b\bar{b}: &\qquad \frac{\cos\alpha}{\cos\beta}\frac{1}{1+\Delta_b}\left[1+\frac{\Delta_b\tan\alpha}{\tan\beta}\right],\\
A^0 b\bar{b}: &\qquad \gamma_5\frac{\tan\beta}{1+\Delta_b},
\end{aligned}
\tag{29}
$$

where $\Delta_b \propto \tan\beta$ [see Eq. (25)]. Similarly, the neutral Higgs couplings to $\tau^+\tau^-$ are modified by replacing Δ_b in Eq. (29) with Δ_τ [70, 71]. The corresponding radiative corrections to the couplings of the neutral Higgs bosons to $t\bar{t}$ are not $\tan\beta$-enhanced (see [73] for a useful approximation for these corrections). One can also derive radiatively corrected couplings of the charged Higgs boson to fermion pairs [74, 75]. The tree-level couplings of the charged Higgs boson to fermion pairs are modified accordingly by replacing $m_b \to m_b/(1+\Delta_b)$ and $m_\tau \to m_\tau/(1+\Delta_\tau)$, respectively.

One consequence of the above results is that the neutral Higgs coupling to $b\bar{b}$ (which is expected to be the dominant decay mode over nearly all of the MSSM Higgs parameter space), can be significantly suppressed at large $\tan\beta$ [76, 77, 78] if $\Delta_b \simeq \mathcal{O}(1)$. For example, Eq. (29) implies that $g_{h^0 b\bar{b}} \simeq 0$ if $\tan\alpha = \Delta_b/\tan\beta$. Inserting this result into the corresponding expression for the $h^0\tau^+\tau^-$ coupling, it follows that

$$
g_{h^0\tau^+\tau^-} \simeq \frac{\sqrt{2}m_\tau\cos\alpha}{v\sin\beta}\left(\frac{\Delta_\tau-\Delta_b}{1+\Delta_\tau}\right), \qquad [\text{if } g_{h^0 b\bar{b}} \simeq 0]\,. \tag{30}
$$

Similarly, $g_{H^0 b\bar{b}} \simeq 0$ if $\tan\alpha = -\tan\beta/\Delta_b$. Inserting this result into the corresponding expression for the $H^0\tau^+\tau^-$ coupling, it follows that

$$
g_{H^0\tau^+\tau^-} \simeq \frac{\sqrt{2}m_\tau\sin\alpha}{v\sin\beta}\left(\frac{\Delta_\tau-\Delta_b}{1+\Delta_\tau}\right), \qquad [\text{if } g_{H^0 b\bar{b}} \simeq 0]\,. \tag{31}
$$

In both cases, we see that although the Higgs coupling to $b\bar{b}$ can be strongly suppressed for certain parameter choices, the corresponding Higgs coupling

[f] A better approximation in which non-leading terms at large $\tan\beta$ are kept can be found in [73].

to $\tau^+\tau^-$ may be unsuppressed. In such cases, the $\tau^+\tau^-$ decay mode can be the dominant Higgs decay channel for the CP-even Higgs boson with SM-like couplings to gauge bosons.

Near the decoupling limit, $\cot\alpha\cot\beta = -1 + \mathcal{O}(m_Z^2/m_{A^0}^2)$ [after an appropriate manipulation of Eq. (16)]. Inserting this result into Eq. (29), one can check that the $h^0 b\bar{b}$ coupling does indeed approach its Standard Model value. However, because $\Delta_b \propto \tan\beta$, the deviation of the $h^0 b\bar{b}$ coupling from the corresponding SM result is of $\mathcal{O}(m_Z^2 \tan\beta/m_{A^0}^2)$. That is, at large $\tan\beta$, the approach to decoupling may be "delayed" [79], depending on the values of other MSSM parameters that enter the radiative corrections.

5.3. *MSSM Higgs Boson Decay Modes*

In this section, we consider the decay properties of the three neutral Higgs bosons (h^0, H^0 and A^0) and of the charged Higgs pair ($H^\pm$). Let us start with the lightest state, h^0. When $m_{A^0} \gg m_Z$, the decoupling limit applies, and the couplings of h^0 to SM particles are nearly indistinguishable from those of $h_{\rm SM}$. If some superpartners are light, there may be some additional decay modes, and hence the h^0 branching ratios would be different from the corresponding Standard Model values, even though the partial widths to Standard Model particles are the same. Furthermore, loops of light charged or colored superpartners could modify the h^0 coupling to photons and/or gluons, in which case the one-loop gg and $\gamma\gamma$ decay rates would also be different. On the other hand, if all superpartners are heavy, all the decay properties of h^0 are essentially those of the SM Higgs boson, and the discussion of Section 3.1 applies.

For $m_{A^0} \gg m_Z$, the heavier Higgs states, H^0, A^0 and $H^\pm$, are roughly mass-degenerate and have negligible couplings to vector boson pairs. In particular, $\Gamma(H^0 \to VV) \ll \Gamma(h_{\rm SM} \to VV)$, while the couplings of A^0 and $H^\pm$ to the gauge bosons are loop-suppressed. The couplings of H^0, A^0 and $H^\pm$ to down-type (up-type) fermions are significantly enhanced (suppressed) relative to those of $h_{\rm SM}$ if $\tan\beta \gg 1$. Consequently, the decay modes $H^0, A^0 \to b\bar{b}$, $\tau^+\tau^-$ dominate the neutral Higgs decay modes for moderate-to-large values of $\tan\beta$ below the $t\bar{t}$ threshold, while $H^+ \to \tau^+\nu$ dominates the charged Higgs decay below the $t\bar{b}$ threshold.

For values of m_{A^0} of order m_Z, all Higgs boson states lie below 200 GeV in mass and would be accessible at the LC. In this parameter regime, there is a significant area of the parameter space in which none of the neutral Higgs boson decay properties approximates those of $h_{\rm SM}$. For example, when $\tan\beta$

is large, supersymmetry-breaking effects can significantly modify the $b\bar{b}$ and/or the $\tau^+\tau^-$ decay rates with respect to those of $h_{\rm SM}$. Additionally, the heavier Higgs bosons can decay into lighter Higgs bosons. Examples of such decay modes are: $H^0 \to h^0h^0$, A^0A^0, and ZA^0, and $H^\pm \to W^\pm h^0$, $W^\pm A^0$ (although in the MSSM, the Higgs branching ratio into vector boson–Higgs boson final states, if kinematically allowed, rarely exceeds a few percent). The decay of the heavier Higgs boson into two lighter Higgs bosons can provide information about Higgs self-couplings. For values of $\tan\beta \lesssim 5$, the branching ratio of $H^0 \to h^0h^0$ is dominant for a Higgs mass range of 200 GeV $\lesssim m_{H^0} \lesssim 2m_t$. The dominant radiative corrections to this decay arise from the corrections to the self-interaction $\lambda_{H^0h^0h^0}$ in the MSSM and are large [80].

The phenomenology of charged Higgs bosons is less model-dependent, and is governed by the values of $\tan\beta$ and $m_{H^\pm}$. Because the $H^\pm$ couplings are proportional to fermion masses, the decays to third-generation quarks and leptons are dominant. In particular, for $m_{H^\pm} < m_t + m_b$ (so that the channel $H^+ \to t\bar{b}$ is closed), $H^+ \to \tau^+\nu_\tau$ is favored if $\tan\beta \gtrsim 1$, while $H^+ \to c\bar{s}$ is favored if $\tan\beta$ is small. Indeed, BR$(H^+ \to \tau^+\nu_\tau) \simeq 1$ if $\tan\beta \gtrsim 5$. These results apply generally to Type-II two-Higgs doublet models. For $m_{H^\pm} \gtrsim 200$ GeV (the precise value depends on $\tan\beta$), the decay $H^+ \to t\bar{b} \to W^+b\bar{b}$ is the dominant decay mode.

In addition to the above decay modes, there exist new Higgs decay channels that involve supersymmetric final states. Higgs decays into charginos, neutralinos and third-generation squarks and sleptons can become important, once they are kinematically allowed [81]. For Higgs masses below 130 GeV, the range of supersymmetric parameter space in which supersymmetric decays are dominant is rather narrow when the current bounds on supersymmetric particle masses are taken into account. One interesting possibility is a significant branching ratio of $h^0 \to \widetilde{\chi}^0\widetilde{\chi}^0$, which could arise for values of m_{h^0} near its upper theoretical limit. Such an invisible decay mode could be detected at the LC by searching for the missing mass recoiling against the Z in $e^+e^- \to h^0Z$.

5.4. *MSSM Higgs Boson Production at the LC*

For $m_{A^0} \gtrsim 150$ GeV, Fig. 10 shows that the MSSM Higgs sector quickly approaches the decoupling limit, where the properties of h^0 approximately coincide with those of $h_{\rm SM}$. In contrast, $|\cos(\beta-\alpha)| \ll 1$ implies that the H^0VV couplings are highly suppressed [see Eq. (17)], and the A^0VV

couplings are loop-suppressed. Thus, the Higgsstrahlung and vector-boson-fusion cross-sections for $h_{\rm SM}$ production also apply to h^0 production, but these are not useful for H^0 and A^0 production. For the heavier neutral Higgs bosons, the most robust production mechanism is $e^+e^- \to Z^* \to H^0A^0$, which is not suppressed since the ZH^0A^0 coupling is proportional to $\sin(\beta-\alpha)$, as indicated in Eq. (18). Radiatively corrected cross-sections for $e^+e^- \to Zh^0$, ZH^0, H^0A^0, and h^0A^0 have been obtained in [82]. The charged Higgs boson is also produced in pairs via s-channel photon and Z exchange. However, since $m_{H^0} \simeq m_{A^0} \simeq m_{H^\pm}$ in the decoupling limit, H^0A^0 and H^+H^- production are kinematically allowed only when $m_{A^0} \lesssim \sqrt{s}/2$.[g] In $\gamma\gamma$ collisions, one can extend the Higgs mass reach for the neutral Higgs bosons. As described in Section 11, the s-channel resonant production of H^0 and A^0 (due primarily to the top and bottom-quark loops in the one-loop Higgs–$\gamma\gamma$ triangle) can be detected for some choices of m_{A^0} and $\tan\beta$ if the heavy Higgs masses are less than about 80% of the initial $\sqrt{s}$ of the primary e^+e^- system. The corresponding cross sections are a few fb [14, 83].

If $m_{A^0} \lesssim 150$ GeV, deviations from the decoupling limit become more pronounced, and H^0 can now be produced via Higgsstrahlung and vector boson fusion at an observable rate. In addition, the factor of $\cos(\beta-\alpha)$ in the Zh^0A^0 coupling no longer significantly suppresses h^0A^0 production. Finally, if $m_{H^\pm} \lesssim 170$ GeV, the charged Higgs boson will also be produced in $t \to H^+b$. In the non-decoupling regime, all non-minimal Higgs states can be directly produced and studied at the LC.

Higgs boson production in association with a fermion-antifermion pair can also be considered. Here, the new feature is the possibility of enhanced Higgs–fermion Yukawa couplings. Consider the behavior of the Higgs couplings at large $\tan\beta$, where some of the Higgs couplings to down type fermion pairs (denoted generically by $b\bar{b}$) can be significantly enhanced.[h] Let us examine two particular regions with large $\tan\beta$. In the decoupling limit (where $m_{A^0} \gg m_Z$ and $|\cos(\beta-\alpha)| \ll 1$), it follows from Eq. (19) that the $b\bar{b}H^0$ and $b\bar{b}A^0$ couplings have equal strength and are significantly enhanced by a factor of $\tan\beta$ relative to the $b\bar{b}h_{\rm SM}$ coupling, while the $b\bar{b}h^0$ coupling

[g] The pair production of scalars is P-wave suppressed near threshold, so in practice the corresponding Higgs mass reach is likely to be somewhat lower than $\sqrt{s}/2$.

[h] We do not consider the possibility of $\tan\beta \ll 1$, which would lead to enhanced Higgs couplings to up-type fermions. As previously noted, models of low-energy supersymmetry tend to favor the parameter regime of $1 \lesssim \tan\beta \lesssim m_t/m_b$. Moreover, some of the low $\tan\beta$ region is already ruled out by the MSSM Higgs search (see Section 6.1).

is given by the corresponding Standard Model value. If $m_{A^0} \lesssim m_Z$ and $\tan\beta \gg 1$, then $|\sin(\beta-\alpha)| \ll 1$, as implied by Fig. 10, and $m_{h^0} \simeq m_{A^0}$. In this case, the $b\bar{b}h^0$ and $b\bar{b}A^0$ couplings have equal strength and are significantly enhanced (by a factor of $\tan\beta$)[i] relative to the $b\bar{b}h_{\rm SM}$ coupling. Note that in both cases above, only two of the three neutral Higgs bosons have enhanced couplings to $b\bar{b}$. If ϕ is one of the two neutral Higgs bosons with enhanced $b\bar{b}\phi$ couplings, then the cross-section for $e^+e^- \to f\bar{f}\phi$ ($f=b$ or τ) will be significantly enhanced relative to the corresponding Standard Model cross-section by a factor of $\tan^2\beta$. The phase-space suppression is not as severe as in $e^+e^- \to t\bar{t}\phi$ (see Fig. 5), so this process could extend the mass reach of the heavier neutral Higgs states at the LC given sufficient luminosity. The production of the charged Higgs boson via $e^+e^- \to t\bar{b}H^-$ is also enhanced by $\tan^2\beta$, although this process has a more significant phase-space suppression because of the final state top quark. If any of these processes can be observed, it would allow for a direct measurement of the corresponding Higgs–fermion Yukawa coupling.

6. MSSM Higgs Boson Searches before the LC

6.1. *Direct Search Limits from LEP*

Although no direct experimental evidence for the Higgs boson yet exists, there are both experimental as well as theoretical constraints on the parameters of the MSSM Higgs sector. Experimental limits on the charged and neutral Higgs masses have been obtained at LEP. For the charged Higgs boson, $m_{H^\pm} > 78.6$ GeV [84]. This is the most model-independent bound. It is valid for more general non-supersymmetric two-Higgs doublet models and assumes only that the H^+ decays dominantly into $\tau^+\nu_\tau$ and/or $c\bar{s}$. The LEP limits on the masses of h^0 and A^0 are obtained by searching simultaneously for $e^+e^- \to Z \to Zh^0$ and $e^+e^- \to Z \to h^0A^0$. Radiative corrections can be significant, as shown in Section 5.2, so the final limits depend on the choice of MSSM parameters that govern the radiative corrections. The third generation squark parameters are the most important of these. The LEP Higgs working group [85] quotes limits for the case of $M_{\rm SUSY} = 1$ TeV in the maximal-mixing scenario, which corresponds to the choice of third generation squark parameters that yields the largest correc-

[i]However in this case, the value of the $b\bar{b}H^0$ coupling can differ from the corresponding $b\bar{b}h_{\rm SM}$ coupling when $\tan\beta \gg 1$, since in case (ii), where $|\sin(\beta-\alpha)| \ll 1$, the product $\tan\beta\sin(\beta-\alpha)$ need not be particularly small.

tions to m_{h^0}. The present LEP 95% CL lower limits are $m_{A^0} > 91.9$ GeV and $m_{h^0} > 91.0$ GeV. The theoretical upper bound on m_{h^0} as a function of $\tan\beta$, exhibited in Fig. 11, can then be used to exclude a region of $\tan\beta$ in which the predicted value of m_{h^0} lies below the experimental bound. Under the same MSSM Higgs parameter assumptions stated above, the LEP Higgs search excludes the region $0.5 < \tan\beta < 2.4$ at 95% CL.

In discussing Higgs discovery prospects at the Tevatron and LHC, we shall quote limits based on the assumption of $M_{\text{SUSY}} = 1$ TeV and maximal squark mixing. This tends to be a conservative assumption; that is, other choices give sensitivity to *more* of the $[m_{A^0}, \tan\beta]$ plane. However, there are other parameter regimes where certain Higgs search strategies become more difficult. While these issues are of vital importance to the Tevatron and LHC Higgs searches, they are much less important at the LC.

6.2. *MSSM Higgs Searches at the Tevatron*

The Tevatron SM Higgs search can be reinterpreted in terms of the search for the CP-even Higgs boson of the MSSM. Since the theoretical upper bound was found to be $m_{h^0} \lesssim 135$ GeV (for $M_{\text{SUSY}} < 2$ TeV), only the Higgs search of the low-mass region, 100 GeV $\lesssim m_{h^0} \lesssim 135$ GeV, applies. In the MSSM at large $\tan\beta$, the enhancement of the $A^0b\bar{b}$ coupling (and a similar enhancement of either the $h^0b\bar{b}$ or $H^0b\bar{b}$ coupling) provides a new search channel: $q\bar{q}$, $gg \to b\bar{b}\phi$, where ϕ is a Higgs boson with enhanced couplings to $b\bar{b}$. Combining both sets of analyses, the Tevatron Higgs Working Group [47] obtained the expected 95% CL exclusion and 5σ Higgs discovery contours for the maximal mixing scenario as a function of total integrated luminosity per detector (combining both CDF and D0 data sets).

With 5 fb^{-1} of integrated luminosity per experiment, it is possible to test nearly the entire MSSM Higgs parameter space, and obtain a 95% CL exclusion limit if nò signal is observed. To assure discovery of a CP-even Higgs boson at the 5σ level, the luminosity requirement becomes very important. Figure 13 shows that a total integrated luminosity of about 20 fb^{-1} per experiment is necessary in order to assure a significant, although not exhaustive, coverage of the MSSM parameter space. If the anticipated 15 fb^{-1} integrated luminosity is achieved, the discovery reach will significantly extend beyond that of LEP. Nevertheless, the MSSM Higgs boson could still evade capture at the Tevatron. One would then turn to the LHC to try to obtain a definitive Higgs boson discovery.

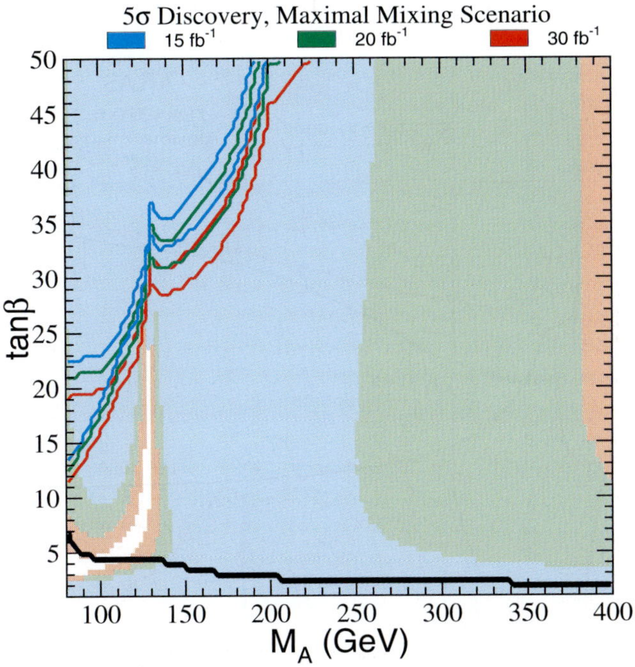

Fig. 13. The 5σ discovery region in the m_{A^0}–$\tan\beta$ plane, for the maximal mixing scenario and two different search channels: $q\overline{q} \to V\phi$ ($\phi = h^0$, H^0), $\phi \to b\overline{b}$ (shaded regions) and gg, $q\overline{q} \to b\overline{b}\phi$ ($\phi = h^0$, H^0, A^0), $\phi \to b\overline{b}$ (region in the upper left-hand corner bounded by the solid lines). Different integrated luminosities are explicitly shown by the color coding. The two sets of lines (for a given color) correspond to the CDF and DØ simulations, respectively. The region below the solid black line near the bottom of the plot is excluded by the absence of observed $e^+e^- \to Z\phi$ events at LEP. Taken from [47].

6.3. *MSSM Higgs Searches at the LHC*

The potential of the LHC to discover one or more of the MSSM Higgs bosons has been exhaustively studied for the minimal and maximal mixing scenarios described above. One of the primary goals of these studies has been to demonstrate that at least one of the MSSM Higgs bosons will be observable by ATLAS and CMS for any possible choice of $\tan\beta$ and m_{A^0} consistent with bounds coming from current LEP data. In order to establish such a "no-lose" theorem, an important issue is whether or not the Higgs bosons have substantial decays to supersymmetric particle pairs. It is reasonable to suppose that these decays will be absent or relatively insignificant for the

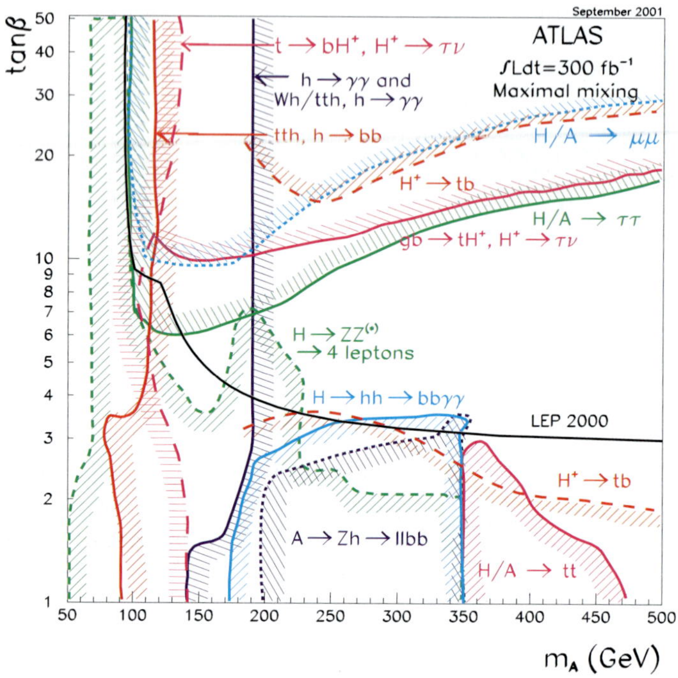

Fig. 14. 5σ discovery contours for MSSM Higgs boson detection in various channels are shown in the $[m_{A^0}, \tan\beta]$ parameter space, assuming maximal mixing and an integrated luminosity of $L = 300$ fb^{-1} for the ATLAS detector. Taken from [86].

light h^0. Current mass limits on supersymmetric particles are such that only $h^0 \to \widetilde{\chi}_1^0\widetilde{\chi}_1^0$ is not kinematically excluded, and this possibility arises only in a very limited class of models. For $m_{A^0} \gtrsim 200$ GeV, decays of the $A^0, H^0, H^\pm$ to supersymmetric particles (especially pairs of light charginos/neutralinos) are possible, but the branching ratios are generally not significantly large. In all such cases, the discovery limits we discuss below would only be slightly weakened. Furthermore, at high $\tan\beta$ the enhancement of the $b\bar{b}$ and $\tau^+\tau^-$ couplings of the heavy A^0 and H^0 imply that supersymmetric decay modes will not be important even when $m_{A^0} \sim m_{H^0} \sim m_{H^\pm} \gg m_Z$. We will summarize the LHC discovery prospects for the MSSM Higgs bosons assuming that supersymmetric decays are not significant.

One of the primary Higgs discovery modes is detection of the relatively SM-like h^0 using the same modes as employed for a light h_{SM}. Based on Fig. 14 (which assumes $L = 300\ \mathrm{fb}^{-1}$) [86], we see that for $m_{A^0} \gtrsim 180$ GeV, the h^0 will be detected via $gg, WW \to h^0$ and $Wh^0, t\bar{t}h^0$ with $h^0 \to \gamma\gamma$, while the $t\bar{t}h^0$ with $h^0 \to b\bar{b}$ mode is viable down to $m_{A^0} \gtrsim 100$–120 GeV, depending on $\tan\beta$. Similar results have been obtained by CMS [87]. There are also many possibilities for detecting the other MSSM Higgs bosons. First, there is a small domain in which $m_{A^0} \lesssim 130$ GeV, but yet m_{A^0} is still large enough for consistency with LEP limits, in which $t \to bH^\pm$ discovery will be possible. However, the most interesting alternative detection modes are based on $gg \to A^0, H^0$ and $gb \to H^\pm t$ production. We focus first on the former. For low-to-moderate $\tan\beta$ values, the channels $H^0 \to ZZ^{(*)} \to 4\ell$, $H^0 \to h^0h^0 \to b\bar{b}\gamma\gamma$ and $A^0 \to Zh^0 \to \ell\ell b\bar{b}$ are viable when $m_{A^0} \lesssim 2m_t$, whereas the $A^0, H^0 \to t\bar{t}$ modes are viable for $m_{A^0} > 2m_t$. For large enough $\tan\beta$, the $gg \to A^0, H^0 \to \tau^+\tau^-, \mu^+\mu^-$ discovery modes become viable. For the $gb \to H^\pm t$ process, the $H^\pm \to tb$ decays provide a 5σ signal both for low-to-moderate $\tan\beta \lesssim 2$–3 and for high $\tan\beta \gtrsim 15$–25, depending upon mass. In addition, the $H^\pm \to \tau^\pm\nu$ decay mode yields a viable signal for $\tan\beta \gtrsim 7$–12. Of course, if the plot were extended to higher m_{A^0}, the minimum $\tan\beta$ value required for H^0, A^0 or $H^\pm$ detection would gradually increase.

We noted earlier that the present LEP limits imply that $\tan\beta > 2.4$ in the case of maximal mixing and $M_{SUSY} = 1$ TeV (with even more stringent limits possible in other regions of the supersymmetric parameter space). Thus, it is very likely that $\tan\beta$ and m_{A^0} will be in one of two regions: (a) the increasingly large (as m_{A^0} increases) wedge of moderate $\tan\beta > 3$ in which only the h^0 will be detected; or (b) the high $\tan\beta$ region for which the $gg \to H^0, A^0 \to \tau^+\tau^-, \mu^+\mu^-$ and $gb \to H^\pm t \to \tau^\pm\nu t, tbt$ modes are viable as well. If the $H^0, A^0, H^\pm$ are heavy and cannot be detected either at the LHC (because $\tan\beta$ is not large enough) or at the LC (because they are too heavy to be pair-produced), precision measurements of the h^0 branching ratios and other properties will be particularly crucial. The precision measurements might provide the only means for constraining or approximately determining the value of m_{A^0} aside from possible direct detection in $\gamma\gamma \to H^0, A^0$ production. Expected LC precisions are such that deviations of h^0 branching ratios from the predicted SM values can be detected (in the maximal mixing scenario) for $m_{A^0} \lesssim 600$ GeV [73].

At the LHC there is another important possibility for h^0 detection. Provided that the mass of the second-lightest neutralino exceeds that of the lightest neutralino (the LSP) by at least m_{h^0}, gluino and squark production will lead to chain decays in which $\widetilde{\chi}_2^0 \to h^0 \widetilde{\chi}_1^0$ occurs with substantial probability. In this way, an enormous number of h^0's can be produced, and the $h^0 \to b\bar{b}$ decay mode will produce a dramatic signal [88].

7. Non-exotic Extended Higgs Sectors

In this section, we consider the possibility of extending only the Higgs sector of the SM, leaving unchanged the gauge and fermionic sectors of the SM. We will also consider extensions of the two-doublet Higgs sector of the MSSM.

The simplest extensions of the minimal one-doublet Higgs sector of the SM contain additional doublet and/or singlet Higgs fields. Such extended Higgs sectors will be called non-exotic (to distinguish them from exotic Higgs sectors with higher representations, which will be considered briefly in Section 8). Singlet-only extensions have the advantage of not introducing the possibility of charge violation, since there are no charged Higgs bosons. In models with more than one Higgs doublet, tree-level Higgs-mediated flavor-changing neutral currents are present unless additional symmetries (*e.g.*, discrete symmetries or supersymmetry) are introduced to restrict the form of the tree-level Higgs-fermion interactions [89]. Extensions containing additional doublet fields allow for spontaneous and explicit CP violation within the Higgs sector. These could be the source of observed CP-violating phenomena. Such models require positive squared-masses for the charged Higgs boson(s) in order to avoid spontaneous breaking of electric charge conservation.

Extensions of the SM Higgs sector containing doublets and singlets can certainly be considered on a purely *ad hoc* basis. But there are also many dynamical models in which the effective low-energy sector below some scale Λ of order 1 to 10 TeV, or higher, consists of the SM fermions and gauge bosons plus an extended Higgs sector. Models with an extra doublet of Higgs fields include those with new strong forces, in which the effective Higgs doublet fields are composites containing new heavier fermions. The heavy fermions should be vector-like to minimize extra contributions to precision electroweak observables. In many of these models, the top quark mixes with the right-handed component of a new vector-like fermion. The top quark could also mix with the right-handed component of a Kaluza-Klein (KK) excitation of a fermion field, so that Higgs bosons would be composites of the top quark and fermionic KK excitations. (For a review and references to the literature, see [90].) Although none of these (non-perturbative) models have been fully developed, they do provide significant motivation for studying the Standard Model with a Higgs sector containing extra doublets and/or singlets if only as the effective low-energy theory below a scale Λ in the TeV range.

When considering Higgs sectors in the context of a dynamical model with strong couplings at scale Λ, restrictions on Higgs self-couplings and Yukawa couplings, which would arise by requiring perturbativity for such couplings up to some large GUT scale, do not apply. At most, one should only demand perturbativity up to the scale Λ at which the new (non-perturbative) dynamics enter and the effective theory breaks down.

The minimal Higgs sector of the MSSM is a Type-II two-doublet model, where one neutral Higgs doublet (Φ_d^0) couples at tree level only to down quarks and leptons while the other (Φ_u^0) couples only to up quarks. Non-minimal extended Higgs sectors are also possible in low-energy supersymmetric models. Indeed, string theory realizations of low-energy supersymmetry often contain many extra singlet, doublet and even higher representations, some of which can yield light Higgs bosons (see, *e.g.*, [91]). However, non-singlet Higgs representations spoil gauge coupling unification, unless additional intermediate-scale matter fields are added to restore it. A particularly well-motivated extension is the inclusion of a single extra complex singlet Higgs field, often denoted S. Including S, the superpotential for the theory can contain the term $\lambda_S H_u H_d S$, which can then provide a natural source of a weak scale value for the μ parameter appearing in the bilinear superpotential form $\mu H_u H_d$ required in the MSSM. A weak-scale value for $s \equiv \langle S^0 \rangle$, where S^0 is the scalar component of the superfield S, is natural and yields an effective $\mu = \lambda_S s$. This extension of the MSSM is referred to as the next-to-minimal supersymmetric model, or NMSSM, and has received considerable attention. For an early review and references, see [1].

7.1. *The Decoupling Limit*

In many extended Higgs sector models, the most natural parameter possibilities correspond to a decoupling limit in which there is only one light Higgs boson, with Yukawa and vector boson couplings close to those of the SM Higgs boson. In contrast, all the other Higgs bosons are substantially heavier (than the Z) with negligibly small relative mass differences, and with suppressed vector boson couplings (which vanish in the exact limit of decoupling). By assumption, the decoupling limit assumes that all Higgs self-couplings are kept fixed and perturbative in size. [j] In the MSSM, such a decoupling limit arises for $m_{A^0} \gg m_Z$, and quickly becomes a very good approximation for $m_{A^0} \gtrsim 150$ GeV.

The decoupling limit can be evaded in special cases, in which the scalar potential exhibits a special form (*e.g.*, a discrete symmetry can forbid certain terms). In such models, there could exist regions of parameter space in which all but one Higgs boson are significantly heavier than the Z, but the light scalar state does *not* possess SM-like properties [92]. A complete exposition regarding the decoupling limit in the two-Higgs doublet model, and special cases that evade the limit can be found in [62].

7.2. *Constraints from Precision Electroweak Data and LC Implications*

In the Standard Model, precision electroweak constraints require $m_{h_{\rm SM}} \lesssim 193$ GeV at 95% CL. This is precisely the mass region preferred in the MSSM and its extensions. However, in models of general extended Higgs sectors with only weak doublet and singlet representations, there are more complicated possibilities. First, it could be that there are several, or even many, Higgs bosons that couple to vector bosons and it is only their average mass weighted by the square of their VV coupling strength (relative to the SM strength) that must obey the SM Higgs mass limit. Second, there can be weak isospin violations either within the Higgs sector itself or involving extra dynamics (for example related to the composite Higgs approach) can compensate for the excessive deviations predicted if there is a SM-like Higgs boson with mass substantially above 200 GeV.

A particularly simple example of this latter possibility arises in the context of the two-Higgs doublet model (2HDM) [92] and is illustrated in Fig. 15. Consider a 2HDM in which one of the CP-even neutral Higgs bosons has SM-like couplings but has mass just above a particular presumed value of $\sqrt{s}$ (500 or 800 GeV) for the linear collider. In addition, focus on cases in which there is a lighter A^0 or h^0 with no VV coupling (for either, we use the notation $\widehat{h}$) and in which all other Higgs bosons have mass larger than $\sqrt{s}$. Next, isolate mass and $\tan\beta$ choices for which detection of the $\widehat{h}$ will also be impossible at the LC. Finally, scan over masses of the heavy Higgs bosons so as to achieve the smallest precision electroweak $\Delta\chi^2$

[j]In the decoupling limit, the heavier Higgs bosons may have enhanced couplings to fermions (*e.g.*, at large $\tan\beta$ in the two-Higgs doublet model). We assume that these couplings also remain perturbative.

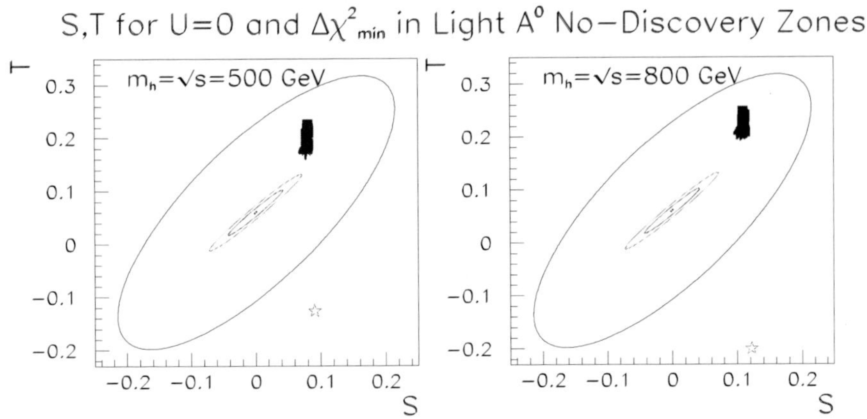

Fig. 15. The outer ellipses show the 90% CL region from current precision electroweak data in the S,T plane for $U=0$ relative to a central point defined by the SM prediction with $m_{h_{\rm SM}}=115$ GeV. The blobs of points show the S,T predictions for 2HDM models with a light A^0 and with $\tan\beta$ such that the A^0 cannot be detected in $b\bar{b}A^0$ or $t\bar{t}A^0$ production at either the LC or the LHC; the mass of the SM-like h^0 is set equal to $\sqrt{s}=500$ GeV (left) or 800 GeV (right) and $m_{H^\pm}$ and m_{H^0} have been chosen to minimize the χ^2 of the full precision electroweak fit. The innermost (middle) ellipse shows the 90% (99.9%) CL region for $m_{h_{\rm SM}}=115$ GeV after the Giga-Z operation *and* a $\Delta m_W \lesssim 6$ MeV threshold scan measurement. The stars to the bottom right show the S,T predictions in the case of the SM with $m_{h_{\rm SM}}=500$ GeV (left) or 800 GeV (right). This figure is from [93].

relative to that found in the Standard Model for $m_{h_{\rm SM}}=115$ GeV [93]. For $\widehat{h}=A^0$, the blobs of overlapping points in Fig. 15 indicate the S,T values for the optimal choices and lie well within the current 90% CL ellipse. The heavy Higgs boson with SM couplings gives a large positive contribution to S and large negative contribution to T, and in the absence of the other Higgs bosons would give the S,T location indicated by the star. However, there is an additional positive contribution to T arising from a slight mass non-degeneracy among the heavier Higgs bosons. For instance, for the case of a light $\widehat{h}=A^0$, the h^0 is heavy and SM-like and $\Delta\rho\equiv\alpha\Delta T$, where

$$\Delta T = \frac{1}{16\pi m_W^2\cos^2\theta_W}\left\{\frac{\cos^2\theta_W}{\sin^2\theta_W}\left(\frac{m_{H^\pm}^2-m_{H^0}^2}{2}\right)\right.$$
$$\left. -3m_W^2\left[\ln\frac{m_{h^0}^2}{m_W^2}+\frac{1}{6}+\frac{1}{\sin^2\theta_W}\ln\frac{m_W^2}{m_Z^2}\right]\right\} \quad (32)$$

can be adjusted by an appropriate choice of $m^2_{H^\pm} - m^2_{H^0}$ to place the prediction in the S–T plane at the location of the blob in Fig. 15. Indeed, even if the "light" decoupled Higgs boson is not so light, but rather has mass equal to $\sqrt{s}$ (and is therefore unobservable), one can still obtain adequate agreement with current precision electroweak data. Fortunately, one can only push this scenario so far. To avoid moving beyond the current 90% ellipse (and also to maintain perturbativity for the Higgs self-couplings), the Higgs with SM-like VV coupling must have mass $\lesssim 1$ TeV.

In composite Higgs models with extra fermions, there are similar non-degeneracies of the fermions that can yield a similar positive contribution to ΔT. As reviewed in [28], consistency with current precision electroweak data inevitably constrains parameters so that some type of new physics (including a possible heavy scalar sector) would again have to lie below a TeV or so. Future Giga-Z and W-threshold measurements could provide much stronger constraints on these types of models, as discussed in Section 10. Moreover, such measurements would become a priority if a heavy SM Higgs boson is found at the LHC, but no other new physics (needed for a consistent explanation of the precision electroweak data) is discovered.

7.3. *Constraints on Higgs Bosons with VV Coupling*

In the MSSM, we know that the Higgs boson(s) that carry the VV coupling must be light: if m_{A^0} is large (the decoupling limit) then it is the mass-bounded h^0 that has all the VV coupling strength; if $m_{A^0} \lesssim 2m_Z$, then the H^0 can share the VV coupling with the h^0, but then m_{H^0} cannot be larger than about $2m_Z$. In the NMSSM, assuming Higgs-sector CP conservation, there are 3 neutral CP-even Higgs bosons, $h_{1,2,3}$ ($m_1 < m_2 < m_3$), which can share the VV coupling strength. One can show (see [94] for a recent update) that the masses of the h_i with substantial VV coupling are strongly bounded from above. This result generalizes to the most general supersymmetric Higgs sector as follows. Labeling the neutral Higgs bosons by i with masses m_{h_i} and denoting the ZZ squared-coupling relative to the SM by K_i, it can be shown that [95]

$$\sum_i K_i \geq 1\,, \qquad \sum_i K_i m^2_{h_i} \leq m^2_B\,. \tag{33}$$

where the value of m_B depends somewhat on the low-energy supersymmetric model. That is, the aggregate strength of the VV squared-coupling of all the neutral Higgs bosons is at least that of the SM, and the squared-masses of the neutral h_i weighted by the squared-couplings must lie below a certain bound. A value of $m_B \sim 200$ GeV in Eq. (33) is obtained [66] by assuming that the low-energy supersymmetric theory remains perturbative up to the GUT scale of order 10^{19} GeV. This value of m_B applies for the most general possible Higgs representations (including triplets) in the supersymmetric Higgs sector and for arbitrary numbers of representations. If only doublet and singlet representations are allowed, the bound would be lower. The value of $m_B \sim 200$ GeV also applies to general Higgs-sector-only extensions of the SM by requiring consistency with precision electroweak constraints *and* assuming the absence of a large contribution to T from the Higgs sector itself or from new physics, such as discussed in Section 7.2.

7.4. *Detection of Non-exotic Extended Higgs Sector Scalars at the Tevatron and LHC*

In the case of extended Higgs sectors, all of the same processes as discussed for the SM and MSSM will again be relevant. However, one can no longer guarantee Higgs discovery at the Tevatron and/or LHC. In particular, if there are many Higgs bosons sharing the WW, ZZ coupling, Higgs boson discovery based on processes that rely on the VV coupling could be much more difficult than in models with just a few light Higgs bosons with substantial VV coupling. This is true even if the sum rule of Eq. (33) applies. For example, in the NMSSM, the addition of one singlet to the minimal two-doublet Higgs structure allows for parameter choices such that no Higgs boson can be discovered at the LHC [96, 97] using any of the processes considered for SM Higgs and MSSM Higgs detection. The problematical regions of parameter space originally discussed in [96] were those in which the $\gamma\gamma$ decay channel signals for the CP-even Higgs bosons are decreased (because of decreased W-loop contribution to the coupling) and a moderate value of $\tan\beta$ was chosen so that $t\bar{t}$+Higgs processes are weak and $b\bar{b}$+Higgs processes are insufficiently enhanced. However, as shown in [97], the recent addition of the $t\bar{t}h \to t\bar{t}b\bar{b}$ and, especially, the $W^*W^* \to h \to \tau^+\tau^-$ discovery modes to the list of viable channels by the CMS and ATLAS collaborations means that discovery of at least one NMSSM Higgs bosons would be possible in all the "bad" parameter regions found in [96]. However, [97] also finds that there are parameter choices (explicitly excluded in the study of [96]) such that the CP-even Higgs bosons decay primarily to a pair of CP-odd Higgs bosons. Discovery techniques for this type of final state have not been developed for the LHC.

However, in other cases, the Tevatron and LHC could observe signals not expected in an approximate decoupling limit. For example, in the 2HDM model discussed earlier the light $\widehat{h}$ with no VV couplings decays via $\widehat{h} \to b\bar{b}, \tau^+\tau^-$ and discovery in $t\bar{t}\widehat{h}$, $b\bar{b}\widehat{h}$ and even $gg \to \widehat{h}$ [98] is possible, though certainly not guaranteed. Further, in these models there is a heavy neutral Higgs boson having the bulk of the VV coupling and (for consistency with current precision electroweak constraints or with perturbativity) a mass less than about 1 TeV. This latter Higgs boson would be detected at the LHC using gg, WW fusion production and $ZZ \to 4\ell, WW \to 2j\ell\nu, \ldots$ decay modes, just like a heavy SM Higgs boson.

7.5. *LC Production Mechanisms for Non-exotic Extended Higgs Sector Scalars*

Any physical Higgs eigenstate with substantial WW and ZZ coupling will be produced in Higgsstrahlung and WW fusion at the LC. Although there could be considerable cross section dilution and/or resonance peak overlap, the LC will nonetheless always detect a signal. This has been discussed for the MSSM in Section 5.4. In the NMSSM, if one of the heavier CP-even h_i has most of the VV coupling, the strong bound on its mass [94] noted earlier implies that it will be detected at any LC with $\sqrt{s} > 350$ GeV within a small fraction of a year when running at planned luminosities [99]. The worst possible case is that in which there are many Higgs bosons with VV coupling with masses spread out over a large interval with separation smaller than the mass resolution. In this case, the Higgs signal becomes a kind of continuum distribution. Still, in [95] it is shown that the sum rule of Eq. (33) guarantees that the Higgs continuum signal will still be detectable for sufficient integrated luminosity, $L \gtrsim 200$ fb^{-1}, as a broad excess in the recoil mass spectrum of the $e^+e^- \to ZX$ process. (In this case, WW fusion events do *not* allow for the reconstruction of Higgs events independently of the final state Higgs decay channel.) As already noted, the value of $m_B \sim 200$ GeV appearing in Eq. (33) can be derived from the constraints of perturbative evolution up to some high energy (GUT or Planck) scale for the most general Higgs sector in supersymmetric theories. The same bound on m_B is also required by precision electroweak data for general SM Higgs sector extensions, at least in theories that do not have a large positive contribution to T from a non-decoupling structure in the Higgs sector or from new physics not associated with the Higgs sector.

Other production modes of relevance include Higgs pair production [100], single Higgs production in association with $t\bar{t}$ [44, 101, 102, 45,

103, 104, 106, 105] and with $b\bar{b}$ [107, 104, 106, 105]. In multi-doublet models, $t\bar{b}H^-/b\bar{t}H^+$ [44, 108, 109] and $\tau\nu H^\pm$ [108, 110, 111] production are also present (and in some cases $c\bar{b}H^-/b\bar{c}H^+$ production is competitive [111]). The associated production of a charged Higgs boson and gauge boson has also been studied [112]. However, none of these modes are guaranteed to be either kinematically accessible at the LC or to have a sufficiently high event rate to be observed (if kinematically allowed). Regardless of the production process, relevant decay channels could include cases where heavier Higgs bosons decay to lighter ones. If observed, such decays would provide vital information on Higgs self-couplings [113].

It is also important to consider the production processes that are most relevant for those Higgs bosons (denoted $\widehat{h}$) which do not have substantial VV couplings. Such processes have particular relevance in the non-decoupling scenario for the general 2HDM model discussed earlier. In this scenario, $\widehat{h}$ is the only Higgs boson light enough to be produced at an LC with $\sqrt{s} \lesssim 1$ TeV, but it cannot be produced and detected in WW fusion or Higgsstrahlung. Since the other Higgs bosons are heavy, the $\widehat{h}$ also cannot be produced in association with another Higgs boson. As shown in [105, 92], the $b\bar{b}\widehat{h}$ and $t\bar{t}\widehat{h}$ processes will also not be detectable at the LC if $\tan\beta$ is moderate in value. The most interesting tree-level processes are then those based on the quartic couplings $WW\widehat{h}\widehat{h}$ and $ZZ\widehat{h}\widehat{h}$ required by gauge invariance [114]. These couplings allow for $WW \to \widehat{h}\widehat{h}$ fusion and $Z^* \to Z\widehat{h}\widehat{h}$ production, respectively. The exact cross sections for these processes are only mildly sensitive to the masses of the other heavier Higgs bosons via 2HDM Higgs self-couplings. Of course, phase space restrictions imply an upper limit on the $\widehat{h}$ masses that can be probed in this way.

The case of $\widehat{h} = A^0$ is illustrated in Figs. 16 and 17 [115]. Figure 16 shows the wedge region in which fewer than 20 events per 1000 fb^{-1} are obtained in $b\bar{b}A^0$ and $t\bar{t}A^0$ production at $\sqrt{s} = 800$ GeV. Figure 17 shows the m_{A^0} values for which fewer than 20 events are obtained in $Z^* \to ZA^0A^0$ and $W^*W^* \to A^0A^0$ production at $\sqrt{s} = 500$ GeV and 800 GeV. Thus, single A^0 discovery is not possible via any of the above processes in the region of the wedge with $m_{A^0} \gtrsim 300$ GeV. With the $\gamma\gamma$ collider option at the LC, we note that $\gamma\gamma \to A^0$ can provide a signal for the decoupled A^0 over a significant portion of the wedge region. The results from a realistic study of [17] are illustrated in Fig. 16, which focuses on $m_{A^0} \geq 250$ GeV. In particular, the operation of the $\gamma\gamma$ collider for two years using a polarization configuration for the electron beams and laser photons yielding a broad $E_{\gamma\gamma}$ spectrum and one year using a configuration yielding a peaked spectrum is assumed. The

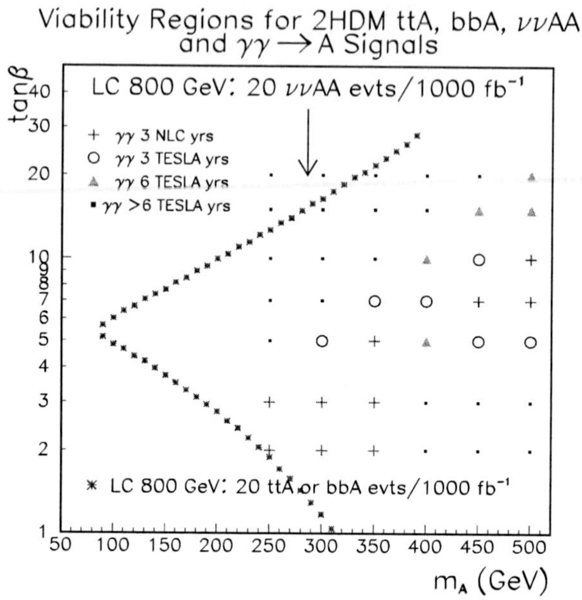

Fig. 16. The stars form an outline of a wedge of $[m_{A^0}, \tan\beta]$ parameter space inside which the LC operating at $\sqrt{s} = 800$ GeV yields fewer than 20 events per 1000 fb^{-1} in both the $t\bar{t}A^0$ and $b\bar{b}A^0$ production modes. Also shown by the arrow is the m_{A^0} value above which the process $e^+e^- \to \nu\bar{\nu}A^0A^0$ yields fewer than 20 events per 1000 fb^{-1}. The + symbols on the grid of $[m_{A^0}, \tan\beta]$ values show the points for which a 4σ signal for $\gamma\gamma \to A^0$ would be achieved using LC operation at $\sqrt{s} = 630$ GeV after three years of operation assuming running conditions and strategies as specified in [17] using the NLC-based design of the $\gamma\gamma$ collider. (The nominal e^+e^- luminosity for the underlying NLC design employed is 220 fb^{-1} per year at $\sqrt{s} = 500$ GeV.) The circles (triangles) show the additional points that would yield a 4σ signal for twice (four times) the integrated luminosity of the current NLC $\gamma\gamma$ interaction region design. The small squares show the additional points sampled in the study of [17]. This figure was taken from [115].

pluses indicate 4σ discovery points after three years of operation (assuming that the machine operates for 10^7 sec during one calendar year) in the appropriate configurations at the NLC. The results of Fig. 16 employ the $E_{\gamma\gamma}$ luminosity spectra as computed in [17] based on the particular laser and interaction region design that is discussed in more detail in Section 11. A factor of two higher $\gamma\gamma$ luminosity, as might be achievable at TESLA or through further design improvement at the NLC, would allow 4σ discovery for the additional points indicated by the circles. The corresponding results for a decoupled CP-even $\widehat{h} = h^0$ are similar.

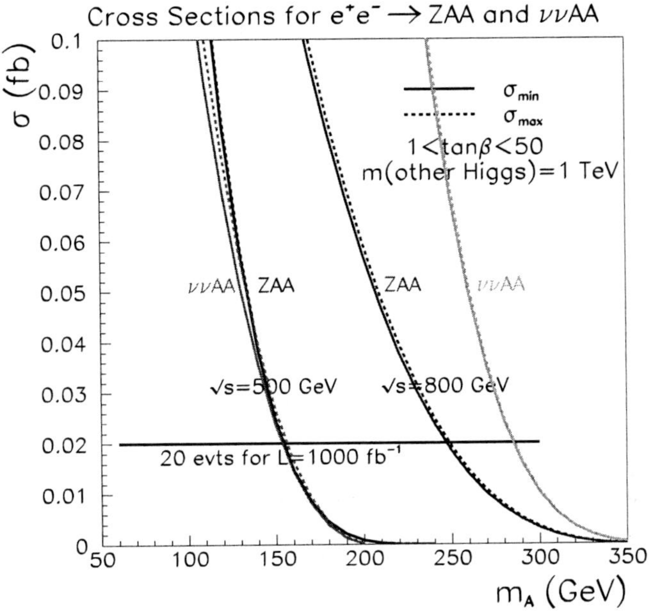

Fig. 17. The cross sections for $e^+e^- \to ZA^0A^0$ and $e^+e^- \to \nu\overline{\nu}A^0A^0$ as a function of m_{A^0} are shown, assuming a 2HDM model with a heavy SM-like h^0. We have taken $m_{h^0} = m_{H^0} = m_{H^\pm} = 1$ TeV. Maximum and minimum values found after scanning $1 \leq \tan\beta \leq 50$ are shown for $\sqrt{s} = 500$ GeV and 800 GeV. The variation with $\tan\beta$ arises from small contributions associated with exchanges of the heavy Higgs bosons. The 20 event level for $L = 1000$ fb^{-1} is indicated. Taken from [115].

8. Exotic Higgs Sectors and Other Possibilities

As we have seen, there are many scenarios and models in which the Higgs sector is more complicated than the one-Higgs-doublet of the Standard Model. Supersymmetry requires at least two Higgs doublets. Even in the absence of supersymmetry, a two-doublet Higgs sector allows for CP-violating phenomena. Singlets can also be added without altering the tree-level prediction of $\rho = 1$. However, the possibility of Higgs representations with still higher weak isospin (left-handed, denoted by L) and/or hypercharge should not be ignored. Indeed, for judicious choices of the numbers and types of such representations, gauge coupling unification (although at scales below 10^{15} GeV) is possible without introducing supersymmetry [116]. The main drawback of introducing scalars of this type is that, for triplets and most higher representations, if the vacuum expectation value of the neutral Higgs field member of the representation is non-zero ($v_L \neq 0$) then ρ becomes infinitely renormalized and can no longer be computed [117]. In this case, ρ becomes an independent input parameter to the theory. Triplets have received the most attention, as they arise naturally in left-right symmetric extensions of the Standard Model gauge group [118]. (These and other models that utilize Higgs triplets are reviewed in [1].) In this section we will also briefly consider the Higgs-like pseudo-Nambu–Goldstone bosons that arise in generic technicolor theories.

8.1. *A Triplet Higgs Sector*

Including a single complex SU(2)-triplet Higgs representation, in addition to some number of doublets and singlets, results in six additional physical Higgs eigenstates: $H^{--,++}$, $H^{-,+}$, H^0 and $H^{0\prime}$. All but the doubly-charged states can mix with the doublet/singlet Higgs states under some circumstances. Even if $v_L \neq 0$ for the neutral field, $\rho = 1$ can be preserved at tree level if, in addition, a real triplet field is also included [119, 120]. However, ρ will still be infinitely renormalized at one-loop unless $v_L = 0$ is chosen. Left-right (L-R) symmetric models capable of yielding the see-saw mechanism for neutrino mass generation *require* two triplet Higgs representations (an L-triplet and an R-triplet). The large see-saw mass entry, M, arises from a "Majorana coupling" which L-R symmetry requires to be present for both the L-triplet and R-triplet representations. [k] Again, ρ will not be altered if $v_L = 0$, but v_R must be non-zero and large for large M. We will briefly discuss the phenomenology of an L-triplet; the corresponding phenomenology of the R-triplet is quite different. (See [1] for a review.)

The resulting Higgs phenomenology can be very complex. We focus on the most unequivocal signal for a triplet representation, namely observation of a doubly-charged Higgs boson. Pair production, $Z^* \to H^{++}H^{--}$ would be visible at the LHC for $m_{H^{--}} \lesssim 1$ TeV [116], but has limited mass reach, $m_{H^{--}} < \sqrt{s}/2$, at the LC. Fortunately, single production of a doubly-charged Higgs boson at the LC is also generally possible. In particular, the generically-allowed Majorana coupling leads to an $e^-e^- \to H^{--}$ coupling and the possibility of s-channel resonance production of the H^{--} in e^-e^- collisions. Observation of this process would provide a dramatic

[k] The so-called Majorana couplings correspond to the terms in the Yukawa interactions in which a fermion bilinear of lepton number +2 or −2 (*e.g.*, the latter contains $\nu\nu$ and e^-e^-) couples to the triplet fields. When the neutral component of the R-triplet acquires a vacuum expectation value, a Majorana mass term for the right-handed neutrino is generated, and lepton number is spontaneously broken. In this model B − L is a spontaneously broken gauged U(1) symmetry.

confirmation of the presence of the Majorana coupling and, in many cases, the ability to actually measure its magnitude. For a discussion and review, see [121]. If the H^{--} is heavy *and* has significant W^-W^- coupling (requiring $v_L \neq 0$), then it can become broad and the s-channel resonant production cross section is suppressed (see, *e.g.*, [122]) and might not be observable. Another production mechanism sensitive to the $e^-e^- \to H^{--}$ coupling that might be useful in such an instance is $e^-e^- \to H^{--}Z$, and $e^-e^- \to H^-W^-$ will be sensitive to the $e^-\nu_e \to H^-$ coupling that would be present for the H^- member of the triplet representation [123]. Using just the Majorana coupling, doubly-charged Higgs bosons can also be produced via $e^-\gamma \to e^+H^{--}$ and $e^+e^- \to e^+e^+H^{--}$ [124] and the singly-charged members of the same representation can be produced in $e^-e^- \to H^-W^-$ [123].

Despite the loss of predictivity for ρ, it could be that non-zero v_L is Nature's choice. In this case, the e^-e^- collider option again has some unique advantages. The neutral, singly-charged and doubly-charged Higgs bosons of the triplet representation can *all* be produced (via ZZ fusion, W^-Z fusion and W^-W^- fusion, respectively). For example, [125] studies $W^-W^- \to H^{--}$ fusion.

8.2. *Pseudo Nambu Goldstone Bosons*

In the context of technicolor and related theories, the lowest-mass states are typically a collection of pseudo-Nambu–Goldstone bosons, of which the lightest is very possibly a state P^0 which can have mass below 200 GeV and couplings and other properties not unlike those of a light SM-like Higgs boson. Typically, its WW, ZZ coupling is very small (arising via loops or anomalies), while its $b\bar{b}$ coupling can be larger. The phenomenology of such a P^0 was studied in [126, 127]. The best modes for detection of the P^0 at the LC are $e^+e^- \to \gamma P^0 \to \gamma b\bar{b}$ and $\gamma\gamma \to P^0 \to b\bar{b}$. Since the P^0 is likely to be discovered at the LHC in the $\gamma\gamma$ final state, we will know ahead of time of its existence, and precision measurements of its properties would be a primary goal of the LC. High integrated luminosity would be required.

9. LC Measurements of Higgs Boson Properties

In addition to the observation of one or more Higgs boson(s), an essential part of the LC physics program consists of measuring the mass, width, and couplings of the Higgs boson(s) with precision, and the determination of the scalar potential that gives rise to electroweak symmetry breaking.

Determinations of the Higgs couplings are needed to demonstrate that a Higgs boson generates mass for vector bosons, charged leptons, and up- and down-type quarks. Whether there is only one Higgs doublet can be checked if the ratios of measured branching ratios of directly coupled particles are indeed proportional to the ratios of the squared-masses. Small variations can distinguish between a SM Higgs and h^0 of the MSSM with couplings close to the SM Higgs boson. Couplings are determined through measurements of Higgs branching ratios and cross sections. Higgs bosons are also expected to couple to themselves, and this self-coupling λ can only be explored through the direct production of two or more Higgs bosons. It is in this category of *direct* and *model independent* determination of many absolute couplings (and not just their ratios) that the strength of the LC Higgs physics program really stands out.

Details of some of the studies can be found in [128] and a comprehensive description of European studies using results of the simulated TESLA detector can be found in [129]. North American studies consider simulations of detectors with capabilities as described elsewhere [10]. The program of measurements of Higgs boson properties strongly impacts detector design. Measurement of branching ratios into fermions requires sophisticated vertex detectors to separate b from c (and gluon) jets. Precise recoil mass measurements opposite leptons need excellent momentum resolution (particularly for $\mu^+\mu^-$) from charged particle tracking. The performance of the combined tracking and calorimetry systems needs to result in precise jet-jet invariant masses, missing mass measurements, and the ability to separate hadronic W from hadronic Z decays.

The specific measurements used to determine the Higgs couplings to vector bosons, fermions and scalars are significantly different depending on the mass of the Higgs boson. A generic neutral CP-even Higgs boson will be denoted by h in this section. We treat three cases separately: a light Higgs boson ($m_h < 2m_W$), an intermediate mass Higgs boson ($2m_W \leq m_h < 2m_t$), and a heavy Higgs boson ($m_h \geq 2m_t$).

9.1. *Mass*

In the Standard Model, the Higgs mass determines all its other properties. Thus, the precision of the mass measurement affects the comparison of theory and experiment, for example, in a global fit of cross sections, branching ratios, and precision electroweak data. Similarly, in the MSSM or other models with extended Higgs sectors, the masses of all the Higgs bosons are an important input to determining the underlying model parameters.

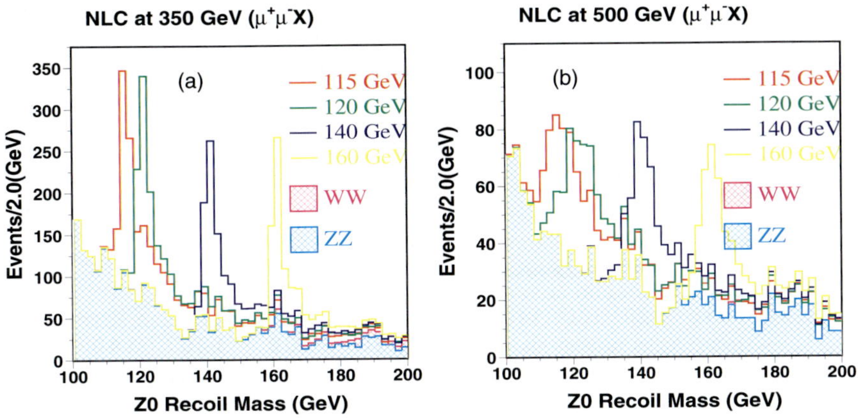

Fig. 18. Recoil mass from a pair of leptons for different Higgs masses at (a) $\sqrt{s} = 350$ GeV and (b) 500 GeV simulated in the LCD Large detector.

For this fundamental mass measurement, the LC can exploit recoils against a Z (independently of Higgs decay), and event reconstruction plus kinematic constraints can improve resolution and clean up mass tails. For a light or intermediate mass Higgs boson, optimal running conditions would have smaller center-of-mass energy (*e.g.*, $\sqrt{s} = 350$ GeV for better momentum resolution), and possess as small a level of beamstrahlung as possible. Under such conditions, one can precisely measure the recoil mass in $e^+e^- \to Zh$ events opposite to the reconstructed leptonic decay $Z \to e^+e^-$ or $\mu^+\mu^-$, with the additional advantage of decay mode independence. Accuracy can be improved by reconstructing specific decay modes, leading for example to a four-jet topology where effective (5-C) kinematic constrained fits can be employed.

Fig. 18 shows the distribution of the recoil mass,

$$M_{\text{recoil}} = \sqrt{s - 2\sqrt{s} \cdot E_{\ell^+\ell^-} + M^2_{\ell^+\ell^-}}\,, \tag{34}$$

in a simulation of the "Large" Linear Collider Detector (LCD) [130] for Higgs masses between 115 and 160 GeV [131]. Using fits of Monte Carlo shape templates and an integrated luminosity of 500 fb^{-1}, precisions of $\Delta m_{h_{\text{SM}}} \simeq 60$ MeV at $\sqrt{s} = 350$ GeV and $\Delta m_{h_{\text{SM}}} \simeq 120$ MeV at $\sqrt{s} =$ 500 GeV have been estimated for either the e^+e^- or $\mu^+\mu^-$ mode.

Realistic simulations have also been made of the LCD Large detector for the process $Zh \to q\bar{q}h$ resulting in four jets. Figure 19(a) shows the jet-jet

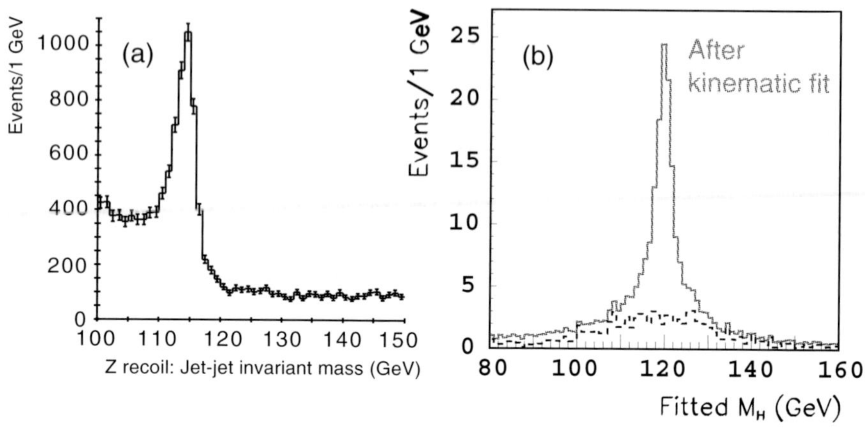

Fig. 19. (a) Jet-jet invariant mass of the jets recoiling from a Z reconstructed hadronically simulated in the LCD Large detector, $m_{h_{\rm SM}} = 115$ GeV. (b) Direct reconstruction of the four-jet $q\bar{q}h_{\rm SM}$ state simulated in the LCD Large detector after fitting with full kinematic constraints, $m_{h_{\rm SM}} = 120$ GeV.

invariant mass distribution for pairs of jets for Higgs with $m_{h_{\rm SM}} = 115$ GeV recoiling against a Z that has been reconstructed from its hadronic decay mode [132]. A clean Higgs signal with a mass resolution of approximately 2 GeV is observed. The central Higgs mass is shifted down due to the loss of low energy charged and neutral particles in the simulated event reconstruction. A low mass tail of the Higgs signal arises from missing neutrinos in semi-leptonic b and c quark decays. Using neural net tags and full kinematic fitting [133], the mass peak shown in Fig. 19(b) is obtained for $m_{h_{\rm SM}} = 120$ GeV, $\sqrt{s} = 500$ GeV, and 500 fb^{-1} resulting in $\Delta m_{h_{\rm SM}} \simeq 50$ MeV. With the possibility of a second lower-energy interaction point, scans across the Zh threshold may be attractive. With a total integrated luminosity of 100 fb^{-1}, $\Delta m_{h_{\rm SM}} \simeq 100$ MeV at $m_{h_{\rm SM}} = 150$ GeV is achievable [134], competitive with the methods above.

More work is necessary to confirm analogous precisions for heavier Higgs bosons and MSSM Higgs bosons with different decay modes and possible near-mass degeneracies. The number of Zh events with $Z \to \ell^+\ell^-$ for an intermediate mass ($m_h > 2m_W$) or heavy Higgs ($m_h > 2m_t$) with SM coupling plummets quickly [135]. In this case, and for the decays $h \to ZZ$, hadronic decays of the Z would have to be considered to gain sufficient statistics. For the heavier MSSM Higgs boson states, European studies [136]

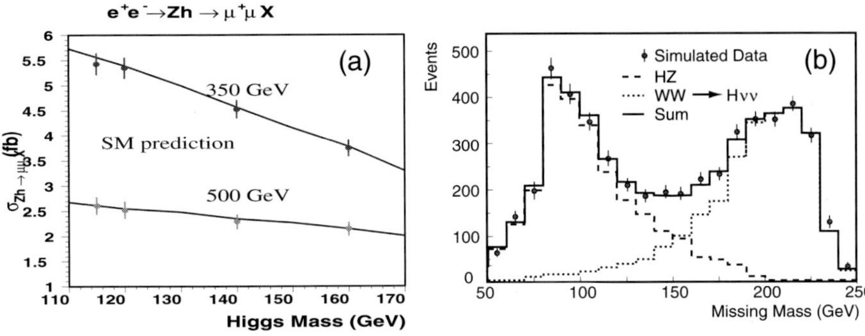

Fig. 20. (a) Cross section measurement for 500 fb^{-1} and (b) separation of Higgsstrahlung and WW fusion ($\sqrt{s}$ = 350 GeV) through a fit (after background subtraction), both simulated in the LCD Large detector.

have shown typical mass precisions of $\Delta m_{H^\pm}$ and $\Delta m_{A^0,H^0}$ of around 1 GeV for 500 fb^{-1}, but at $\sqrt{s} = 800$ GeV.

9.2. *Coupling Determinations — Light Higgs Boson*

9.2.1. *Cross Sections*

For Higgs masses below $2m_W$, the couplings g_{hZZ} and g_{hWW} are best measured through measurements of the Higgsstrahlung and WW fusion cross sections, respectively. These cross sections are also critical in the extraction of branching ratios since the experimental measurement will be a product of cross section and branching ratio.

Measurement of the Higgsstrahlung cross section is best addressed via the recoil mass method outlined above [131]. To reduce the contribution from the WW fusion process, it is best to run at a lower energy, *i.e.*, $\sqrt{s} = 350$ GeV, and to examine recoil against $\mu^+\mu^-$ to avoid large Bhabha backgrounds. A simulation based on the LCD Large detector with 500 fb^{-1} finds $\Delta\sigma/\sigma \simeq 3\%$ [4.7%] at $\sqrt{s} = 350$ [500] GeV, as shown in Fig. 20(a). These results agree roughly with estimates from European studies [137].

With efficient and pure b-jet tagging, events due to the process $e^+e^- \to W^+W^-\nu\overline{\nu} \to \nu\overline{\nu}h \to \nu\overline{\nu}b\overline{b}$ can be separated from those due to Higgsstrahlung, $Zh \to \nu\overline{\nu}h \to \nu\overline{\nu}b\overline{b}$ by examining the missing mass distribution and fitting to the expected shapes of a peak at m_Z from Higgsstrahlung and the higher missing masses from WW fusion. This technique has been confirmed with simulations of the LCD Large detector as shown

Table 1. Predicted branching ratio precisions in the LCD Large detector and typical vertex detector configuration for 500 fb^{-1} and $\sqrt{s} = 500$ GeV [142].

	$m_{h_{\rm SM}} = 120$ GeV		$m_{h_{\rm SM}} = 140$ GeV	
	BR	δBR/BR	BR	δBR/BR
$h_{\rm SM} \to b\bar{b}$	$(69 \pm 2.0)\%$	2.9%	$(34 \pm 1.3)\%$	3.8%
$h_{\rm SM} \to WW^*$	$(14 \pm 1.3)\%$	10%	$(51 \pm 1.5)\%$	3.0%
$h_{\rm SM} \to c\bar{c}$	$(2.8 \pm 1.1)\%$	39%	$(1.4 \pm 0.64)\%$	44%
$h_{\rm SM} \to gg$	$(5.2 \pm 0.93)\%$	18%	$(3.5 \pm 0.79)\%$	23%
$h_{\rm SM} \to \tau^+\tau^-$	$(7.1 \pm 0.56)\%$	8.0%	$(3.6 \pm 0.38)\%$	10%

in Fig. 20(b) [138]. With 500 fb^{-1} and a precision of $\delta{\rm BR}(h_{\rm SM} \to b\bar{b})/{\rm BR} \simeq 3\%$ (see below), the fusion-process cross section with this analysis can be found with a precision $\Delta\sigma/\sigma = 3.5\%$ for $m_{h_{\rm SM}} = 120$ GeV.

9.2.2. *Branching Ratios*

A key advantage of the linear collider in Higgs studies is the attractive situation of identifying Higgsstrahlung Zh events through the tag of the Z decays. This selection is largely independent of the decay mode of the h and simplifies the measurement of Higgs boson branching ratios.

Small beam sizes, the possibility of a first measurement as close as 1 cm from the beam axis, and sophisticated pixel vertex detectors allow for efficient and clean separation of quark flavors. Separate tagging of b and c jets is possible. By assumption, jets that are not identified as heavy flavor are assumed to be gluon jets, since the Higgs partial width into gg dominates the width into light quark pairs.

In a study [139] of vertexing in a CCD vertex detector in a standard configuration of a LCD detector (C1 in [140]), topological vertexing [141] with neural net selection [see Fig. 21(a)] was used for flavor (or anti-flavor, *i.e.*, veto WW^*) tagging. Assuming 500 fb^{-1} and 80% polarization, results shown in Table 1 were obtained (and include the most recent updated branching ratio precisions of [142]). These results have been checked to scale roughly with other studies [143, 144, 145] as $(\sigma \int \mathcal{L} dt)^{-1/2}$, with the results of [145] (that includes $h_{\rm SM}\nu\bar{\nu}$) being noticeably better for $c\bar{c}$ and gg with errors as shown in Fig. 21. As described in more detail in Section 9.2.5, these branching ratio measurements can then be used to either distinguish a SM Higgs boson from an MSSM Higgs boson, or to probe higher mass states and extract MSSM parameters such as m_{A^0} even if A^0 is not accessible.

With enough data and good charged particle momentum resolution, even the rare decay mode into $\mu^+\mu^-$ can be observed. As shown in Fig. 22,

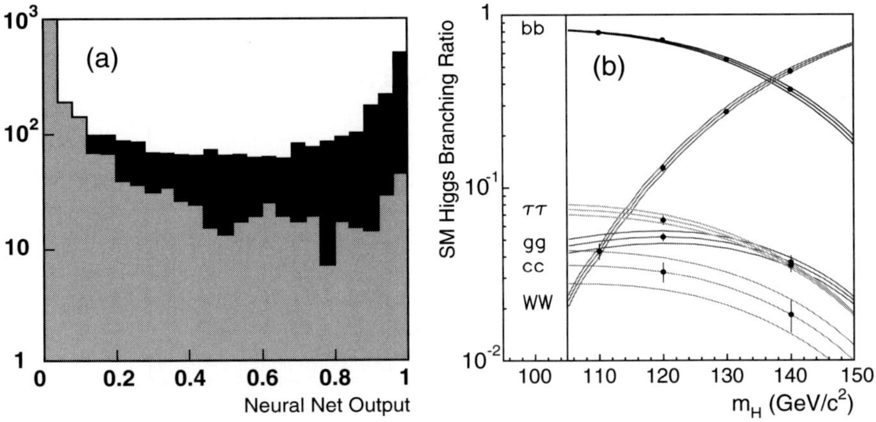

Fig. 21. (a) For the simulated LCD Large detector with CCD vertex detector, neural net $h_{SM} \to c\bar{c}$ output for $h_{SM} \to c\bar{c}$ events (dark) compared to output for $h_{SM} \to b\bar{b}$ events (gray). (b) Variation of branching ratios with SM Higgs mass (bands are 1σ uncertainties on the predictions) and measurement precisions in the TESLA detector (points with error bars).

for a Higgs boson mass of 120 GeV where the Standard Model branching ratio is predicted to be only 3×10^{-4}, the signal can be extracted from the underlying background with more than 5σ significance with 1000 fb^{-1} [146]. For this channel, there are clear advantages to running at multi-TeV energies where precisions could be reached to extend the test of the Higgs mechanism of mass generation in the lepton sector by checking if $g_{h_{SM}\mu\mu}/g_{h_{SM}\tau\tau} = m_\mu/m_\tau$.

For lighter Higgs bosons, the coupling to top quarks is still accessible via the radiative process $t\bar{t}h$ described below or indirectly through BR$(h \to gg)$. An accessible decay mode for light Higgs bosons is $h \to \gamma\gamma$ requiring excellent electromagnetic calorimetry in the detector. For a SM Higgs boson in a typical LCD detector, this is a difficult measurement requiring a great deal of luminosity, which is best for masses around 120 GeV [147], as shown in Fig. 23. A higher luminosity study [148] with 1000 fb^{-1} and $m_{h_{SM}} = 120$ GeV for the TESLA detector finds δBR/BR $= 14\%$. A gamma-gamma collider, discussed in Section 11, would be a more powerful tool for determining the Higgs coupling to photons.

Invisible Higgs boson decay channels, if significant, pose a difficult challenge for the LHC Higgs search. [149]. Possible invisible final states include neutralinos [150], majorons [151], heavy neutrinos [152], or the disappear-

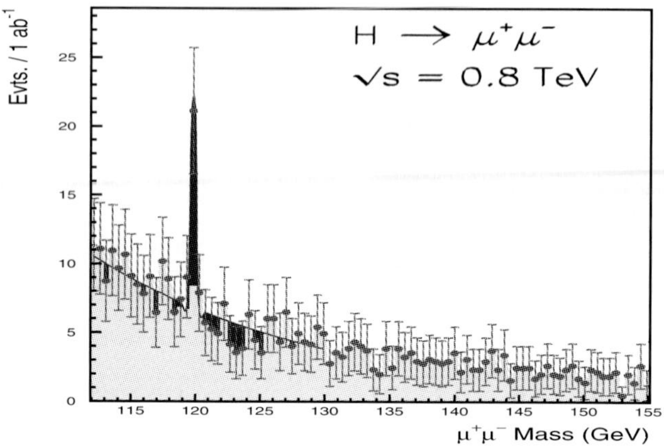

Fig. 22. Distribution of the $\mu^+\mu^-$ invariant mass for $m_{h_{\rm SM}} = 120$ GeV. The points with error bars represent 1000 fb^{-1} of data.

ance of the Higgs boson due to Higgs-radion mixing [153]. Likewise, the LHC Higgs search will be difficult if the $b\bar{b}$, WW^*, ZZ^* and $\gamma\gamma$ decay channels are significantly suppressed (in which case, *e.g.*, Higgs decay into light hadronic jets could be dominant [154]). The LC can close these loopholes and measure the branching ratio easily even for branching ratios as small as 5% as shown in Fig. 24 for a relatively narrow Higgs state using the recoil mass method and demanding no detector activity opposite the Z [155], or by comparing the number of events tagged with $Z \to \ell^+\ell^-$ with the total number of observed Higgs decays into known states.

9.2.3. *Radiative Production, $t\bar{t}h$*

For a light Higgs boson, production through radiation off a top quark is feasible [44, 101, 102, 103, 45, 104, 106, 105], resulting in a final state of $t\bar{t}h$, thus allowing for a determination of the Yukawa top quark coupling g_{htt}. For a SM-like Higgs boson with $m_h = 120$ GeV, the $t\bar{t}h$ cross section is roughly 10 times larger at $\sqrt{s} = 700$–800 GeV than at 500 GeV. At $\sqrt{s} = 800$ GeV, a statistical error of $\delta g_{htt}/g_{htt} \sim 5\%$ was estimated [102] for $L = 500\ \text{fb}^{-1}$ on the basis of an optimal-observable analysis [156]. At $\sqrt{s} = 500$ GeV, a statistical error of $\delta g_{htt}/g_{htt} \simeq 21\%$ is estimated [103, 104] using a total integrated luminosity of 1000 fb^{-1}. A more sophisticated

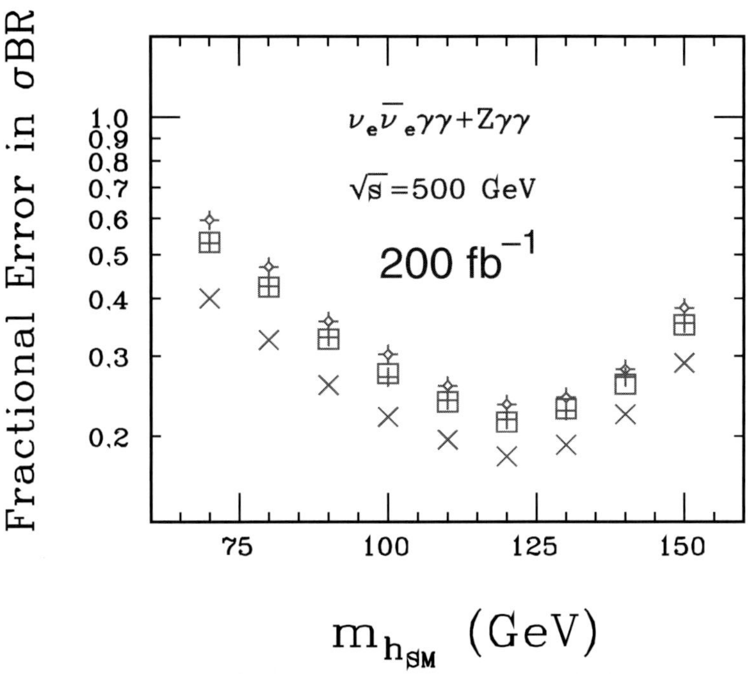

Fig. 23. Fractional error on the branching ratio BR($h_{\rm SM} \to \gamma\gamma$). The open squares are for a typical LCD detector electromagnetic energy resolution of $\Delta E/E = 10\%/\sqrt{E} \oplus 1.0\%$.

analysis using neural net selections, full simulation, and the same integrated luminosity at $\sqrt{s} = 800$ GeV finds a total error of 6% on the coupling [157].

9.2.4. *Self-Coupling*

To fully delineate the Higgs sector, it is essential to measure the shape of the Higgs potential. The cross section for double Higgs production (*e.g.*, Zhh) is related to the triple Higgs coupling g_{hhh}, which in turn is related to the spontaneous symmetry breaking shape of the Higgs potential [158]. In the MSSM, a variety of double Higgs production processes [113] would be required to determine g_{hhh}, g_{hAA}, *etc.*

These cross sections are low, and high integrated luminosity is needed, bolstered by polarization and neural net selections. Experimental studies [159, 160] indicate that for a SM-like Higgs boson with $m_h = 120$ GeV at $\sqrt{s} = 500$ GeV and 1000 fb^{-1}, a precision of $\delta g_{hhh}/g_{hhh} = 23\%$ is

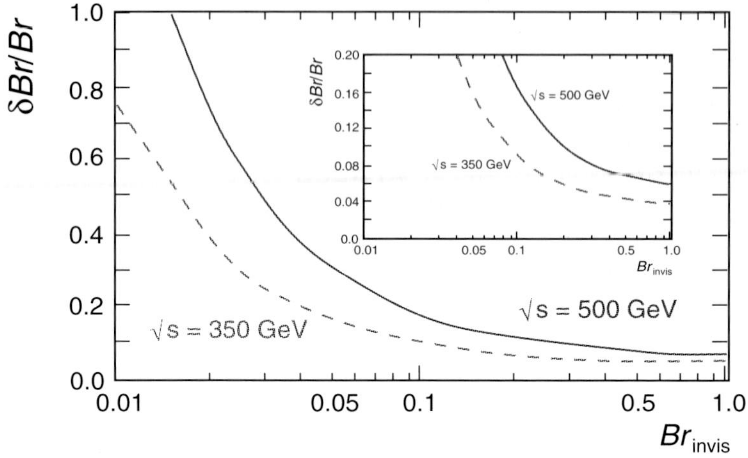

Fig. 24. Fractional error on a Higgs invisible branching ratio for $m_h = 120$ GeV and 500 fb^{-1} of data. The asymptotic value as the branching ratio (Br) approaches unity is due solely to the statistical uncertainty in $\sigma(hZ)$. A blow-up of the region of $\delta\mathrm{Br}/\mathrm{Br} \leq 0.2$ is also shown.

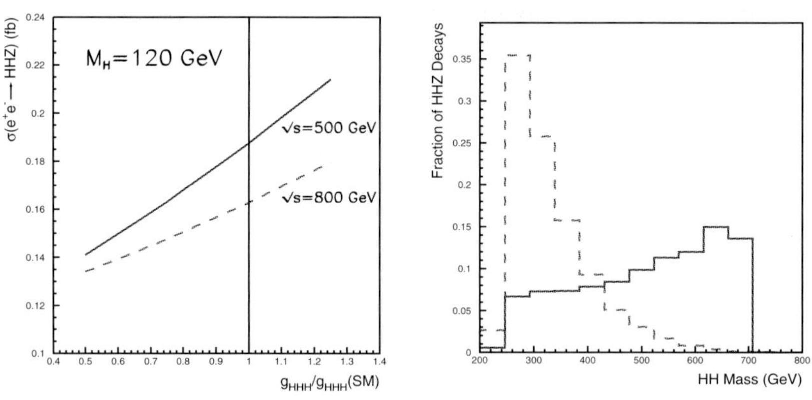

Fig. 25. (a) Dependence of the hhZ cross section on the triple Higgs coupling, normalized to its SM value for indicated mass and $\sqrt{s}$ values. (b) Distribution of the hh invariant mass in hhZ events from diagrams containing the triple Higgs vertex (solid line) compared to that from other diagrams (dashed line).

possible. A measurement of the double Higgs production cross section to extract g_{hhh} has to deal with dilution due to the existence of diagrams that lead to the same final states but that do not include a triple Higgs vertex [see Fig. 25(a)], so additional kinematical variables can be considered to enhance sensitivity to the signal. A study considering the h decay angle in the hh rest frame and the hh invariant mass [Fig. 25(b)] has shown improved separation and an increased precision of $\delta g_{hhh}/g_{hhh} = 20\%$ [161] for the same mass, energy, and integrated luminosity as above. The regions of accessibility in the MSSM parameter space for the MSSM Higgs self-couplings have also been performed [113, 162].

With the cross section for SM triple Higgs production so low [$\sigma(Zhhh) < 0.001$ fb], measurement of the quartic coupling g_{hhhh} is hopeless with currently envisioned luminosities.

9.2.5. *Implications for the MSSM Higgs Sector*

The discussion of light Higgs coupling determinations has been based on the assumption that the actual Higgs couplings to fermions, vector bosons and scalars are close to the corresponding Standard Model expectations. In Section 7.1, it was argued that such an expectation is rather generic, and applies to the decoupling limit of models of Higgs physics beyond the Standard Model. In particular, the decoupling limit of the MSSM Higgs sector sets in rather rapidly once $m_{A^0} \gtrsim 150$ GeV [see Section 5.1]. Since $m_{h^0} \lesssim 135$ GeV in the MSSM [Eq. (22)], the precision study of h^0 using the techniques discussed above can distinguish between h^0 and $h_{\rm SM}$ with a significance that depends on how close the model is to the decoupling limit. Said another way, the detection of deviations in the Higgs couplings from their Standard Model predictions would yield evidence for the existence of the non-minimal Higgs sector, and in the context of the MSSM would provide constraints on the value of m_{A^0} (with some dependence on $\tan\beta$ and other MSSM parameters that enter in the Higgs radiative corrections).

In [73], the potential impact of precision Higgs measurements at the LC on distinguishing h^0 from $h_{\rm SM}$ was examined. The fractional deviation of the h^0 branching ratios into a given final state from the corresponding result for $h_{\rm SM}$ (assuming the same Higgs mass in both cases) is defined as: $\delta{\rm BR} \equiv ({\rm BR}_{\rm MSSM} - {\rm BR}_{\rm SM})/{\rm BR}_{\rm SM}$. For the MSSM Higgs boson decay, both m_{h^0} and the corresponding branching ratios were computed including the radiative corrections due to the virtual exchange of Standard Model and supersymmetric particles, as described in Section 5.2. Thus, the h^0 branching ratios depend on m_{A^0} and $\tan\beta$ (which fix the tree-level MSSM Higgs sector properties) and a variety of MSSM parameter that govern the loop corrections. Four scenarios were considered: the minimal and maximal top-

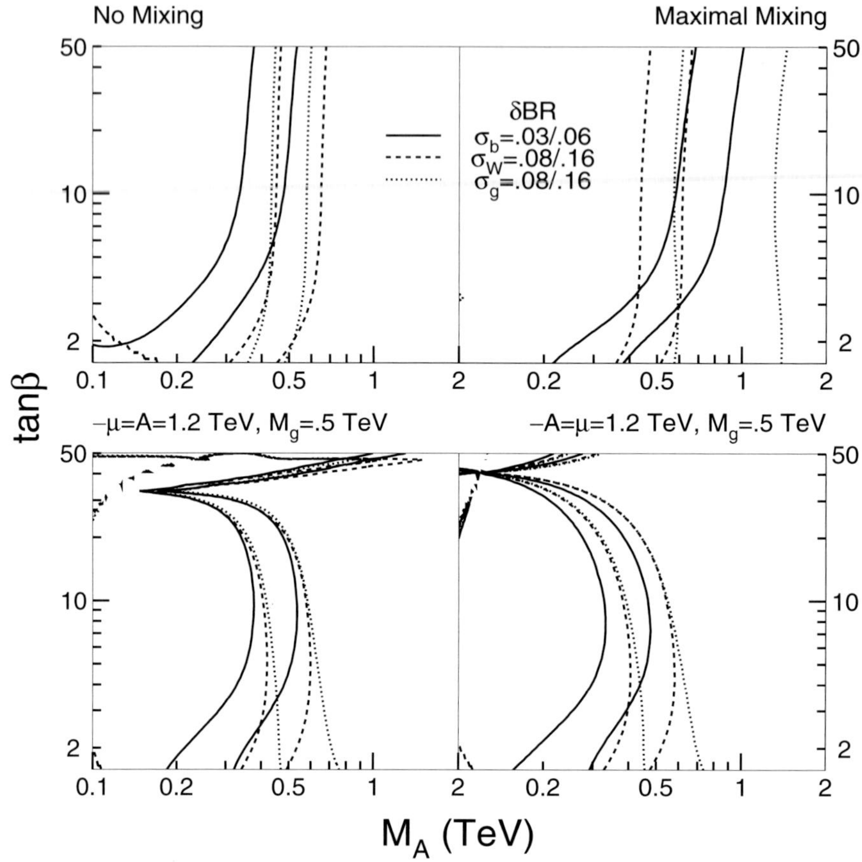

Fig. 26. Contours of $\delta\text{BR}(b\bar{b}) = 3$ and 6% (solid), $\delta\text{BR}(WW^*) = 8$ and 16% (dashed) and $\delta\text{BR}(gg) = 8$ and 16% (dotted) [BR deviations are defined in the text] in the no (*i.e.* minimal) mixing scenario (top left), the maximal mixing scenario (top right), and the large μ and A_t scenario with $\mu = -A_t = 1.2$ TeV (bottom left) and $\mu = -A_t = -1.2$ TeV (bottom right). Taken from [73].

squark mixing cases [see Eq. (22) and surrounding text], and two additional cases with large $|\mu| = |A_t|$ (for $\mu A_t < 0$ and two possible sign choices of μ), where μ and A_t control the top-squark mixing. In the latter two scenarios, significant renormalization of the CP-even Higgs mixing angle α and Δ_b [see Eq. (29)] can arise.

In Fig. 26, contours of δBR are plotted for three h^0 decay modes: $b\bar{b}$, WW^* and gg. The contours shown correspond roughly to the 1σ and 2σ

measurements claimed by [145], rescaled for the LC at $\sqrt{s} = 500$ GeV (see also the $b\bar{b}$ and WW^* branching ratio precisions given in Table 1). In the minimal and maximal scenarios, the dependence on m_{A^0} is nearly independent of $\tan\beta$, and demonstrates that one can achieve sensitivity to values of m_{A^0} that lie significantly beyond $\sqrt{s}/2$ where direct production at the LC via $e^+e^- \to H^0A^0$ is kinematically forbidden. However in some MSSM parameter regimes [*e.g.*, the case of large $\mu = -A_t$ as shown in Fig. 26(c) and (d)], a region of $\tan\beta$ may yield almost no constraint on m_{A^0}. This is due to the phenomenon of m_{A^0}-independent (or premature) decoupling [73] in which $\cos(\beta-\alpha)$ [which controls the departure from the decoupling limit] vanishes at a particular value of $\tan\beta$ independently of the value of m_{A^0}. Thus, a measured deviation of Higgs branching ratios that distinguishes h^0 from $h_{\rm SM}$ can place significant constraints on the heavier non-minimal Higgs states, although the significance of the resulting constraints can depend in a nontrivial way on the value of the MSSM parameters that control the Higgs radiative corrections.

9.3. *Coupling Determinations — Intermediate Mass Higgs Boson*

For $m_h < 2m_W$, there are many possible branching ratio measurements at the LC (see Table 1). This rich phenomenology allows for the determination of Higgs couplings to many of the fermions and bosons. For larger masses, decays to $f\bar{f}$ become rarer until the threshold for decays into top is crossed. In this intermediate mass range, the LC can measure the W and Z couplings more precisely than the LHC both through Higgs production rates and via branching ratios for decays into these bosons. Whether the observed Higgs boson fully generates the W and Z mass can then be checked.

We have noted in Section 2 that precision electroweak measurements in the framework of the Standard Model imply that $m_{h_{\rm SM}} \lesssim 193$ GeV at 95% CL [24]. Thus, any Higgs boson observed with a mass much greater than this would imply new physics. At this point, measurements from a Giga-Z dataset would be particularly useful to probe this new sector.

9.3.1. *Cross Sections*

Techniques described earlier [131, 138] for cross section measurements of both the Higgsstrahlung and W-fusion processes, with subsequent Higgs decays into $b\bar{b}$, can still be performed for the lower portion of the intermediate mass range below $m_{h_{\rm SM}} = 160$ GeV. Even in this intermediate mass range, it is beneficial to run at the peak of the cross section at roughly $m_h + m_Z + 50$ GeV. Typical precisions of $\Delta\sigma(Zh_{\rm SM})/\sigma(Zh_{\rm SM}) \simeq 5\%$ and $\Delta\sigma(\nu\bar{\nu}h_{\rm SM})/\sigma(\nu\bar{\nu}h_{\rm SM}) \simeq 17\%$ for $m_{h_{\rm SM}} = 160$ GeV at $\sqrt{s} = 350$ GeV with 500 fb^{-1} can be obtained.

For heavier Higgs bosons in this mass range, cross sections for both Higgsstrahlung and W-fusion will need to be extracted from these processes followed by $h \to WW^*$ decays for example as described in [144]. Couplings determined from $t\bar{t}h$ and Zhh production would clearly need higher center-of-mass energy.

9.3.2. *Branching Ratios*

Using Higgsstrahlung events at an optimal $\sqrt{s}$, the statistical error on BR($h_{\rm SM} \to b\bar{b}$) is still only 6.5% at $m_{h_{\rm SM}} = 160$ GeV [145]. At $\sqrt{s} = 500$ GeV, with leptonic decays of the Z only, the statistical error on this branching ratio reaches 25% at $m_{h_{\rm SM}} \simeq 165$ GeV with 250 fb^{-1} and remains below 30% for $m_{h_{\rm SM}} < 200$ GeV with 2000 fb^{-1} [135]. However, in addition to the leptonic decays of the Z, hadronic decays can also be used to effectively tag the associated Z. Extrapolating from full LCD detector simulations, it is conservatively estimated that including the hadronic decays of the Z results in an increase in signal statistics above background by a factor of four. With these assumptions and 500 fb^{-1}, again with the optimal $\sqrt{s} \simeq 350$ GeV, the error on the $b\bar{b}$ branching ratio can then be estimated to reach 25% at $m_{h_{\rm SM}} \simeq 200$ GeV. Measurement of branching ratios to $c\bar{c}$, $\tau^+\tau^-$, gg, and $\gamma\gamma$ does not seem feasible in this mass range.

Branching ratios into vector bosons can be measured with good precision in this intermediate mass range. For $m_{h_{\rm SM}} = 160$ GeV and 500 fb^{-1}, a predicted excellent precision of 2.1% on BR($h_{\rm SM} \to WW$), has been reported [144] and extrapolated estimated precisions of better than 7% over the mass range of 150 to 200 GeV [135].

To measure BR($h \to ZZ$), it will be necessary to distinguish hadronic Z decays from hadronic W decays, serving as an important benchmark for electromagnetic and hadronic calorimetry. With the same luminosity, and assuming that this separation allows one to identify one of the two Z's in the Higgs decays (through leptons or $b\bar{b}$) 40% of the time, the statistical uncertainty of this branching ratio would be approximately 8% for $m_{h_{\rm SM}} \simeq 210$ GeV [135] degrading to 17% for $m_{h_{\rm SM}} = 160$ GeV [129] where the branching ratio into Z's is still small.

9.4. *Coupling Determinations — Heavy Higgs Boson*

If a SM-like Higgs boson is heavy (*i.e.*, $m_h > 2m_t$), then new physics beyond the Standard Model must exist (*e.g.*, see Section 7.2; other examples of possible new physics effects are surveyed in [28, 163]). High-statistics measurements at the Z peak would again be useful to elucidate these effects.

9.4.1. *Cross Sections*

As a specific case, for $m_h = 500$ GeV at $\sqrt{s} = 800$ GeV, a SM-like Higgs boson would have a width of 70 GeV and dominant decay modes into W^+W^- (55%), ZZ (25%), and $t\bar{t}$ (20%). The production cross section for Zh would be 6 fb, but production would be dominated by $\nu\bar{\nu}h$ production at 10 fb. With 1000 fb^{-1}, one expects 400 Zh events where the Z decays to e^+e^- or $\mu^+\mu^-$. With reasonable selection and acceptance cuts, a measurement of $\sigma(Zh)$ to better than 7% should be feasible. See [163] for an independent assessment of the LC discovery reach for a heavy Higgs boson.

9.4.2. *Branching Ratios*

The LHC will have great difficulty distinguishing $h \to t\bar{t}$ decays due to huge QCD $t\bar{t}$ backgrounds. On the other hand, this mode should be observable at the LC. In the SM, the important coupling $g^2_{tth_{\rm SM}} \simeq 0.5$ can be compared to $g^2_{bbh_{\rm SM}} \simeq 4 \times 10^{-4}$. If the Higgs boson is heavier than 350 GeV, it will be possible obtain a good determination of the top-Higgs Yukawa coupling. Full simulations are needed for heavy Higgs decays into top, but with reasonable assumptions, one can expect a statistical error of δBR/BR $\simeq 14\%$ with 500 fb^{-1} [135]. Simulations using the TESLA detector of the $W^+W^- \to h_{\rm SM} \to t\bar{t}$ process with 1000 fb^{-1} and 6-jet final states show impressive signal significance for $\sqrt{s} = 1000$ GeV and reasonably good significance at $\sqrt{s} = 800$ GeV [164]. These studies find that a relative error in the top quark Yukawa coupling measurement better than 10% can be achieved for Higgs masses in the 350–500 GeV and 350–650 GeV ranges at $\sqrt{s} = 800$ GeV and 1000 GeV, respectively.

Again assuming detector performance allowing for separation of hadronic W and Z decays, and using production through W-fusion, similar techniques for extracting BR($h_{\rm SM} \to t\bar{t}$) can be applied resulting in estimates on BR($h_{\rm SM} \to W^+W^-$) and BR($h_{\rm SM} \to ZZ$) as shown in Table 2.

Table 2. Summary of measurement precisions for the properties of a SM-like Higgs boson, h, for a range of Higgs boson masses. Unless otherwise noted (see footnotes below the table), $\sqrt{s} = 500$ GeV and the total integrated luminosity is taken to be 500 fb^{-1}.

Δm_h	$\simeq$ 120 MeV (recoil against leptons from Z) $\simeq$ 50 MeV (direct reconstruction)				
m_h (GeV)	120	140	160	200	400–500
$\sqrt{s}$ (GeV)	500				800
$\Delta\sigma(Zh)/\sigma(Zh)$	4.7%	6.5%	6%	7%	10%
$\Delta\sigma(\nu\bar{\nu}h)\mathrm{BR}(b\bar{b})/\sigma\mathrm{BR}$	3.5%	6%	17%	–	–
$\delta g_{hxx}/g_{hxx}$ (from BR's)					
$t\bar{t}$	6–21% †	–	–	–	10%
$b\bar{b}$	1.5%	2%	3.5%	12.5%	–
$c\bar{c}$	20%	22.5%	–	–	–
$\tau^+\tau^-$	4%	5%	–	–	–
$\mu^+\mu^-$	15% ‡	–	–	–	–
$WW^{(*)}$	4.5%	2%	1.5%	3.5%	8.5%
$ZZ^{(*)}$	–	–	8.5%	4%	10%
gg	10%	12.5%	–	–	–
$\gamma\gamma$	7%	10%	–	–	–
g_{hhh}	20% §	–	–	–	–
Γ_{tot} ††	10.1%	8.2%	12.9%	10.6%	22.3%

† The range of $ht\bar{t}$ couplings is obtained from $e^+e^- \to t\bar{t}h$, with $\sqrt{s} = 500$–800 GeV and 1000 fb^{-1} of data.
‡ based on $\sqrt{s} = 800$ GeV and 1000 fb^{-1} of data.
§ based on $\sqrt{s} = 500$ GeV and 1000 fb^{-1} of data.
†† indirect determination from $\Gamma(VV^*)/\mathrm{BR}(VV^*)$, $V = W, Z$.

9.5. *Summary of Couplings*

The relative measurement errors for a SM Higgs at various masses are summarized in Table 2. As much as possible, the entries in this compilation have been collected from LCD detector simulations, particularly for lighter masses. For uniformity, the entries have been scaled to 500 fb^{-1}, except where noted otherwise. The significant measurements of many branching ratios, couplings and the total width demonstrates the strength of the LC Higgs program.

Just as the computer program ZFITTER [165] is used with Z mass, widths, asymmetries and branching ratios to make global fits for Z couplings, a program `HFITTER` [128] is now available that performs a global fit taking into account correlations between measurements of Higgs boson properties. Individual couplings of the Higgs boson can then be extracted optimally, for example through the correct combination of cross section and branching ratio measurements for such couplings as g_{hWW} and g_{hZZ}. Such precision fits can be used to probe indirectly for higher-mass states.

9.6. *Total Width*

For light Higgs bosons, the predicted SM width is far too small to be measured directly (and any anomalously wide state will indicate new physics), but the combination of branching ratios and coupling measurements allows the indirect and *model-independent* measurement of the total width through $\Gamma_{\text{tot}} = \Gamma(h \to X)/\text{BR}(h \to X)$. For $m_{h_{\text{SM}}} < 115$ GeV, the total width measurement would very likely require a $\gamma\gamma$ collider, an e^+e^- LC, and input from the LHC [2]. However, limits from LEP indicating $m_{h_{\text{SM}}} \gtrsim 115$ GeV would result in a non-negligible branching ratio to WW^* with the exciting and attractive prospect of an indirect model-independent measurement of the total width using LC measurements alone.

First, measurements of $\sigma(h\nu\nu) \times \text{BR}(h \to b\bar{b})$ and $\text{BR}(h \to b\bar{b})$ independently (through recoil in Higgsstrahlung) gives $\Gamma(h \to WW^*)$. Using a similar independent measurement of $\text{BR}(h \to WW^*)$ then gives the total width through the relation $\Gamma_{\text{tot}} = \Gamma(h \to WW^*)/\text{BR}(h \to WW^*)$. Similarly, both $\Gamma(h \to ZZ^*)$ and $\text{BR}(h \to ZZ^*)$ can be determined from the Higgsstrahlung process and used to compute $\Gamma_{\text{tot}} = \Gamma(h \to ZZ^*)/\text{BR}(h \to ZZ^*)$. These two independent results for Γ_{tot} can then be combined. The results are summarized in Table 2. Even with as little as 200 fb^{-1}, Γ_{tot} can be found to approximately 10% for $m_{h_{\text{SM}}} = 120$ GeV, improving to $\sim 7\%$ for $m_{h_{\text{SM}}} \sim 150$ GeV. Even better precision can be attained with the introduction of some model assumptions in the value used for $\Gamma(h_{\text{SM}} \to WW^*)$, *i.e.*, assuming the SU(2) relation between W and Z couplings along with $\sigma_{meas}(Zh_{\text{SM}})$, or else by using its SM value directly as a consistency check.

For $m_{h_{\text{SM}}} \gtrsim 205$ GeV, $\Gamma_{\text{tot}}(h_{\text{SM}}) > 2$ GeV and the total width could be directly resolved with typical LCD detector resolutions. The variations of precision for indirect and direct measurements for different values of $m_{h_{\text{SM}}}$ and inputs from different machines are examined in [2, 166]. The jet-jet mass resolution assumed in [2] has been verified by full simulations [132] in the LCD Large Detector resulting in estimated direct measurements of the total width reaching minimum values of $\simeq 6\%$ in the mass range of 240–280 GeV with 200 fb^{-1}. A more complete analysis for $m_{h_{\text{SM}}} = 240$ GeV, where the predicted SM Higgs width is 3.4 GeV, fits to a convolution of resolution and a Breit-Wigner as shown in Fig. 27 to find $\Delta\Gamma_{h_{\text{SM}}}/\Gamma_{h_{\text{SM}}} = 12\%$ [167]. The indirect determination described above can also still be pursued and a combination would allow even better precisions.

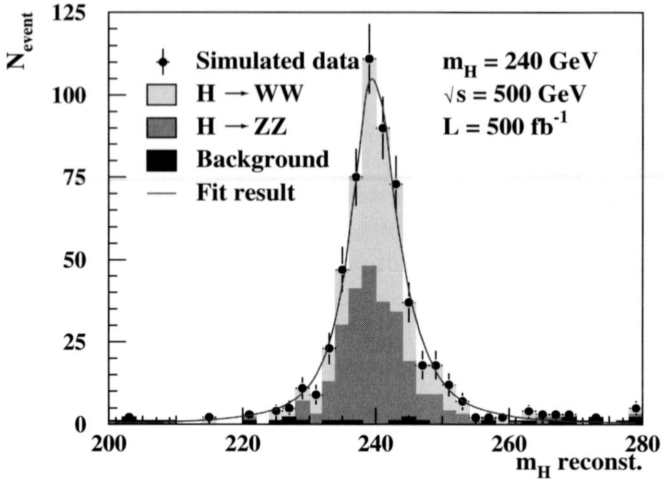

Fig. 27. Fit to the reconstructed $h_{\rm SM}$ mass to directly extract $\Gamma_{h_{\rm SM}}$.

9.7. *Quantum Numbers*

The spin, parity, and charge conjugation quantum numbers J^{PC} of a Higgs boson candidate, generically denoted by ϕ, can potentially be determined in a model-independent way. Useful ingredients include the following:

(1) If ϕ is produced in $\gamma\gamma$ collisions, then it cannot have $J = 1$ and it must have positive C [168].

(2) The behavior of the $Z\phi$ Higgsstrahlung cross section at threshold constrains the possible values of $J^{\rm PC}$ of the state. If the spin of the ϕ is 2 or less, a cross section growing as β indicates a CP-even object, whereas a cross section growing as β^3 signals a CP-odd state [169] with differences as shown in Fig. 28(a).

3) The angular dependence of the $e^+e^- \to Z\phi$ cross section depends upon whether the ϕ is CP-even, CP-odd, or a mixture [169, 170, 171, 172]. Following [172] we parameterize the $ZZ\phi$ vertex as

$$\Gamma_{\mu\nu}(k_1, k_2) = a g_{\mu\nu} + b\,\frac{k_{1\mu}k_{2\nu} - g_{\mu\nu}k_1 \cdot k_2}{m_Z^2} + \tilde{b}\,\frac{\epsilon_{\mu\nu\alpha\beta}k_1^\alpha k_2^\beta}{m_Z^2}\,, \tag{35}$$

where k_1 and k_2 are the momenta of the two Zs. The first term arises from a Standard-Model-like $ZZ\phi$ coupling, and the last two from effective

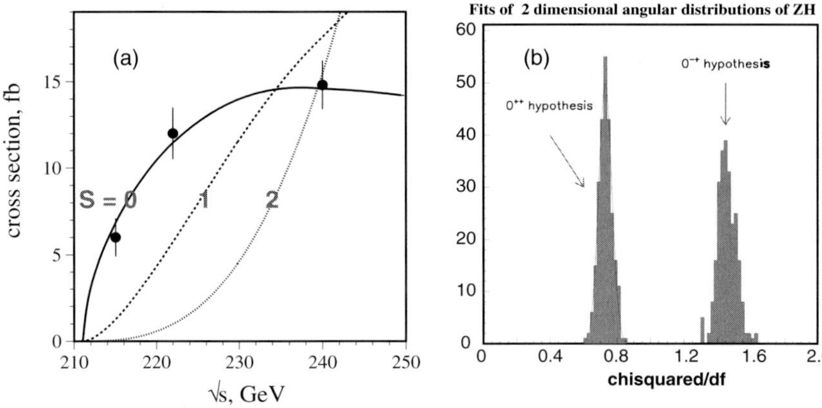

Fig. 28. (a) Behavior of Higgsstrahlung threshold for various spin states along with typical measurement precisions on the cross section. (b) Fit to the double-differential angular distribution in $Z\phi$ events (see text) to distinguish CP-even and CP-odd states.

interactions that could be induced by high-mass virtual particles. With this vertex the Higgsstrahlung cross section becomes

$$\frac{d\sigma}{d\cos\theta_Z} \propto 1 + \frac{p_Z^2}{m_Z^2}\sin^2\theta_Z - 4\,\mathrm{Im}\left[\frac{\tilde{b}}{\tilde{a}}\right]\frac{v_e a_e}{v_e^2 + a_e^2}\frac{p_z\sqrt{s}}{m_Z^2}\cos\theta_Z + \left|\frac{\tilde{b}}{\tilde{a}}\right|^2 \frac{p_z^2 s}{2m_Z^4}(1+\cos^2\theta_Z)\,, \tag{36}$$

where θ_Z, p_Z, and E_Z are the scattering angle, momentum, and energy of the final-state Z boson; v_e and a_e are the vector and axial-vector Ze^+e^- couplings; and $\tilde{a} \equiv a - bE_Z\sqrt{s}/m_Z^2$. The term in Eq. (36) proportional to $\cos\theta_Z$ arises from interference between the CP-even and CP-odd couplings in Eq. (35). If the CP-odd coupling $\tilde{b}$ is large enough, it can be extracted from the forward-backward asymmetry. Even upper limits on this asymmetry would be interesting. Note that the CP-even component of a Higgs boson will typically couple at tree level whereas the CP-odd component will only couple via one-loop diagrams (typically dominated by the t quark loop). As a result, one expects the coupling strength $\tilde{b}$ to be proportional to m_Z^2/s times a loop suppression factor. Thus, an asymmetry measurement may be able to provide a crude determination of the $\tilde{b}/a$ term. If ϕ is a purely CP-odd state with a one-loop coupling, the resulting ZA^0 cross section will simply be too small to provide a useful measurement of the asymmetry.

(4) The angular distribution of the fermions in the $Z \to f\overline{f}$ decays in $Z\phi$ production also reflects the CP nature of the state ϕ [170, 171]. For the decay $Z \to e^+e^-$ or $\mu^+\mu^-$, the following angles can be defined: the angle between the initial e^- and the Z; the angle between the final state e^- or μ^- and the direction of motion of the Z, in the rest frame of the Z; and the angle between the Z production plane and Z decay plane. Correlations between these angles can be exploited, *e.g.*, a fit to the double-differential angular distribution of the first two of these angles results in a 14σ separation between the 0^{++} (CP-even, scalar) and the 0^{-+} (CP-odd, pseudoscalar) [135], assuming that the $Z\phi$ cross section is independent of the CP nature of ϕ [see Fig. 28(b)]. Even more powerful are fits to the triple-differential angular distribution where sufficient luminosity can uncover non-standard $ZZ\phi$ couplings. However, this technique again suffers from the difficulty described in the previous item; namely, the CP-odd part of the state ϕ is typically so weakly coupled to ZZ that there is little sensitivity to the CP-odd component (if there is any significant CP-even component in ϕ) or little cross section (if ϕ is almost purely CP-odd).

(5) If ϕ has significant branching ratios to either $\tau^+\tau^-$ or $t\bar{t}$, channels which are 'self-analyzing', then the decay distribution structure can provide a direct determination of the ratio b_f/a_f in the $y_f\overline{f}(a_f + ib_f\gamma_5)f\phi$ ($f = \tau$ or t) Yukawa coupling structure of ϕ [173, 174, 175]. Detector simulation studies have been made of the specific decay $\phi \to \tau^+\tau^- \to \rho^+\overline{\nu}_\tau\rho^-\nu_\tau$ where likelihood fits were made to angular correlations between decay products of each τ [176] resulting in distinguishing the CP nature of the ϕ to a confidence level greater than 95% with 500 fb^{-1} at $\sqrt{s} = 500$ GeV.

(6) The angular distributions in the $t\bar{t}\phi$ final state, which has adequate cross section for $\sqrt{s} \gtrsim 800$ GeV for modest values of $m_\phi \lesssim 200$ GeV, assuming Yukawa coupling $y_t\overline{t}(a_t+ib_t\gamma_5)t\phi$ comparable to SM values, appears to provide an excellent means for determining the CP nature of ϕ by allowing one to probe the ratio b_t/a_t [177, 102]. An observable CP-violating asymmetry in $t\bar{t}\phi$ production (via a T-odd triple correlation product proposed in [101]) would provide evidence for a CP-violating Higgs sector.

(7) It is likely that the CP properties of the ϕ can be well determined using photon polarization asymmetries in $\gamma\gamma \to \phi$ collisions [178, 179, 175] (see Section 11).

(8) If the ϕ has substantial ZZ coupling, then $e^-e^- \to ZZe^-e^- \to \phi e^-e^-$ can be used to probe its CP nature [180] via the energy distributions of the ϕ and the final electrons, which are much harder in the case of a CP-odd state than for a CP-even state. Certain correlations are also useful probes of the CP properties of the ϕ. However, if the CP-odd portion of ϕ couples at one-loop (as expected for a Higgs boson), there will be either little sensitivity to this component or little cross section.

9.8. *Precision Studies of Non-SM-like Higgs Bosons*

We confine our remarks to the two-doublet Higgs model (either the MSSM Higgs sector or a more general 2HDM). In the MSSM, we noted in Section 5.4 that for $m_{A^0} \lesssim \sqrt{s}/2$, as long as one is not too close to threshold, it is possible to observe all Higgs scalars of the non-minimal Higgs sector. In particular, in parameter regions away from the decoupling limit, none of the CP-even Higgs scalars may resemble the SM Higgs boson. Precision studies of all the Higgs bosons will provide a detailed profile of the non-minimal Higgs sector. Once $m_{A^0} \gtrsim \sqrt{s}/2$, only the h^0 will be visible at the LC in an approximate decoupling regime (although there may be some possibilities for observing the heavier Higgs states produced singly either in association with a $b\bar{b}$ pair at large $\tan\beta$ where the coupling to $b\bar{b}$ is enhanced, or by s-channel resonance production at a $\gamma\gamma$ collider). In the more general 2HDM, the simplest mechanisms for producing the heavier neutral Higgs bosons are $e^+e^- \to H^0A^0$, $e^+e^- \to H^0b\bar{b}, A^0b\bar{b}$ and $e^+e^- \to H^0t\bar{t}, A^0t\bar{t}$. Charged Higgs bosons can be produced via $e^+e^- \to H^+H^-$ and $e^+e^- \to H^+\bar{t}b + H^-t\bar{b}$. We have seen that none of these processes are guaranteed to be either kinematically accessible or have a useful rate.

Values of m_{A^0} and m_{H^0} in excess of 500 GeV to 1 TeV are certainly possible. In such cases, a substantial increase of energy for the LC will be required to observe these states directly, either in association with $b\bar{b}$ (at large $\tan\beta$) or via H^0A^0 production. Measuring the former will provide a crucial determination of the $b\bar{b}$ couplings, which in the given model context will provide a determination of $\tan\beta$, with accuracy determined by the production rates. Moreover, if H^0 and A^0 can be produced at a high rate (by whatever process), a detailed study of their branching ratios has the potential for providing vital information regarding the model parameters.

In low-energy supersymmetric models, the heavy H^0, A^0 and $H^\pm$ would generally decay to various pairs of supersymmetric particles as well as to b's

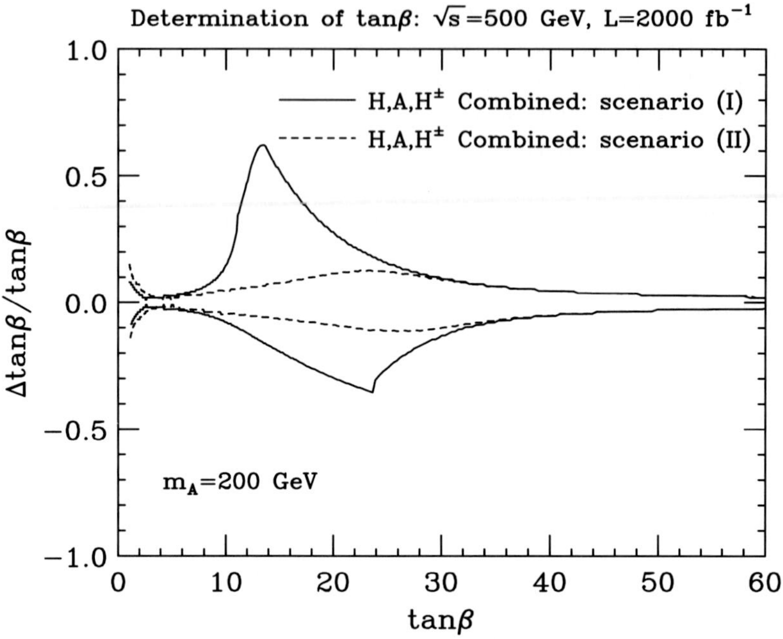

Fig. 29. The accuracy with which $\tan\beta$ can be measured, assuming $m_{A^0} = 200$ GeV, $\sqrt{s} = 500$ GeV and integrated luminosity of $L = 2000$ fb^{-1}, after combining in quadrature the errors on $\tan\beta$ from: a) the rate for $b\bar{b}A^0$, $b\bar{b}H^0 \to b\bar{b}b\bar{b}$; b) the $H^0A^0 \to b\bar{b}b\bar{b}$ rate; c) a measurement of the average H^0, A^0 total width in H^0A^0 production; d) the $H^+H^- \to t\bar{b}\bar{t}b$ rate; and e) the total $H^\pm$ width measured in $H^+H^- \to t\bar{b}\bar{t}b$ production. Results are shown for two supersymmetric scenarios: in (I), the H^0 and A^0 do not decay to supersymmetric particles; in (II), there are substantial decays of the H^0 and A^0 to a pair of the lightest supersymmetric particles ($\widetilde{\chi}_1^0\widetilde{\chi}_1^0$). Taken from [185].

and t's. A study of the relative branching ratios and the rates for specific processes would provide powerful determinations of $\tan\beta$ and many of the soft-supersymmetry-breaking parameters [181, 182, 183, 184, 185]. In addition, at large $\tan\beta$ the absolute widths of H^0, A^0 (H^+) become large enough to be directly measured in the $b\bar{b}$ ($t\bar{b}$) final state and provide an excellent determination of $\tan\beta$ [185].[1] The accuracy in the $\tan\beta$ measurement that can be achieved after combining all these $\tan\beta$ determinations in quadra-

[1]More precisely, what is being measured via the above techniques is the relevant Yukawa coupling, which is determined by $\tan\beta$ at tree level. At one-loop, radiative corrections must be included in order to convert from the Yukawa coupling to a uniform definition of $\tan\beta$ [186].

ture are illustrated in Fig. 29 assuming $\sqrt{s} = 500$ GeV, $L = 2000$ fb^{-1}, and $m_{A^0} = 200$ GeV. Excellent precision is achieved at low and high $\tan\beta$ regardless of the specific supersymmetric scenario. However, in the range $10 < \tan\beta < 30$, good accuracy in the $\tan\beta$ measurement is only achieved if there are some supersymmetric decays of H^0 and A^0 against which the corresponding $b\bar{b}$ decay modes have to compete.

10. The Giga-Z Option — Implications for Higgs Physics

Measurements of the effective leptonic mixing angle and the W boson mass to precisions of $\delta\sin^2\theta_{\rm eff} \simeq 10^{-5}$ and $\delta m_W \approx 6$ MeV at Giga-Z can be exploited in many ways. The size of the resulting 90% CL ellipses were shown in Fig. 15. These measurements imply that the SM Higgs boson mass can be determined indirectly to a precision of about 7%. A deviation between the directly observed value and the value implied by a global SM fit to Giga-Z data would require new physics beyond the SM.

If new physics is present then it will be possible to obtain information about new high mass scales beyond the direct reach of the collider. For example, in the MSSM such information would be of particular importance if the heavier scalar top quark, $\tilde{t}_2$, and the heavy Higgs bosons A^0, H^0 and $H^\pm$ were beyond the kinematical reach of the LC and background problems precluded their observation at the LHC. Similar considerations apply to more general extended Higgs sector models.

10.1. *Giga-Z and the MSSM*

In the MSSM, the relation between m_W and $\sin^2\theta_{\rm eff}$ is affected by the parameters of the supersymmetric sector, especially the $\tilde{t}$ sector. At the LC, the mass of the light $\tilde{t}$, $m_{\tilde{t}_1}$, and the $\tilde{t}$ mixing angle, $\theta_{\tilde{t}}$, should be measurable very well if the process $e^+e^- \to \tilde{t}_1\overline{\tilde{t}_1}$ is accessible [187].

In Fig. 30 [21], it is demonstrated how *upper* bounds on m_{A^0} and $m_{\tilde{t}_2}$ can be derived from measurements of m_{h^0}, m_W and $\sin^2\theta_{\rm eff}$, supplemented by precise determinations of $m_{\tilde{t}_1}$ and $\theta_{\tilde{t}}$. The most restrictive upper bounds are obtained if $\tan\beta > 10$ (in this case, one assumes that the value of $\tan\beta$ would have already been determined from measurements in the gaugino sector [188]). The other parameters values are assumed to have the uncertainties as expected from LHC [189] and the LC [129].

For low $\tan\beta$ (where the prediction for m_{h^0} depends sensitively on $\tan\beta$) the heavier $\tilde{t}$ mass, $m_{\tilde{t}_2}$, can be restricted to 760 GeV $\lesssim m_{\tilde{t}_2} \lesssim 930$ GeV from the m_{h^0}, m_W and $\sin^2\theta_{\rm eff}$ precision measurements. The mass m_{A^0}

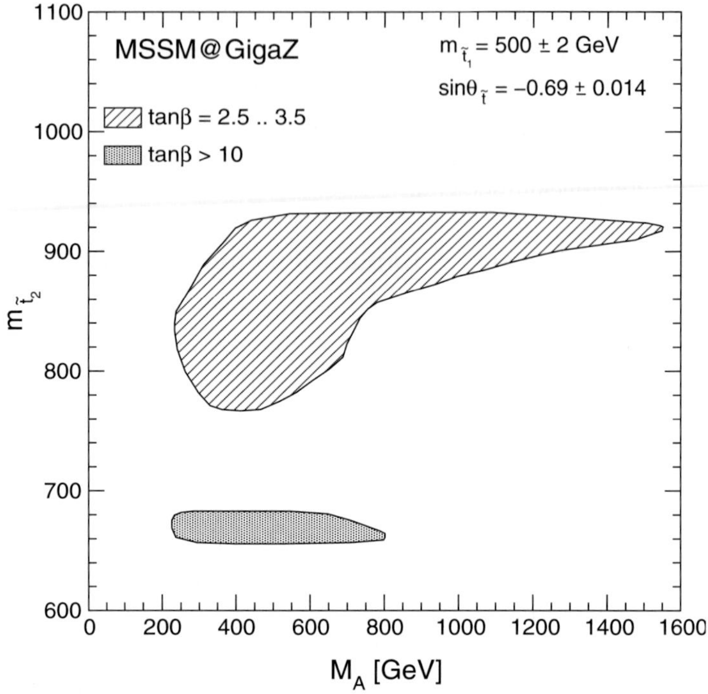

Fig. 30. The region in the $[m_{A^0}, m_{\tilde{t}_2}]$ parameter space, allowed by $1\,\sigma$ errors obtained from the Giga-Z measurements of m_W and $\sin^2\theta_{\rm eff}$: $m_W = 80.400 \pm 0.006$ GeV, $\sin^2\theta_{\rm eff} = 0.23140 \pm 0.00001$, and from the LC measurement of m_{h^0}: $m_{h^0} = 115 \pm 0.05\,(\text{exp.}) \pm 0.5\,(\text{theo.})$ GeV. $\tan\beta$ is assumed to be $\tan\beta = 3 \pm 0.5$ or $\tan\beta > 10$. The other parameters are given by $m_{\tilde{t}_1} = 500{\pm}2$ GeV, $\sin\theta_{\tilde{t}} = -0.69{\pm}0.014$, $A_b = A_t{\pm}10\%$, $m_{\tilde{g}} = 500 \pm 10$ GeV, $\mu = -200 \pm 1$ GeV and $M_2 = 400 \pm 2$ GeV. Taken from [21].

varies between 200 GeV and 1600 GeV. If $\tan\beta \geq 10$ (where m_{h^0} has only a mild dependence on $\tan\beta$), the allowed region for the $\tilde{t}_2$ is much smaller, 660 GeV $\lesssim m_{\tilde{t}_2} \lesssim$ 680 GeV, and m_{A^0} is restricted to $m_{A^0} \lesssim 800$ GeV.

In deriving the bounds on the heavier $\tilde{t}$ mass, $m_{\tilde{t}_2}$, the constraints from m_{h^0}, $\sin^2\theta_{\rm eff}$ and m_W play an important role. For the bounds on m_{A^0}, the main effect comes from $\sin^2\theta_{\rm eff}$. The assumed value of $\sin^2\theta_{\rm eff} = 0.23140$ differs slightly from the corresponding value obtained in the SM limit. For this value the (logarithmic) dependence on m_{A^0} is still large enough [21] so that from the high precision in $\sin^2\theta_{\rm eff}$ at Giga-Z an *upper limit* on m_{A^0} can be set. With an error in $\sin^2\theta_{\rm eff}$ that would be at least ten times larger at the LC without the Giga-Z mode, no bound on m_{A^0} could be inferred.

10.2. *Giga-Z and Non-exotic Extended Higgs Sectors*

Building on the discussion of the general 2HDM given earlier, one can imagine many situations for which the very small Giga-Z 90% CL ellipses illustrated in Fig. 15 would provide crucial (perhaps the only) constraints. For example, suppose the LHC observes a 1 TeV Higgs boson with very SM-like properties and no other new physics below the few-TeV scale. We have seen that this is possible in the 2HDM scenarios consistent with current precision electroweak constraints. Suppose further that it is not immediately possible to increase $\sqrt{s}$ sufficiently so that h^0A^0 production is allowed (typically requiring $\sqrt{s} > 1.5$ TeV in these models). Giga-Z measurements would provide strong guidance as to the probable masses of the non-SM-like Higgs bosons of any given non-minimal Higgs sector. However, it must be accepted that a particular Giga-Z result for S, T might have other non-Higgs interpretations as well.

11. The $\gamma\gamma$ Collider Option

Higgs production in $\gamma\gamma$ collisions offers a unique capability to measure the two-photon width of the Higgs and to determine its CP composition through control of the photon polarization.

The $\gamma\gamma$ coupling of a SM-like Higgs boson $h_{\rm SM}$ of relatively light mass receives contributions from loops containing any particle whose mass arises in whole or part from the vacuum expectation value of the corresponding neutral Higgs field. A measurement of $\Gamma(h_{\rm SM} \to \gamma\gamma)$ provides the possibility of revealing the presence of arbitrarily heavy particles that acquire mass via the Higgs mechanism.[m] However, because such masses are basically proportional to some coupling times v, if the coupling is perturbative the masses of these heavy particles are unlikely to be much larger than 0.5–1 TeV. Since ${\rm BR}(h_{\rm SM} \to X)$ is entirely determined by the spectrum of light particles, and is thus not affected by heavy states, $N(\gamma\gamma \to h_{\rm SM} \to X) \propto \Gamma(h_{\rm SM} \to \gamma\gamma){\rm BR}(h_{\rm SM} \to X)$ will provide an extraordinary probe for such heavy states. Even if there are no new particles that acquire mass via the Higgs mechanism, a precision measurement of $N(\gamma\gamma \to \widehat{h} \to X)$ for specific final states X ($X = b\bar{b}, WW^*, \ldots$) can allow one to distinguish between a $\widehat{h}$ that is part of a larger Higgs sector and the SM $h_{\rm SM}$.[n] The deviations from

[m] Loop contributions from particles that acquire a large mass M from some other mechanism will decouple as M^{-2} and $\Gamma(h_{\rm SM} \to \gamma\gamma)$ will not be sensitive to their presence.
[n] It may also be possible to detect some rare Higgs decay modes at the $\gamma\gamma$ collider. For example, [19] argues that the detection of $\gamma\gamma \to h_{\rm SM} \to \gamma\gamma$ will be possible.

the SM predictions typically exceed 5% if the other heavier Higgs bosons have masses below about 400 GeV.

The predicted rate for Higgs boson production followed by decay to final state X can be found in [14]. This rate depends strongly on $d\mathcal{L}_{\gamma\gamma}/dy$, the differential $\gamma\gamma$ collider luminosity, where $y = m_{\hat{h}}/\sqrt{s}$ and $\sqrt{s}$ is the ee collider center-of-mass energy. An important parameter to maximize peak luminosity is $\langle\lambda\lambda'\rangle$, the average value of the product of the helicities of the two colliding photons after integration over their momentum fractions z and z'. Larger values of this parameter also suppress the dominant $J_z = \pm 2$, $\gamma\gamma \to b\bar{b}(g)$ background, which is proportional to $(1 - \langle\lambda\lambda'\rangle)$. The computation of $d\mathcal{L}_{\gamma\gamma}/dy$ was first considered in [11, 12]. More realistic determinations [190] including beamstrahlung, secondary collisions between scattered electrons and photons from the laser beam, and other non-linear effects result in a substantial enhancement of the luminosity in the low-$E_{\gamma\gamma}$ region as shown in Fig. 31.

The choice of parameters that gives a peaked spectrum is well suited for light Higgs studies. Using the spectrum of Fig. 31 as an example, the di-jet invariant mass distributions for the Higgs signal and for the $b\bar{b}(g)$ background for $m_{h_{\rm SM}} = 120$ GeV are shown in Fig. 32 [17]. After a nominal year of operation (10^7 sec), $\Gamma(h_{\rm SM} \to \gamma\gamma){\rm BR}(h_{\rm SM} \to b\bar{b})$ could be measured with an accuracy of about 2.9%. (The more optimistic error of close to 2%, quoted in [15, 191] for $m_{h_{\rm SM}} = 120$ GeV, is based upon a significantly higher peak luminosity.) The error for this measurement increases to about 10% for $m_{h_{\rm SM}} = 160$ GeV, primarily due to the decrease of the Higgs di-jet branching fraction by a factor of 18.

In many scenarios, it is possible that by combining this result with other types of precision measurements for the SM-like Higgs boson, small deviations can be observed indicating the possible presence of heavier Higgs bosons. For a 2HDM (either the MSSM or a two-Higgs-doublet model with partial decoupling), if $m_{H^0} \sim m_{A^0} > \sqrt{s}/2$ then $e^+e^- \to H^0 A^0$ is not possible. Further, as discussed earlier in association with Fig. 16, the alternatives of $b\bar{b}H^0$ and $b\bar{b}A^0$ production will only allow H^0 and A^0 detection if $\tan\beta$ is large [105]. Thus, $\gamma\gamma \to H^0, A^0$ may be the only option allowing their discovery (without a collider with higher $\sqrt{s}$). The back-scattering kinematics are such that a LC for which the maximum energy is $\sqrt{s} = 630$ GeV can potentially probe Higgs masses as high as 500 GeV in $\gamma\gamma$ collisions. If m_{H^0} and m_{A^0} are known to within roughly 50 GeV on the basis of precision h^0 data, then they will be easily detected by choosing the machine $\sqrt{s}$ and polarization/helicity configurations that yield a $E_{\gamma\gamma}$ spectrum peaked

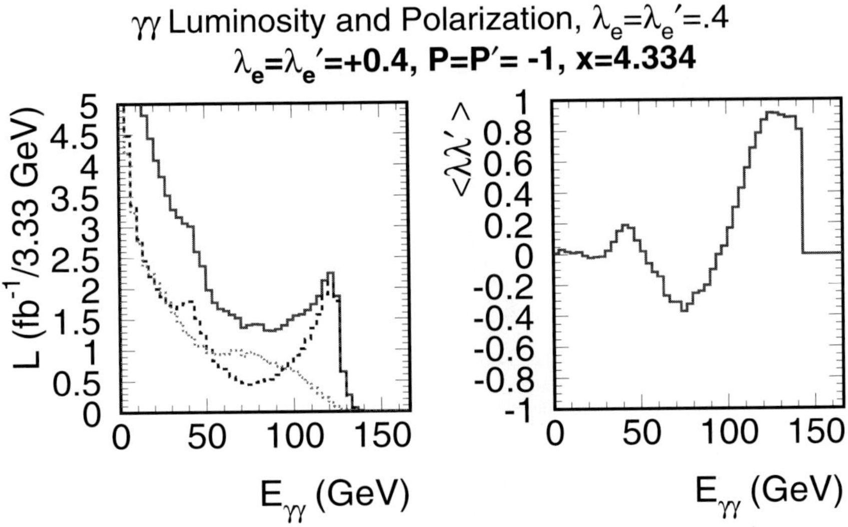

Fig. 31. The CAIN [190] predictions for the $\gamma\gamma$ luminosity, $L = d\mathcal{L}/dE_{\gamma\gamma}$, are plotted in units of fb^{-1} per 3.33 GeV bin size, for circularly polarized photons assuming the NLC-based design for the $\gamma\gamma$ collider [10, 17], with $\sqrt{s} = 160$ GeV, 80% electron beam polarization, and a 1.054/3 micron laser wavelength. The directions of the photon polarizations and electron helicities have been chosen to produce a peaked $E_{\gamma\gamma}$ spectrum. Beamstrahlung and other effects are included. The dashed (dotted) curve gives the component of the total luminosity that derives from the $J_z = 0$ ($J_z = 2$) two-photon configuration. Also plotted is the corresponding value of $\langle\lambda\lambda'\rangle$ [given by $\langle\lambda\lambda'\rangle = (L_{J_z=0} - L_{J_z=2})/(L_{J_z=0} + L_{J_z=2})$]. Taken from [17].

near the upper edge of the expected mass range [83, 17]. But, if it happens that no constraints have been placed on the H^0, A^0 masses, then it is most appropriate to run entirely at the maximal $\sqrt{s}$ and to consider how best to cover the portion of the LHC wedge [o] with $m_{H^0} \sim m_{A^0} \gtrsim \sqrt{s}/2$ (for which H^0A^0 pair production is not possible). Assuming a maximum of $\sqrt{s} = 630$ GeV, the goal would then be to roughly uniformly cover the 300 GeV $\lesssim m_{A^0}, m_{H^0} \lesssim 500$ GeV region of the LHC wedge. For this purpose, it is best to run part of the time with polarization/helicity directions that yield a peaked (Type-II) spectrum (with peak at $E_{\gamma\gamma} \sim 500$ GeV for $\sqrt{s} = 630$ GeV) and roughly three times as long in a polarization/helicity

[o]The LHC wedge, discussed in Section 6.3, corresponds to the moderate $\tan\beta$ large m_{A^0} wedge-like region in the $[m_{A^0}, \tan\beta]$ parameter space (see Fig. 14) where only the h^0 of the MSSM Higgs sector can be observed.

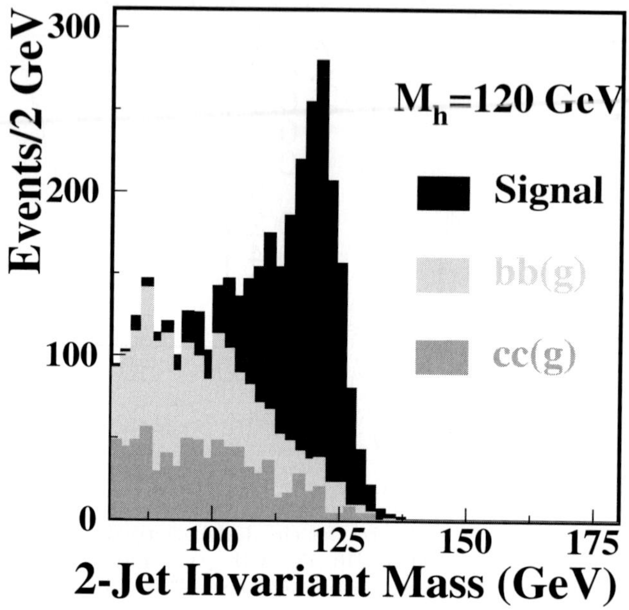

Fig. 32. Higgs signal and heavy quark backgrounds for $\gamma\gamma \to h$ in units of events per 2 GeV for a Higgs mass of 120 GeV after one year of running, with the $\gamma\gamma$ luminosity spectrum of Fig. 31. Results are from [17].

configuration that yields a broad (Type-I) $E_{\gamma\gamma}$ spectrum with good luminosity and substantial $\langle\lambda\lambda'\rangle$ in the $350 \lesssim E_{\gamma\gamma} \lesssim 450$ GeV region.

For the NLC-based design of the $\gamma\gamma$ collider as specified in [10, 17], after roughly four years of total running time (one year with the peaked spectrum and three years with the broad spectrum, assuming that the machine operates 10^7 sec per year) one can achieve $> 4\sigma$ signals (99% exclusion) in $\gamma\gamma \to H^0, A^0 \to b\bar{b}$ for most (all) of the $[m_{A^0}, \tan\beta]$ LHC wedge region. With twice the luminosity of the nominal NLC design (as might be achieved at TESLA or by using round beams at the NLC), essentially all of the wedge region is covered at the 5σ level except for the lowest $\tan\beta$ values in the $m_{A^0} > 2m_t$ part of the wedge (where it is anticipated that $\gamma\gamma \to H^0, A^0 \to t\bar{t}$ will yield a viable signal). This is summarized in Fig. 33. In fact, for H^0, A^0 discovery, the $\gamma\gamma$ collider is almost perfectly complementary to the LHC (and also the LC operation in the e^+e^- collision mode).

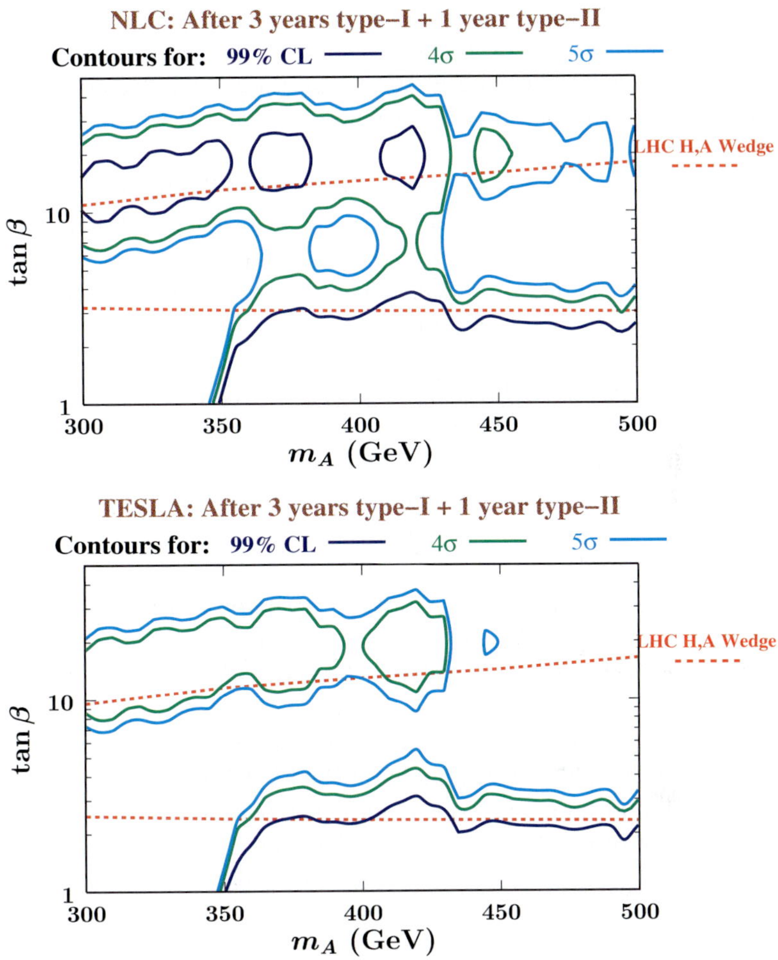

Fig. 33. Assuming a machine energy of $\sqrt{s} = 630$ GeV and employing only $\gamma\gamma \to H^0, A^0 \to b\bar{b}$, we show the $[m_{A^0}, \tan\beta]$ regions for which three years of operation using the type-I $P\lambda_e, P'\lambda'_e > 0$ polarization configuration and one year of operation using the type-II $P\lambda_e, P'\lambda'_e < 0$ configuration will yield $S/\sqrt{B} \geq 5$, $S/\sqrt{B} \geq 4$ or exclude the H^0, A^0 at 99% CL. The dashed curves indicate the wedge region from the LHC plot of Fig. 14—the lower curve gives a rough approximation to the LEP (maximal-mixing) limits while the upper curve is that above which $H^0, A^0 \to \tau^+\tau^-$ can be directly detected at the LHC. The upper plot is for the NLC-based $\gamma\gamma$ collider design and the lower plot is that achieved with a factor of 2 increase in luminosity, as might be achieved at TESLA or by using round e^-e^- beams instead of flat beams at the NLC. Results are from [17] as summarized in [19].

The corresponding results for $\gamma\gamma \to \widehat{h} \to b\bar{b}$ for the wedge region of a 2HDM model with a light decoupled $\widehat{h}$, and all other Higgs bosons heavier than $\sqrt{s}$, were given earlier in Fig. 16. Again, the $\gamma\gamma$ collider would allow $\widehat{h}$ discovery in a substantial portion of the wedge region in which its discovery would not otherwise be possible.

Once one or more Higgs bosons have been detected, precision studies can be performed including: determination of CP properties; a detailed scan to separate the H^0 and A^0 in the decoupling limit of a 2HDM; and branching ratios measurements. The branching ratios to supersymmetric final states are especially important for determining the basic supersymmetry breaking parameters [192, 182, 183, 83].

The CP properties can be determined for any spin-0 Higgs $\widehat{h}$ produced in $\gamma\gamma$ collisions. Since $\gamma\gamma \to \widehat{h}$ is of one-loop order, whether $\widehat{h}$ is CP-even, CP-odd or a mixture, the CP-even and CP-odd parts of $\widehat{h}$ have $\gamma\gamma$ couplings of similar size. However, the structure of the couplings is very different:

$$\mathcal{A}_{\text{CP-even}} \propto \vec{\epsilon}_1 \cdot \vec{\epsilon}_2 \,, \quad \mathcal{A}_{\text{CP-odd}} \propto (\vec{\epsilon}_1 \times \vec{\epsilon}_2) \cdot \widehat{p}_{\text{beam}} \,. \tag{37}$$

By adjusting the orientation of the initial laser photon polarization vectors with respect to one another, it is possible to determine the relative CP-even and CP-odd content of the resonance $\widehat{h}$ [178]. If $\widehat{h}$ is a mixture, one can use helicity asymmetries for this purpose [178, 175]. However, if $\widehat{h}$ is either purely CP-even or purely CP-odd, then one must employ transverse linear polarizations [179, 175]. Substantial luminosity with transverse polarization can be obtained, although the spectrum is not peaked (see [17]).

One measure of the CP nature of a Higgs boson is the asymmetry for parallel *vs.* perpendicular orientation of the linear polarizations of the initial laser beams,

$$\mathcal{A} \equiv \frac{N_{\parallel} - N_{\perp}}{N_{\parallel} + N_{\perp}} \,, \tag{38}$$

which is positive (negative) for a CP-even (odd) state. Since 100% linear polarization for the laser beams translates into only partial linear polarization for the colliding photons, both $N_{\parallel}$ and $N_{\perp}$ will be non-zero for the signal. In addition, the heavy quark background contributes to both. The expected value of $\mathcal{A}$ must be carefully computed for a given model. For the SM Higgs boson with $m_{h_{\rm SM}} = 120$ GeV, it is found [17] that $\mathcal{A}$ can be measured with an accuracy of about 20% in one year of operation using linear polarizations for the two lasers. This measurement would thus provide a moderately strong test of the even-CP nature of $h_{\rm SM}$. Of course, the linear polarization configuration is not ideal for the most accurate determination of $\Gamma(h_{\rm SM} \to \gamma\gamma){\rm BR}(h_{\rm SM} \to b\bar{b})$, but would allow measuring this product with an accuracy of about 8%.

We end by noting that the $e^-\gamma$ and e^-e^- collider options are most relevant to exotic Higgs scenarios, as discussed in Section 8.

12. Concluding Remarks

The physical origin of electroweak symmetry breaking is not yet known. In all theoretical approaches and models, the dynamics of electroweak symmetry breaking must be revealed at the TeV-scale or below. This energy scale will be thoroughly explored by hadron colliders, starting with the Tevatron and followed later in this decade by the LHC. Even though the various theoretical alternatives can only be confirmed or ruled out by future collider experiments, a straightforward interpretation of the electroweak precision data suggests that electroweak symmetry breaking dynamics is weakly-coupled, and a Higgs boson with mass between 100 and 200 GeV must exist. With the supersymmetric extension of the Standard Model, this interpretation opens the route to grand unification of all the fundamental forces, with the eventual incorporation of gravity in particle physics.

The discovery of the Higgs boson at the Tevatron and/or the LHC is a crucial first step. The measurement of Higgs properties at the LHC will begin to test the dynamics of electroweak symmetry breaking. However, a high-luminosity e^+e^- linear collider, now under development, is needed for a systematic program of precision Higgs measurements. For example, depending on the value of the Higgs mass, branching ratios and Higgs couplings can be determined in some cases at the level of a few percent. In this way, one can extract the properties of the Higgs sector in a comprehensive way, and establish (or refute) the theory of scalar sector dynamics as the mechanism responsible for generating the fundamental particle masses.

Acknowledgments

We are very grateful to Andreas Kronfeld who served with us as a co-convener of the Higgs Working Group of the American Linear Collider Working Group. This chapter is an update of our contribution to the American Linear Collider Working Group Resource Book for Snowmass 2001. We have benefited greatly from Andreas' contributions and counsel. We would also like to thank David Asner, Marcela Carena, Sven Heinemeyer, Heather Logan, Steve Mrenna, Michael Peskin, Carlos Wagner, Georg Weiglein, Dieter Zeppenfeld and Peter Zerwas for fruitful conversations and collaborative efforts that contributed to this work.

J.F.G., H.E.H. and R.V.K. are supported in part by the U.S. Department of Energy under the respective grants DE-FG03-91ER40674, DE-FG03-92ER40689 (Task B) and DE-FG02-91ER40661 (Task A).

References

1. J.F. Gunion, H.E. Haber, G. Kane and S. Dawson, *The Higgs Hunter's Guide* (Perseus Publishing, Reading, MA, 1990).
2. J.F. Gunion, L. Poggioli, R. Van Kooten, C. Kao, P. Rowson *et al.*, "Higgs boson discovery and properties," in *New Directions for High Energy Physics*, Proceedings of the 1996 DPF/DPB Summer Study on High Energy Physics, Snowmass '96, edited by D.G. Cassel, L.T. Gennari and R.H. Siemann (Stanford Linear Accelerator Center, Stanford, CA, 1997) pp. 541–587 [hep-ph/9703330].
3. H.E. Haber, T. Han, F.S. Merritt, J. Womersley *et al.*, "Weakly coupled Higgs bosons and precision electroweak physics," in *New Directions for High Energy Physics*, Proceedings of the 1996 DPF/DPB Summer Study on High Energy Physics, Snowmass '96, edited by D.G. Cassel, L.T. Gennari and R.H. Siemann (Stanford Linear Accelerator Center, Stanford, CA, 1997) pp. 482–498 [hep-ph/9703391].
4. H. Murayama and M.E. Peskin, Ann. Rev. Nucl. Part. Sci. **46**, 533 (1996) [hep-ex/9606003].
5. E. Accomando *et al.* [ECFA/DESY LC Physics Working Group], Phys. Rep. **299**, 1 (1998) [hep-ph/9705442].
6. M. Carena and H.E. Haber, FERMILAB-Pub-02/114-T and SCIPP 02/07 [hep-ph/0208209], to appear in Progress in Particle and Nuclear Physics, edited by A. Faessler.
7. C. Adolphsen *et al.* [International Study Group Collaboration], "International study group progress report on linear collider development," SLAC-R-559 and KEK-REPORT-2000-7 (April, 2000).
8. N. Akasaka *et al.*, "JLC design study," KEK-REPORT-97-1.
9. R. Brinkmann, K. Flottmann, J. Rossbach, P. Schmuser, N. Walker and H. Weise [editors], "TESLA: The superconducting electron positron linear collider with an integrated X-ray laser laboratory. Technical design report, Part 2: The Accelerator," DESY-01-011 (March, 2001), http://tesla.desy.de/tdr/.
10. T. Abe *et al.* [American Linear Collider Working Group Collaboration], SLAC-R-570 (2001) [hep-ex/0106055–58].
11. I.F. Ginzburg, G.L. Kotkin, V.G. Serbo and V.I. Telnov, Nucl. Instrum. Meth. **A205**, 47 (1983).
12. I.F. Ginzburg, G.L. Kotkin, S.L. Panfil, V.G. Serbo and V.I. Telnov, Nucl. Instrum. Meth. **A219**, 5 (1984).
13. D.L. Borden, D.A. Bauer and D.O. Caldwell, Phys. Rev. **D48**, 4018 (1993).

14. J.F. Gunion and H.E. Haber, Phys. Rev. **D48**, 5109 (1993).
15. S. Söldner-Rembold and G. Jikia, Nucl. Instrum. Meth. **A472**, 133 (2001) [hep-ex/0101056];
16. E. Boos *et al.*, Nucl. Instrum. Meth. **A472**, 100 (2001) [hep-ph/0103090]; M. Melles, Nucl. Instrum. Meth. A **472**, 128 (2001) [hep-ph/0008125].
17. D. Asner, J. Gronberg and J.F. Gunion, hep-ph/0110320, Phys. Rev. **D67** (2003) in press.
18. M.M. Velasco *et al.*, "Photon-Photon and Electron-Photon Colliders with Energies Below a TeV," in *Proceedings of the APS/DPF/DPB Summer Study on the Future of Particle Physics* (Snowmass 2001), edited by R. Davidson and C. Quigg, E3005 [hep-ph/0111055].
19. D. Asner, B. Grzadkowski, J.F. Gunion, H.E. Logan, V. Martin, M. Schmitt and M.M. Velasco, hep-ph/0208219.
20. U. Baur *et al.*, Report of the P1-WG1 subgroup, "Present and Future Electroweak Precision Measurements and the Indirect Determination of the Mass of the Higgs Boson," in *Proceedings of the APS/DPF/DPB Summer Study on the Future of Particle Physics* (Snowmass 2001), edited by R. Davidson and C. Quigg, P1WG1 [hep-ph/0202001].
21. J. Erler, S. Heinemeyer, W. Hollik, G. Weiglein and P.M. Zerwas, Phys. Lett. **B486**, 125 (2000) [hep-ph/0005024]; J. Erler and S. Heinemeyer, in *Proceedings of the 5th International Symposium on Radiative Corrections* (RADCOR 2000), 11–15 September 2000, edited by H.E. Haber, SLAC-R-579, eConf C000911 [hep-ph/0102083].
22. H.E. Haber and Y. Nir, Phys. Lett. **B306**, 327 (1993) [hep-ph/9302228]; H.E. Haber, in *Physics From the Planck Scale to the Electroweak Scale*, Proceedings of the US–Polish Workshop, Warsaw, Poland, September 21–24, 1994, edited by P. Nath, T. Taylor, and S. Pokorski (World Scientific, Singapore, 1995) pp. 49–63 [hep-ph/9501320].
23. For a recent review, see R.N. Mohapatra, in *Supersymmetry, Supersymmetry and Supercolliders*, edited by J. Bagger (World Scientific, Singapore, 1998) [hep-ph/9911272].
24. D. Abbaneo *et al.* [LEP Electroweak Working Group] and N. de Groot *et al.* [SLD Heavy Flavor Group], CERN-EP-2002-091 [hep-ex/0212036]; and additional updates at `http://lepewwg.web.cern.ch/LEPEWWG/`.
25. ALEPH, DELPHI, L3 and OPAL Collaborations, The LEP working group for Higgs boson searches], LHWG Note 2002-01 (July 2002), contributed paper for ICHEP'02 (Amsterdam, July 2002), and additional updates at http://lephiggs.web.cern.ch/LEPHIGGS/www/Welcome.html; P.A. McNamara and S.L. Wu, Rept. Prog. Phys. **65** (2002) 465.
26. E. Farhi and L. Susskind, Phys. Rep. **74**, 277 (1981); R.K. Kaul, Rev. Mod. Phys. **55**, 449 (1983). For a recent review, see C.T. Hill and E.H. Simmons, FERMI-PUB-02/045-T [hep-ph/0203079].
27. M.E. Peskin and T. Takeuchi, Phys. Rev. Lett. **65**, 964 (1990); Phys. Rev. **D46**, 381 (1992).
28. M.E. Peskin and J.D. Wells, Phys. Rev. **D64**, 093003 (2001) [hep-ph/0101342].

29. See, *e.g.*, A.G. Cohen, in *The Building Blocks of Creation—From Microfermis to Megaparsecs*, Proceedings of the Theoretical Advanced Study Institute (TASI 93) in Elementary Particle Physics, Boulder, CO, 6 June—2 July 1993, edited by S. Raby and T. Walker (World Scientific, Singapore, 1994); C.P. Burgess, in *Proceedings of the 4th International Symposium on Radiative Corrections (RADCOR 98): Applications of Quantum Field Theory to Phenomenology*, Barcelona, Catalonia, Spain, 8–12 September 1998, edited by J. Sola (World Scientific, Singapore, 1999) pp. 471–488 [hep-ph/9812470].
30. R. Dashen and H. Neuberger, Phys. Rev. Lett. **50**, 1897 (1983).
31. T. Hambye and K. Riesselmann, Phys. Rev. **D55**, 7255 (1997) [hep-ph/9610272].
32. See, *e.g.*, G. Altarelli and G. Isidori, Phys. Lett. **B337**, 141 (1994); J.A. Casas, J.R. Espinosa and M. Quirós, Phys. Lett. **B342**, 171 (1995) [hep-ph/9409458]; **B382**, 374 (1996) [hep-ph/9603227].
33. W. Langguth, I. Montvay and P. Weisz, Nucl. Phys. **B277**, 11 (1986); A. Hasenfratz, K. Jansen, J. Jersak, C.B. Lang, T. Neuhaus and H. Yoneyama, Nucl. Phys. **B317**, 81 (1989); J. Kuti, L. Lin and Y. Shen, Phys. Rev. Lett. **61**, 678 (1988).
34. U.M. Heller, M. Klomfass, H. Neuberger and P. Vranas, Nucl. Phys. **B405**, 555 (1993).
35. M.Lüscher and P. Weisz, Phys. Lett. **B212**, 472 (1988), Nucl. Phys. **B318**, 705 (1989).
36. J. Shigemitsu, Phys. Lett. **B226**, 364 (1989); W. Bock, A.K. De, C. Frick, K. Jansen and T. Trappenberg, Nucl. Phys. **B371**, 683 (1992); W. Bock, C. Frick, J. Smit and J.C. Vink, Nucl. Phys. **B400**, 309 (1993) [hep-lat/9212026].
37. K. Riesselmann, hep-ph/9711456.
38. G. Isidori, G. Ridolfi and A. Strumia, Nucl. Phys. **B609**, 387 (2001) [hep-ph/0104016].
39. G. 't Hooft, in *Recent Developments in Gauge Theories,* Proceedings of the NATO Advanced Summer Institute, Cargese, 1979, edited by G. 't Hooft *et al.* (Plenum, New York, 1980) p. 135; E. Witten, Nucl. Phys. **B188**, 513 (1981); L. Susskind, Phys. Rep. **104**, 181 (1984).
40. H.P. Nilles, Phys. Rep. **110**, 1 (1984).
41. H.E. Haber and G.L. Kane, Phys. Rep. **117**, 75 (1985); S.P. Martin, hep-ph/9709356.
42. A. Djouadi, J. Kalinowski and M. Spira, Comput. Phys. Commun. **108**, 56 (1998) [hep-ph/9704448]. The HDECAY program is available at `http://home.cern.ch/~mspira/proglist.html`.
43. T. Han, Int. J. Mod. Phys. **A11**, 1541 (1996) [hep-ph/9601393].
44. A. Djouadi, J. Kalinowski and P.M. Zerwas, Z. Phys. **C54**, 155 (1992).
45. S. Dittmaier, M. Kramer, Y. Liao, M. Spira and P. M. Zerwas, Phys. Lett. **B441**, 383 (1998) [hep-ph/9808433].
46. A. Freitas, W. Hollik, W. Walter and G. Weiglein, Phys. Lett. **B495**, 338 (2000) [hep-ph/0007091]; A. Freitas, S. Heinemeyer, W. Hollik, W. Walter and G. Weiglein, Nucl. Phys. Proc. Suppl. **89**, 82 (2000) [hep-ph/0007129];

and in *Proceedings of the 5th International Symposium on Radiative Corrections* (RADCOR 2000), 11–15 September 2000, edited by H.E. Haber, SLAC-R-579, eConf C000911 [hep-ph/0101260].
47. M. Carena, J.S. Conway, H.E. Haber and J.D. Hobbs *et al.*, FERMILAB-CONF-00/270-T [hep-ph/0010338].
48. D. Costanzo, "Higgs Physics at the Large Hadron Collider," presented at the 36th Rencontres de Moriond on Electroweak Interactions and Unified Theories, Les Arcs, France, 10–17 March 2001 [hep-ex/0105033]; V.A. Mitsou, "Search for New Physics with ATLAS at the LHC," in *From Particles to the Universe*, Proceedings of the 15th Lake Louise Winter Institute, Alberta, Canada, 20–26 February 2000, edited by A. Astbury, B.A. Campbell, F.C. Khanna and M.G. Vincter (World Scientific, Singapore, 2001) [hep-ph/0004161].
49. K. Lassila-Perini, ETH Dissertation thesis No. 12961 (1998). [CERN-THESIS-99-3]; "Higgs Physics at the LHC," presented at the III International Symposium on LHC Physics and Detectors, Chia, 25–27 October 2001, CMS Conference Report CMS CR 2001/018.
50. M. Dittmar, Pramana **55**, 151 (2000); M. Dittmar and A.-S. Nicollerat, "High Mass Higgs Studies using $gg \to h_{\rm SM}$ and $qq \to qqh_{\rm SM}$ at the LHC," CMS NOTE 2001/036.
51. M. Hohlfeld, "On the determination of Higgs parameters in the ATLAS experiment at the LHC," ATL-PHYS-2001-004.
52. D. Zeppenfeld, R. Kinnunen, A. Nikitenko and E. Richter-Wąs, Phys. Rev. **D62**, 013009 (2000) [hep-ph/0002036].
53. D. Zeppenfeld, "Higgs Couplings at the LHC," in *Proceedings of the APS/DPF/DPB Summer Study on the Future of Particle Physics* (Snowmass 2001), edited by R. Davidson and C. Quigg, SNOWMASS-2001-P123 [hep-ph/0203123].
54. A. De Roeck, presentations for the ATLAS Collaboration at the International Workshop on Linear Colliders, *Workshop on Physics and Experiments with Future Electron-Positron Linear Colliders*, 26–30 August, 2002, Jeju Island, Korea; and private communications.
55. O.J. Éboli and D. Zeppenfeld, Phys. Lett. **B495**, 147 (2000) [hep-ph/0009158].
56. L.J. Hall and M.B. Wise, Nucl. Phys. **B187**, 397 (1981).
57. For a review of the two-Higgs-doublet model, see Chapter 4 of [1].
58. K. Inoue, A. Kakuto, H. Komatsu, and S. Takeshita, Prog. Theor. Phys. **67**, 1889 (1982); R. Flores and M. Sher, Ann. Phys. (NY) **148**, 95 (1983).
59. J.F. Gunion and H.E. Haber, Nucl. Phys. **B272**, 1 (1986); **B278**, 449 (1986) [E: **B402**, 567 (1993)].
60. V.D. Barger, J.L. Hewett and R.J.N. Phillips, Phys. Rev. **D41** (1990) 3421.
61. B. Schrempp and M. Wimmer, Prog. Part. Nucl. Phys. **37** (1996) 1 [hep-ph/9606386].
62. J.F. Gunion and H.E. Haber, SCIPP-02/10 and UCD-2002-10 [hep-ph/0207010], Phys. Rev. D, in press.

63. H.E. Haber and R. Hempfling, Phys. Rev. Lett. **66**, 1815 (1991); Y. Okada, M. Yamaguchi and T. Yanagida, Prog. Theor. Phys. **85**, 1 (1991); J. Ellis, G. Ridolfi and F. Zwirner, Phys. Lett. **B257**, 83 (1991).
64. S. Heinemeyer, W. Hollik and G. Weiglein, Phys. Rev. **D58**, 091701 (1998) [hep-ph/9803277]; Phys. Lett. **B440**, 296 (1998) [hep-ph/9807423]; Eur. Phys. J. **C9**, 343 (1999) [hep-ph/9812472]; M. Carena, H.E. Haber, S. Heinemeyer, W. Hollik, C.E.M. Wagner and G. Weiglein, Nucl. Phys. **B580**, 29 (2000) [hep-ph/0001002]; R.-J. Zhang, Phys. Lett. **B447**, 89 (1999) [hep-ph/9808299]; J.R. Espinosa and R.-J. Zhang, JHEP **0003**, 026 (2000) [hep-ph/9912236]; Nucl. Phys. **B586**, 3 (2000) [hep-ph/0003246]; J.R. Espinosa and I. Navarro, Nucl. Phys. **B615**, 82 (2001) [hep-ph/0104047]; G. Degrassi, P. Slavich and F. Zwirner, Nucl. Phys. **B611**, 403 (2001) [hep-ph/0105096]; A. Brignole, G. Degrassi, P. Slavich and F. Zwirner, Nucl. Phys. **B631**, 195 (2002) [hep-ph/0112177]; Nucl. Phys. **B643**, 79 (2002) [hep-ph/0206101]; G. Degrassi, S. Heinemeyer, W. Hollik, P. Slavich and G. Weiglein, hep-ph/0212020; Computer programs developed for computing the radiatively-corrected Higgs masses, FeynHiggs and FeynHiggsFast, are described respectively in S. Heinemeyer, W. Hollik and G. Weiglein, Comput. Phys. Commun. **124**, 76 (2000) [hep-ph/9812320]; and CERN-TH-2000-055 [hep-ph/0002213].
65. J. Ellis, J.F. Gunion, H.E. Haber, L. Roszkowski and F. Zwirner, Phys. Rev. **D39**, 844 (1989). For a recent treatment and guide to the literature see: C. Panagiotakopoulos and A. Pilaftsis, Phys. Rev. **D63**, 055003 (2001) [hep-ph/0008268]. A detailed discussion of NMSSM phenomenology can be found in U. Ellwanger, M. Rausch de Traubenberg and C.A. Savoy, Nucl. Phys. **B492**, 21 (1997) [hep-ph/9611251].
66. J.R. Espinosa and M. Quirós, Phys. Rev. Lett. **81**, 516 (1998) [hep-ph/9804235]; M. Masip, Phys. Lett. **B444**, 352 (1998) [hep-ph/9810303].
67. K. Tobe and J.D. Wells, Phys. Rev. **D66**, 013010 (2002) [hep-ph/0204196].
68. J.A. Coarasa, R.A. Jiménez and J. Solà, Phys. Lett. **B389**, 312 (1996) [hep-ph/9511402]; R.A. Jiménez and J. Solà, Phys. Lett. **B389**, 53 (1996) [hep-ph/9511292]; A. Bartl, H. Eberl, K. Hikasa, T. Kon, W. Majerotto and Y. Yamada, Phys. Lett. **B378**, 167 (1996) [hep-ph/9511385].
69. R. Hempfling, Phys. Rev. **D49**, 6168 (1994); L. Hall, R. Rattazzi and U. Sarid, Phys. Rev. **D50**, 7048 (1994) [hep-ph/9306309].
70. M. Carena, M. Olechowski, S. Pokorski and C.E.M. Wagner, Nucl. Phys. **B426**, 269 (1994) [hep-ph/9402253].
71. D. Pierce, J. Bagger, K. Matchev, and R. Zhang, Nucl. Phys. **B491**, 3 (1997) [hep-ph/9606211].
72. S. Heinemeyer, W. Hollik and G. Weiglein, Eur. Phys. J. **C16**, 139 (2000) [hep-ph/0003022].
73. M. Carena, H.E. Haber, H.E. Logan and S. Mrenna, Phys. Rev. **D65**, 055005 (2002) [E: **D65**, 099902 (2002)] [hep-ph/0106116].
74. M. Carena, D. Garcia, U. Nierste and C.E.M. Wagner, Nucl. Phys. **B577**, 88 (2000) [hep-ph/9912516].
75. H. Eberl, K. Hidaka, S. Kraml, W. Majerotto and Y. Yamada, Phys. Rev. **D62**, 055006 (2000) [hep-ph/9912463].

76. M. Carena, S. Mrenna and C.E.M. Wagner, Phys. Rev. **D60**, 075010 (1999) [hep-ph/9808312]; Phys. Rev. **D62**, 055008 (2000) [hep-ph/9907422].
77. H. Baer and J.D. Wells, Phys. Rev. **D57**, 4446 (1998) [hep-ph/9710368]; W. Loinaz and J. D. Wells, Phys. Lett. **B445**, 178 (1998) [hep-ph/9808287].
78. J.L. Diaz-Cruz, H.-J. He, T. Tait and C.P. Yuan, Phys. Rev. Lett. **80**, 4641 (1998) [hep-ph/9802294]; C. Balázs, J.L. Diaz-Cruz, H.-J. He, T. Tait and C.P. Yuan, Phys. Rev. **D59**, 055016 (1999) [hep-ph/9807349].
79. H.E. Haber, M.J. Herrero, H.E. Logan, S. Peñaranda, S. Rigolin and D. Temes, Phys. Rev. **D63**, 055004 (2001) [hep-ph/0007006].
80. P.H. Chankowski, S. Pokorski and J. Rosiek, Nucl. Phys. **B423**, 497 (1994).
81. A. Djouadi, P. Janot, J. Kalinowski and P.M. Zerwas, Phys. Lett. **B376**, 220 (1996) [hep-ph/9603368]; A. Djouadi, J. Kalinowski, P. Ohmann and P.M. Zerwas, Z. Phys. **C74**, 93 (1997) [hep-ph/9605339].
82. S. Heinemeyer, W. Hollik, J. Rosiek and G. Weiglein, Eur. Phys. J. **C19**, 535 (2001) [hep-ph/0102081].
83. M.M. Mühlleitner, M. Krämer, M. Spira and P.M. Zerwas, Phys. Lett. **B508**, 311 (2001) [hep-ph/0101083].
84. ALEPH, DELPHI, L3 and OPAL Collaborations, The LEP working group for Higgs boson searches], LHWG Note 2001-05 (July 2001), [hep-ex/0107031].
85. ALEPH, DELPHI, L3 and OPAL Collaborations, The LEP working group for Higgs boson searches], LHWG Note 2001-04 (July 2001), [hep-ex/0107030].
86. Fig. 14 was provided by E. Richter-Was and F. Gianotti on behalf of the ATLAS collaboration.
87. D. Denegri *et al.*, "Summary of the CMS Discovery Potential for the MSSM SUSY Higgses," CMS NOTE 2001/032 [hep-ph/0112045].
88. I. Hinchliffe, F.E. Paige, M.D. Shapiro, J. Soderqvist and W. Yao, Phys. Rev. **D55**, 5520 (1997) [hep-ph/9610544]; S. Abdullin *et al.* [CMS Collaboration], J. Phys. **G28**, 469 (2002) [hep-ph/9806366].
89. S.L. Glashow and S. Weinberg, Phys. Rev. **D15**, 1958 (1977); E.A. Paschos, Phys. Rev. **D15**, 1966 (1977).
90. B.A. Dobrescu, FERMILAB-CONF-99-051 [hep-ph/9903407]; and in *Physics and experiments with future linear e^+e^- colliders*, Proceedings of the 5th International Linear Collider Workshop, Batavia, IL, USA, 2000, edited by A. Para and H.E. Fisk (American Institute of Physics, New York, 2001), pp. 113–121 [hep-ph/0103038].
91. M. Cvetic, H. Lu, C.N. Pope, A. Sadrzadeh and T.A. Tran, Nucl. Phys. **B586**, 275 (2000) [hep-th/0003103].
92. P. Chankowski, T. Farris, B. Grzadkowski, J.F. Gunion, J. Kalinowski and M. Krawczyk, Phys. Lett. **B496**, 195 (2000) [hep-ph/0009271].
93. J.F. Gunion, in *Physics and experiments with future linear e^+e^- colliders*, Proceedings of the 5th International Linear Collider Workshop, Batavia, IL, USA, 2000, edited by A. Para and H.E. Fisk (American Institute of Physics, New York, 2001), pp. 231–236 [hep-ph/0012199].
94. U. Ellwanger and C. Hugonie, Eur. Phys. J. **C25** (2002) 297 [hep-ph/9909260].

95. J.R. Espinosa and J.F. Gunion, Phys. Rev. Lett. **82**, 1084 (1999) [hep-ph/9807275].
96. J.F. Gunion, H.E. Haber and T. Moroi, "Will at least one of the Higgs bosons of the next-to-minimal supersymmetric extension of the Standard Model be observable at LEP2 or the LHC?", in *New Directions for High Energy Physics*, Proceedings of the 1996 DPF/DPB Summer Study on High Energy Physics, Snowmass '96, edited by D.G. Cassel, L.T. Gennari and R.H. Siemann (Stanford Linear Accelerator Center, Stanford, CA, 1997) [hep-ph/9610337].
97. U. Ellwanger, J.F. Gunion and C. Hugonie, hep-ph/0111179.
98. J.F. Gunion and B. Grzadkowski, work in progress.
99. J. Kamoshita, Y. Okada and M. Tanaka, Phys. Lett. **B328**, 67 (1994) [hep-ph/9402278].
100. J.F. Gunion, L. Roszkowski, A. Turski, H.E. Haber, G. Gamberini, B. Kayser, S.F. Novaes, F.I. Olness and J. Wudka, Phys. Rev. **D38**, 3444 (1988).
101. S. Bar-Shalom, D. Atwood, G. Eilam, R.R. Mendel and A. Soni, Phys. Rev. **D53**, 1162 (1996) [hep-ph/9508314].
102. J.F. Gunion, B. Grzadkowski and X. He, Phys. Rev. Lett. **77** (1996) 5172 [hep-ph/9605326].
103. S. Dawson and L. Reina, Phys. Rev. **D57** (1998) 5851 [hep-ph/9712400]; Phys. Rev. **D59**, 054012 (1999) [hep-ph/9808443]; H. Baer, S. Dawson, and L. Reina, Phys. Rev. **D61**, 0113002 (2000) [hep-ph/9906419].
104. S. Dawson and L. Reina, Phys. Rev. **D60**, 015003 (1999) [hep-ph/9812488].
105. B. Grzadkowski, J.F. Gunion and J. Kalinowski, Phys. Lett. **B480**, 287 (2000) [hep-ph/0001093].
106. S. Dittmaier, M. Kramer, Y. Liao, M. Spira and P.M. Zerwas, Phys. Lett. **B478**, 247 (2000) [hep-ph/0002035].
107. U. Cotti, A. Gutierrez-Rodriguez, A. Rosado and O.A. Sampayo, Phys. Rev. **D59**, 095011 (1999) [hep-ph/9902417].
108. S. Kanemura, S. Moretti and K. Odagiri, JHEP **0102**, 011 (2001) [hep-ph/0012030].
109. B.A. Kniehl, F. Madricardo and M. Steinhauser, Phys. Rev. **D66**, 054016 (2002) [hep-ph/0205312].
110. A. Gutierrez-Rodriguez and O. A. Sampayo, Phys. Rev. **D62**, 055004 (2000); S. Moretti, Eur. Phys. J. direct **C4**, 15 (2002) [hep-ph/0206208].
111. H.J. He and C.-P. Yuan, Phys. Rev. Lett. **83**, 28 (1999) [hep-ph/9810367]; H.J. He, S. Kanemura and C.-P. Yuan, Phys. Rev. Lett. **89**, 101803 (2002) [hep-ph/0203090]; and hep-ph/0209376.
112. S.H. Zhu, hep-ph/9901221; A. Arhrib, M. Capdequi Peyranère, W. Hollik and G. Moultaka, Nucl. Phys. **B581**, 34 (2000) [hep-ph/9912527]; S. Kanemura, Eur. Phys. J. **C17**, 473 (2000) [hep-ph/9911541]; H.E. Logan and S. Su, Phys. Rev. **D66**, 035001 (2002) [hep-ph/0203270]; **D67**, 017703 (2003) [hep-ph/0206135].
113. A. Djouadi, H.E. Haber and P.M. Zerwas, Phys. Lett. **B375**, 203 (1996) [hep-ph/9602234]; P. Osland and P.N. Pandita, Phys. Rev. **D59**, 055013

(1999) [hep-ph/9806351]; A. Djouadi, W. Kilian, M. Mühlleitner and P.M. Zerwas, Eur. Phys. J. **C10**, 27 (1999) [hep-ph/9903229]; F. Boudjema and A. Semenov, Phys. Rev. **D66**, 095007 (2002) [hep-ph/0201219].

114. H.E. Haber and Y. Nir, Phys. Lett. **B306**, 327 (1993) [hep-ph/9302228].
115. T. Farris, J.F. Gunion and H.E. Logan, "Higgs sectors in which the only light Higgs boson is CP-odd and linear collider strategies for its discovery," in *Proceedings of the APS/DPF/DPB Summer Study on the Future of Particle Physics* (Snowmass 2001), edited by R. Davidson and C. Quigg, SNOWMASS-2001-P121 [hep-ph/0202087].
116. J.F. Gunion, C. Loomis and K.T. Pitts, "Searching for doubly-charged Higgs bosons at future colliders," in *New Directions for High Energy Physics*, Proceedings of the 1996 DPF/DPB Summer Study on High Energy Physics, Snowmass '96, edited by D.G. Cassel, L.T. Gennari and R.H. Siemann (Stanford Linear Accelerator Center, Stanford, CA, 1997), pp. 603–607 [hep-ph/9610237].
117. J.F. Gunion, R. Vega and J. Wudka, Phys. Rev. **D43**, 2322 (1991).
118. R.N. Mohapatra, Fortsch. Phys. **31**, 185 (1983); J.F. Gunion, J. Grifols, A. Mendez, B. Kayser and F. Olness, Phys. Rev. **D40**, 1546 (1989); N.G. Deshpande, J.F. Gunion, B. Kayser and F. Olness, Phys. Rev. **D44**, 837 (1991).
119. H. Georgi and M. Machacek, Nucl. Phys. **B262**, 463 (1985).
120. M.S. Chanowitz and M. Golden, Phys. Lett. **B165**, 105 (1985).
121. J.F. Gunion, Int. J. Mod. Phys. **A11**, 1551 (1996) [hep-ph/9510350]; **A13**, 2277 (1998) [hep-ph/9803222]; F. Cuypers and M. Raidal, Nucl. Phys. **B501**, 3 (1997) [hep-ph/9704224]; P.H. Frampton, Int. J. Mod. Phys. **A15**, 2455 (2000) [hep-ph/0002017].
122. J. Gluza, Phys. Lett. **B403**, 304 (1997) [hep-ph/9704202].
123. R.A. Alanakian, Phys. Lett. **B436**, 139 (1998) [hep-ph/9706383].
124. G. Barenboim, K. Huitu, J. Maalampi and M. Raidal, Phys. Lett. **B394**, 132 (1997) [hep-ph/9611362].
125. V.D. Barger, J.F. Beacom, K. Cheung and T. Han, Phys. Rev. **D50**, 6704 (1994) [hep-ph/9404335].
126. R. Casalbuoni, A. Deandrea, S. De Curtis, D. Dominici, R. Gatto and J.F. Gunion, Nucl. Phys. **B555**, 3 (1999) [hep-ph/9809523].
127. K. Lane, K.R. Lynch, S. Mrenna and E.H. Simmons, Phys. Rev. **D66**, 015001 (2002) [hep-ph/0203065].
128. K. Desch, and M. Battaglia, in *Physics and experiments with future linear e^+e^- colliders*, Proceedings of the 5th International Linear Collider Workshop, Batavia, IL, USA, 2000, edited by A. Para and H.E. Fisk (American Institute of Physics, New York, 2001), pp. 312–316; M. Battaglia and K. Desch, in *ibid.*, pp. 163–182 [hep-ph/0101165].
129. R.D. Heuer, D.J. Miller, F. Richard and P.M. Zerwas [editors], "TESLA: The superconducting electron positron linear collider with an integrated X-ray laser laboratory. Technical design report, Part 3: Physics at an e^+e^- Linear Collider," DESY-01-011 (March, 2001), http://tesla.desy.de/tdr/, [hep-ph/0106315].

130. LCD refers to "Linear Collider Detector" as defined by the North American simulation studies, see `http://www-sldnt.slac.stanford.edu/nld/`.
131. H. Yang and K. Riles, "Impact of Tracker Design on Higgs Mass and Cross Section Resolutions," in *Proceedings of the APS/DPF/DPB Summer Study on the Future of Particle Physics* (Snowmass 2001), edited by R. Davidson and C. Quigg, SNOWMASS-2001-E3040.
132. M. Ronan, "Jet-Jet Reconstruction in Full Detector Simulations," presented at Workshop on Physics and Detectors for Future e^+e^- Linear Colliders, Johns Hopkins University, Baltimore, 19–21 March 2001.
133. A. Juste, "Higgs mass determination from direct reconstruction at a linear e^+e^- collider," in *Physics and Experiments with Future Linear e^+e^- Colliders*, edited by E. Fernández and A. Pacheco (Universitat Autònoma de Barcelona, Bellaterra, Spain, 1999) [hep-ex/9912041].
134. V.D. Barger, M.S. Berger, J.F. Gunion and T. Han, Phys. Rev. Lett. **78**, 3991 (1997) [hep-ph/9612279].
135. P.F. Derwent *et al.*, "Linear Collider Physics," FERMILAB-FN-701 [hep-ex/0107044].
136. A. Küskinen, M. Battaglia, P. Pöyhoönen, "Study of $e^+e^- \to H^+H^-$ at a 800 GeV Linear Collider", in *Physics and experiments with future linear e^+e^- colliders*, Proceedings of the 5th International Linear Collider Workshop, Batavia, IL, USA, 2000, edited by A. Para and H.E. Fisk (American Institute of Physics, New York, 2001); "Study of HA Production in e^+e- Collisions at $\sqrt{s}$ = 800 GeV, DESY-123-E, p. 417; "Pair production of charged Higgs bosons at future linear e^+e^- colliders," in *Proceedings of the APS/DPF/DPB Summer Study on the Future of Particle Physics* (Snowmass 2001), edited by R. Davidson and C. Quigg, SNOWMASS-2001-E3017 [hep-ex/0112015]; A. Andreazza and C. Troncon, "Study of HA Production in e^+e- Collisions at $\sqrt{s} = 800$ GeV," DESY-123-E, p. 417.
137. P. Garcia-Abia and W. Lohmann, Eur. Phys. J. direct **C2**, 1 (2000) [hep-ex/9908065].
138. R. Van Kooten, "Separation of the Higgsstrahlung and WW-fusion process," presented at Workshop on Physics and Detectors for Future e^+e^- Linear Colliders, Johns Hopkins University, Baltimore, 19–21 March 2001.
139. C.T. Potter, J.E. Brau and M. Iwasaki, "Standard Model Higgs Boson Branching Ratio Measurements at a Linear Collider," in *Proceedings of the APS/DPF/DPB Summer Study on the Future of Particle Physics* (Snowmass 2001), edited by R. Davidson and C. Quigg, SNOWMASS-2001-P118.
140. J. Brau, C. Potter, and M. Iwasaki, "Linear Collider Vertex Detector Optimization for Higgs Branching Ratio Measurements," in *Physics and experiments with future linear e^+e^- colliders*, Proceedings of the 5th International Linear Collider Workshop, Batavia, IL, USA, 2000, edited by A. Para and H.E. Fisk (American Institute of Physics, New York, 2001).
141. D.J. Jackson, Nucl. Instrum. Meth. **A388**, 247 (1997).
142. C.T. Potter, J.E. Brau and M. Iwasaki, "Standard Model Higgs boson branching ratio measurements at a linear collider," to appear in the Proceedings of the International Workshop on Linear Colliders, *Workshop on*

Physics and Experiments with Future Electron-Positron Linear Colliders, 26–30 August, 2002, Jeju Island, Korea.
143. M.D. Hildreth, T.L. Barklow, and D.L. Burke, Phys. Rev. **D49**, 3441 (1994); K. Kawagoe, in *Proceedings of the 2nd International Workshop on Physics and Experiments with Linear e^+e^- Colliders*, Volume II, Waikoloa, HI, 26–30 April 1993, edited by F.A. Harris, S.L. Olsen, S. Pakvasa and X. Tata (World Scientific, Singapore, 1993) pp. 660–665.
144. G. Borisov and F. Richard, Precise measurement of Higgs decay rate into WW^* at future e^+e^- linear colliders and theoretical consequences, in *Physics and Experiments with Future Linear e^+e^- Colliders*, edited by E. Fernández and A. Pacheco (Universitat Autoònoma de Barcelona, Bellaterra, Spain, 1999) [hep-ph/9905413].
145. M. Battaglia, Nucl. Instrum. Meth. **A473**, 75 (2001) [hep-ex/0012021]; "Measuring Higgs branching ratios and telling the SM from a MSSM Higgs boson at the e^+e^- linear collider," in *Physics and Experiments with Future Linear e^+e^- Colliders*, edited by E. Fernández and A. Pacheco (Universitat Autoònoma de Barcelona, Bellaterra, Spain, 1999) [hep-ph/9910271].
146. M. Battaglia and A. De Roeck, "Determination of the muon Yukawa coupling at high energy e^+e^- linear colliders," in *Proceedings of the APS/DPF/DPB Summer Study on the Future of Particle Physics*, (Snowmass 2001), edited by R. Davidson and C. Quigg, SNOWMASS-2001-E3066 [hep-ph/0111307].
147. J.F. Gunion and P.C. Martin, hep-ph/9610417; J.F. Gunion and P.C. Martin, Phys. Rev. Lett. **78**, 4541 (1997) [hep-ph/9607360].
148. E. Boos, J.C. Brient, D.W. Reid, H.J. Schreiber and R. Shanidze, Eur. Phys. J. **C19**, 455 (2001) [hep-ph/0011366].
149. O.J. Eboli and D. Zeppenfeld, Phys. Lett. **B495**, 147 (2000) [hep-ph/0009158].
150. F. Boudjema, G. Belanger and R.M. Godbole, CERN-TH-2002-141 [hep-ph/0206311]; K. Griest and H.E. Haber, Phys. Rev. **D37**, 719 (1988).
151. L. F. Li, Y. Liu and L. Wolfenstein, Phys. Lett. **B159**, 45 (1985); S. Bertolini and A. Santamaria, Phys. Lett. **B213**, 487 (1988); A.S. Joshipura and S.D. Rindani, Phys. Rev. Lett. **69**, 3269 (1992); A.S. Joshipura and J.W.F. Valle, Nucl. Phys. **B397**, 105 (1993); M.A. Diaz, M.A. Garcia-Jareno, D.A. Restrepo and J.W.F. Valle, Nucl. Phys. **B527**, 44 (1998) [hep-ph/9803362].
152. A. Pilaftsis, Z. Phys. **C55**, 275 (1992) [hep-ph/9901206]; V.A. Khoze, hep-ph/0105069; K. Belotsky, D. Fargion, M. Khlopov, R. Konoplich and K. Shibaev, hep-ph/0210153.
153. G.F. Giudice, R. Rattazzi and J.D. Wells, Nucl. Phys. **B595**, 250 (2001) [hep-ph/0002178]; K.M.. Cheung, Phys. Rev. **D63**, 056007 (2001) [hep-ph/0009232]; Phys. Lett. **B524**, 161 (2002) [hep-ph/0110035]; J.L. Hewett and T.G. Rizzo, SLAC-PUB-9132 [hep-ph/0202155].
154. E.L. Berger, C.W. Chiang, J. Jiang, T.M. Tait and C.E.M. Wagner, Phys. Rev. **D66**, 095001 (2002) [hep-ph/0205342].

155. R. Van Kooten, "Measurement of the Invisible Branching Ratio of the Higgs," presented at Chicago LC Workhsop, 7–9 January 2002, Gleacher Center, University of Chicago, Chicago.
156. D. Atwood and A. Soni, Phys. Rev. **D45**, 2405 (1992); M. Diehl and O. Nachtmann, Z. Phys. **V62**, 397 (1994).
157. A. Juste and G. Merino, hep-ph/9910301.
158. F. Boudjema and E. Chopin, Z. Phys. **C73**, 85 (1996) [hep-ph/9507396]; D.J. Miller and S. Moretti, Eur. Phys. J. **C13**, 459 (2000) [hep-ph/9906395].
159. D.J. Miller and S. Moretti, hep-ph/0001194.
160. C. Castanier, P. Gay, P. Lutz and J. Orloff, hep-ex/0101028.
161. M. Battaglia, E. Boos and W.M. Yao, "Studying the Higgs potential at the e^+e^- linear collider," in *Proceedings of the APS/DPF/DPB Summer Study on the Future of Particle Physics* (Snowmass 2001), edited by R. Davidson and C. Quigg, SNOWMASS-2001-E3016 [hep-ph/0111276].
162. A. Djouadi, W. Kilian, M.M. Mühlleitner and P.M. Zerwas, hep-ph/0001169. R. Lafaye, D.J. Miller, M.M. Mühlleitner and S. Moretti, hep-ph/0002238;
163. D. Choudhury, T.M. Tait and C.E.M. Wagner, Phys. Rev. **D65**, 115007 (2002) [hep-ph/0202162].
164. J. Alcaraz and E. Ruiz Morales, Phys. Rev. Lett. **86**, 3726 (2001) [hep-ph/0012109].
165. D. Bardin, P. Christova, M. Jack, L. Kalinovskaya, A. Olchevski, S. Riemann and T. Riemann, Comput. Phys. Commun. **133**, 229 (2001) [hep-ph/9908433].
166. V. Drollinger and A. Sopczak, in *Physics and experiments with future linear e^+e^- colliders*, Proceedings of the 5th International Linear Collider Workshop, Batavia, IL, USA, 2000, edited by A. Para and H.E. Fisk (American Institute of Physics, New York, 2001) pp. 226–230 [hep-ph/0102342].
167. N. Meyer, "Properties of a 240 GeV SM Higgs," presented at Chicago LC Workshop, 7-9 January 2002, Gleacher Center, University of Chicago, Chicago.
168. L.D. Landau, Dokl. Akad. Nauk Ser. Fiz. **60**, 207 (1948); C.N. Yang, Phys. Rev. **77**, 242 (1950); J.J. Sakurai, *Invariance Principles and Elementary Particles* (Princeton University Press, 1964).
169. D.J. Miller, S.Y. Choi, B. Eberle, M.M. Mühlleitner and P.M. Zerwas, Phys. Lett. **B505**, 149 (2001) [hep-ph/0102023].
170. K. Hagiwara and M.L. Stong, Z. Phys. **C62**, 99 (1994) [hep-ph/9309248]; K. Hagiwara, S. Ishihara, J. Kamoshita and B.A. Kniehl, Eur. Phys. J. **C14**, 457 (2000) [hep-ph/0002043].
171. V.D. Barger, K. Cheung, A. Djouadi, B.A. Kniehl and P.M. Zerwas, Phys. Rev. **D49**, 79 (1994) [hep-ph/9306270].
172. T. Han and J. Jiang, Phys. Rev. **D63**, 096007 (2001) [hep-ph/0011271].
173. B. Grzadkowski and J.F. Gunion, Phys. Lett. **B350**, 218 (1995) [hep-ph/9501339].
174. B. Grzadkowski and J.F. Gunion, hep-ph/9503409.
175. M. Krämer, J. Kühn, M.L. Stong and P.M. Zerwas, Z. Phys. **C64**, 21 (1994) [hep-ph/9404280].

176. G.R. Bower, T. Pierzchala, Z. Was and M. Worek, Phys. Lett. **B543**, 227 (2002) [hep-ph/0204292].
177. J.F. Gunion and X.G. He, "Discriminating between Higgs boson models using $e^+e^- \to t\bar{t}h$ and Zh at the NLC," in *New Directions for High Energy Physics*, Proceedings of the 1996 DPF/DPB Summer Study on High Energy Physics, Snowmass '96, edited by D.G. Cassel, L.T. Gennari and R.H. Siemann (Stanford Linear Accelerator Center, Stanford, CA, 1997), pp. 627–631 [hep-ph/9609453].
178. B. Grzadkowski and J.F. Gunion, Phys. Lett. **B294**, 361 (1992) [hep-ph/9206262].
179. J.F. Gunion and J.G. Kelly, Phys. Lett. **B333**, 110 (1994) [hep-ph/9404343].
180. C.A. Boe, O.M. Ogreid, P. Osland and J. Zhang, Eur. Phys. J. **C9**, 413 (1999) [hep-ph/9811505].
181. J.F. Gunion and J. Kelly, "Using Higgs pair production in e^+e^- or $\mu^+\mu^-$ collisions to probe GUT-scale boundary conditions in the minimal supersymmetric model," in *New Directions for High Energy Physics*, Proceedings of the 1996 DPF/DPB Summer Study on High Energy Physics, Snowmass '96, edited by D.G. Cassel, L.T. Gennari and R.H. Siemann (Stanford Linear Accelerator Center, Stanford, CA, 1997), pp. 632–636 [hep-ph/9610421].
182. J.F. Gunion and J. Kelly, Phys. Rev. **D56**, 1730 (1997) [hep-ph/9610495].
183. J.L. Feng and T. Moroi, Phys. Rev. **D56**, 5962 (1997) [hep-ph/9612333].
184. V.D. Barger, T. Han and J. Jiang, Phys. Rev. **D63**, 075002 (2001) [hep-ph/0006223].
185. J.F. Gunion, T. Han, J. Jiang and A. Sopczak, hep-ph/0212151; J.F. Gunion, T. Han, J. Jiang, S. Mrenna and A. Sopczak, "Determination of tan beta at a future e^+e^- linear collider," in *Proceedings of the APS/DPF/DPB Summer Study on the Future of Particle Physics* (Snowmass 2001), edited by R. Davidson and C. Quigg, SNOWMASS-2001-P120 [hep-ph/0112334].
186. A. Freitas and D. Stöckinger, Phys. Rev. **D66**, 095014 (2002) [hep-ph/0205281].
187. A. Bartl, H. Eberl, S. Kraml, W. Majerotto, W. Porod, and A. Sopczak, Z. Phys. **C76**, 549 (1997) [hep-ph/9701336]; hep-ph/9909378; M. Berggren, R. Keränen, H. Nowak and A. Sopczak, hep-ph/9911345.
188. See, *e.g.*, S.Y. Choi, A. Djouadi, M. Guchait, J. Kalinowski and P.M. Zerwas, Eur. Phys. J. **C14**, 535 (2000) [hep-ph/0002033].
189. ATLAS Collaboration, "Detector and Physics Performance Technical Design Report", CERN/LHCC/99-15 (1999); CMS Collaboration, Technical Design Reports, CMS TDR 1-5 (1997/98).
190. P. Chen, G. Horton-Smith, T Ohgaki, A.W. Weidemann and K. Yokoya, Nucl. Instrum. Meth. **A355**, 107 (1995). See `.http://www-acc-theory.kek.jp/members/cain/cain21b.manual/main.html`.
191. P. Niezurawski, A.F. Zarnecki and M. Krawczyk, Acta Phys. Polon. **B34**, 177 (2003) [hep-ph/0208234].
192. J.F. Gunion, J.G. Kelly and J. Ohnemus, Phys. Rev. **D51**, 2101 (1995) [hep-ph/9409357].

CHAPTER 3

Top Quark Physics

Y. Sumino
Department of Physics, Tohoku University
Sendai, 980-8578 Japan

We review physics associated with the production and decay processes of top quarks at a future linear e^+e^- collider. We focus on rich physics to be covered there with an emphasis on key theoretical ideas.

1. Introduction

Among all the fermions included in the Standard Model (SM), the top quark plays a very unique role. The mass of the top quark is by far the largest and approximates the electroweak (EW) symmetry breaking scale. In fact the top quark is the heaviest of all the elementary particles discovered up to now. It means that in the SM Lagrangian the top quark mass term breaks the $SU(2)_L \times U(1)_Y$ symmetry maximally. This fact suggests that the top quark couples strongly to the physics that breaks the EW symmetry. It is therefore important to investigate properties of the top quark in detail, for the purpose of probing the symmetry breaking physics as well as to gain deeper understanding of the origin of the flavor structure. The standard procedures for investigating top quark properties are: measurements of fundamental quantities such as its mass and decay width; detailed examinations of various interactions of top quark to see if there are signs of new physics.

Recently studies of various properties of the top quark have been started at Tevatron. For example, polarization of W bosons from the decay of top quarks has been measured. Detailed properties will be investigated further in future experiments at Tevatron Run II, at LHC and at a future e^+e^- linear collider. Above all, the most fundamental parameter of the top quark is its mass. It is now measured to around ± 5 GeV accuracy. The recent reported values are

$$m_t = \begin{cases} 175.9 \pm 4.8 \pm 4.9 \text{ GeV} & \text{(CDF [1])} \\ 173.3 \pm 5.6 \pm 6.2 \text{ GeV} & \text{(D0 [2])} \end{cases} . \tag{1}$$

This value will be pinned down to an unmatched accuracy in an experiment in the top quark threshold region at a future linear collider.

The second fundamental parameter of the top quark is its decay width Γ_t. Within the SM the top quark decays almost 100% to the b quark and W. The width can be predicted as a function of m_t, α_s, M_W, etc. Already a fairly precise theoretical prediction at the level of a few percent accuracy is available [3]. Here, we only note that $\Gamma_t \simeq 1.5$ GeV for the above top quark mass range. This value deviates with simple extensions of the SM. For example, $\Gamma_t \propto |V_{tb}|^2$ can be smaller if there is a fourth generation, whereas it will be greater if there are additional decay modes such as $t \to bH^+$ or $t \to \tilde{t}\tilde{\chi}$. Thus, it is important to measure Γ_t. A linear collider will be capable to measure also this quantity more accurately than the hadron colliders.

An important property of the top quark, beside it is heavy, is that it decays so quickly that no top-hadrons will be formed. Therefore all the spin information of the top quark will be transferred to its decay daughters in its decay processes [4], and the energy-angular distributions of the decay products are calculable as purely partonic processes. In fact we may take full advantage of the spin information in studying the top quark properties through its decay processes in a clean environment expected at an e^+e^- collider [5].

We discuss separately physics of the top quarks in the threshold region $\sqrt{s} \simeq 2m_t$ (Section 2) and those in the open top region $\sqrt{s} \gg 2m_t$ (Section 3). In this article, particular emphasis is placed on the former, where we present an extensive review of the analyses performed so far. On the other hand, only limited subjects will be covered for the latter; we include a study on some strategical issue which has not been discussed in detail up to now.

2. Top Quark Threshold Region

An experiment in the top quark threshold region ($\sqrt{s} \approx 2m_t$) is considered as a serious candidate for the first stage operation of a future linear e^+e^- collider. In this section we review the subjects that can be covered in experiments in this region. It is shown that we will be able to extract rich physics and test properties of the top quark in detail.

2.1. *Physics Motivations and Goals*

The major three goals that should be achieved in experiments in the $t\bar{t}$ threshold region will be as follows:

1) A precision measurement of the top quark mass.

2) To elucidate dynamics involved in formations and decays of (remnant of) the toponium resonances.

3) Tests of the various top quark interactions: search for anomalies in the $t\bar{t}\gamma$, $t\bar{t}Z$, tbW, ttg and ttH couplings and possibly for other non-SM interactions.

Here, we briefly explain the physics motivation for each of these goals.

2.1.1. *Top Mass Determination*

The top quark mass m_t is one of the fundamental parameters of the SM. It is conceivable that, in the future, to know its precise value will be crucial to achieve a high predictive power in precision physics. In fact some important observables depend on m_t strongly through radiative corrections, when m_t acts as a non-decoupling parameter. Then it is required that m_t should be known more precisely than other parameters of the theory. For example, suppose that a light Higgs boson is discovered in the future, and suppose that we want to test predictions of its mass by the Minimal Supersymmetric Model within 50 MeV accuracy (comparable to the error of the Higgs mass measurement; see the chapter on Higgs Physics). Then we need to know the top quark mass with a similar accuracy due to the large dependence of the radiative correction on the top quark mass: $\delta m_H^2 \sim G_F m_t^4 \ln(m_{\tilde{t}_1} m_{\tilde{t}_2}/m_t^2)$.

2.1.2. *Testing Dynamics of $t\bar{t}$ Resonances*

We expect to observe intriguing phenomena of the (remnant of) $t\bar{t}$ resonances which are quite different from the phenomenology of charmonium and bottomonium. This is because: (1) The top quark decays via the EW interaction very quickly, and the decay daughters carry direct information on the dynamics of t and $\bar{t}$ inside the resonances. (2) Interplays of QCD and EW interactions induce unique phenomena. On the other hand, high precision theoretical predictions based on the first principles are available, due to the large mass and the large decay width of the top quark. Therefore, we will be able to make detailed tests of the dynamics of the boundstate formation and decays. This aspect is worth emphasizing, since predictability of the theory on the quarkonium physics has improved dramatically recently. For example, so far the quarkonium wave functions calculated in perturbative QCD have never been tested experimentally, and this will be possible for the $t\bar{t}$ resonances.

2.1.3. *Examinations of Various Top Quark Interactions*

Studies of various top quark interactions will be pursued both in the open top region, where we can explore more deeply into the structures of many of the interactions, and in the $t\bar{t}$ threshold region. There are specific advantages in each case. The advantages in the $t\bar{t}$ threshold region are as follows. First, to study the decay properties of the top quark, the threshold region will be optimal. This is because: the top quark can be polarized maximally by polarizing the electron beam; we are guaranteed to be (almost) at the rest frame of the top quark without need to reconstruct its momentum; we do not gain resolving power by raising the CM energy. In particular, the top quark polarization and the top quark decay width can be measured most precisely in the threshold region. Secondly, we can test not only the EW interactions of the top quark but also the QCD interactions. Since the threshold cross section is enhanced by QCD boundstate effects, it has a high sensitivity to the anomalous ttg couplings.[a]

2.2. *Top Quark Mass*

2.2.1. *How to Determine the Mass*

The top quark mass will be determined to high accuracy from the shape of the $t\bar{t}$ total production cross section in the threshold region [6]. If the top quark were stable, the cross section would show distinct resonance peaks in this region. Due to the large decay width of the top quark, $\Gamma_t \simeq$ 1.5 GeV, however, these peaks will be smeared out. The resonances merge with one another, leading to a broad enhancement of the cross section over the threshold region; see Fig. 1a below. Still, the location of a sharp rise of the cross section is determined mainly from the mass of the lowest-lying ($1S$) $t\bar{t}$ resonance. Since the mass of the resonance can be calculated as a

[a]The cross section includes an infinite number of ttg couplings at the leading order of the non-relativisitic expansion.

function of the top quark mass and α_s from perturbative QCD, we will be able to extract the top quark mass from the measurement of the $1S$ peak position of the cross section.

In early days, people planned to extract the pole mass of the top quark from the cross section. This is because the position of the "threshold" is determined as twice of the top quark pole mass, and it seemed to be most natural to relate this mass parameter directly to the threshold cross section or the resonance spectrum. On the other hand, existence of the pole mass of a quark contradicts the confinement picture of QCD and this mass is most likely to be ill-defined beyond perturbation theory; even within perturbative QCD we face a problem in defining this quantity more precisely than $\mathcal{O}(\Lambda_{QCD}) = \mathcal{O}(300\text{ MeV})$ accuracy due to the renormalon problem. Recently there has been significant progress in our theoretical understanding of the heavy quarkonia such as Υ's and (remnant of) toponium resonances. Developments in technologies of higher-order calculations [7–9] and the subsequent discovery of renormalon cancellation [10] enabled extractions of the top quark mass with high accuracy in future experiments [11]. It became clear that we have to relate the threshold cross section to a short-distance mass (e.g. the $\overline{\text{MS}}$ mass), which renormalizes only ultraviolet part of the radiative corrections, in order to extract a precise and physically meaningful mass parameter.

The upshot of the theoretical studies up to now is as follows. The series expansion of the mass of (remnant of) the $1S$ $t\bar{t}$ resonance in α_s becomes much more convergent when we express the series in terms of the $\overline{\text{MS}}$ mass of the top quark instead of the pole mass. At the present situation, theoretical uncertainty of the $\overline{\text{MS}}$ mass $\overline{m}_t \equiv m_t^{\overline{\text{MS}}}(m_t^{\overline{\text{MS}}})$ is about 150 MeV when we determine it from the threshold scan. Also, concerning future prospects, there seems to be a good chance to reduce the theoretical uncertainty to about 50 MeV by the time of the linear collider operation.

2.2.2. *Simulation Studies on Expected Precision*

There have been a number of simulation studies which incorporate realistic experimental conditions expected at future linear colliders [6, 12–14]. Here, we quote the result of the study [6, 12] on the determination of the peak position of the $t\bar{t}$ production cross section, from which the $\overline{\text{MS}}$ mass of the top quark $\overline{m}_t$ can be extracted. The effects of the beam energy spread and the beamstrahlung[b] as well as those of the initial state radiation are important when studying the threshold cross section. In particular the beam energy spread smears out the $1S$ peak of the cross section. In order to determine the peak position of the original cross section, we convolute the original cross section with the radiator function and with the beam energy spectrum, and then we make a fit to the measured cross section. Selection of top quark events can be performed with high efficiency and good background suppression. (More details will be explained in Section 2.4.) Taking these into account, a simulation study of the energy scan of the cross section was performed. Using 11 energy points with 1 fb^{-1} each, we can determine the peak position with a statistical error ≈ 200 MeV, which translates to a statistical error on the top quark $\overline{\text{MS}}$ mass of $\Delta\overline{m}_t \approx 100$ MeV.

In the remaining part of this subsection we review the theoretical background on the determination of the top quark mass.

2.2.3. *Renormalon Problem and Renormalon Cancellation*

As one may naively expect, generally convergence property of perturbative expansions become worse when there are contributions from soft gluons. This causes a poorly convergent behavior of the perturbative expansion of the resonance spectrum, if the pole mass is used.

The quarkonium spectrum is calculated by solving the non-relativistic Schrödinger equation

$$\hat{H}\,\psi_n(r) = E_n\,\psi_n(r), \tag{2}$$

where the quantum mechanical Hamiltonian is determined from perturbative QCD order by order in expansion in $1/c$ (inverse of the speed of light):

$$\hat{H} = \hat{H}_0 + \frac{1}{c}\,\hat{H}_1 + \frac{1}{c^2}\,\hat{H}_2 + \cdots . \tag{3}$$

Since the quark and antiquark inside a heavy quarkonium system are non-relativistic, the expansion in $1/c$ leads to a reasonable systematic approximation. The "renormalon problem" stems from the static QCD potential contained in the Hamiltonian:

$$V_{\text{QCD}}(r) \simeq -\int \frac{d^3\mathbf{q}}{(2\pi)^3}\, e^{i\mathbf{q}\cdot\mathbf{r}}\, C_F\, \frac{4\pi\alpha_s(|\mathbf{q}|)}{|\mathbf{q}|^2}. \tag{4}$$

[b]The beamstrahlung is a bremsstrahlung from beam particles due to the strong electromagnetic fields produced by opposing beam bunches.

The potential is essentially the Fourier transform of the Coulomb gluon propagator exchanged between quark and antiquark, where the coupling constant is replaced by the running coupling evaluated at the scale of the gluon momentum $|\mathbf{q}|$. When we examine the large-order behavior of the perturbative expansion of $V_{\rm QCD}(r)$ in $\alpha_s(\mu)$, we find that it is an asymptotic series and has an intrinsic uncertainty

$$\delta V_{\rm QCD}(r) \sim \Lambda_{\rm QCD} = \mathcal{O}(300 \text{ MeV}). \tag{5}$$

The uncertainty originates from the fact that the running coupling constant $\alpha_s(\mu)$ becomes large at $\mu \sim \Lambda_{\rm QCD}$ in the integrand of Eq. (4).

If we want to extract the quark mass from the spectrum of boundstates, this uncertainty in the potential is directly reflected to the uncertainty in the quark mass. It is because the quarkonium mass is determined as twice of the quark pole mass minus the binding energy, and an uncertainty in the potential means an uncertainty in the binding energy. This implies that we cannot determine the quark pole mass to an accuracy better than $\mathcal{O}(300 \text{ MeV})$.

This problem can be circumvented, however. Consider the total energy of a color-singlet non-relativistic quark-antiquark pair:

$$E_{\rm tot}(r) \simeq 2m_{\rm pole} + V_{\rm QCD}(r). \tag{6}$$

It was found [10] that the leading renormalon contained in the potential $V_{\rm QCD}(r)$ gets cancelled in the total energy $E_{\rm tot}(r)$ if the pole mass $m_{\rm pole}$ is expressed in terms of the $\overline{\rm MS}$ mass:

$$m_{\rm pole} \simeq \overline{m} + \frac{1}{2} \int\limits_{|\mathbf{q}|<\overline{m}} \frac{d^3\mathbf{q}}{(2\pi)^3} C_F \frac{4\pi\alpha_s(|\mathbf{q}|)}{|\mathbf{q}|^2}, \tag{7}$$

where $\overline{m} \equiv m_{\overline{\rm MS}}(m_{\overline{\rm MS}})$ is the renormalization-group invariant $\overline{\rm MS}$ mass. The difference of $m_{\rm pole}$ and $\overline{m}$ is essentially the infrared portion of the quark self-energy. The signs of the renormalon contributions are opposite between $V_{\rm QCD}(r)$ and $m_{\rm pole}$ because the color charges are opposite between q and $\overline{q}$ while the self-enregy is proportional to the square of the same charge. Their magnitudes differ by a factor of two because both the quark and antiquark propagator poles contribute in the calculation of the potential whereas only one of the two contributes to the mass-relation (7). Expanding the Fourier factor $e^{i\mathbf{q}\cdot\mathbf{r}}$ in a Taylor series for small $\mathbf{q}$, the leading renormalon contributions cancel in $E_{\rm tot}(r)$ [cf. Eqs. (4) and (7)].

As a result of this cancellation, the series expansion of the total energy $E_{\rm tot}(r)$ in $\alpha_s(\mu)$ became much more convergent if we use the $\overline{\rm MS}$

mass instead of the pole mass. This suggests that the $\overline{\rm MS}$ mass has a more natural relation to physical quantities of a static (or non-relativistic) quark-antiquark system. Residual uncertainties due to the uncancelled singularity in Eq. (4) is suppressed as

$$\Lambda_{\rm QCD} \times \langle (\mathbf{q}\cdot\mathbf{r})^2 \rangle \sim \Lambda_{\rm QCD} \times \left(\frac{\Lambda_{\rm QCD}}{\alpha_s m_t} \right)^2 , \tag{8}$$

which is much smaller than the original uncertainty.[c]

2.2.4. *Theoretical Prediction for the* 1*S* *Energy Level*

To see how well the renormalon cancellation works, we examine the perturbation series of the energy level of the (remnant of) toponium $1S$ state. Taking the input parameter as $m_{t,\rm pole} = 174.79$ GeV/$\overline{m}_t = 165.00$ GeV and setting $\mu = \overline{m}_t$ [i.e. expansion parameter is $\alpha_s(\overline{m}_t) = 0.1092$], we obtain

$$M_{1S} = 2 \times \begin{cases} (174.79 - 0.46 - 0.39 - 0.28 - 0.19^*) \text{ GeV} & \text{(Pole-mass scheme)} \\ (165.00 + 7.21 + 1.24 + 0.22 + 0.052^*) \text{ GeV} & (\overline{\rm MS} \text{ scheme}). \end{cases}$$

The numbers with stars $(*)$ are calculated in the large-β_0 approximation, which can be regarded as reasonable order-of-magnitude estimates of the corresponding terms [15]. In Figs. 1 are shown the convergence properties of the above series together with the corresponding total cross sections. The series expansion converges very slowly when the $1S$ energy level is expressed using the pole mass. On the other hand, when the energy level is expressed by the $\overline{\rm MS}$ mass, the series shows a healthy convergence behavior. Non-perturbative uncertainties are estimated to be much smaller than the present perturbative theoretical uncertainty [16]. Thus, from the above series we estimate that $\overline{m}_t$ can be determined to about 100 MeV accuracy from the $1S$ energy level with the present theoretical accuracy, and to below 50 MeV accuracy if the computation of the term with the star is completed.

Let us also examine the dependence of the $1S$ energy level $M_{1S}(\alpha_s(\mu), \overline{m}_t)$ on the scale μ. As shown in Fig. 2 the scale dependence decreases as we include more terms of the series expansion. The scale dependences of the series are also consistent with the estimates of the theoretical uncertainties given above. The convergence property of the series becomes optimal with a choice of the scale $\mu \sim 50$ GeV.

[c] Typical size of the $t\bar{t}$ resonances is given by the Bohr radius of the system, $\langle r \rangle \sim 1/(\alpha_s m_t) \sim 1/(20 \text{ GeV})$.

a)

b)

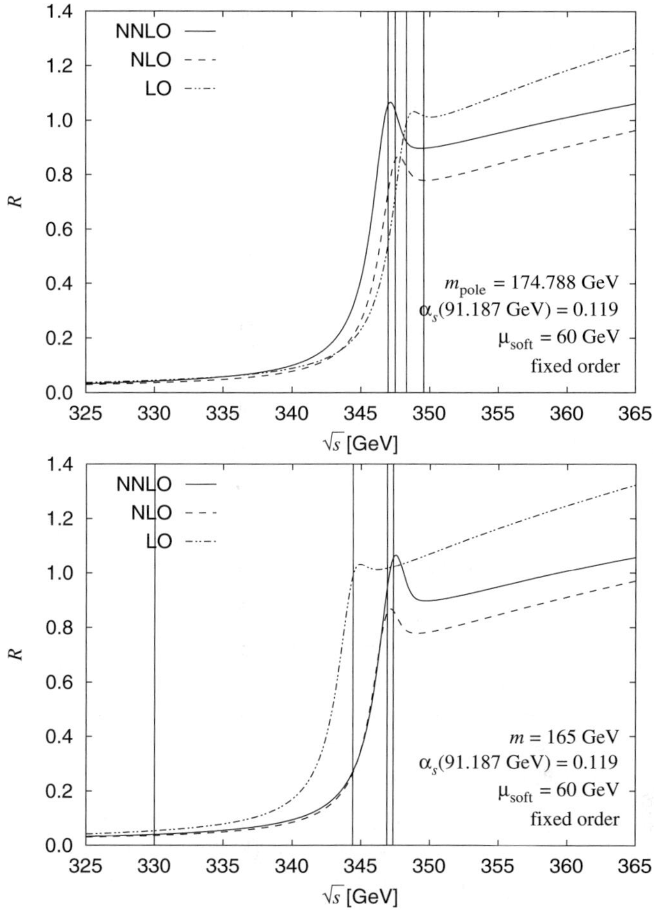

Fig. 1. The total production cross sections for $e^+e^- \to t\bar{t}$ in the threshold region, where the leading-order (LO), next-to-leading-order (NLO) and next-to-next-to-leading-order (NNLO) curves are shown. The vertical lines represent the locations of the corresponding $1S$ resonances when the top width is artificially taken to zero; also the position of $2m_t$ is shown by vertical lines. The two figures correspond to (a) the pole-mass scheme, and (b) the $\overline{\text{MS}}$-scheme.

2.2.5. *Physical Implications*

Let us discuss some physical implications of the renormalon cancellation in the heavy quarkonium system. Firstly, we expect that gluons with wavelength much longer than the size of the quarkonium cannot couple to this system. Hence, we expect that infrared gluons with momentum transfer

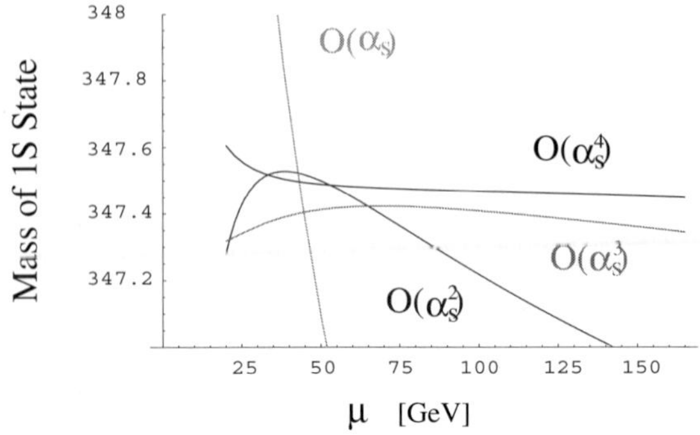

Fig. 2. μ-dependences of the sums of the series expansion for the $1S$ energy level. The $\mathcal{O}(\alpha_s^4)$ term is evaluated in the large-β_0 approximation.

$q \ll \alpha_s m$ should decouple from the expression of $E_{\rm tot}$. This is a naive expectation based on classical dynamics. Such understanding should be valid when it is described by the bare QCD Lagrangian without large quantum corrections. The $\overline{\rm MS}$ mass is closely related to the bare mass of a quark; only ultraviolet divergences are subtracted. On the other hand, the pole mass has much more intricate relation to the bare mass, because the relation includes in addition infrared dynamics of the quantum corrections to the quark self-energy. In this sense it would be natural to expect the decoupling to be realized when the $\overline{\rm MS}$ mass is used to express $E_{\rm tot}$.

Secondly, the pole mass of a quark is ill-defined beyond perturbation theory. It can be determined only when the quark can propagate an infinite distance. Generally accepted belief is that when a quark and an antiquark are separated beyond a distance $\sim \Lambda_{\rm QCD}^{-1}$ the color flux is spanned between the two charges due to non-perturbative effects and the free quark picture is no longer valid. On the other hand, it is natural to consider the total energy (or the mass) of the quarkonium which is a color-singlet state. It can propagate for a long time and the notion of its mass is not limited by the hadronization scale.

Thirdly, renormalon cancellation seems to be a universal feature which occurs in many processes. The same phenomenon was known e.g. in the QCD corrections to the ρ-parameter [17] and B decays [18]. Therefore, the $\overline{\rm MS}$ mass of the top quark, which is determined accurately from the threshold scan, is more suited than the pole mass for an input parameter in describing other physical processes.

2.2.6. *Remaining Theoretical Uncertainties*

At the present situation, the largest theoretical uncertainty originates from the uncertainty of the strong coupling constant $\alpha_s(M_Z)$: it induces an uncertainty in the relation between the $1S$ energy level and $\overline{m}_t$. The present theoretical uncertainty $\Delta\alpha_s(M_Z) = 0.0020$ corresponds to $\Delta\overline{m}_t \approx 150$ MeV. So, we are already in a position to determine the top quark mass with this accuracy once the threshold scan will be performed.

Concerning future prospects, the simulation studies show that the statistical error on $\overline{m}_t$ can be reduced to 50 MeV with a moderate integrated luminosity of about 40 fb^{-1}. On the theoretical side, there are corrections that should be computed before we achieve 50 MeV accuracy. These are:

- The 4-loop relation between the pole mass and the $\overline{\text{MS}}$ mass.
- The final-state interaction corrections (non-factorizable corrections) to the $e^+e^- \to t\bar{t}$ total cross section at $\mathcal{O}(1/c^2)$.
- The EW corrections to the cross section. These include the contributions from the tree-level non-resonant-type diagrams and the one-loop resonant-type corrections.

In addition, there remains a large uncertainty, 10–15%, in the normalization of the cross section, originating from the large next-to-next-to-leading order corrections to it; [11, 19, 20]; see (Fig. 1). The uncertainty in the normalization can affect the measurement of the top quark mass when it is energy dependent. The smearing effects by the beam energy spread necessitates a precise knowledge of the energy dependence of the threshold cross section in order to reconstruct the $1S$ peak position. Therefore, reduction of this theoretical uncertainty is needed. Last but not least, the strong coupling constant needs to be determined at the level of $\Delta\alpha_s(M_Z) = 0.0007$ from other experiments.

2.3. *Dynamics and Observables*

In this section we discuss the dynamics of t and $\bar{t}$ within SM close to the top quark threshold. Also we survey the observables that contain interesting information. In this section, we restrict ourselves to qualitative arguments. Quantitative analyses will be given in the next section.

2.3.1. *Time Evolution*

It is illuminating to consider the time evolution of this system: a $t\bar{t}$ pair produced in e^+e^- annihilation; (see Fig. 3). The t and $\bar{t}$ are produced

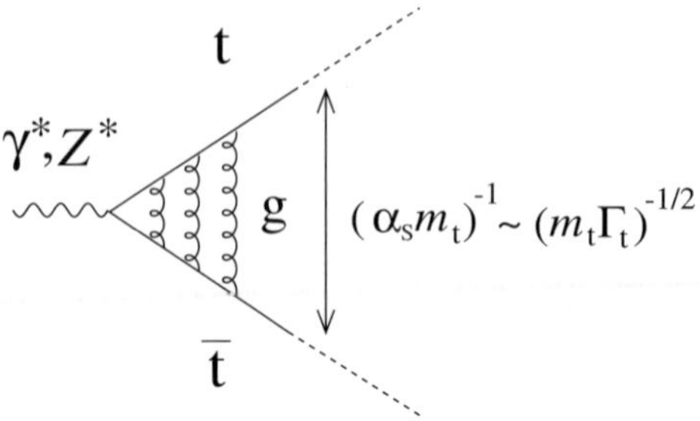

Fig. 3. The time evolution of the $t\bar{t}$ system.

close to each other, and they start to spread apart. Since they are slow near the threshold, they cannot escape even the relatively weak attractive force mediated by the exchange of gluons; t and $\bar{t}$ are bound to form Coulombic resonances when they reach the distance of the Bohr radius $(\alpha_s m_t)^{-1} \sim 0.1\ \text{GeV}^{-1}$. At this stage, the size of the coupling of t and $\bar{t}$ to gluons is $\alpha_s(\mu = \alpha_s m_t) \sim 0.15$. If they could continue to spread apart even further to the distance $\Lambda_{\text{QCD}}^{-1} \sim$ a few GeV^{-1}, there would occur hadronization effects as the coupling becomes very strong, since gluons with wavelength $\sim \Lambda_{\text{QCD}}^{-1}$ would be able to resolve the color charge of each constituent. For a realistic top quark, however, the $t\bar{t}$ pair will decay at the distance $(m_t\Gamma_t)^{-1/2} \sim 0.1\ \text{GeV}^{-1}$ into energetic b and $\bar{b}$ jets and W's well before hadronization effects become important. Thus, the toponium can be regarded as a Coulombic resonance state (with reasonably weak coupling) due to the large mass and the large width of the top quark. Since infrared gluons decouple from the resonances, the strong interaction participating in the formation of the boundstate is dictated by the perturbative domain of QCD. Due to this reason, we are able to compute the amplitude from the first principles with order 5% accuracy or better, even though the QCD boundstates are involved.[d] The spin and PC of the dominantly produced boundstate are $J^{PC} = 1^{--}$. Inside this boundstate: t and $\bar{t}$ are in the S-wave state ($L = 0$); the spins of t and $\bar{t}$ are aligned to each other and pointing to e^- beam direction $|\uparrow\uparrow\rangle$ or to e^+ beam direction $|\downarrow\downarrow\rangle$ or they are in a linear combination of the two states ($S = 1$).

[d]At the moment only one exception is the normalization of the total $t\bar{t}$ cross section, where we still have 10–15% theoretical uncertainties [11]. On the other hand, the resonance masses can be predicted most accurately, at the level of 10^{-3} accuracy or better.

The toponium resonances decay dominantly via the EW interaction. We may regard, in the first approximation, the decays of the resonances as partonic decays of the individual t and $\bar{t}$ inside the resonances. Thus, the momenta and spins of t and $\bar{t}$ inside the resonances determine the distributions of the momenta and spins of the decay products.

How do we describe the time evolution of the system theoretically? We first determine the quantum mechanical Hamiltonian of the system, such that it generates (within the framework of quantum mechanics) the $t\bar{t}$ scattering amplitude which coincides the scattering amplitude computed from perturbative QCD in the non-relativistic region. Presently the Hamiltonian is known to $\mathcal{O}(1/c^2)$ accuracy [7, 21]:

$$\hat{H} = \hat{H}_0 + \frac{1}{c}\,\hat{H}_1 + \frac{1}{c^2}\,\hat{H}_2 + \cdots\,, \tag{9}$$

$$\hat{H}_0 = \frac{\mathbf{p}^2}{m_t} - C_F\,\frac{\alpha_s}{r}, \tag{10}$$

$$\hat{H}_1 = -\,C_F\,\frac{\alpha_s}{r}\left(\frac{\alpha_s}{4\pi}\right)\Big\{\beta_0\,\log(\mu'^2 r^2) + a_1\Big\}, \tag{11}$$

$$\begin{aligned}\hat{H}_2 =& -\frac{\mathbf{p}^4}{4m_t^3} - C_F\,\frac{\alpha_s}{r}\left(\frac{\alpha_s}{4\pi}\right)^2\Big\{\beta_0^2\,[\log^2(\mu'^2 r^2) + \frac{\pi^2}{3}] \\ &\qquad + (\beta_1 + 2\beta_0 a_1)\log(\mu'^2 r^2) + a_2\Big\} \\ &+ \frac{\pi C_F\alpha_s}{m_t^2}\,\delta^3(\mathbf{r}) + \frac{3C_F\alpha_s}{2m_t^2 r^3}\,\mathbf{L}\cdot\mathbf{S} - \frac{C_F\alpha_s}{2m_t^2 r}\left(\mathbf{p}^2 + \frac{1}{r^2}r_i r_j p_j p_i\right) \\ &- \frac{C_A C_F\alpha_s^2}{2m_t r^2} - \frac{C_F\alpha_s}{2m_t^2}\Big\{\frac{S^2}{r^3} - 3\frac{(\mathbf{S}\cdot\mathbf{r})^2}{r^5} - \frac{4\pi}{3}(2S^2-3)\delta^3(\mathbf{r})\Big\},\end{aligned} \tag{12}$$

where m_t denotes the pole mass; $\alpha_s \equiv \alpha_s(\mu)$; $C_F = 4/3$, $C_A = 3$ are color factors; $\mu' = \mu\, e^{\gamma_E}$. The lowest-order Hamiltonian $\hat{H}_0$ represents interaction of non-relativistic t and $\bar{t}$ via the Coulomb potential.

In quantum field theory the total $t\bar{t}$ production cross section is given, via the optical theorem, by the imaginary part of the top quark contribution to the photon or Z vacuum polarization function:

$$\sigma_{t\bar{t}}(s) \propto \text{F.T. Im}\,\langle 0|\,T\,\bar{t}(x)\gamma^\mu(\gamma_5)t(x)\,\bar{t}(0)\gamma_\mu(\gamma_5)t(0)\,|0\rangle\,, \tag{13}$$

where F.T. stands for the Fourier transform. In the threshold region, we replace the vacuum polarization function by the Green function of the non-relativistic Schrödinger equation determined from the above Hamiltonian:

$$G(\mathbf{0},\mathbf{0};E) = \text{F.T.} \lim_{\mathbf{r},\mathbf{r}'\to\mathbf{0}} \langle\mathbf{r}|\, e^{-i(\hat{H}-i\Gamma_t)T}\,|\mathbf{r}'\rangle \tag{14}$$

$$= \lim_{\mathbf{r},\mathbf{r}'\to\mathbf{0}} \langle\mathbf{r}|\, \frac{-1}{E-\hat{H}+i\Gamma_t}\,|\mathbf{r}'\rangle\,, \tag{15}$$

where $\mathbf{r},\mathbf{r}'$ denote the relative coordinate between t and $\bar{t}$. The Green function $G(\mathbf{0},\mathbf{0};E)$ represents the propagation of t and $\bar{t}$, which are pair created at a same point at time $t=0$, and which pair annihilate at some point at time $t=T>0$. The ultraviolet divergences that arise when taking the limit $\mathbf{r},\mathbf{r}'\to\mathbf{0}$ are absorbed by renormalization of the $t\bar{t}\gamma$ and $t\bar{t}Z$ vertices. The above Green function incorporates the boundstate effects in the threshold region, $\beta \lesssim \alpha_s \ll 1$. ($\beta$ is the velocity of t or $\bar{t}$.) More precisely, in this way we resum systematically the Coulomb singularities, α_s^n/β^n, α_s^{n+1}/β^n, α_s^{n+2}/β^n, ..., which consist in the $t\bar{t}$ production cross section.

Also, we can show that the $t\bar{t}$ production vertex is proportional to the Green function in momentum space

$$\Gamma^\mu_{t\bar{t}} \propto \tilde{G}(\mathbf{p}_t;E) = \int d^3\mathbf{r}\; e^{-i\mathbf{p}_t\cdot\mathbf{r}}\, G(\mathbf{0},\mathbf{r};E) \tag{16}$$

$$= -\sum_n \frac{\phi_n(\mathbf{p}_t)\psi_n^*(\mathbf{0})}{E-E_n+i\Gamma_t}, \tag{17}$$

where

$$\hat{H}\,|n\rangle = E_n\,|n\rangle\,, \qquad \begin{aligned} \phi_n(\mathbf{p}) &= \langle\mathbf{p}|n\rangle \\ \psi_n(\mathbf{x}) &= \langle\mathbf{x}|n\rangle \end{aligned} \tag{18}$$

defines the wave functions of the energy eigenstate of the Hamiltonian. The above Green function represents the propagation of t and $\bar{t}$, from time $t=0$ when they are pair created at a same point, until time $t=T$ when they are separated by $\mathbf{r}$ and either of the two particles decays.

2.3.2. *Production Process of Top Quarks*

Total Cross Section

The first observable we measure in the $t\bar{t}$ threshold region will be the total cross section. The total cross section can be written as [22, 23]

$$\sigma_{\text{tot}}(e^+e^-\to t\bar{t}) \propto -\text{Im}\sum_n \frac{|\psi_n(\mathbf{0})|^2}{E-E_n+i\Gamma_t}. \tag{19}$$

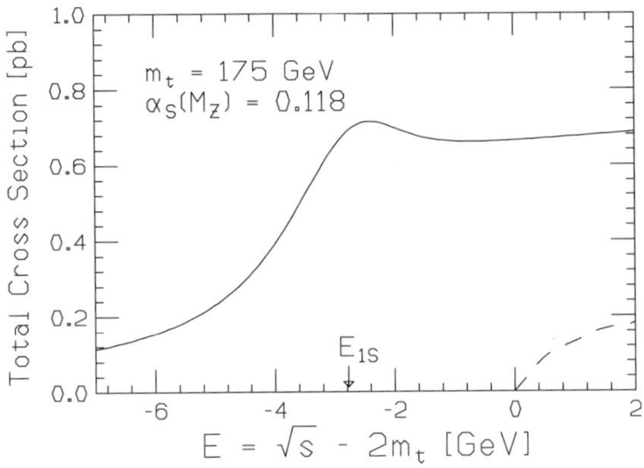

Fig. 4. The total cross section vs. energy, $E = \sqrt{s} - 2m_t$. The solid curve is calculated from the Green function. The dashed curve shows the tree-level total cross section for a stable top quark.

One sees that the energy dependence of the total cross section is determined by the resonance spectra. Due to the large width Γ_t of the top quark, however, distinct resonance peaks are smeared out; (see Fig. 4). (We will show explicitly the resonance spectra below.) In the same figure, the tree level cross section is also shown as a dashed curve. Despite the disappearance of each resonance peak, one sees that the cross section is indeed largely enhanced by the QCD interaction, and that inclusion of the QCD binding effect is mandatory for a proper account of the cross section in the threshold region.

Top Momentum Distribution

Next we consider the top-quark momentum ($|\mathbf{p}_t|$) distribution near $t\bar{t}$ threshold [24, 25]. It has been shown that experimentally it will be possible to reconstruct the top-quark momentum $\mathbf{p}_t$ from its decay products with reasonable resolution and detection efficiency. Figure 5(a) shows a comparison of reconstructed top momenta (solid circles) with that of generated ones (histogram), where the events are generated by a Monte Carlo generator and are reconstructed from the data from detector simulators with appropriate selection cuts; see Section 2.4 for details. The figure demonstrates that the agreement is fairly good.

Theoretically, the top-quark momentum distribution is given by

$$\frac{d\sigma}{d|\mathbf{p}_t|} \propto \left| \sum_n \frac{\phi_n(\mathbf{p}_t)\psi_n^*(\mathbf{0})}{E - E_n + i\Gamma_t} \right|^2 + (\text{sub-leading}). \tag{20}$$

The $|\mathbf{p}_t|$-distribution is thus governed by the momentum-space wave functions of the resonances. By measuring the momentum distribution, essentially we measure (a superposition of) the wave functions of the toponium resonances. Shown in Fig. 5(b) are the top momentum distributions for various energies. One may also vary the magnitude of α_s and confirm that the distribution is indeed sensitive to the resonance wave functions [24, 25]. Hence, the momentum distribution provides information independent of that from the total cross section.

Note that the toponium states will be the first quarkonium resonances whose wave functions can be measured experimentally. For comparison, consider $\Upsilon(4S)$, which decays into B and $\overline{B}$. Since the mass of $B(\overline{B})$ is fixed, its momentum is fixed by the on-shell condition; the momentum distribution of $B(\overline{B})$ is a δ-function in this case. Meanwhile, in the case of toponium, the invariant mass distribution of top quarks has a large width, and accordingly the top-quark three momenta have a distribution that is just sufficiently broad for probing the wave functions of the resonances.

Forward-Backward Asymmetry

Another observable that can be measured experimentally is the forward-backward asymmetry of the top quark [26]. Generally in a fermion pair production process, a forward-backward asymmetric distribution originates from an interference of the vector and axial-vector $f\overline{f}$ production vertices at tree level of electroweak interaction. One can show from the spin-parity argument that in the threshold region the $t\bar{t}$ vector vertex creates S-wave resonance states, while the $t\bar{t}Z$ axial-vector vertex creates P-wave states. Therefore, by observing the forward-backward asymmetry of the top quark, we observe an interference of the S-wave and P-wave states.

In general, S-wave resonance states and P-wave resonance states have different energy spectra. So if the CM energy is fixed at either of the spectra, there would be no contribution from the other. However, the widths of resonances are large for the toponium in comparison to their level splittings, which permit sizable interferences of the S-wave and P-wave states.[e]

[e]Note that no forward-backward asymmetry is observed for charmonium or bottomonium states because the widths of the resonances are too small compared to their level splittings. Thus, the asymmetry reveals to be another observable unique to the toponium states.

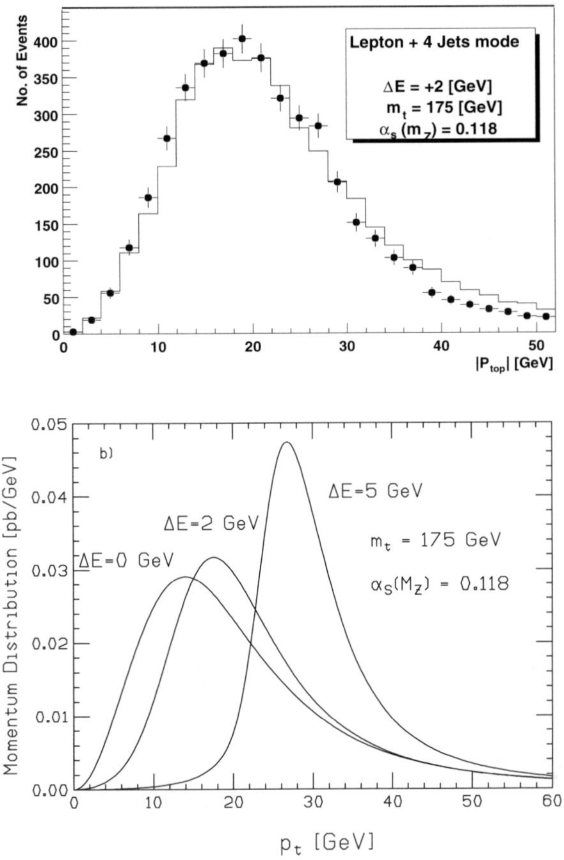

Fig. 5. (a) Reconstructed momentum distribution (solid circles) for the lepton-plus-4-jet mode, compared with the generated distribution (histogram). The Monte Carlo events were generated with $\alpha_s(M_Z) = 0.118$ and $m_t = 175$ GeV. (b) Top-quark momentum distributions $d\sigma/d|\mathbf{p}_t|$ for various CM energies measured from the lowest lying resonance, $\Delta E = \sqrt{s} - M_{1S}$.

Figure 6 shows the pole position $E_n - i\Gamma_t$ of these states on the complex energy plane. One sees that the widths of the resonances are comparable to the mass difference between the lowest lying S-wave and P-wave states, and exceeds by far the level spacings between higher S-wave and P-wave states. This gives rise to a forward-backward asymmetry even below threshold, and

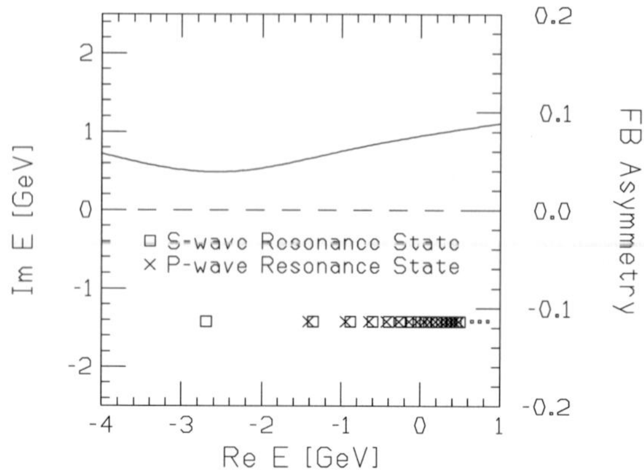

Fig. 6. The positions of poles of the S-wave and P-wave states on the complex energy plane, together with the forward-backward asymmetry as a function of the energy, taking $\alpha_s(M_Z) = 0.118$ and $m_t = 175$ GeV. The right-axis is for the forward-backward asymmetry.

provides information on the resonance level structure which is concealed in the total cross section. Shown on the same figure is the forward-backward asymmetry as a function of the energy. It is seen that the asymmetry takes its minimum value at around the lowest lying S-wave state, where the interference is smallest, and increases up to $\sim$ 10% with energy as the resonance spectra appear closer to each other. One may also confirm that essentially the forward-backward asymmetry measures the degree of overlap of the S-wave and P-wave states by varying the coupling constant α_s or the top quark decay width Γ_t.

The resonance level structure is determined by QCD, whereas the resonance widths are determined by electroweak interaction. The interplay of the two interactions generates the forward-backward asymmetry.

Polarization

Top quarks are highly polarized in the threshold region due to kinematical constraints specific to this region. The intermediate vector bosons (γ, Z) are in $J = 1$ state, pointing to the e^- or e^+ beam direction. Close to threshold, $t\bar{t}$ pairs are produced dominantly in the S-wave ($L = 0$) state, since there is only small available phase space. Thus, the total angular-momentum $J = 1$ is transferred to the sum of the t and $\bar{t}$ spins, so they are parallel to each other and pointing to the e^- beam direction $|\uparrow\uparrow\rangle$ or to the e^+ beam direction $|\downarrow\downarrow\rangle$, or they are in a linear combination of the

Table 1. The polarization of top quarks in the $t\bar{t}$ threshold region for a linearly polarized e^- beam and an unpolarized e^+ beam ($P_{e^+} = 0$).

P_{e^-}	P_t
+80 %	$\sim$ +60 %
0 %	$\sim$ −40 %
−80 %	$\sim$ −90 %

two states. Assuming a linearly polarized e^- beam and an unpolarized e^+ beam, it is possible to raise the polarization of top quarks higher than the e^- polarization (see Table 1). The asymmetry of P_t with respect to left- and right-handed electrons in the table can be understood as follows: the couplings of e^-_L and e^-_R to the Z boson are different, so the weight factors of the states $|\uparrow\uparrow\rangle$ and $|\downarrow\downarrow\rangle$ are different even if $P_{e^-} = 0$.

As we have seen already, there is about 10% admixture of the P-wave states in the top quark production. This modifies the top quark polarization vector by about 10% through S–P interferences. It is convenient to express the top quark polarization vector $\mathbf{S}$ in components:

$$\mathbf{S} = S_{\|}\,\mathbf{n}_{\|} + S_{\perp}\,\mathbf{n}_{\perp} + S_{\mathrm{N}}\,\mathbf{n}_{\mathrm{N}}, \tag{21}$$

where the orthonormal basis is defined from the e^- beam direction and the top quark momentum direction as shown in Fig. 7

$$\mathbf{n}_{\|} = \frac{\mathbf{p}_e}{|\mathbf{p}_e|}, \quad \mathbf{n}_{\mathrm{N}} = \frac{\mathbf{p}_e \times \mathbf{p}_t}{|\mathbf{p}_e \times \mathbf{p}_t|}, \quad \mathbf{n}_{\perp} = \mathbf{n}_{\mathrm{N}} \times \mathbf{n}_{\|}. \tag{22}$$

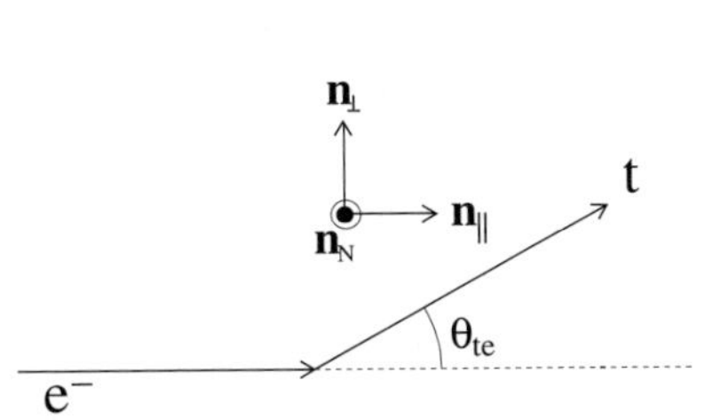

Fig. 7. The definition of the orthonormal basis for the top quark spin. The normal component ($\mathbf{n}_{\mathrm{N}}$) points out of the plane.

The components are given by [27, 28]

$$S_{\parallel} = C_{\parallel}^{0} + C_{\parallel}^{1}\,\mathrm{Re}\Big[\tilde{F}/\tilde{G}\Big]\,\beta\cos\theta_{te}, \tag{23}$$

$$S_{\perp} = C_{\perp}\,\mathrm{Re}\Big[\tilde{F}/\tilde{G}\Big]\,\beta\sin\theta_{te}, \tag{24}$$

$$S_{\mathrm{N}} = C_{\mathrm{N}}\,\mathrm{Im}\Big[\tilde{F}/\tilde{G}\Big]\,\beta\sin\theta_{te}. \tag{25}$$

Here, $C_{\parallel}^{i}$, $C_{\perp}$ and C_{N} denote combinations of the electroweak couplings of e^- and t as well as of $e^{\pm}$ beam polarization. As can be understood from the above argument, $C_{\parallel}^{0} = \pm 1$ when $P_{e^-} = \pm 100\%$. $\tilde{G}$ and $\tilde{F}$ denote, respectively, the S-wave and P-wave Green functions which incorporate the QCD boundstate effects. The ratio $\tilde{F}/\tilde{G}$ represents the degree of S–P interference (with respect to the leading-order cross section $\propto |\tilde{G}|^2$): its real part in $S_{\parallel}$ and $S_{\perp}$ provides the same information that can be extracted from the forward-backward asymmetry; the imaginary part in S_{N} provides new information, which essentially measures the rate of gluon exchanges between t and $\bar{t}$ inside the resonances.[f] The above components of the top quark polarization vector can be extracted using distributions of the decay products of top quarks, as we will see below.

Higgs Effects

If the Higgs particle turns out to be relatively light, $M_H \lesssim 150$ GeV, there is a chance to extract the top-Higgs Yukawa coupling of the top quark from the threshold cross section [29]. Qualitatively, the effect of Higgs exchange between t and $\bar{t}$ can be incorporated by adding the Yukawa potential

$$V_{\mathrm{Yuk}}(r) = -\frac{g_{tH}^2}{4\pi}\,\frac{e^{-M_H r}}{r} \tag{26}$$

to the quantum mechanical Hamiltonian (9). The effect is larger for smaller M_H and larger g_{tH}. Note that the Higgs mass should not be taken as a parameter, since if the Higgs effect is of observable size, it will already have been found and M_H will have been determined; see the chapter on Higgs physics. As the Yukawa potential is much short-ranged compared to the toponium resonance size, it does not affect the resonance spectra but makes the wave function at the origin significantly larger. Thus, only the normalization of the total cross section is changed, while the momentum distribution of top quark is quite insensitive to the Yukawa coupling.

[f] S_{N} is odd under $\mathcal{T}_N$ (naive time-reversal tranformation). Hence, it vanishes at tree level.

At the moment, theoretical uncertainties to the normalization of the $t\bar{t}$ cross section are large, of the order of 10–15%, which makes extraction of the Higgs effects difficult.

2.3.3. *Decay of Top Quarks and Final-State Interactions*

We turn to decay processes of the top quark near threshold. The top quarks produced via $e^+e^- \to t\bar{t}$ in the threshold region will be highly polarized [4] (see Table 1). Therefore, the threshold region can be an ideal place for studying properties of the top quark through decay processes, using the highly polarized top quark sample.

Decay of Free Polarized Top Quarks

Detailed studies of the decay of *free* polarized top quarks have already been available including the full $\mathcal{O}(\alpha_s)$ corrections [30–32]. The best example is the energy-angular distribution of charged leptons l^+ in the semi-leptonic decay of the top quark. At leading order, the l^+ distribution has a form where the energy and angular dependences are factorized [33, 34]:

$$\frac{d\Gamma_{t\to bl^+\nu}(\mathbf{S})}{dE_l d\Omega_l} = h(E_l)\,(1+|\mathbf{S}|\cos\theta_l) \quad + \quad [\,\mathcal{O}(\alpha_s)\text{ correction}\,]. \tag{27}$$

Here E_l, Ω_l, and θ_l denote, respectively, the l^+ energy, the solid angle of l^+, and the angle between the l^+ direction and the top polarization vector $\mathbf{S}$, all of which are defined in the top-quark rest frame. Hence, we may measure the top-quark polarization with maximal sensitivity using the l^+ angular distribution.

Effects of Final-State Interactions

Close to threshold, the above precise analyses of the free top-quark decays do not apply directly because of the existence of corrections unique to this region. Namely, these are the final-state interactions due to gluon exchange between t and $\bar{b}$ ($\bar{t}$ and b) or between b and $\bar{b}$. (Fig. 8) The size of the corrections is at 10% level in the threshold region, hence it is necessary to incorporate their effects in precision studies of top-quark production and decay near threshold.

Before presenting the formula for these final-state interaction corrections, let us first see what kind of effects we expect from physics ground [28].

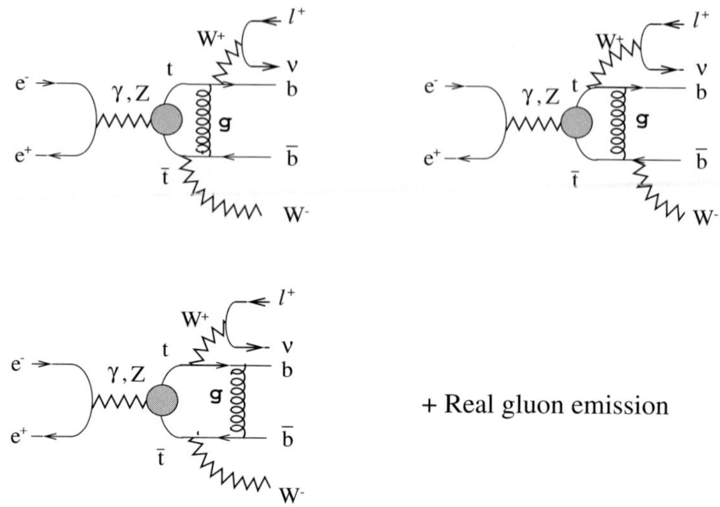

Fig. 8. Diagrams for the final-state interactions for $e^+e^- \to t\bar{t} \to bl^+\nu\bar{b}W^-$.

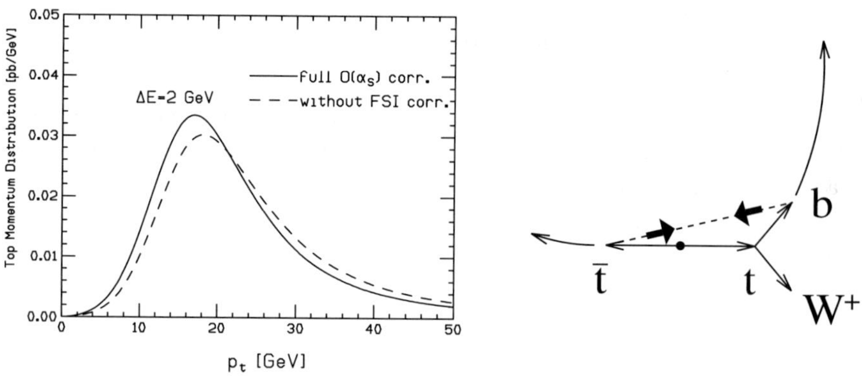

Fig. 9. (a) The top momentum distribution with (solid) and without (dashed) the final-state interaction corrections for $\alpha_s(M_Z) = 0.118$ and $m_t = 175$ GeV. (b) Attractive force between $\bar{t}$ and b (from t decay). The momentum transfer $\delta\mathbf{p}_b = -\delta\mathbf{p}_{\bar{t}}$ due to the attraction is indicated by thick arrows.

• *Top Momentum Distribution*

Perhaps it is easiest to understand the effect of final-state interactions on the top momentum distribution. Figure 9(a) shows the momentum distribution with (solid) and without (dashed) the final-state interactions. We see that the average momentum is reduced due to the interaction. To understand this, consider for example the case where t decays first. Figure 9(b) shows the attractive force between $\bar{t}$ and b, which deflects the trajectory of b. Since $\mathbf{p}_t$ is reconstructed from the bW^+ momenta at time $t \to \infty$, it is obvious that the reconstructed momentum $|\mathbf{p}_t| = |\mathbf{p}_b + \mathbf{p}_{W^+}|$ is decreased by the attraction.

- *Forward-Backward Asymmetric Distribution*

Next we consider the $\cos\theta_{te}$ distribution of the top quark. (θ_{te} denotes the angle between t and e^- in the $t\bar{t}$ CM frame.) We consider the case where $\bar{t}$ decays first and examine the interaction between t and $\bar{b}$. The t and $\bar{t}$ pair-produced near threshold in e^+e^- collisions have their spins approximately parallel or anti-parallel to the e^- beam direction ($\hat{\mathbf{n}}_\parallel$) and the spins are always oriented parallel to each other. On the other hand, the decay of $\bar{t}$ occurs via $V{-}A$ coupling, and $\bar{b}$ is emitted preferably in the spin direction of the parent $\bar{t}$, see Fig. 10. More precisely, the excess of the $\bar{b}$'s emitted in the $\bar{t}$ spin direction over those emitted in the opposite direction is given by $\kappa = (m_t^2 - 2M_W^2)/(m_t^2 + 2M_W^2)$. Now suppose t and $\bar{t}$ have their spins in the $\hat{\mathbf{n}}_\parallel$ direction. Then $\bar{b}$ will be emitted dominantly in the $\hat{\mathbf{n}}_\parallel$ direction. One can see from Fig. 11(a) that in this case t is always attracted to the forward direction due to the attractive force between t and $\bar{b}$. The direction of the attractive force will be opposite if t and $\bar{t}$ have their spins in the $-\hat{\mathbf{n}}_\parallel$ direction (Fig. 11(b)). Thus, polarized top quarks will be pulled in a definite (forward or backward) direction, and we may expect that a forward-backward asymmetric distribution of the top quark $\sim \kappa\, S_\parallel \cos\theta_{te}$ is generated by the final-state interaction.

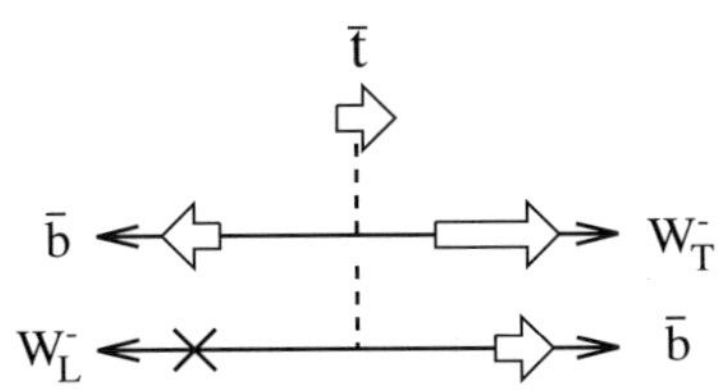

Fig. 10. Typical configurations in the decay of $\bar{t}$ with definite spin orientation. Transverse W^- (W_T^-) tends to be emitted in the direction of the $\bar{t}$ spin orientation, while longitudinal W^- (W_L^-) is emitted in the opposite direction due to helicity conservation. For $m_t \simeq 175$ GeV, $\bar{t}$ decays mainly to W_L^-, hence $\bar{b}$ is emitted more in the $\bar{t}$ spin direction.

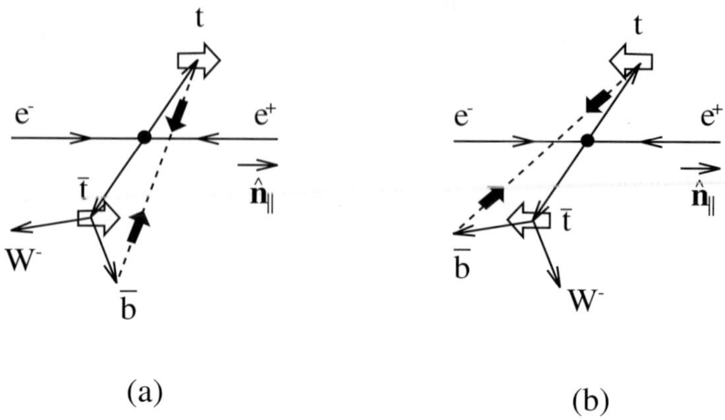

Fig. 11. Attractive force between t and $\bar{b}$ when the t and $\bar{t}$ spins are oriented in the (a) $\hat{\mathbf{n}}_{\|}$ direction, and in the (b) $-\hat{\mathbf{n}}_{\|}$ direction. The momentum transfer $\delta \mathbf{p}_b = -\delta \mathbf{p}_{\bar{t}}$ due to the attraction is indicated by thick black arrows.

- *Top-Quark Polarization Vector*

From Fig. 11 we can also learn the effect of the final-state interaction on the top polarization vector. We have seen that if the t and $\bar{t}$ spins are oriented in the $\hat{\mathbf{n}}_{\|}$ direction, t will be attracted to the forward direction due to the attraction by $\bar{b}$, and oppositely attracted to the backward direction if the t and $\bar{t}$ spins are in the $-\hat{\mathbf{n}}_{\|}$ direction. This means that in the forward region ($\cos\theta_{te} \simeq 1$) the number of t's with spin in $\hat{\mathbf{n}}_{\|}$ direction increases, whereas in the backward region the number of those with spin in the opposite direction increases. Or equivalently, the $\hat{\mathbf{n}}_{\|}$-component of the top-quark polarization vector increases in the forward region and decreases in the backward region. We may thus conjecture that the top-quark polarization vector is modified as $\delta S_{\|} \sim \kappa \cos\theta_{te}$ due to the interaction between t and $\bar{b}$.

- l^+ *Energy-Angular Distribution*

Finally let us examine the effect of the attraction between b and $\bar{t}$ on the l^+ energy-angular distribution in the semi-leptonic decay of t. The b-quark from t decay will be attracted in the direction of $\bar{t}$ (and *vice versa*) due to the gluon exchange between these two particles. We show schematically typical configurations of the particles in the top-quark semileptonic decay in Fig. 12. It can be seen that if the probability for b being emitted in the $\bar{t}$ direction increases, correspondingly the probability for particular l^+ energy-angular configurations increases. These configurations are either "E_l is small and l^+ emitted in $-\mathbf{p}_t$ direction" or "E_l is large and l^+ emitted in $\mathbf{p}_t$ direction".

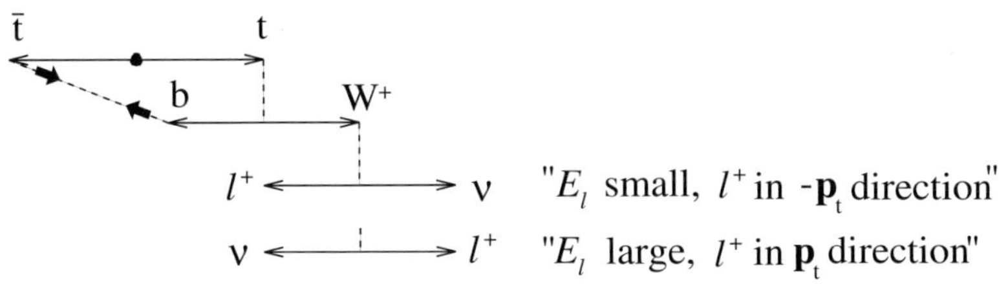

Fig. 12. Typical configurations of the particles in semileptonic decay of t when the b-quark is emitted in the $\bar{t}$ direction. Due to the boost by W^+, there will be an energy-angle correlation of l^+.

Lepton Energy-Angular Distribution Near $t\bar{t}$ Threshold

Here, we present the formula for the charged lepton energy-angular distribution in the decay of top quarks that are produced via $e^+e^- \to t\bar{t}$ near threshold.

First, without including the final-state interactions, the differential distribution of t and l^+ has a form where the production and decay processes of the top quark are factorized:

$$\frac{d\sigma(e^+e^- \to t\bar{t} \to bl^+\nu\bar{b}W^-)}{d^3\mathbf{p}_t dE_l d\Omega_l} = \frac{d\sigma(e^+e^- \to t\bar{t})}{d^3\mathbf{p}_t} \times \frac{1}{\Gamma_t} \frac{d\Gamma_{t\to bl^+\nu}(\mathbf{S})}{dE_l d\Omega_l}. \quad (28)$$

Namely, the cross section is given as a product of the production cross section of unpolarized top quarks and the differential decay distribution of l^+ from polarized top quarks. The above formula holds true even including all $\mathcal{O}(\alpha_s)$ corrections other than the final-state interactions.

Including the final-state interactions, the factorization of production and decay processes is destroyed. The formula including the full $\mathcal{O}(\alpha_s)$

corrections is given by [28]

$$\frac{d\sigma(e^+e^-\to t\bar{t}\to bl^+\nu\bar{b}W^-)}{d^3\mathbf{p}_t dE_l d\Omega_l} = \frac{d\sigma(e^+e^-\to t\bar{t})}{d^3\mathbf{p}_t}\times(1+\delta_0+\delta_1\cos\theta_{te})$$
$$\times\frac{1}{\Gamma_t}\frac{d\Gamma_{t\to bl^+\nu}(\mathbf{S}+\delta\mathbf{S})}{dE_l d\Omega_l}\times\left[1+\xi(|\mathbf{p}_t|,E,E_l,\cos\theta_{lt})\right]. \tag{29}$$

Here, the first line on the right-hand-side shows that there are corrections to the top-quark production cross section, while the second line shows that the correction to the decay distribution of l^+ is accounted for by a modification of the parent top-quark polarization vector, and finally there is a non-factorizable correction ξ which cannot be assigned either to the production or the decay process alone.

We have already seen in Fig. 9(a) that the top momentum distribution is modified by δ_0 to take a lower average momentum. The forward-backward asymmetric distribution and the top polarization vector get corrections as

$$\delta_1\cos\theta_{te} = \kappa S_{\parallel}\cos\theta_{te}\times\frac{1}{2}\psi_{\rm R}, \tag{30}$$

$$\delta\mathbf{S} = \left[1-(S_{\parallel})^2\right]\times\kappa\cos\theta_{te}\times\frac{1}{2}\psi_{\rm R}\cdot\hat{\mathbf{n}}_{\parallel} \tag{31}$$

with

$$\psi_{\rm R}(|\mathbf{p}_t|,E) = -\,C_F\cdot 4\pi\alpha_s$$
$$\times\,\mathrm{Pr.}\int\frac{d^3\mathbf{q}}{(2\pi)^3}\,\frac{1}{|\mathbf{q}-\mathbf{p}_t|^3}\,\frac{\mathbf{p}_t\cdot(\mathbf{q}-\mathbf{p}_t)}{|\mathbf{p}_t|\,|\mathbf{q}-\mathbf{p}_t|}\,2\,\mathrm{Re}\left[\frac{\tilde{G}(\mathbf{q};E)}{\tilde{G}(\mathbf{p}_t;E)}\right]. \tag{32}$$

The above formulas (30) and (31) have exactly the forms that we anticipated in the previous subsection if $\psi_{\rm R}$ is positive. Indeed, the numerical evaluation in Ref. [27] shows that $\psi_{\rm R}(|\mathbf{p}_t|,E)\gtrsim 0$ holds in the entire threshold region.[g]

We show the $\cos\theta_{lt}$ and E_l dependences of the non-factorizable correction ξ as a 3-dimensional plot in Fig. 13. One can see that ξ takes comparatively large positive values for either "small E_l and $\cos\theta_{lt}\simeq -1$" or "large E_l and $\cos\theta_{lt}\simeq +1$". Oppositely, in the other two corners of the E_l–$\cos\theta_{lt}$ plane ξ becomes negative. These features are consistent with our previous qualitative argument. Typical magnitude of ξ is 10–20%.

[g]It shows that the force between b and $\bar{t}$ ($\bar{b}$ and t) is attractive in the entire threshold region. Note that the sign of $\psi_{\rm R}$ will be reversed if the force is repulsive.

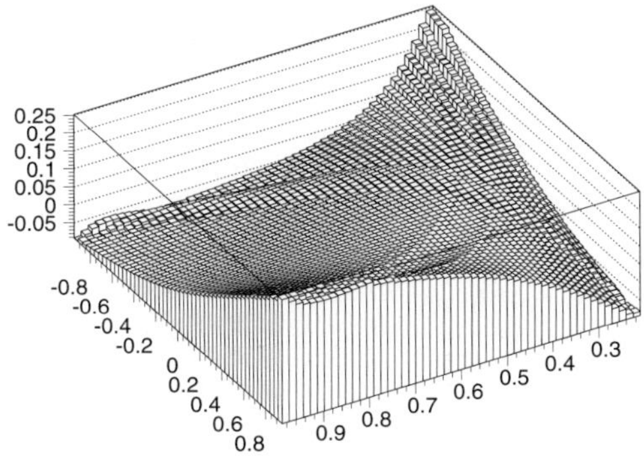

Fig. 13. A three-dimensional plot of ξ as a function of $2E_l/m_t$ (x-axis) and $\cos\theta_{lt}$ (y-axis).

Thus, the theoretical prediction for the distribution of l^+ in the top decay is under good control in the $t\bar{t}$ threshold region, together with a good qualitative understanding.

Prior to the calculation of the l^+ differential distribution Eq. (29), an inclusive quantity, the mean value $\langle n\ell\rangle$ of the charged lepton four-momentum projection on an arbitrarily chosen four-vector n, was proposed as an observable sensitive to the top quark polarization, and this quantity was calculated including the final-state interactions [27].

Observable Proper to Decay Processes

We have seen that the final-state interactions destroy the factorization of the production and decay cross sections of the top quark. In order to study the decay of the top quark in a clean environment in the threshold region, it would be useful if we could find an observable which depends only on the decay process of free polarized top quarks, $d\Gamma_{t\to bl^+\nu}(\mathbf{S})/dE_l d\Omega_l$. In fact, such an observable can be constructed, which at the same time preserves most of the differential information of the l^+ energy-angular distribution.

It is possible to show the invariance of the non-factorizable correction factor $\xi(|\mathbf{p}_t|, E, E_l, \cos\theta_{lt})$ under a transformation of the l^+ kinematical variables

$$E_l \to E_l' = 2m_t \left(\frac{m_t^2 + M_W^2}{M_W^2} - \frac{2m_t}{E_l} \right)^{-1}, \tag{33}$$

$$\mathbf{p}_l \to \mathbf{p}_l' \quad \text{such that} \quad \cos\theta_{lt} \to -\cos\theta_{lt}. \tag{34}$$

Using this symmetry, it is possible to cancel out not only the non-factorizable correction but also the top production cross section by taking an appropriate ratio of cross sections.

Let us define an observable as

$$\overline{A}(E_l, a = \mathbf{S}\cdot\mathbf{p}_l) \equiv \frac{\int d^3\mathbf{p}_t d\Omega_l \; \delta(\mathbf{S}\cdot\mathbf{p}_l - a) \left[\frac{d\sigma(e^+e^- \to t\bar{t} \to bl^+\nu\bar{b}W^-)}{d^3\mathbf{p}_t dE_l d\Omega_l} \right]_{E_l, \mathbf{p}_l}}{\int d^3\mathbf{p}_t d\Omega_l \; \delta(\mathbf{S}\cdot\mathbf{p}_l + a) \left[\frac{d\sigma(e^+e^- \to t\bar{t} \to bl^+\nu\bar{b}W^-)}{d^3\mathbf{p}_t dE_l d\Omega_l} \right]_{E_l', \mathbf{p}_l}}. \tag{35}$$

Here, the top-quark polarization vector $\mathbf{S}$ in the delta functions depends on $\mathbf{p}_t$ [cf. Eqs. (23)–(25)]. The numerator and denominator, respectively, depend on two external kinematical variables (the lepton energy and the lepton angle from the parent top-quark polarization vector), and all other variables are integrated out before taking the ratio.

Then substituting the differential distribution Eq. (29), one can show that theoretically $\overline{A}$ is determined solely from the decay distribution of free polarized top quarks:

$$\overline{A}(E_l, a) = \left[\frac{d\Gamma_{t\to bl^+\nu}(\mathbf{S})}{dE_l d\Omega_l} \right]_{E_l, \mathbf{S}\cdot\mathbf{p}_l = a} \Bigg/ \left[\frac{d\Gamma_{t\to bl^+\nu}(\mathbf{S})}{dE_l d\Omega_l} \right]_{E_l', \mathbf{S}\cdot\mathbf{p}_l = -a}. \tag{36}$$

This is a general formula that is valid even if the decay vertices of the top quark deviate from the standard-model forms.

This quantity will be useful from the theoretical point of view. If one claims that he calculates $\overline{A}(E_l, a)$ defined in Eq. (35) in the threshold region, it can be calculated without including any bound-state effects or final-state interaction corrections but only from a decay distribution of free polarized top quarks via Eq. (36).

2.4. *Measurements of Top Quark Couplings*

As we have seen so far, studying various top quark properties in the $t\bar{t}$ threshold region is promising and interesting. Regarding the energy upgrading scenario of a future linear collider, it is possible that the machine operates first in the $t\bar{t}$ threshold region for a significant amount of time, while measuring the top quark mass precisely, etc., before the beam energies will be increased to the open-top region. Therefore it would be desirable that measurements of the various top quark couplings can be performed concurrently with other unique measurements near threshold, with sensitivities comparable to those in the open-top region.

Thus, systematic simulation studies on sensitivities to all the top quark couplings or the form factors in the $t\bar{t}$ threshold region are requested, which are not yet complete. Up to now, serious sensitivity studies in the threshold region were performed on the top quark width [6], on the top-Higgs Yukawa coupling [6, 35] and on the strong coupling constant α_s. Also, there exist theoretical studies on the CP violating couplings of the top quark, as well as on the top quark production and decay vertices. We review the present status of these studies.

2.4.1. *Measurements of* Γ_t, g_{tH}, α_s

Among various observables in the top quark threshold region, the height (normalization) of the $1S$ peak of the threshold cross section is most sensitive to the top quark decay width Γ_t, the top-Higgs Yukawa coupling g_{tH}, and the strong coupling constant α_s. This is because: (1) according to Eq. (19), the peak height is inversely proportional to Γ_t in the narrow width limit; (2) the effect of Higgs exchange enhances the cross section [cf. Eq. (26)]; (3) the wave function at the origin $|\psi_n(\mathbf{0})|^2$ is proportional to α_s^3; when we take into account the smearing effect by the large decay width the dependence reduces to α_s^2. Nevertheless, extracting these quantities requires a precise theoretical prediction for the normalization of the threshold cross section. In view of the present large theoretical uncertainty, accurate measurements of g_{tH} and α_s seem to be difficult, whereas we will still be able to measure Γ_t relatively accurately. Also, the top momentum distribution and forward-backward asymmetry of the top quark depend on Γ_t and α_s. Theoretical predictions for these observables are expected to be more stable, but some more calculations are needed for accurate predictions.

Total $t\bar{t}$ Cross Section

Let us first discuss the results of the simulation studies, which set a benchmark of measurements in future experiments.

It is well known that the initial state radiation has to be properly taken into account in the vicinity of a resonance state. Also, in order to study experimental feasibilities in a realistic environment, we need to include the effects due to the beam energy spread and beamstrahlung (a bremsstrahlung from beam particles due to the strong electromagnetic fields produced by opposing beam bunches). These effects can be incorporated by convoluting the original cross section with the radiator function and the beam spectrum. The beam spectrum is machine dependent, and in our study below we use the parameters given in Ref. [36].

The effects on the cross section can be summarized as follows (see Fig. 14).

(i) The initial state radiation and the beamstrahlung reduce usable luminosity in the threshold region. The usable part is essentially restricted to the δ-function part of the beam spectrum. Therefore, we need to know the height of the δ-function part accurately for precision measurements.

(ii) The beam energy spread is the major source of the peak smearing. When the energy spread is less than 0.4%, the cross section shape is practically independent of the structure inside the δ-function part. If we cannot achieve such a narrow band beam, it is important to measure the spectrum inside the δ-function part with a high-energy resolution.

The signature of a $t\bar{t}$ pair production is two b quarks and two W bosons in the final state. The two W bosons decay into either $q\bar{q}'$ or $\ell\nu$. Therefore the final state configurations are

(i) two b jets and four jets from W's (45%),

(ii) two b jets, two jets, and one charged lepton (44%), and

(iii) two b jets and two charged leptons (11%).

All three cases can be used for the measurement of the total cross section. Case (i) and (ii) are useful for the measurements of the top momentum distribution and top quark forward-backward asymmetry.[h] The basic cuts used in the event selection can be classified into four groups: (a) event shape cuts such as those on the number of charged particles, the number of jets, and thrust, (b) mass cuts to select W's and t's by jet-invariant-mass method, (c) requirements of leptons in cases (ii) and (iii), and (d) b taggings.

In order to measure the total $t\bar{t}$ cross section reliably, we need to select $t\bar{t}$ events with a good signal-to-background ratio. The largest background comes from W^+W^- pair productions, whose cross section is larger by a

[h] A recent study shows that the charge of b can be identified efficiently from a measurement of the vertex charge of b jets. In this way t and $\bar{t}$ can be distinguished even in case (i) [37].

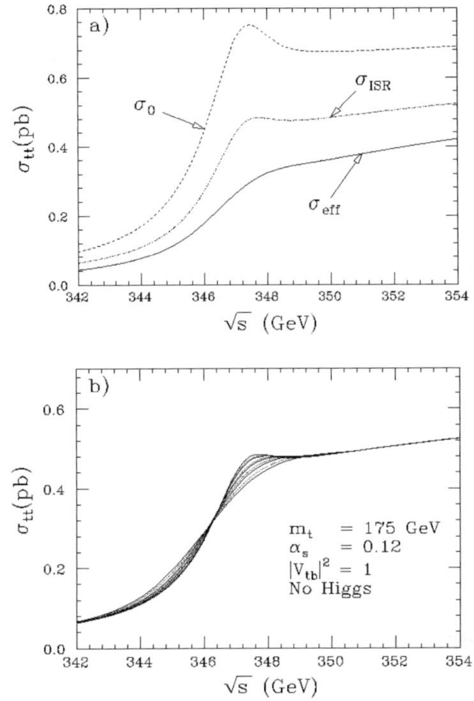

Fig. 14. (a) Effects of the initial state radiation (ISR) and the beam energy spread and beamstrahlung on the threshold shape for $m_t = 175$ GeV and $\alpha_s(M_Z) = 0.12$: with no ISR and no beam effects (dash), with ISR but without beam effects (dotdash), and with both ISR and beam effects (solid). (b) Threshold cross section including the effects of ISR and natural beam energy spread but leaving out the effect of beamstrahlung. Two kinds of spectra, flat-top (dot) and double-peaked (solid), which is more realistic, are examined for various beam energy spread: 0.1%, 0.4%, 0.7%, 1.0%, 1.4%.

factor 10^3. We take case (i) as an example and show the result of a Monte Carlo simulation study. We require 20 or more charged tracks and impose cuts on the visible energy (the energy sum of all the detected particles including neutral ones) and the total transverse momentum: $E_{\text{vis}} \geq 200$ GeV and $p_T \leq 50$ GeV. The remaining events must contain six or more hadronic jets after jet clustering with $y_{\text{cut}} = 5 \times 10^{-3}$, where y_{cut} is the cut on jet-invariant-masses normalized by the visible energy. If there are more than six jets, we increase the y_{cut} value so as to make the event yield exactly six jets. Two pairs of jets out of these six jets must have 2-jet invariant masses consistent with M_W. These two W candidates are then required to make invariant masses consistent with m_t, when combined with one of the remaining two jets. Finally we impose a cut on the event thrust: thrust ≤ 0.75. After the final cut, the detection efficiency is 29%, while the signal-to-background ratio is greater than 10. This detection efficiency translates to about 63% when the branching ratio for $W \to q\bar{q}'$ of 67% is taken into account. We can select events with similar efficiencies and signal-to-background ratios for cases (ii) and (iii) as well.

Using the $t\bar{t}$ samples so obtained, we can determine the parameters Γ_t, g_{tH}, α_s. A total of 11 points with 1 fb^{-1} each were generated with the inputs $m_{t,\text{pole}} = 170$ GeV, $\alpha_s(M_Z) = 0.12$, $m_H = \infty$ (no Higgs effects) and the SM value for Γ_t. By fitting these data points to the theoretical prediction for the total $t\bar{t}$ production cross section, convoluted with the beam spectrum, we obtained the following estimates of the statistical errors on the above parameters: $\Delta\Gamma_t/\Gamma_t^{\text{SM}} = 0.18$, $\Delta g_{tH}/g_{tH}^{\text{SM}} = 0.3$ in the case $M_H = 115$ GeV, and $\Delta\alpha_s(M_Z) = 0.002$. Each error is given for a measurement of each parameter, assuming that all the other parameters are known precisely from other sources.

While the above errors serve as measures for potential sensitivities of the threshold measurements, as long as the theoretical uncertainty on the normalization of the cross section remains at the present level, theoretical errors on these parameters can be much larger.[i] For example, the induced theoretical error on the strong coupling constant is $\Delta\alpha_s(M_Z) \simeq 0.01$. As for Γ_t, we may consider a strategy to determine it to a good accuracy, nonetheless. Since the effect of the variation of Γ_t appears only in the vicinity of the $1S$ peak[j] (see Fig. 15), we may use, for instance, the ratio of the $t\bar{t}$ total cross section at $\sqrt{s} \simeq M_{1S}$ and at $\sqrt{s} \simeq M_{1S} + 3$ GeV. This ratio is much less affected by the normalization uncertainty, and the statistical error on $\Delta\Gamma_t/\Gamma_t$ will be determined by the luminosity we invest at $\sqrt{s} \simeq M_{1S}$ and $\sqrt{s} \simeq M_{1S} + 3$ GeV. We expect that an accuracy better than 15% (both theoretical and statistical) would be achievable with a moderate integrated luminosity. On the other hand, extraction of the Yukawa coupling seems to be difficult. This is because its effect is to enhance the cross section almost energy independently in the threshold region, and disentanglement from the normalization uncertainty is difficult. For the same reason, a precise measurement of α_s will also be difficult. A better strategy would be to de-

[i] An attempt at solving this problem has been given recently in Ref. [38].

[j] In the limit $\sqrt{s} \gg 2m_t$, the normalization of the cross section becomes independent of Γ_t.

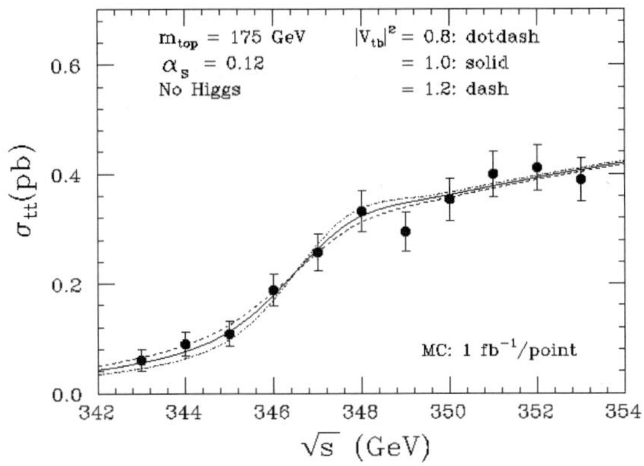

Fig. 15. The dependence of the cross section as $\Gamma_t \propto |V_{tb}|^2$ is varied. An example of energy scan is superimposed, where each point corresponds to 1 fb^{-1}.

termine the value of $\alpha_s(M_Z)$ in other experiments and to use it as an input of the theoretical predictions in the top quark threshold region.

Top Momentum Distribution

By virtue of the large top quark width, we can measure the top quark momentum by reconstructing the 3-jet decay of a top quark, even below the $t\bar{t}$ threshold. This is in contrast to the charmonium and bottomonium cases, where the pair annihilation modes dominate. We assume that the program of a linear collider operation is such that the machine will first scan the threshold shape with a relatively low integrated luminosity and determines the top quark mass accurately, then the CM energy is fixed at some optimal point in the threshold region and we invest more luminosity to perform the measurements of the top momentum distribution, etc. The choice of the CM energy should be made from the viewpoint of insensitivity to theoretical ambiguities and beam effects and of sensitivities to physical parameters to be determined. Since the momentum distribution depends heavily on the CM energy, the initial state radiation and beam effects must be properly taken into account. From the study in Ref. [6], we find that the position of the peak momentum $|\mathbf{p}|_{\text{peak}}$ is fairly stable against these effects, although the distribution itself is quite affected. An optimal CM energy, which maximizes the sensitivity to $\alpha_s(M_Z)$ and Γ_t, while keeping small the beam effects and theoretical uncertainties, is chosen as 2 GeV above the 1S peak energy, $\Delta E \equiv \sqrt{s} - M_{1S} = 2$ GeV. For simplicity, we assume that M_{1S} is known in what follows.

In order to measure the top momentum, we need to select top quarks which are well-reconstructed as 3-jet systems from $t \to bW$ decays. This requires final states with at least one t or $\bar{t}$ quark decaying into three jets: the configurations (i) and (ii) discussed above. The 6-jet mode [case (i)] is advantageous from the statistical point of view, since both t and $\bar{t}$ quarks in a single event can be used for the momentum measurement. This mode, however, tends to suffer from combinatorial background: even if we have a perfect b-tagging capability, some ambiguity would still remain in assigning other four jets to two W's. The lepton-plus-4-jet mode [case (ii)] is cleaner in this respect.

The selection of top quark events goes similarly to the one discussed in the total cross section analysis. In the present case, however, we have to make sure that the momentum of a reconstructed top quark is well measured. Rejection of events with energetic missing neutrinos is thus necessary. The main differences are the cut values which must be much tighter to require four-momentum balance, well reconstructed W's and t's. The b tagging and the cut on the b-W angle are necessary in addition. (See Ref. [6] for details.) In case (ii) Monte Carlo simulation study resulted in a detection efficiency of 4% including the branching fraction to the lepton-plus-4-jet mode. The systematic shift between the peak momentum of the generated events and that of the reconstructed events is less than 1 GeV. The momentum resolution is 3–4 GeV in the relevant range, which is much smaller than the width of the momentum distribution. In case (i) the detection efficiency is 2%, including the branching fraction to the 6-jet mode. Since both t and $\bar{t}$ can be used for the momentum measurement, effective selection efficiency is twice this value. The agreement of the reconstructed momentum distribution with the generated one is not as good as in the lepton-plus-4-jet case, because of the combinatorial background. A further study to gain in the detection efficiency using a kinematical fitting procedure is now underway [39].

We can estimate the statistical error on $|\mathbf{p}|_{\text{peak}}$ from the width of the expected momentum distribution and the expected number of reconstructed top quarks. Using the events reconstructed from the lepton-plus-4-jet mode at $\Delta E = 2$ GeV with $m_{t,\text{pole}} = 170$ GeV, the 1σ bound corresponding to an integrated luminosity of 100 fb^{-1} is $\Delta|\mathbf{p}|_{\text{peak}} \simeq 200$ MeV. This value translates to sensitivities to Γ_t and $\alpha_s(M_Z)$ of $\Delta\Gamma_t/\Gamma_t = 0.03$ and $\Delta\alpha_s(M_Z) = 0.002$, when each parameter is varied while the other is fixed and assuming M_{1S} is known.

This indicates that the momentum distribution can be useful for a precise determination of Γ_t, once $\alpha_s(M_Z)$ is known precisely from other sources. The sensitivity of $|\mathbf{p}|_{\text{peak}}$ to Γ_t arises due to the following reason. If the width becomes larger, t and $\bar{t}$ decay when they are closer to each other, so the momentum distribution is reduced in the low momentum region. Accordingly $|\mathbf{p}|_{\text{peak}}$ shifts toward the higher momentum side.

In actual experiments, we need to examine various systematic errors of both theoretical and experimental nature. One of the most serious theoretical ambiguities comes from the fact that the top quark has color. In the threshold region, the color neutralization is expected to take place between b and $\bar{b}$ quarks. Hence, an additional uncertainty from possible top-hadron formations is absent. Nevertheless the b-quark fragmentation must be well understood when we group particles in the final state into t and $\bar{t}$ sides. Within the framework of perturbative QCD, the full next-to-leading order corrections are known, including the gluon exchange effects between the top decay products, as we have seen in Sec. 2.3. Full calculations of the next-to-next-to-leading order corrections are necessary (and probably sufficient) for the prediction of $|\mathbf{p}|_{\text{peak}}$ with 200 MeV accuracy. Part of the corrections has been computed and shown to be of the expected size [20], in contrast to the large correction to the normalization of the cross section [11]. As for the non-perturbative contributions, studies on the color reconnection effects are in progress [40]. Furthermore, we will be able to measure the effects from various comparisons of the top-jet distributions with the theoretical predictions, where for the latter we will have a fairly precise value of the $\overline{\text{MS}}$ mass of the top quark as input, as determined from the threshold scan. This should be useful in reducing systematic uncertainties.

Forward-Backward Asymmetry

As we have seen in Section 2.3, interference of the S and P-wave resonances produces a forward-backward asymmetry of the top quark. The spectrum of the would-be toponium resonances is such that the lowest-lying S-wave resonance ($1S$ state) stands alone, while the higher S-wave resonances are accompanied by nearly degenerate P-wave resonances. The level spacings are determined by α_s while the resonance widths are determined mainly by the top quark width. The relative size of the two determines the size of the interference. Therefore, we expect that the forward-backward asymmetry provides additional information on the top quark width.

The CM energy is chosen according to the criterion similar to that of the momentum distribution measurement. The asymmetry depends fairly strongly on the CM energy. As a consequence, the beam effects are sizable in the entire threshold region, and a good understanding or control of these effects will be mandatory in the asymmetry measurement. In consideration of the sensitivity to Γ_t, we chose the CM energy as $\Delta E = 1$ GeV in the simulation study. Since the asymmetry is quite small, the systematic error due to the detector asymmetry must be carefully studied in actual experiments. We expect collider operations at the Z pole (Giga Z) will be very useful, since the experimental limit on the measurable asymmetry will be determined by the number of events available for the detector calibration. We assume that a sample of 40,000 reconstructed events is given and estimate the expected statistical errors on α_s and Γ_t. The sample corresponds to 200 fb^{-1} even if the detection efficiency is as high as 40%. For $m_{t,\text{pole}} = 170$ GeV, the sensitivity to the strong coupling constant is $\Delta\alpha_s(M_Z) = 0.004$ if M_{1S} and Γ_t are known. On the other hand, the sensitivity to the top quark width is $\Delta\Gamma_t/\Gamma_t = 0.06$ if M_{1S} and $\alpha_s(M_Z)$ are known [12].

2.4.2. *CP Violating Couplings*

Here, we review the recent theoretical study on probing the anomalous CP violating interactions of the top quark in the threshold region. Besides its own interest, it also serves as a case study for the analyses of the top quark properties using its production-decay chain. (See the chapter on CP violation for a general analysis of CP violating couplings.)

Among various interactions of the top quark, testing the CP-violating interactions is particularly interesting due to the following reasons:

- Within the SM, CP-violation in the top quark sector is extremely small. If any CP-violating effect is detected in the top quark sector in a near-future experiment, it immediately signals new physics.

- In relatively wide class of models beyond the SM, CP violation emerges especially sizably in the top quark sector.

In this analysis we consider the *CP*-violating interactions of the top quark with γ, Z, and g:

$$\mathcal{L}_{CP\text{-odd}} = -\frac{ed_{t\gamma}}{2m_t}(\bar{t}i\sigma^{\mu\nu}\gamma_5 t)\partial_\mu A_\nu - \frac{g_Z d_{tZ}}{2m_t}(\bar{t}i\sigma^{\mu\nu}\gamma_5 t)\partial_\mu Z_\nu$$
$$-\frac{g_s d_{tg}}{2m_t}(\bar{t}i\sigma^{\mu\nu}\gamma_5 T^a t)\partial_\mu G^a_\nu, \qquad \sigma^{\mu\nu} \equiv \tfrac{i}{2}[\gamma^\mu, \gamma^\nu], \tag{37}$$

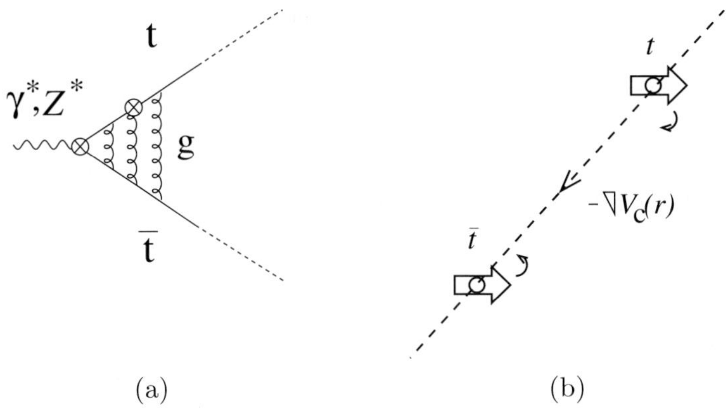

Fig. 16. (a)The time evolution of the $t\bar{t}$ system when the effects of $\mathcal{L}_{\text{CP-odd}}$ are included. ⊗ represent the anomalous couplings. (b)The potential $V_{\text{CP-odd}}$ tends to align the t spin in the direction of the color-electric field and the $\bar{t}$ spin in the opposite direction.

where $e = g_W \sin\theta_W$ and $g_Z = g_W/\cos\theta_W$. These represent the interactions of γ, Z, g with the electric dipole moment (EDM), Z-EDM, chromo-EDM of the top quark, respectively.[k] Each of the interactions has $C = +1$ and $P = -1$. The SM contributions to these couplings are extremely small, $d_{t\gamma}^{(\text{SM})}, d_{tZ}^{(\text{SM})}, d_{tg}^{(\text{SM})} \sim 10^{-14}$. Since we will not be able to detect them in near-future collider experiments, we neglect the SM contributions below. Our concern is in the anomalous couplings which are induced from some new physics. Generally the couplings $d_{t\gamma}$, d_{tZ}, d_{tg} are complex where their imaginary parts may be induced from some absorptive processes.

When studying CP violation of the top quark, unique aspects of the $t\bar{t}$ threshold region may be put as follows:

- The QCD interaction is enhanced in this region, hence the cross section is sensitive to the top-gluon (tg) couplings. We can study anomalous tg couplings in a clean environment in comparison to hadron colliders.

- In certain models (e.g. those in which a neutral Higgs boson is exchanged between t and $\bar{t}$ [41]), induced top quark EDM and Z-EDM are enhanced near the $t\bar{t}$ threshold.

Time evolution of the $t\bar{t}$ system

Let us consider effects of the anomalous interactions Eq. (37) on the time evolution of the $t\bar{t}$ system as described in Section 2.3; [see Fig. 16(a)]. CP-

[k]The magnitudes of these EDMs are given by $ed_{t\gamma}/m_t$, $g_Z d_{tZ}/m_t$, $g_s d_{tg}/m_t$, respectively. $d_{t\gamma} = 1$ corresponds to $e/m_t \sim 10^{-16}\, e\,\text{cm}$, etc.

violation originating from the $t\gamma$ or tZ coupling occurs at the stage of the pair creation, *i.e.* when t and $\bar{t}$ are very close to each other. The generated boundstate has $J^{PC} = 1^{+-}$, so t and $\bar{t}$ are in the P-wave ($L = 1$) and spin-0 state $|\uparrow\downarrow\rangle - |\downarrow\uparrow\rangle$. On the other hand, CP-violation originating from the tg coupling takes place after the boundstate formation when multiple gluons are exchanged between t and $\bar{t}$, i.e. when t and $\bar{t}$ are separated at a distance of the Bohr radius. The anomalous top-gluon coupling generates effectively a spin-dependent potential between t and $\bar{t}$

$$V_{\text{CP}-\text{odd}} = \frac{d_{tg}}{m_t}\,(\mathbf{s}_t - \bar{\mathbf{s}}_t)\cdot\nabla V_{\text{C}}(r). \qquad (38)$$

Here, $\mathbf{s}_t$ and $\bar{\mathbf{s}}_t$ denote the spins of non-relativistic t and $\bar{t}$, respectively; $V_{\text{C}}(r) = -C_F\alpha_s/r$ is the Coulomb potential with the color factor $C_F = 4/3$. When $d_{tg} > 0$, the potential $V_{CP\text{-odd}}$ tends to align both chromo-EDMs in the direction of chromo-electric field, or, align $\bar{\mathbf{s}}_t$ in the direction of $\mathbf{r} = \mathbf{r}_t - \bar{\mathbf{r}}_t$ and $\mathbf{s}_t$ in the direction of $-\mathbf{r}$; [see Fig. 16(b)]. Therefore, first the boundstate is formed in $J^{PC} = 1^{--}$ ($L = 0$ and $S = 1$) state and after interacting via the potential $V_{CP\text{-odd}}$ it turns into $J^{PC} = 1^{+-}$ ($L = 1$ and $S = 0$) state, i.e. the t and $\bar{t}$ spins are aligned into antiparallel directions.

We can disentangle the effects of the three couplings, $d_{t\gamma}$, d_{tZ}, d_{tg}, on the amplitude using the differences in the dependences on energy and $e^\pm$ polarization in the threshold region. Firstly when the CM energy is raised the CP-violating effects due to $d_{t\gamma}$ and d_{tZ} increase proportionally to the velocity of the top quark, since these effects are induced directly by the dimension-five operators. On the other hand, the effect of d_{tg} does not increase so rapidly. The enhancement of the tg-coupling due to multiple exchanges of gluons will be lost when the energy is raised and t and $\bar{t}$ spread apart quickly without forming boundstates. Secondly, one may vary the relative weight of the photon-induced *CP*-violating effect and the Z-induced effect by varying the $e^\pm$ longitudinal polarization. This is because e_L and e_R couple differently to γ and Z, and the relative weight of virtual γ and Z changes [42].

CP-odd observables

For the process $e^+e^- \to t\bar{t}$, we may conceive of following expectation values of kinematical variables for CP-odd observables:

$$\begin{gathered} \langle\,(\mathbf{p}_e - \bar{\mathbf{p}}_e)\cdot(\mathbf{s}_t - \bar{\mathbf{s}}_t)\,\rangle\,, \\ \langle\,(\mathbf{p}_t - \bar{\mathbf{p}}_t)\cdot(\mathbf{s}_t - \bar{\mathbf{s}}_t)\,\rangle\,, \\ \langle\,[(\mathbf{p}_e - \bar{\mathbf{p}}_e)\times(\mathbf{p}_t - \bar{\mathbf{p}}_t)]\cdot(\mathbf{s}_t - \bar{\mathbf{s}}_t)\,\rangle\,, \end{gathered} \qquad (39)$$

where the spins and momenta are defined in the CM frame. (The initial state is CP-even if we assume the SM interactions of $e^{\pm}$ with γ and Z.) These are the three components of the difference of the t and $\bar{t}$ spins. One might say (in a somewhat oversimplified way) that in the SM the t and $\bar{t}$ spins are parallel to each other, so that the SM contributions drop in the difference $\mathbf{s}_t - \bar{\mathbf{s}}_t$, whereas the t and $\bar{t}$ spins become antiparallel to each other by the effects of $\mathcal{L}_{CP\text{-odd}}$, so they remain in $\mathbf{s}_t - \bar{\mathbf{s}}_t$. Thus, we want to measure the difference of the spins of t and $\bar{t}$. It is equivalent to measuring the difference of the polarization vectors of t and $\bar{t}$. All other CP-odd observables for $e^+e^- \to t\bar{t}$ are bilinear in $\mathbf{s}_t$ and $\bar{\mathbf{s}}_t$. Since analyses of spin correlations are complicated, we focus on the difference of the polarization vectors.

Practically we can measure the t and $\bar{t}$ polarization vectors efficiently using $\ell^{\pm}$ angular distributions. It is known that the angular distribution of the charged lepton ℓ^+ from the decay of top quark is maximally sensitive to the top quark polarization vector; [cf. Eq. (27)]. Indeed the ℓ^+ angular distribution is ideal for extracting CP-violation in the $t\bar{t}$ *production process*; the angular distribution is unchanged even if anomalous interactions are included in the tbW decay vertex, up to the terms linear in the decay anomalous couplings and within the approximation $m_b = 0$ [43]. Therefore, if we consider the average of the lepton direction, for instance, we may extract the top quark polarization vector efficiently:

$$\langle \mathbf{n} \cdot \mathbf{n}_\ell \rangle_{\text{Lab}} \simeq \frac{1}{3} \mathbf{n} \cdot \mathbf{S}. \tag{40}$$

The average is to be taken at the top quark rest frame, but in the threshold region, we may take the average in the laboratory frame barely without loss of sensitivities to the anomalous couplings.

We made rough estimates of the statistical errors to the anomalous CP-violating couplings of the top quark with γ, Z and g in the $t\bar{t}$ threshold region: $\delta d_{t\gamma}$, δd_{tZ}, $\delta d_{tg} \sim 10\%$ corresponding to an integrated luminosity of 50 fb^{-1}. Qualitatively, the characteristics of the $t\bar{t}$ region are summarized as follows.

(1) We can measure the three couplings $d_{t\gamma}$, d_{tZ}, d_{tg} simultaneously and we can disentangle each contribution.

(2) Typical sizes of components of the difference of the t and $\bar{t}$ polarizations are given by

$$|\delta S_\perp|,\ |\delta S_{\text{N}}| \sim (5\text{–}20\%) \times (d_{t\gamma}, d_{tZ}, d_{tg}).$$

They can be extracted efficiently from the directions of charged leptons from decays of t and $\bar{t}$:

$$\langle\langle \mathbf{n} \cdot (\mathbf{n}_\ell + \overline{\mathbf{n}}_\ell) \rangle\rangle_{\mathbf{p}_t} \simeq \frac{2}{3} \mathbf{n} \cdot \delta \mathbf{S}.$$

(3) We can measure the complex phases of the couplings $d_{t\gamma}$, d_{tZ}, d_{tg}. Since the strong phases can be modulated at our disposal, a single observable ($\delta S_\perp$ or $\delta S_{\rm N}$) probes the phases of the couplings.

We note that if one of the couplings $d_{t\gamma}$, d_{tZ}, d_{tg} is detected in the future, we would certainly want to measure the others in order to gain deeper understanding of the CP-violating mechanism. This is because one may readily think of various underlying processes which give different contributions to the individual couplings.

2.4.3. *Production and Decay Vertices*

The left–right asymmetry

$$A_{RL} = \frac{\sigma_R - \sigma_L}{\sigma_R + \sigma_L} \tag{41}$$

can be measured with longitudinally polarized beams. (Here R and L correspond to fully polarized right- and lefthanded beams, respectively.) The left-right asymmetry of the top quark production cross section close to the threshold can be measured with 1–2% accuracy, with an integrated luminosity of 100 fb^{-1} and assuming electron and positron longitudinal polarizations of $|P_{e^-}| = 0.8$ and $|P_{e^+}| = 0.6$. This allows us to measure the vector coupling of the top quark with γ and Z with a similar accuracy. The theoretical prediction is well under control, since the uncertainty of the normalization of the total cross section cancels in the left-right asymmetry [44].

On the other hand, by using the high polarization of top quarks close to the threshold, the couplings which dictate the decays of top quarks can be measured accurately. It was shown that using the angular distribution of charged leptons in the top quark decay $t \to b\ell^+\nu$, the anomalous tbW couplings can be constrained. Also, from the angular distribution of the charged Higgs boson in a non-SM decay mode $t \to H^+ b$, one can measure $\tan\beta$ accurately when it is large, or otherwise, confirm the dominance of right- or of lefthanded couplings [44].

3. Open Top Region

Studies of the top quark properties in the energy region far above the threshold extend to wide varieties. Generally speaking, we have more resolving power in studying various top quark properties by carrying out experiments at highest possible energy. Here it is not our purpose to give a comprehensive review of the existing works. Rather we discuss only very limited subjects, which may form the basis of various analyses in the open top region, and which have stronger links to other important physics subjects at a linear collider. After giving a short (and selective) review of previous works in Section 3.1, we discuss a possible strategy for extracting individual top quark form factors in Section 3.2.

3.1. *Short Review*

3.1.1. *Yukawa Coupling*

If there is a light Higgs boson, it is likely to be discovered at Tevatron or at LHC. Then, one of the most important physics goals of the open top region at an e^+e^- collider is to measure the coupling of the top quark to the Higgs boson, i.e. the Yukawa coupling of the top quark. Its measurement serves as a test for the mechanism of fermion mass generation, since in models beyond the Standard Model, such as Supersymmetric Standard Model, the couplings of the Higgs boson to fermions deviate from those of the Standard Model in general. (See also the chapter on Higgs physics.)

For a light Higgs boson of $m_H < 2m_t$, the Yukawa coupling g_{tH} can be measured through a measurement of the cross section for $e^+e^- \to t\bar{t}H$ [45]. There are two types of tree diagrams contributing to this process: the Higgs radiated from the top quark and from the Z boson. The contributions of the former type dominate the cross section in the Standard Model, hence a measurement of the cross section translates directly to a measurement of g_{tH}^2. The cross section grows as the CM energy is raised and saturates at around $\sqrt{s} \simeq 700$ GeV.

There have been Monte-Carlo simulation studies which considered a light Higgs boson decaying dominantly via $H \to b\bar{b}$ [46–48]. All of them report similar sensitivities to the Yukawa coupling. Here, we take the study [48] as an example and summarize the result of the analysis. The process $e^+e^- \to t\bar{t}H$ have 3 types of final states: (i) 8 jets, (ii) 6 jets plus one lepton, and (iii) 4 jets plus 2 leptons. All these final states include 4 b-jets, so that the flavor tagging is crucial for the event identification. Only the cases (i) and (ii) are considered. Background can be classified into two groups: (1) irreducible background containing $e^+e^- \to t\bar{t}Z$ followed by $Z \to b\bar{b}$, and (2) reducible background, the largest of which is $e^+e^- \to t\bar{t}g$ with a cross section larger than that for the signal events by 2 to 3 orders of magnitude. Taking $m_H = 100$ GeV and $\sqrt{s} = 700$ GeV, the cross section for $e^+e^- \to t\bar{t}H$ is about 3 fb, including the initial state radiation and beam effects. Basic cuts to select signal events from the background consist of event shape cuts requiring 8 or 6 jets in the final state and small thrust, mass cuts demanding $m_{2J} \simeq m_W$ or m_H and $m_{3J} \simeq m_t$, and in the case of the 8-jet mode, four momentum balance. In addition, the b-tagging requiring more than two b-jet candidates in the final state is crucial. The total selection efficiency for the above two modes (i) and (ii) is estimated to be 0.39. On the other hand, the effective background cross sections (those which have been taken into account) amount to $\sigma_{t\bar{t}Z} \simeq 4.2$ fb and $\sigma_{t\bar{t}} \simeq 402$ fb, including the initial state radiation and beam effects. The overall detection efficiencies for the background events are $\epsilon_{t\bar{t}Z} = 0.14$ and $\epsilon_{t\bar{t}} = 0.0019$, respectively. The expected number of the signal events and background events are $S = 114$ and $B = 133$, respectively, for 100 fb^{-1}, resulting in $S/N = 0.86$. This is translated to a statistical error on the Yukawa coupling of $\Delta g_{tH}/g_{tH} \sim 14\%$ (4%) for 100 fb^{-1} (1000 fb^{-1}).

In the case of a heavy Higgs of $m_H > 2m_t$, the Yukawa coupling can be extracted from a measurement of the decay branching ratio of the Higgs boson into $t\bar{t}$, combined with the direct measurement of the decay width of the Higgs boson. From the simulation study [49], the sensitivity to the Yukawa coupling of $\Delta g_{tH}/g_{tH} \sim 5\%$ (12%) is expected for $m_H = 400$ (500) GeV and the integrated luminosity of $\int d\mathcal{L} \sim 1000$ fb^{-1}.

3.1.2. *Form Factors*

Another study which is important in the open top region at an e^+e^- collider is a systematic and detailed analysis of the various interactions of the top quark. In contrast to the study of top quark properties close to the threshold, where the top quark is strongly influenced by QCD boundstate effects, in the open-top region we may regard the top quark to be a free parton in the zeroth-order approximation. In fact, hadronization effects are negligible in its production and decay processes, and perturbative evaluations of the various differential cross sections for the production-decay chain are reliable. In particular, the clean environment in comparison to the hadron colliders enables precise measurements of the form factors of the gauge interactions of the top quark ($t\bar{t}V$ vertices with $V = W^\pm, \gamma, Z, g$). Top quarks will be produced with a relatively high polarization, so that the angular distributions of the decay products can be used to disentangle different helicity components of top quarks efficiently.

Already radiative corrections to a number of differential as well as inclusive cross sections are available within the SM, both for the QCD [30, 50–55], and EW corrections [56, 57], which allow for the theoretical predictions typically at 1–5% accuracy. The 1-loop radiative corrections in the Supersymmetric Standard Model are also known [58]. Thanks to these calculations, precision measurements of the form factors of the top quark at a few percent accuracy are meaningful. Enhanced signals of new physics may be visible clearly in these observables if the EW symmetry breaking is driven by a strongly-interacting Higgs sector. Indeed, some models for strongly-coupled Higgs sector predict 5–10% deviations of the CP-conserving form factors from the SM predictions [59–61].

Sensitivity studies on the individual form factors have been performed based on Monte-Carlo simulations [62–65]. Here, we only mention that these studies have shown that, at $\sqrt{s} = 500$ GeV, they can be determined typically with 1–5% statistical accuracy corresponding to an integrated luminosity of 300 fb^{-1}. Further analysis seems to be needed in order to find an optimized strategy for disentangling individual form factors systematically. We discuss this issue in Section 3.2.

3.1.3. *Top Mass Reconstruction*

As we have discussed in Section 2.2, the top quark mass can be measured precisely in the threshold region. Theoretically it is an interesting but challenging subject to extract the top quark mass far above the threshold and compare it with that determined in the threshold region. Experimentally, the peak position of the jet invariant-mass distribution of the decay products of the top quark can be measured to an accuracy of order 100 MeV or better [66]. The question is: What is the mass extracted from the peak position of the Breit-Wigner distribution? Naively we may consider it the pole mass, since the peak position is determined from the position of the pole of the top quark propagator in perturbative QCD. On the other hand, as we have seen, the pole mass suffers from a theoretical uncertainty of order 300 MeV due to renormalon. Thus, we will face this serious conceptual question on what we measure. The main problem lies in the fact that we have no reliable theoretical method at hand to calculate the invariant mass distribution of realistic color-singlet final states. Rather we calculate the invariant mass of final partons which do not combine to a color-singlet state. Then we cannot expect the renormalon cancellation to take place. This will result in a rather large uncertainty of the top quark mass even if we will be able to compute higher-order perturbative corrections. It was also shown that from a full kinematical reconstruction of the dilepton events $e^+e^- \to t\bar{t} \to b\bar{b}\ell^+\ell^-\nu\bar{\nu}$, the top quark mass can be "determined" to an accuracy of order 200 MeV [67]. In this case, however, determination of the momenta of colored partons play an essential role as well. We may not expect the leading renormalons to cancel in the perturbative calculation of this observable. Observables less ambiguous from a theoretical point of view are those which can be obtained from color-singlet final states, such as the invariant mass distribution of $\ell^+\ell^-$. If there is a strong correlation between such an observable and the top quark mass, we will be able to determine the top quark mass accurately, provided we attain computations of radiative corrections to a sufficient accuracy. In any case, further theoretical studies are mandatory to take full advantage of expected experimental capabilities.

3.2. *Use of Inclusive Observables in Studying Form Factors*

In most of the previous sensitivity studies of the top quark form factors in the open top region, strategy for extracting individual form factors is based on reconstruction of the spins of the top quark and W boson (statistically) from the energy-angular distributions of their decay products. Our purpose here is to propose an alternative method for extracting individual form factors using inclusive observables. We discuss a systematic algorithm which may be useful. But we have not carried out any simulation studies yet, so this analysis should be taken as a hint for further analysis.

We follow the convention of [43] in defining the form factors of the top quark vertices. There are eight independent form factors in the $t\bar{t}$ production vertices ($V = \gamma,\ Z$):

$$\Gamma^{\mu}_{t\bar{t}V} = \frac{g}{2}\,\overline{u}(p_t)\Big[\gamma^{\mu}\Big\{A_V + \delta A_V - (B_V + \delta B_V)\gamma_5\Big\} + \frac{(p_t - p_{\bar{t}})^{\mu}}{2m_t}(\delta C_V - \delta D_V\gamma_5)\Big]v(p_{\bar{t}}). \qquad (42)$$

The tree-level SM contributions are given by

$$A_\gamma = \tfrac{4}{3}\sin\theta_W, \qquad B_\gamma = 0,$$
$$A_Z = \tfrac{1}{2\cos\theta_W}\Big(1 - \tfrac{8}{3}\sin^2\theta_W\Big), \qquad B_Z = \tfrac{1}{2\cos\theta_W}. \qquad (43)$$

The loop corrections within the SM as well as non-SM contributions are included in the anomalous form factors δA_V, δB_V, δC_V and δD_V. Among these, $\delta D_{\gamma,Z}$ are the CP-odd form factors, whereas all the other form factors are CP-even. The decay vertices of t and $\bar{t}$, respectively, include four form factors $f_{1,2}^{L,R}$ and $\overline{f}_{1,2}^{L,R}$:

$$\Gamma^{\mu}_{Wtb} = -\frac{g}{\sqrt{2}} V_{tb}\, \overline{u}(p_b) \Big[\gamma^{\mu}(f_1^L P_L + f_1^R P_R) \\ -\frac{i\sigma^{\mu\nu}k_\nu}{M_W}(f_2^L P_L + f_2^R P_R) \Big] u(p_t), \tag{44}$$

$$\overline{\Gamma}^{\mu}_{Wtb} = -\frac{g}{\sqrt{2}} V^*_{tb}\, \overline{v}(p_{\bar{t}}) \Big[\gamma^{\mu}(\overline{f}_1^L P_L + \overline{f}_1^R P_R) \\ -\frac{i\sigma^{\mu\nu}k_\nu}{M_W}(\overline{f}_2^L P_L + \overline{f}_2^R P_R) \Big] v(p_{\bar{b}}), \tag{45}$$

where $P_L = (1-\gamma_5)/2$ and $P_R = (1+\gamma_5)/2$; k denotes the momentum of W; we assume that W is on-shell. At tree level of the SM, $f_1^L = \overline{f}_1^L = 1$, and all the other form factors are zero.

It would be convenient to first separate the CP-odd contributions δD_V, $f_1^L - \overline{f}_1^L$, $f_1^R - \overline{f}_1^R$, $f_2^L - \overline{f}_2^R$ and $f_2^R - \overline{f}_2^L$ using CP-odd observables and to determine them. These form factors may be determined efficiently by using the forward-backward charge asymmetry of the charged lepton or b-quark [43]

$$A^f_{CP} = \frac{\int_{-c_m}^{0} d\cos\theta_f \frac{d\sigma^+(t\to f)}{d\cos\theta_f} - \int_0^{+c_m} d\cos\theta_f \frac{d\sigma^-(t\to \overline{f})}{d\cos\theta_f}}{\int_{-c_m}^{0} d\cos\theta_f \frac{d\sigma^+(t\to f)}{d\cos\theta_f} + \int_0^{+c_m} d\cos\theta_f \frac{d\sigma^-(t\to \overline{f})}{d\cos\theta_f}} \tag{46}$$

for $f = \ell, b$. Also, one may use the energy asymmetry $\langle E_{\ell^+} - E_{\ell^-}\rangle$ or $\langle \mathrm{sign}(E_{\ell^+} - E_{\ell^-})\rangle$, or the observables studied in Refs. [68–71]. See the chapter on CP-asymmetry for more details. In the following we study how to extract the CP-even form factors.

Suppose we are interested in the case where deviations of the form factors from the tree-level SM contributions are small and consider deviations only up to the first order, *i.e.* we neglect the terms quadratic in the anomalous form factors. Then, only the contributions from f_1^L, $\overline{f}_1^L$, f_2^R and $\overline{f}_2^L$ need to be considered among all the decay form factors, if we neglect the mass of the b-quark. (Sensitivity studies based on the direct measurement of f_1^R and $\overline{f}_1^R$ have been performed in Refs. [31] and [65]. These studies incorporated the terms quadratic in these form factors. As expected, the sensitivities are rather low. On the other hand, fairly stringent constraints have already been posed on them indirectly from the process $b \to s\gamma$ [72].)

Thus, we would like to determine the eight CP-even form factors ($V = \gamma, Z$):

$$\delta A_V, \;\; \delta B_V, \;\; \delta C_V, \;\; f_1^L + \overline{f}_1^L, \;\; f_2^R + \overline{f}_2^L. \tag{47}$$

Among these, a deviation of $f_1^L + \overline{f}_1^L$ from the SM value modifies only the overall normalizations of the top quark decay distributions. In other words, essentially its effect is to vary the total decay width of the top quark Γ_t. A precise measurement of the top quark decay width in the open top region will probably be difficult. The best accuracy would be attained in the threshold region from the measurement of the total cross section, the momentum distribution and the forward-backward asymmetry of the top quark.

Separation between the $V = \gamma$ and $V = Z$ components of the production form factors, δA_V, δB_V, δC_V, can be made optimally by varying the parameters of the initial state, the polarizations and the energies of the e^+e^- beams.[1] The reason is as follows. The differences in the structure of the amplitude between the $V = \gamma$ and $V = Z$ contributions consist only in the γ/Z propagators and the way they couple to the electron current $J^\mu_{(5)} = \overline{v}(p_{e^+})\gamma^\mu(\gamma_5)u(p_{e^-})$. Therefore, changing these factors by controlling the beam polarizations and $\sqrt{s}$ is the most straightforward way for varying the relative weight of the $V = \gamma$ and $V = Z$ contributions to the amplitude.

Then our task is to separate the four form factors (without distinguishing between the $V = \gamma$ and $V = Z$ components),

$$\delta A_V, \;\; \delta B_V, \;\; \delta C_V, \;\; f_2^R + \overline{f}_2^L, \tag{48}$$

using the dependences on the kinematical variables of the *final states*. Here, we consider inclusive observables which are insensitive to the top quark decay width.

The production process can be characterized by (1) the production rate of the top quark, (2) the angular distribution of the top quark, and (3) the spins of t and $\bar{t}$. The production rate (1) will be measured from the total cross section $\sigma(e^+e^- \to t\bar{t})$. The leading angular distribution (2) may be characterized by the forward-backward asymmetry of the top quark. Since the charge of the b quarks can be identified with a good efficiency [37],

[1]This point has been emphasized especially for the CP-odd form factor δD_V [42].

the forward-backward asymmetry can be measured using both 4-jet-plus-1-lepton and 6-jet final states. As for the spins of t and $\bar{t}$ (3), it is known that the top quark spin can be analyzed with maximal sensitivity using the angular distribution of the charged lepton in the rest frame of the top quark [see discussion in Section 2.3 C)]. Also, it was shown [43] that the angular distribution of the lepton $d\sigma/d\cos\theta_\ell$, where θ_ℓ is the angle between the charged lepton and the electron beam direction in the CM frame, is independent of the anomalous decay form factors (in the approximation $m_b = 0$ and retaining only the linear terms in the anomalous form factors). Taking these features into account, the spins of t and $\bar{t}$ can be analyzed using the charged lepton angular distributions. As simple inclusive observables, we may use the forward-backward asymmetry of the charged lepton,

$$\langle \mathrm{sign}(\cos\theta_f) \rangle = \frac{\sigma(\cos\theta_f > 0) - \sigma(\cos\theta_f < 0)}{\sigma(\cos\theta_f > 0) + \sigma(\cos\theta_f < 0)} \qquad (f = \ell^+ \text{ or } \ell^-), \quad (49)$$

or the correlation of the charged lepton momenta, $\langle \vec{p}_{\ell^+} \cdot \vec{p}_{\ell^-} \rangle$.

All the above observables considered for extracting the production form factors are insensitive to the decay form factors. In order to extract the remaining decay form factors f_2^R and $\overline{f}_2^L$, we should use the observables sensitive to these anomalous form factors. Such examples are the forward-backward asymmetry of the b-quark or $\bar{b}$-quark, the correlation of the b and $\bar{b}$ momenta (it is necessary to identify the charge of b-quarks in both observables), the mean energy of the charged leptons $\langle E_{\ell^+} + E_{\ell^-} \rangle$, etc.

Previous sensitivity studies of the top quark form factors have put emphasis on reconstructing spins of t and W from the energy-angular distributions of decay products. Here we focused on inclusive observables. In this way, (a) sensitivities to the form factors may be raised; (b) correspondences between the observables and the form factors may be understood more simply.[m]

To summarize our strategy, (1) We first separate and determine the CP-odd form factors using CP-odd observables; (2) We extract and decompose the production form factors δA_V, δB_V and δC_V from the total cross section $\sigma_{\mathrm{tot}}(e^+e^- \to t\bar{t})$, the forward-backward asymmetry of the top quark, the forward-backward asymmetry of the charged lepton $\langle \mathrm{sign}(\cos\theta_{\ell^\pm}) \rangle$ and the correlation of the charged-lepton momenta $\langle \vec{p}_{\ell^+} \cdot \vec{p}_{\ell^-} \rangle$; (3) The $V = \gamma$ and $V = Z$ components are decomposed by varying the polarizations and the

[m]The former idea has been pushed mathematically in the analyses [68, 73] which use optimal observables [74], although that method may be disadvantageous from the latter viewpoint.

energies of the e^+e^- beams; (4) We extract f_1^L and $\overline{f}_1^L$ from the measurement of the top quark width; (5) We extract f_2^R and $\overline{f}_2^L$ from the forward-backward asymmetry of b or $\bar{b}$, the correlation of the b and $\bar{b}$ momenta $\langle \vec{p}_b \cdot \vec{p}_{\bar{b}} \rangle$, and the average energy of the charged lepton $\langle E_{\ell^+} \rangle$ or $\langle E_{\ell^-} \rangle$.

4. Conclusions

We have seen that, at a future e^+e^- linear collider, we will be able to examine detailed properties of the top quark as well as investigate its interactions through various processes in a clean environment. In particular, powerful analyses will be available utilizing the high polarization of the top quark and feasibility of detailed kinematical reconstruction.

With respect to the grand goal of the linear collider: "to uncover the mechanism of electroweak symmetry breaking," we may put the role of Top Quark Physics as follows.

- Precision determination of the top quark $\overline{\rm MS}$ mass — $\overline{m}_t$ is the largest symmetry-breaking mass parameter in the SM. It will become the crucial parameter for an accurate understanding of the symmetry breaking physics (including radiative corrections).
- Detailed tests of various top quark interactions — The top quark is expected to be a sensitive probe for the symmetry breaking physics due to its large coupling to the order parameter of the electroweak gauge symmetry.

There will be other aspects as well, which are not directly related to the symmetry breaking physics:

- Anomalous CP-violating interaction of top quark — Since the SM contributions to the top CP-violating interactions are extremely small, their measurements will be sensitive to new physics.
- Physics of heavy quarkonium: remnant of "toponium" — It is a unique physics target in the $t\bar{t}$ threshold region. The detailed study will advance our understanding of the structure and dynamics of QCD bound-states, which, as yet, is still at a premature stage.

According to the linear collider operation program, Top Quark Physics will be studied both in the $t\bar{t}$ threshold region and in the open top region.

There have been extensive studies on physics that can be covered in experiments in the $t\bar{t}$ threshold region. In particular, theoretical understanding has developed remarkably over the last decade. Also, detailed Monte

Carlo simulation studies were initiated focusing in this energy region. The most important outcome up to now is the establishment of a future prospect that we will be able to measure the top quark $\overline{\rm MS}$ mass precisely. Furthermore, it has been shown that unique studies of various properties of the top quark and (remnant of) $t\bar{t}$ resonances will be possible. We may achieve the following measurements: $\overline{m}_t$ can be measured with 150 MeV accuracy (10^{-3} relative accuracy); Γ_t can be measured with 5–10% accuracy; we will be able to test various form factors of the top quark; we will test dynamics of heavy quarkonium. Furthermore, if we can reduce uncertainties of the present theoretical predictions, we can further gain in the accuracies of the above measurements, as well as measure the top-Higgs Yukawa coupling (provided the Higgs boson is relatively light).

Studies of the top quark properties in the energy region far above the threshold extend to wide varieties. We have reviewed only a part of these studies. Among them, determination of the top-quark Yukawa coupling (if the Higgs boson mass is below 500 GeV) and a systematic analysis of the top quark interactions will provide important information which is useful for other analyses in the whole program of the precision experiments to be carried out at a linear collider. With a high luminosity, we will be able to measure the Yukawa coupling at the level of 5–15% accuracy both in the cases $m_H < 2m_t$ and $m_H > 2m_t$. Also, we will be able to measure the various top quark form factors at 1–5% accuracy.

These studies illustrate that the future linear collider will play an important role in understanding the nature of the top quark and in searching for new physics beyond the SM through the study of the top quark interactions. We invite the readers to further develop theoretical and experimental studies by the time of the collider operation.

References

1. CDF Collaboration, Fermilab preprint, FERMILAB-PUB-97-284-E (1997).
2. D0 Collaboration, Phys. Rev. Lett. **79**, 1197 (1997).
3. M. Jeżabek and J.H. Kühn, Phys. Rev. **D48**, R1910 (1993); erratum Phys. Rev. **D49**, 4970 (1994); and references therein.
4. J.H. Kühn, Acta Phys. Polonica **B12**, 347 (1981).
5. M. Peskin, in *Proceedings of Workshop on Physics and Experiments with Linear Colliders*, Saariselka, Finland, 1991, edited by R. Orava, P. Elrola and M. Nordbey (World Scientific, Singapore, 1992).
6. K. Fujii, T. Matsui and Y. Sumino, Phys. Rev. **D50**, 4341 (1994).
7. A. Hoang and T. Teubner, Phys. Rev. **D58**, 114023 (1998); K. Melnikov and A. Yelkhovsky, Nucl. Phys. **B528**, 59 (1998).

8. O. Yakovlev, Phys. Lett. **B457**, 170 (1999).
9. A. Penin and A. Pivovarov, Nucl. Phys. **B549**, 217 (1999); Phys. Atom. Nucl. **64**, 275 (2001).
10. A. Hoang, M. Smith, T. Stelzer and S. Willenbrock, Phys. Rev. **D59**, 114014 (1999); M. Beneke, Phys. Lett. **B434**, 115 (1998).
11. A. H. Hoang, et al., Eur. Phys. J. direct **C3**, 1 (2000).
12. Y. Sumino, Acta Phys. Polonica **B25**, 1837 (1994).
13. P. Comas, R. Miquel, M. Martinez and S. Orteu, in *"e^+e^- Collisions at TeV Energies"*, ed. P. Zerwas, DESY 96-123D, (1996).
14. D. Peralta, M. Martinez and R. Miquel, in the Proceedings of "*Physics and Experiments with Future Linear e^+e^- Colliders*," edited by E. Fernández and A. Pacheco (Universitat Autònoma de Barcelona); D. Cinabro, in the same proceedings.
15. Y. Kiyo and Y. Sumino, Phys. Lett. **B496**, 83 (2000).
16. V. Fadin and O. Yakovlev, *Sov. J. Nucl. Phys.* **53** (1991) 688.
17. P. Gambino and A. Sirlin, Phys. Lett. **B355**, 295 (1995).
18. M. Beneke, V. Braun and V. Zakharov, Phys. Rev. Lett. **73**, 3058 (1994); M. Luke, A. Manohar and M. Savage, Phys. Rev. **D51**, 4924 (1995); M. Neubert and C. Sachrajda, Nucl. Phys. **B438**, 235 (1995).
19. M. Beneke, A. Signer and V. Smirnov, Phys. Lett. **B454**, 137 (1999).
20. T. Nagano, A. Ota and Y. Sumino, Phys. Rev. **D60**, 114014 (1999); A. Hoang and T. Teubner, Phys. Rev. **D60**, 114027 (1999).
21. A. Pineda and F. Ynduráin, Phys. Rev. **D58**, 094022 (1998).
22. V.S. Fadin and V.A. Khoze, JETP Lett. **46**, 525 (1987) ; Sov. J. Nucl. Phys. **48**, 309 (1988).
23. M. Strassler and M. Peskin, Phys. Rev. **D43**, 1500 (1991).
24. Y. Sumino, K. Fujii, K. Hagiwara, H. Murayama, and C.-K. Ng, Phys. Rev. **D47**, 56 (1993).
25. M. Ježabek, J.H. Kühn and T. Teubner, Z. Phys. **C56**, 653 (1992); M. Ježabek and T. Teubner, Z. Phys. **C59**, 669 (1993).
26. H. Murayama and Y. Sumino, Phys. Rev. **D47**, 82 (1993).
27. R. Harlander, M. Ježabek, J. Kühn, and M. Peter, Z. Phys. **C73**, 477 (1997).
28. M. Peter and Y. Sumino, Phys. Rev. **D57**, 6912 (1998).
29. M. Ježabek and J. Kühn, Phys. Lett. **B316**, 360 (1993).
30. A. Czarnecki, M. Ježabek and J.H. Kühn, Nucl. Phys. **B351**, 70 (1991); A. Czarnecki and M. Ježabek, Nucl. Phys. **B427**, 3 (1994).
31. M. Ježabek, Nucl. Phys. **37 B** *(Proc.Suppl.)*, 197 (1994) .
32. M. Ježabek, Acta Phys. Polonica **B26**, 789 (1995).
33. J. Kühn and K. Streng, Nucl. Phys. **B198**, 71 (1982); M. Ježabek and J. Kühn, Nucl. Phys. **B320**, 20 (1989).
34. M. Ježabek and J.H. Kühn, Nucl. Phys. **B320**, 20 (1989).
35. R. Guth and J. Kühn, Nucl. Phys. **B368**, 38 (1992); R. Harlander, M. Ježabek and J. Kühn, Acta Phys. Polonica **B27**, 1781 (1996).
36. JLC Group, KEK Report 92-16, 1992; K. Yokoya, Proceedings of 1988 Linear Accelerator Conference (CEBAF Report No. 89-08, Geneva, 1989), Vol.1.
37. M. Iwasaki, hep-ex/0102014.

38. A. Hoang, A. Manohar, I. Stewart and T. Teubner, Phys. Rev. Lett. **86**, 1951 (2001).
39. K. Ikematsu, *et al.*, in preparation.
40. T. Sjöstrand and V. Khoze, Eur. Phys. J. direct **C1**, 1 (2000); Proceedings of the 4th International Workshop on Linear Colliders (LCWS 99), Universitat Autonoma de Barcelona, 2000 (hep-ph/9908408).
41. W. Bernreuther, T. Schröder and T. N. Pham, Phys. Lett. **B279**, 389 (1992); A. Soni and R. M. Xu, Phys. Rev. Lett.**69**, 33 (1992).
42. F. Cuypers and S. Rindani, Phys. Lett. **B343**, 333 (1995).
43. B. Grządkowski and Z. Hioki, Phys. Lett. **B476**, 87 (2000).
44. J. Kühn, in 2nd ECFA/DESY Study 1998-2001, p. 1418.
45. A. Djouadi, J. Kalinowski and P. Zerwas, Mod. Phys. Lett. **A7**, 1765 (1992); Z. Phys. **C54**, 255 (1992).
46. A. Juste and G. Merino, hep-ph/9910301.
47. H. Baer, S. Dawson and L. Reina, Phys. Rev. **D61**, 013002 (2000).
48. ACFA Linear Collider Working Group (Koh Abe, *et al.*), "Particle Physics Experiments at JLC", hep-ph/0109166.
49. J. Alcaraz and E. Morales, hep-ph/0012109.
50. J. Jersak, E. Laerman and P. Zerwas, Phys. Rev. **D25**, 1218 (1982); Y. Dokshitzer, V. Khoze and W. Stirling, Nucl. Phys. **B428**, 3 (1994).
51. G. Jikia, Phys. Lett. **B257**, 196 (1991); V. Khoze, *et al.*, Nucl. Phys. **B378**, 413 (1992); Y. Dokshitzer, *et al.*, Nucl. Phys. **B403**, 65 (1993).
52. C. Macesanu and L. Orr, hep-ph/0012177.
53. K. Chetyrkin, J. Kühn and M. Steinhauser, Nucl. Phys. **B482**, 213 (1996); R. Harlander and M. Steinhauser, Eur. Phys. J. **C2**, 151 (1998).
54. C. Schmidt, Phys. Rev. **D54**, 3250 (1996); A. Brandenburg, M. Flesch and P. Uwer, Phys. Rev. **D59**, 014001 (1999).
55. W. Beenakker, F. Berends and A. Chapovsky, Phys. Lett. **B454**, 129 (1999).
56. W. Beenakker, S. van der Marck, W. Hollik, Nucl. Phys. **B365**, 24 (1991); V. Driesen, W/ Hollik and A. Kraft, hep-ph/9603398; W. Beenakker, A. Denner and A. Kraft, Nucl. Phys. **B410**, 219 (1993).
57. A. Akhundov, D. Bardin and A. Leike, Phys. Lett. **B261**, 321 (1991).
58. W. Hollik and C. Schappacher, Nucl. Phys. **B545**, 98 (1999).
59. R. Chivukula, S. Selipsky and E. Simmons, Phys. Rev. Lett. **69**, 575 (1992); R. Chivukula, E. Simmons and J. Terning, Phys. Lett. **B331**, 383 (1994).
60. K. Hagiwara and N. Kitazawa, Phys. Rev. **D52**, 5374 (1995).
61. U. Mahanta, Phys. Rev. **D55**, 5848 (1997); Phys. Rev. **D56**, 402 (1997).
62. American Linear Collider Working Group (T. Abe, *et al.*), Chapter on Top Quark Physics in *"Linear Collider Physics Resource Book for Snowmass 2001"*.
63. M. Schmitt, in *"e^+e^- Collisions at TeV Energies"*, ed. P. Zerwas, DESY 96-123D, (1996).
64. W. Bernreuther, A. Brandenburg and P. Overmann, in *"e^+e^- Collisions at TeV Energies"*, edited by P. Zerwas, DESY 96-123D, (1996); W. Bernreuthe and P. Overmann, Z. Phys. **C61**, 599 (1994); Z. Phys. **C72**, 461 (1996).
65. R. Frey, hep-ph/9606201.

66. C. Bagliesi, et al. in "e^+e^- *Collisions at 500 GeV: The Physics Potential*", edited by P. Zerwas, DESY 92-123A (1992).
67. J. Antoš and G. Yeh, LCWS99, Sitges, Barcelona, Spain (1999), Fermilab-Conf-99/260.
68. D. Atwood and A. Soni, Phys. Rev. **D45**, 2405 (1992).
69. H.-Y. Zhou, Phys. Lett. **B439**, 393 (1998).
70. S.D. Rindani, Pramana J. Phys. **54**, 791 (2000).
71. D. Atwood, S. Bar-Shalom, G. Eilam and A. Soni, Phys. Rep. **347**, 1 (2001).
72. K. Fujikawa and A. Yamada, Phys. Rev. **D49**, 5890 (1994).
73. B. Grzadkowski and Z. Hioki, Nucl. Phys. **B585**, 3 (2000).
74. J. Gunion, B. Grzadkowski and X.-G. He, Phys. Rev. Lett. **77**, 5172 (1996).

CHAPTER 4

Supersymmetry and the Linear Collider

Jonathan L. Feng[a] and Mihoko M. Nojiri[b]

[a] *Department of Physics and Astronomy*
University of California, Irvine, CA 92697, USA

[b] *Yukawa Institute of Theoretical Physics*
Kyoto University, Kyoto 606-8502, Japan

We present a pedagogical introduction to supersymmetry and supersymmetric models and give an overview of the potential of the linear collider for studying them. If supersymmetry is found, its discovery will bring with it many more questions than answers. These include:

Are the newly discovered particles really superpartners?
If not all superpartners are discovered, where are the rest of them?
Do the electromagnetic, weak, and strong forces unify?
Is a supersymmetric particle the dark matter?
How are the supersymmetric flavor and CP problems solved?
What is the scale of supersymmetry breaking?
What are the fundamental interactions at the Planck scale?

We review how the linear collider will provide definitive answers to some of these and may shed light on the rest, even if only one or a few superpartners are kinematically accessible.

1. Introduction

Supersymmetry relates fermions to bosons and postulates the existence of a partner particle for every known particle. Its motivations range from the gauge hierarchy problem and the central role of supersymmetry in quantum theories of gravity to the unification of gauge couplings and the existence of dark matter. The search for supersymmetry in particle physics has been carried out on numerous fronts, and its discovery would mark the culmination of an intense research effort spanning several decades.

At the same time, the discovery of supersymmetry will bring with it many more questions than answers. These include:

- Are the newly discovered particles really superpartners?
- If not all superpartners are discovered, where are the rest of them?
- Do the electromagnetic, weak, and strong forces unify?
- Is a supersymmetric particle the dark matter?
- How are the supersymmetric flavor and CP problems solved?
- What is the scale of supersymmetry breaking?
- What are the fundamental interactions at the Planck scale?

Insights into these and other fundamental questions will have far-reaching implications, transforming not only particle physics, but also astrophysics and cosmology.

Here we present a pedagogical overview of what the linear collider may contribute to answering these questions. The large and growing world-wide interest in linear colliders has motivated hundreds of detailed studies of the potential of linear colliders for measuring supersymmetry parameters. Typically, however, there is little opportunity to review in broad terms the underlying physics questions and overall goals of this program. In this overview, we therefore begin with a brief introduction to supersymmetry in Section 2, focusing on the most salient points for collider physics. We then describe several outstanding successes and puzzles driving work in supersymmetry in Section 3 and some of the resulting models in Section 4. Against this backdrop, we then review what linear colliders may tell us in the remaining sections. An index of key terms is provided at the end. Throughout this overview, we consider an e^+e^- linear collider with center-of-mass energy between $300 - 1000$ GeV, a luminosity of $\mathcal{O}(100)$ fb^{-1}/yr, highly polarizable electron (and possibly also positron) beams, and the capability of running in alternative modes, such as e^-e^-, $e^-\gamma$.

This overview is neither a thorough introduction to supersymmetry nor a comprehensive review of the linear collider literature. Supersymmetry and its phenomenological implications have been systematically presented in a number of beautiful introductions. (See, for example, Refs. [1–8].) Detailed supersymmetry studies at linear colliders and exhaustive bibliographies may be found in recent reports of the large linear collider collaborations [9–11].

2. The Minimal Supersymmetric Standard Model

The *minimal supersymmetric standard model* (MSSM) is the supersymmetric extension of the standard model with minimal field content, conserved R-parity, and no additional theoretical assumptions. It is the underlying framework for most "model-independent" collider studies, and we therefore begin by describing its particle content and interactions.

2.1. *Particle Content*

Supersymmetry is, under general assumptions, the unique extension of the Poincare algebra, the algebra of spacetime translations P, rotations J, and boosts K. In its simplest form, supersymmetry extends the Poincare algebra by introducing two additional operators Q and $\overline{Q}$, which satisfy

$$\{Q_\alpha, \overline{Q}_{\dot{\beta}}\} = 2\sigma^{\mu}_{\alpha\dot{\beta}} P_\mu \ , \quad [Q_\alpha, P^\mu] = [\overline{Q}_{\dot{\beta}}, P^\mu] = 0 \ , \tag{1}$$

where the braces denote anti-commutation, σ^0 is the 2×2 identity matrix, and the σ^i are the Pauli sigma matrices. Q and $\overline{Q}$ are two-component spinors and fermionic, and so supersymmetry transformations relate fermions to bosons.

If supersymmetry is a fundamental symmetry of nature, all particles must lie in representations of the supersymmetry algebra. In conventional quantum field theory, particles are represented by fields, functions of the four spacetime coordinates. The Poincare algebra is represented by differential operators acting on these fields. For example,

$$P_x = i\frac{\partial}{\partial x} \ , \ J_x = -i\left(y\frac{\partial}{\partial z} - z\frac{\partial}{\partial y}\right) \ , \ K_x = i\left(t\frac{\partial}{\partial x} + x\frac{\partial}{\partial t}\right) \ . \tag{2}$$

To formulate supersymmetric field theories, we must extend this formalism to include supersymmetry. Not surprisingly, given the anti-commutator appearing in Eq. (1), this requires an extension of the usual spacetime coordinates to include new anti-commuting, or Grassmanian, coordinates θ_1, θ_2, $\overline{\theta}_1$ and $\overline{\theta}_2$. The supersymmetry operators may then be identified with the differential operators

$$Q_\alpha = \frac{\partial}{\partial\theta^\alpha} - i\sigma^{\mu}_{\alpha\dot{\beta}}\overline{\theta}^{\dot{\beta}}\partial_\mu \ , \ \overline{Q}_{\dot{\beta}} = -\frac{\partial}{\partial\overline{\theta}^{\dot{\beta}}} + i\theta^\alpha\sigma^{\mu}_{\alpha\dot{\beta}}\partial_\mu \ . \tag{3}$$

It is not hard to show that the operators of Eqs. (2) and (3) satisfy the relations of Eq. (1). These operators then act on *superfields*, functions of x^μ, θ, and $\overline{\theta}$, which are the natural representations of the supersymmetry algebra.

To determine the particle content of these theories, we expand superfields in powers of θ and $\overline{\theta}$. Conventional four-dimensional fields then appear as the coefficients in this expansion. This is completely analogous to the case of a field in higher spacetime dimensions, which may be decomposed into four-dimensional fields through a Fourier expansion in the extra dimensional coordinates. Four-dimensional fields emerge as coefficients in this expansion, producing a zero-mode field and a tower of Kaluza-Klein states. The case of supersymmetry is similar, with one important difference: the Grassmanian properties of the θ coordinates imply that the expansion terminates, and so superfields contain only a finite set of four-dimensional particles.

The simplest superfields are *chiral superfields*, independent of $\overline{\theta}$ (and anti-chiral fields, independent of θ). Expanding in θ, ones finds

$$\Psi(x,\theta) = \phi(x) + \sqrt{2}\,\theta^\alpha \psi_\alpha(x) + \theta\theta F(x) \ , \tag{4}$$

where $\theta\theta \equiv \theta^\alpha \theta_\alpha$. A chiral superfield then contains a chiral fermion ψ and a complex scalar ϕ. The θ coordinates have mass dimension $-1/2$. The field F therefore has mass dimension 2 and no renormalizable kinetic term. It contains no physical degrees of freedom and is known as an *auxiliary field.* It may be removed by applying its equation of motion, but it is often convenient to retain it for reasons to be discussed below.

The quarks and leptons of the standard model are contained in chiral superfields, and so supersymmetry requires the existence of scalar superpartners, the squarks and sleptons. Similarly, the Higgs boson of the standard model is in a chiral superfield, and so supersymmetry requires a fermionic superpartner, the Higgsino. Anomaly cancellation requires that this new fermion be accompanied by additional fermions. In the MSSM this is accomplished by introducing an additional Higgs doublet and the accompanying superpartners. The fields of the chiral supermultiplets in the MSSM, along with their quantum numbers, are listed in Table 1. Note that the gauge quantum numbers of particles in the same supermultiplet are identical, since standard model gauge symmetries are 'internal symmetries,' independent of spacetime symmetries.

Vector supermultiplets are superfields that are functions of both θ and $\overline{\theta}$. While there are in general nine terms in the expansion, many of them are not physical, as they may be removed by a supersymmetric generalization of gauge transformations. In the end, we are left with only

$$V(x,\theta,\overline{\theta}) = \theta\sigma^\mu\overline{\theta} v_\mu(x) - i\overline{\theta}\overline{\theta}\theta^\alpha\lambda_\alpha(x) + i\theta\theta\overline{\theta}_{\dot{\alpha}}\overline{\lambda}^{\dot{\alpha}}(x) + \frac{1}{2}\theta\theta\overline{\theta}\overline{\theta}D(x) \ . \tag{5}$$

Table 1. Particles of the Minimal Supersymmetric Standard Model and their SU(3), SU(2), and U(1)$_Y$ quantum numbers. All chiral fermions are left-handed, and the superscript c denotes charge conjugation. We will also use the common alternative notations $\tilde{e}_L \equiv \tilde{e}_L^- \equiv \tilde{e}$, $\tilde{e}_R^* \equiv \tilde{e}_R^+ \equiv \tilde{e}^c$, and so forth.

Chiral supermultiplets	Quarks	Squarks
$Q\ (3,2,1/6)$ $U^c\ (3,1,-2/3)$ $D^c\ (3,1,1/3)$	$q=(u,d)$ u^c d^c	$\tilde{q}=(\tilde{u},\tilde{d})$ $\tilde{u}^c$ $\tilde{d}^c$
	Leptons	Sleptons
$L\ (1,2,-1/2)$ $E^c\ (1,1,1)$	$l=(\nu,e)$ e^c	$\tilde{l}=(\tilde{\nu},\tilde{e})$ $\tilde{e}^c$
	Higgs bosons	Higgsinos
$H_d\ (1,2,-1/2)$ $H_u\ (1,2,1/2)$	(H_d^0, H_d^-) (H_u^+, H_u^0)	$(\tilde{H}_d^0, \tilde{H}_d^-)$ $(\tilde{H}_u^+, \tilde{H}_u^0)$
Vector supermultiplets	Gauge bosons	Gauginos
$(8,1,1)$ $(1,3,1)$ $(1,1,1)$	g $W^\pm, Z$ γ	$\tilde{g}$ (gluino) $\tilde{W}^\pm$, $\tilde{W}^0$ (Winos) $\tilde{B}$ (Bino)
Gravity supermultiplet	Graviton	Gravitino
$(1,1,1)$	$g_{\mu\nu}$	$\tilde{G}$

A vector superfield contains a vector boson v and a Majorana fermion λ. The field D has mass dimension 2 and so, as with F, is an auxiliary field with no physical degrees of freedom. The standard model gauge bosons are contained in vector supermultiplets, and so the MSSM also contains their fermionic partners, the gauginos. These are included in Table 1.

Finally, although not strictly a particle of the standard model, the graviton $g_{\mu\nu}$ is the gauge boson of gravity, coupling to the energy-momentum tensor. Its superpartner, the gravitino $\tilde{G}$, is a spin 3/2 particle that couples to the supercurrent J_Q^μ, which generates supersymmetry. The gravitino plays an important role in the phenomenology of some supersymmetric models.

2.2. *Supersymmetric Matter Interactions*

Given the particles of Table 1, we may then construct the most general supersymmetric Lagrangian. To construct the matter interactions, it is convenient to define the *superpotential*

$$W = \frac{1}{2}\mu_{ij}\Psi_i\Psi_j + \frac{1}{3}y_{ijk}\Psi_i\Psi_j\Psi_k \ , \tag{6}$$

a gauge-invariant function of chiral superfields. The most general matter coupling is then compactly expressed in superfield notation as

$$\mathcal{L}_{\text{matter}} = \Psi_i^\dagger \Psi_i|_{\theta\theta\overline{\theta}\overline{\theta}} + [W|_{\theta\theta} + \text{H.c.}] , \tag{7}$$

where one takes the components in the θ, $\overline{\theta}$ expansion as indicated. The trilinear superpotential terms generate the Yukawa couplings of the standard model with dimensionless couplings y_{ijk}, along with other interactions related by supersymmetry. The parameters μ_{ij} have mass dimension 1.

In component fields, Eq. (7) becomes

$$\begin{aligned}\mathcal{L}_{\text{matter}} &= \partial^\mu \phi_i^* \partial_\mu \phi_i + i\overline{\psi}_i \overline{\sigma}^\mu \partial_\mu \psi_i + F_i^* F_i \\ &+ \left[\mu_{ij}\left(\phi_i F_j - \frac{1}{2}\psi_i\psi_j\right) + y_{ijk}(\phi_i\phi_j F_k - \phi_i \psi_j \psi_k) + \text{H.c.}\right] .\end{aligned} \tag{8}$$

The first line includes canonical kinetic terms for every ψ_i and ϕ_i. As discussed above, however, there are no corresponding terms for the auxiliary fields F_i. These may therefore be removed by their equations of motion

$$F_i^* = -\mu_{ij}\phi_j - y_{ijk}\phi_j\phi_k . \tag{9}$$

With this substitution, the complete matter Lagrangian becomes

$$\begin{aligned}\mathcal{L}_{\text{matter}} &= \partial^\mu \phi_i^* \partial_\mu \phi_i + i\overline{\psi}_i \overline{\sigma}^\mu \partial_\mu \psi_i - \frac{1}{2}\mu_{ij}\psi_i\psi_j - \frac{1}{2}\mu_{ij}^*\overline{\psi}_i\overline{\psi}_j \\ &- y_{ijk}\phi_i\psi_j\psi_k - y_{ijk}^*\phi_i^*\overline{\psi}_j\overline{\psi}_k - \left|\mu_{ij}\phi_j + y_{ijk}\phi_j\phi_k\right|^2 .\end{aligned} \tag{10}$$

Terms such as the last one in Eq. (10) are called *F-terms*, as they arise from $|F_i|^2$.

From Eq. (10) we find that both scalar and fermion masses are determined by μ, and they are degenerate. This is a generic prediction of exact supersymmetry. The absence of a boson with the couplings and mass of the electron implies that supersymmetry must be broken. In addition, we see that there are trilinear and quartic scalar couplings. The quartic couplings are completely determined by the Yukawa couplings y. For this reason, the quadratically-divergent contributions to the Higgs boson mass shown in Fig. 1 exactly cancel, as required for supersymmetry to stabilize the gauge hierarchy.

The superpotential of the MSSM is

$$W = \epsilon_{ab}\left[-y_{ij}^u H_u^a Q_i^b U_j^c + y_{ij}^d H_d^a Q_i^b D_j^c + y_{ij}^e H_d^a L_i^b E_j^c - \mu H_u^a H_d^b\right] , \tag{11}$$

where a, b are SU(2) indices contracted by the anti-symmetric ϵ tensor with $\epsilon_{12} = 1$, and i, j are generational indices. As in Table 1, the superscript c denotes charge conjugation. Gauge invariance forbids bilinear terms in

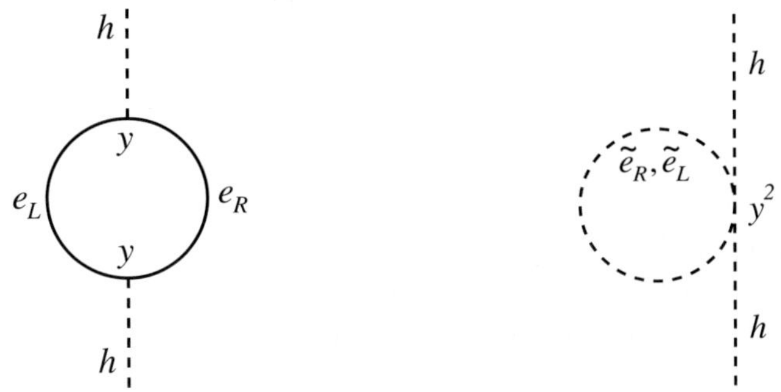

Fig. 1. Quadratically-divergent contributions to the Higgs boson mass in supersymmetry. These cancel exactly in the limit of exact supersymmetry.

the standard model, but the additional Higgs doublet of the MSSM, H_u, may have bilinear couplings. Linear terms are forbidden by gauge invariance in the MSSM, but may be present in extensions of the MSSM that contain complete gauge singlets. Additional non-renormalizable couplings may also be included in extensions of the MSSM with quartic and higher order superpotential terms suppressed by some heavy mass scale.

Equation (11) is *not* the most general gauge invariant superpotential. For example, as evident in Table 1, the superfields L_i and H_d have identical quantum numbers, leading to additional possible terms. The complete list of omitted terms is

$$\begin{aligned} W_L &= \lambda_{ijk} L_i L_j E_k^c + \mu_i L_i H_u \\ W_{LB} &= \lambda'_{ijk} L_i Q_j D_k^c \\ W_B &= \lambda''_{ijk} U_i^c D_j^c D_k^c \ . \end{aligned} \quad (12)$$

W_L violates lepton number L, W_B violates baryon number B, and W_{LB} violates both L and B. If all terms are present, protons decay through renormalizable operators, and present bounds from proton decay are extremely stringent. All of these terms may be eliminated by assigning to each particle an *R-parity* $R = (-1)^{3(B-L)+2S}$, where S is spin, and requiring that R-parity be multiplicatively conserved. All standard model particles have $R = 1$ and all superpartners have $R = -1$, and so all terms of Eq. (12) are eliminated. R-parity conservation implies that the lightest supersymmetric particle (LSP) is stable, with strong implications for collider studies and dark matter. We will focus on R-parity conserving supersymmetry throughout this review.

2.3. *Supersymmetric Gauge Interactions*

The interactions of the MSSM that involve gauge fields fall into several categories. The terms involving only gauge fields are

$$-\frac{1}{4}F^a_{\mu\nu}F^{a\,\mu\nu} + i\overline{\lambda}^a\overline{\sigma}^\mu\partial_\mu\lambda^a + \frac{1}{2}D^{a\,2} \tag{13}$$

for each gauge group, where a labels gauge group generators. The first two terms are the canonical kinetic terms for gauge bosons and gauginos. As with the F field, D has no kinetic term and may be eliminated by its equation of motion. This leads to the quartic scalar couplings

$$V_D = \frac{g_s^2}{2}\Big[\sum_{a,ij}\phi_i^\dagger T^a_{3,ij}\phi_j\Big]^2 + \frac{g^2}{2}\Big[\sum_{a,ij}\phi_i^\dagger T^a_{2,ij}\phi_j\Big]^2 + \frac{g'^2}{2}\Big[\sum_i\phi_i^\dagger Y_i\phi_i\Big]^2 , \tag{14}$$

where T_3 and T_2 are the generators of SU(3) and SU(2), and Y is hypercharge. These are known as *D-terms.* As in Section 2.2, these quartic interactions are completely determined by standard model couplings — in this case, the gauge couplings. A host of quadratically-divergent contributions to the Higgs mass that involve gauge interactions are thereby exactly canceled by supersymmetric contributions, allowing supersymmetry to completely stabilize the gauge hierarchy.

The gauge-matter couplings include $v_\mu\phi\partial^\mu\phi^*$ couplings, as well as four-point $v^\mu v_\mu\phi\phi^*$ interactions, where v^μ denotes gauge bosons. Finally, and perhaps of greatest significance for linear collider studies, there are sfermion-fermion-gaugino couplings, the supersymmetric analogues to the fermion-fermion-gauge boson couplings. These are of the form

$$\sqrt{2}g\phi_i T^a_{ij}\psi_j\lambda^a + \text{H.c.} , \tag{15}$$

where the coupling g is a standard model gauge coupling. Supersymmetry relates the couplings of gauginos to the couplings of gauge bosons. As we will discuss in Section 7, this fact may be exploited to definitively identify new particles as superpartners, as well as to determine the mass scales of undiscovered superpartners.

2.4. *Supersymmetry-breaking Terms*

So far we have been discussing the purely supersymmetric part of the MSSM Lagrangian. However, as noted above, supersymmetry predicts that superpartners have equal mass. This is also evident from the supersymmetry algebra: the supersymmetry generator commutes with the momentum operator, and so

$$(HQ - QH)|\phi\rangle = (E_\psi - E_\phi)|\psi\rangle = 0 \; . \tag{16}$$

Because we have not found superpartners degenerate with the known particles, supersymmetry must be broken.

Supersymmetry may be broken in many ways. However, to preserve the strong phenomenological motivation of solving the gauge hierarchy problem, it is imperative that the quadratic divergences, so elegantly removed by exact supersymmetry, are not re-introduced. When we require this, the allowed terms, known as *soft supersymmetry-breaking terms*, are limited to four classes: gaugino masses $M\lambda\lambda$, scalar masses $m^2\phi^*\phi$, trilinear scalar couplings $A\phi\phi\phi$ (A-*terms*), and bilinear scalar couplings $B\phi\phi$ (B-*terms*). No new dimensionless couplings are introduced, preserving the cancellations discussed in Sections 2.2 and 2.3. Also note, however, that not all dimensionful contributions are allowed; for example, terms of the form $\phi_i^*\phi_j\phi_k$ are not soft.

In the MSSM, the full set of soft terms is

$$\begin{aligned}
-\mathcal{L}_{\text{soft}} = & \frac{1}{2}(M_1\tilde{B}\tilde{B} + M_2\tilde{W}^a\tilde{W}^a + M_3\tilde{g}^a\tilde{g}^a) \\
& + m_{Q\,ij}^2(\tilde{u}_{Li}^*\tilde{u}_{Lj} + \tilde{d}_{Li}^*\tilde{d}_{Lj}) + m_{U\,ij}^2\tilde{u}_{Ri}^*\tilde{u}_{Rj} + m_{D\,ij}^2\tilde{d}_{Ri}^*\tilde{d}_{Rj} \\
& + m_{L\,ij}^2\,(\tilde{\nu}_{Li}^*\tilde{\nu}_{Lj} + \tilde{e}_{Li}^*\tilde{e}_{Lj}) + m_{E\,ij}^2\tilde{e}_{Ri}^*\tilde{e}_{Rj} \\
& + \epsilon_{ab}[-\tilde{A}_{ij}^u H_u^a\tilde{q}_i^b\tilde{u}_{Rj}^* + \tilde{A}_{ij}^d H_d^a\tilde{q}_i^b\tilde{d}_{Rj}^* + \tilde{A}_{ij}^e H_d^a\tilde{l}_i^b\tilde{e}_{Rj}^* \\
& \quad - BH_u^aH_d^b + \text{H.c.}] \; .
\end{aligned} \tag{17}$$

Gauge invariance requires that scalar particles in the same electroweak doublet receive the same soft masses. The A- and B-terms mimic the Yukawa and μ terms of the superpotential. If R-parity is broken, there are additional soft terms corresponding to the superpotential terms of Eq. (12).

What is the origin of these soft terms? At first sight, supersymmetry has just introduced a bewildering array of new parameters. If supersymmetry is to take us closer to a fundamental theory, we would hope that these parameters are not completely arbitrary. In fact, however, blind hope is not required! Experimental constraints already guarantee that, if weak-scale supersymmetry exists, the soft terms cannot be generic. These constraints, some of which will be reviewed in Section 3, imply that some pattern must emerge. Furthermore, we will see in Section 4 that the soft terms contain an imprint of physics at even more microscopic scales, and their determination may shed light on energy scales as high as the Planck scale. Just as discovering the mechanism of spontaneous gauge symmetry breaking is one of the key issues in standard model physics, uncovering the mechanism of supersymmetry breaking is the central issue of supersymmetric physics. The promise of model-independent measurements of the soft parameters is the prime supersymmetric motivation for the linear collider.

2.5. *Sleptons*

Given both the supersymmetric and the soft supersymmetry-breaking terms, we are now ready to determine the mass eigenstates of the MSSM. For lack of space, we focus on those with electroweak quantum numbers only, although squarks and gluinos may also be produced at linear colliders with many interesting results (see, for example, Refs. [12–17]).

We first discuss the scalar superpartners. There are six charged slepton states. These receive masses from soft terms and also, after electroweak symmetry breaking, from Yukawa couplings, F-terms, and D-terms. The full set of slepton mass terms is $\tilde{l}^\dagger \mathcal{M}^2_{\tilde{l}} \tilde{l}$, where $\tilde{l}^T = (\tilde{e}_L, \tilde{\mu}_L, \tilde{\tau}_L, \tilde{e}_R, \tilde{\mu}_R, \tilde{\tau}_R)$, and the 6×6 mass matrix is

$$\mathcal{M}^2_{\tilde{l}} = \begin{pmatrix} m^2_{LL} & m^2_{LR} \\ m^{2\dagger}_{LR} & m^2_{RR} \end{pmatrix} . \tag{18}$$

The diagonal 3×3 blocks are

$$m^2_{LL} = \begin{pmatrix} m^2_{L\,11} + m^2_e + \Delta^2_L & m^2_{L\,12} & m^2_{L\,13} \\ m^2_{L\,21} & m^2_{L\,22} + m^2_\mu + \Delta^2_L & m^2_{L\,23} \\ m^2_{L\,31} & m^2_{L\,32} & m^2_{L\,33} + m^2_\tau + \Delta^2_L \end{pmatrix}$$

$$m^2_{RR} = \begin{pmatrix} m^2_{E\,11} + m^2_e + \Delta^2_R & m^2_{E\,12} & m^2_{E\,13} \\ m^2_{E\,21} & m^2_{E\,22} + m^2_\mu + \Delta^2_R & m^2_{E\,23} \\ m^2_{E\,31} & m^2_{E\,32} & m^2_{E\,33} + m^2_\tau + \Delta^2_R \end{pmatrix} , \tag{19}$$

where $\Delta^2_L = m^2_Z(-\frac{1}{2} + \sin^2\theta_W)\cos 2\beta$ and $\Delta^2_R = m^2_Z(-\sin^2\theta_W)\cos 2\beta$. The $\Delta^2_{L,R}$ contributions originate from gauge interactions, the D-terms of Eq. (14), and are therefore flavor-diagonal. While they contain no supersymmetry-breaking parameters, they require electroweak symmetry breaking, which in turn relies on supersymmetry-breaking soft masses for the Higgs scalars. The off-diagonal blocks are given by

$$m^2_{LR} = -\begin{pmatrix} \tilde{A}^e_{11} v_d + \mu y^e_{11} v_u & \tilde{A}^e_{12} v_d + \mu y^e_{12} v_u & \tilde{A}^e_{13} v_d + \mu y^e_{13} v_u \\ \tilde{A}^e_{21} v_d + \mu y^e_{21} v_u & \tilde{A}^e_{22} v_d + \mu y^e_{22} v_u & \tilde{A}^e_{23} v_d + \mu y^e_{23} v_u \\ \tilde{A}^e_{31} v_d + \mu y^e_{31} v_u & \tilde{A}^e_{32} v_d + \mu y^e_{32} v_u & \tilde{A}^e_{33} v_d + \mu y^e_{33} v_u \end{pmatrix} . \tag{20}$$

Finally, the sneutrino masses are $\tilde{\nu}^\dagger \mathcal{M}_{\tilde{\nu}} \tilde{\nu}$, where $\tilde{\nu}^T = (\tilde{\nu}_e, \tilde{\nu}_\mu, \tilde{\nu}_\tau)$, and

$$\mathcal{M}_{\tilde{\nu}}^2 = \begin{pmatrix} m_{\tilde{L}\,11}^2 + \Delta_\nu^2 & m_{\tilde{L}\,12}^2 & m_{\tilde{L}\,13}^2 \\ m_{\tilde{L}\,21}^2 & m_{\tilde{L}\,22}^2 + \Delta_\nu^2 & m_{\tilde{L}\,23}^2 \\ m_{\tilde{L}\,31}^2 & m_{\tilde{L}\,32}^2 & m_{\tilde{L}\,33}^2 + \Delta_\nu^2 \end{pmatrix}, \tag{21}$$

where $\Delta_\nu^2 = m_Z^2 \frac{1}{2} \cos 2\beta$.

In general, we see that the sleptons masses are complicated. Note that, while it is always possible to rotate to a basis where the Yukawa couplings y^e are diagonal, there is no reason for these rotations to simultaneously diagonalize the soft term matrices $m_{\tilde{L}}^2$, $m_{\tilde{E}}^2$, and $\tilde{A}^e$. Studies of slepton flavor mixing may shed light on fundamental issues, such as the masses of standard model quarks and leptons, and will be discussed in Section 5.4.

Ultimately one hopes to study the sleptons in full generality at colliders. However, in studies so far, simplifying assumptions are typically made to reduce the problem to a more manageable form. A common assumption is flavor conservation, so that y^e, $m_{\tilde{L}}^2$, $m_{\tilde{E}}^2$, and $\tilde{A}^e$ are all simultaneously diagonalizable. The 6×6 matrix then reduces to three 2×2 blocks of the form

$$m_{\tilde{\tau}}^2 = \begin{pmatrix} m_{\tilde{L}\,33}^2 + m_\tau^2 + \Delta_L^2 & -m_\tau(A_\tau + \mu \tan\beta) \\ -m_\tau(A_\tau + \mu \tan\beta) & m_{\tilde{E}\,33}^2 + m_\tau^2 + \Delta_R^2 \end{pmatrix}, \tag{22}$$

and similarly for the first and second generations. The $\tilde{A}^e$ parameters have been replaced by A parameters, following convention. A common assumption is that A_e, A_μ, and A_τ are of the same order. This is motivated by some forms of supersymmetry breaking, but is by no means a universal requirement. With this assumption, bounds on $|A_\tau|$ from the requirement that there be no tachyonic staus typically imply that left-right mixing in selectrons and smuons is insignificant. However, left-right mixing may be important for the staus. In fact, although suppressed by m_τ, the factor $\mu \tan\beta$ leads to large mixing in many models. In general, the stau mass eigenstates and eigenvalues are

$$\begin{aligned} \tilde{\tau}_1 &= \tilde{\tau}_L \cos\theta_\tau + \tilde{\tau}_R \sin\theta_\tau \\ \tilde{\tau}_2 &= -\tilde{\tau}_L \sin\theta_\tau + \tilde{\tau}_R \cos\theta_\tau \end{aligned} \tag{23}$$

and

$$m_{\tilde{\tau}_1,\tilde{\tau}_2}^2 = \frac{1}{2}\left[m_{\tilde{\tau}\,LL}^2 + m_{\tilde{\tau}\,RR}^2 \mp \sqrt{(m_{\tilde{\tau}\,LL}^2 - m_{\tilde{\tau}\,RR}^2)^2 + 4(m_{\tilde{\tau}\,LR}^2)^2} \right]. \tag{24}$$

As we will see, when the slepton masses are assumed to be unified at a high scale, renormalization group effects imply that staus are the lightest sleptons and so are of great importance in collider studies.

Finally, in the flavor conserving case, Eqs. (19) and (21) imply

$$m_{\tilde{e}_L}^2 - m_{\tilde{\nu}_e}^2 = -m_W^2 \cos 2\beta \ . \tag{25}$$

This mass relation is an extremely robust prediction, resulting only from gauge symmetry and supersymmetry in the MSSM. It therefore provides a non-trivial test that newly-discovered scalars must pass to be identified as left-handed sleptons.

2.6. *Charginos and Neutralinos*

We now turn to the fermionic superpartners. The charged fermions are Winos and Higgsinos. Without electroweak breaking, the Winos have soft mass M_2 from Eq. (17), and the Higgsinos get mass through the superpotential term $-\mu H_u H_d$ of Eq. (11). Once electroweak symmetry is broken, however, the Higgs–Wino–Higgsino interactions of Eq. (15) generate Wino–Higgsino mixing. In the standard model, proper electroweak symmetry breaking requires a Higgs vacuum expectation value (vev) $\langle h \rangle^2 = v^2/2$, where $(g^2 + g'^2)v^2/2 = m_Z^2$. As noted above, however, the MSSM is a two Higgs doublet model, and so the vev v may be shared between both neutral Higgses: $\langle H_u^0 \rangle^2 + \langle H_d^0 \rangle^2 = v^2/2$. This freedom is typically parametrized by $\tan\beta$, where

$$\langle H_u^0 \rangle = \frac{v}{\sqrt{2}} \sin\beta \ , \quad \langle H_d^0 \rangle = \frac{v}{\sqrt{2}} \cos\beta \ . \tag{26}$$

With this parametrization, the charged mass terms are $\psi^{-T} \mathcal{M}_C \psi^+ + \text{H.c.}$, where $\psi^{\pm T} = (-i\tilde{W}^\pm, \tilde{H}^\pm)$ and

$$\mathcal{M}_C = \begin{pmatrix} M_2 & \sqrt{2}\, m_W \sin\beta \\ \sqrt{2}\, m_W \cos\beta & \mu \end{pmatrix} . \tag{27}$$

$\mathcal{M}_C$ may be diagonalized by unitary matrices U and V through $M_D = U^* \mathcal{M}_C V^{-1}$. If $\mathcal{M}_C$ is real, U and V may also be chosen real and may be parametrized as

$$U = \begin{pmatrix} \cos\phi_- & \sin\phi_- \\ -\sin\phi_- & \cos\phi_- \end{pmatrix} , \quad V = \begin{pmatrix} \cos\phi_+ & \sin\phi_+ \\ -\sin\phi_+ & \cos\phi_+ \end{pmatrix} . \tag{28}$$

The resulting mass eigenstates are called *charginos* and are $\tilde{\chi}_i^- = U_{ij}\psi_j^-$ and $\tilde{\chi}_i^+ = V_{ij}\psi_j^+$. In the limit where $M_2, |\mu| \gg m_W$ the mass eigenstates become nearly pure Winos and Higgsinos.[a] The mixing angles are sensitive to $\tan\beta$ when $\tan\beta \sim 1$.

The neutral fermionic superpartners are the Bino, the neutral Wino, and the two neutral Higgsinos. As with charginos, electroweak symmetry breaking generates gaugino–Higgsino mixing. The resulting mass eigenstates are called *neutralinos*. The neutralino mass matrix is $\frac{1}{2}\psi^{0T}\mathcal{M}_N\,\psi^0$+H.c., where $\psi^{0T} = (-i\tilde{B}, -i\tilde{W}^3, \tilde{H}_d^0, \tilde{H}_u^0)$ and

$$\mathcal{M}_N = \begin{pmatrix} M_1 & 0 & -m_Z c_\beta\, s_W & m_Z s_\beta\, s_W \\ 0 & M_2 & m_Z c_\beta\, c_W & -m_Z s_\beta\, c_W \\ -m_Z c_\beta\, s_W & m_Z c_\beta\, c_W & 0 & -\mu \\ m_Z s_\beta\, s_W & -m_Z s_\beta\, c_W & -\mu & 0 \end{pmatrix}, \tag{29}$$

and we have introduced the shorthand notation $s_W \equiv \sin\theta_W$, $c_W \equiv \cos\theta_W$, $s_\beta \equiv \sin\beta$, and $c_\beta \equiv \cos\beta$. The neutralino mass eigenstates are $\tilde{\chi}_i^0 = N_{ij}\psi_j^0$, where N diagonalizes $\mathcal{M}_N$.

In order of increasing mass, the two charginos are labeled $\tilde{\chi}_1^\pm$ and $\tilde{\chi}_2^\pm$, and the four neutralinos are $\tilde{\chi}_1^0$, $\tilde{\chi}_2^0$, $\tilde{\chi}_3^0$, and $\tilde{\chi}_4^0$. There are several phenomenologically interesting limits for the chargino and neutralino mass spectrum:

- $M_1 < M_2 < |\mu|$: "Bino LSP scenarios." As we will discuss in Section 4.1, this case is realized in much, but not all, of minimal supergravity parameter space. The lightest state is $\tilde{\chi}_1^0 \approx \tilde{B}$, followed by a nearly degenerate triplet, $\tilde{\chi}_2^0 \approx \tilde{W}^0$ and $\tilde{\chi}_1^\pm \approx \tilde{W}^\pm$. There is typically a large spacing between the lightest state and all others.
- $|\mu| < M_1, M_2$: "Higgsino LSP scenarios." $\tilde{\chi}_1^0$, $\tilde{\chi}_2^0$, $\tilde{\chi}_1^+$ are all fairly degenerate in mass and Higgsino-like. Radiative corrections are important in determining the mass differences.
- $M_2 < M_1, |\mu|$: "Wino LSP scenarios." These are realized in a variety of models, including some anomaly-mediation models. (See Section 4.5.) $\tilde{\chi}_1^0$ and $\tilde{\chi}_1^\pm$ are Wino-like and extremely mass degenerate, and the charginos may have an observable lifetime before decaying to the neutralino. In this case, radiative corrections are essential in determining the mass difference.

3. Successes and Puzzles

The discovery of weak-scale supersymmetry will be accompanied by many fundamental questions, such as those listed in Section 1. Here we expand

[a] We assume that gaugino masses M_i are positive and real, unless otherwise stated.

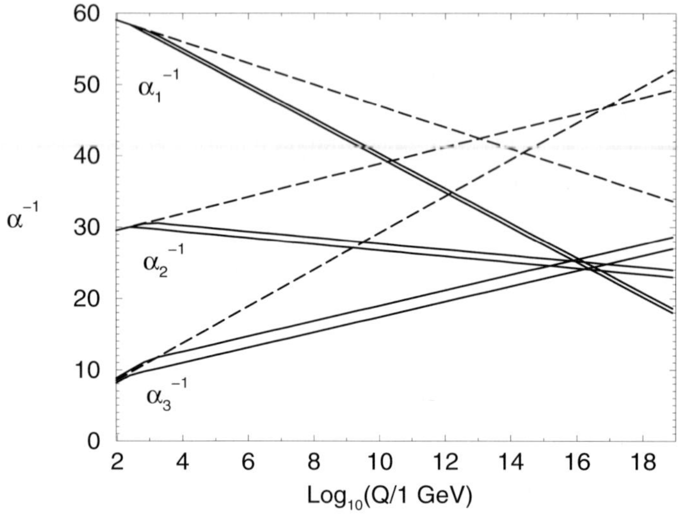

Fig. 2. The renormalization group evolution of the three gauge coupling constants in the standard model (dashed) and in the MSSM (solid) [6].

upon a number of these, both to highlight some important goals of the linear collider program and also to motivate the array of models to be described in Section 4.

3.1. *Unification*

The unification of forces requires that gauge couplings unify at some scale. The renormalization group evolutions of gauge couplings in the standard model and the MSSM are shown in Fig. 2. For the MSSM, all sparticles are assumed to lie between 250 GeV and 1 TeV. In the standard model, the gauge couplings obviously do not unify. In the MSSM, however, they unify at the 1% level. This is widely regarded as the strongest quantitative motivation for any framework for physics beyond the standard model proposed to date.

The fact that the three couplings meet in supersymmetry is significant, but so is where they meet. The couplings unify with value $\alpha_{\text{unif}} \approx 1/24$ at $M_{\text{GUT}} \simeq 2 \times 10^{16}$ GeV. If the couplings had unified at $\alpha \gtrsim 1$, unification would occur beyond the perturbative regime. At the same time, unification at scales above $M_{\text{Pl}} \simeq 10^{19}$ GeV would be invalidated by quantum gravity effects, while unification at scales $Q \lesssim 10^{16}$ GeV would imply proton decay at levels in conflict with experiment. (See Section 3.5.) In fact, the proximity of $M_{\rm GUT}$ to $M_{\rm Pl}$ gives hope that the effective gravitational coupling, $\alpha_{\rm grav} \sim E^2/M_{\rm Pl}^2$ also unifies with the standard model gauge couplings near $Q \sim M_{\rm Pl}$, as is motivated in string theory. The discrepancy between $M_{\rm GUT}$ and $M_{\rm Pl}$ may be resolved by a variety of mechanisms [18, 19].

The unification of gauge couplings has immediate implications for the linear collider. Simple gauge group unification requires also gaugino mass unification. The quantities M_i/g_i^2 are invariants of one-loop renormalization group evolution. (Two-loop effects typically lead to small deviations [6].) If all gauge couplings and gaugino masses are unified at any high scale, one therefore expects

$$M_1 : M_2 : M_3 = g_1^2 : g_2^2 : g_3^2 \approx 1 : 2 : 7 \tag{30}$$

at lower scales, where g_1, g_2, and g_3 are the hypercharge, weak, and strong couplings. The model-independent measurement of gaugino mass parameters is then of great importance, and the verification of $2M_1 \approx M_2$ would provide strong supporting evidence for unification.

Gauge group unification also implies that the particle content of the MSSM lies in representations of a larger gauge group. There is already tantalizing evidence for this — although the quantum numbers of the standard model particles shown in Table 1 appear rather arbitrary, they are consistent with their placement in $\overline{\mathbf{5}} + \mathbf{10}$ representations of SU(5), or, with the addition of a right-handed neutrino, in the spinor representation **16** of SO(10). These unifications imply relations between Yukawa couplings and soft supersymmetry-breaking terms of particles in the same grand unified theory (GUT) multiplets, providing additional motivations for model-independent measurements.

3.2. *Dark Matter*

As noted in Section 2.2, the most general gauge-invariant supersymmetric standard model allows proton decay through renormalizable operators, in gross violation of proton lifetime limits. R-parity conservation is one elegant way to prevent this. This has the immediate consequence that the lightest supersymmetric particle (LSP) is stable. In many models, this LSP is a neutralino, a weakly-interacting, weak-scale mass particle that is a natural candidate for non-baryonic cold dark matter [20–23].

If supersymmetry is discovered at colliders, studies of the superpartner spectrum will shed light on this hypothesis. Of course, collider experiments

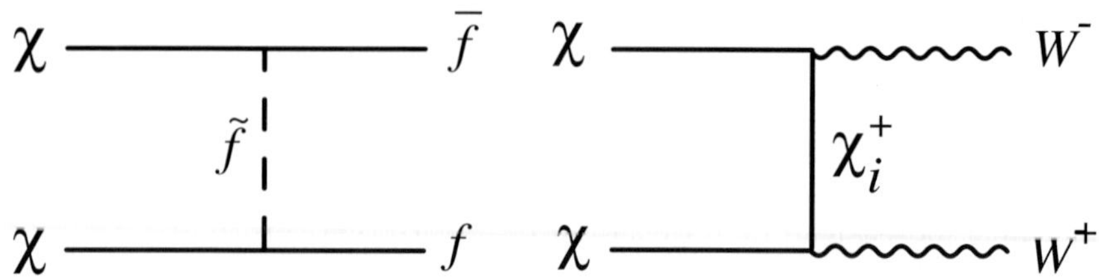

Fig. 3. Neutralino dark matter annihilation processes.

will never be able to prove that the LSP is dark matter — this requires knowing that the LSP is stable on cosmological time scales. However, if the LSP is found to be charged, colored, or unstable, supersymmetric dark matter will be largely disfavored. On the other hand, if the LSP is found to be a seemingly stable neutralino, there will be much work to do. One interesting test will be to determine the neutralino's thermal relic density. The neutralino's relic abundance is determined by its annihilation cross sections through processes such as those of Fig. 3. The processes $\chi\chi \to f\bar{f}$ are chirality suppressed, and so the most relevant t-channel sfermions are typically stops, sbottoms, and staus. If the neutralino's mass and composition are known, as well as the masses and compositions of these sfermions, its thermal relic density may be accurately computed, assuming a standard thermal history for the universe. Values near the favored dark matter density of $\Omega \approx 0.3$ will be strong evidence in favor of neutralino dark matter. Collider and conventional dark matter experiments will also play complementary roles. The interactions of neutralinos with matter are given by interactions such as those in Fig. 4. By determining the relevant supersymmetry parameters, collider experiments may guide dark matter searches, or confirm their signals as being due to supersymmetry. We will discuss the relation of linear collider experiments to cosmology in more detail in Section 10.

3.3. *Flavor Violation*

Weak-scale supersymmetry generically induces gross violations of constraints on flavor-violating observables [24]. Among the best known is the supersymmetric contribution to the K_L–K_S mass splitting. Supersymmetry contributes to this through box diagrams, such as the one involving squarks and gluinos shown in Fig. 5. The crosses on the squark propagators are flavor-violating mass insertions $m^2_{\tilde{q}_{12}}$. As noted in Section 2.5, there

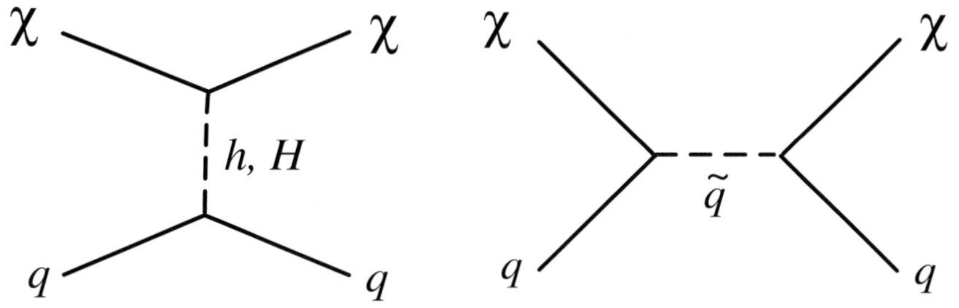

Fig. 4. Neutralino dark matter scattering processes.

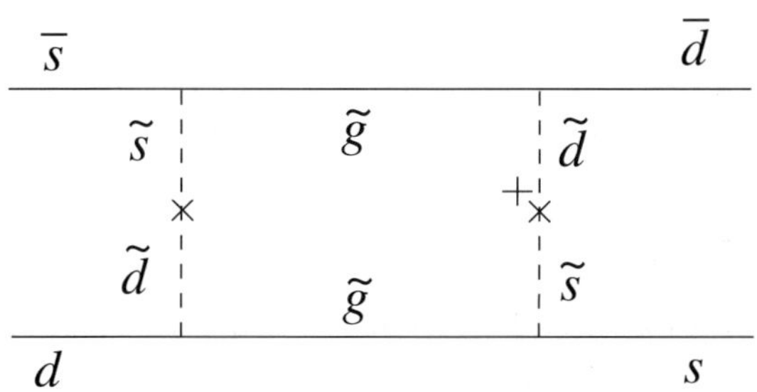

Fig. 5. Supersymmetric contribution to the K_L–K_S mass splitting.

is no reason *a priori* to assume that squark and quark mass matrices are simultaneously diagonalizable, and so such flavor-violating mass insertions generically exist.

The bound from the measured Δm_K implies [25]

$$\left[\frac{10\ \text{TeV}}{m_{\tilde{q},\tilde{g}}}\right]^2 \left[\frac{\Delta m^2_{\tilde{q}_{12}}/m^2_{\tilde{q}}}{0.1}\right]^2 \lesssim 1\ . \tag{31}$$

For TeV-scale squarks and gluinos, the flavor-violating parameters must be very small so that the supersymmetric generalization of the GIM mechanism applies. Alternatively, for $\mathcal{O}(1)$ flavor mixing, squark and gluino masses must be of order 10 TeV, well above the weak scale.

The supersymmetric flavor problem is among the most pressing phenomenological problems of supersymmetry, and motivates many models that naturally produce degenerate squarks, very heavy squarks and gluinos, or quark-squark alignment. Note that this is not simply a problem in the hadronic sector — bounds on lepton flavor violation (LFV) from μ–e conver-

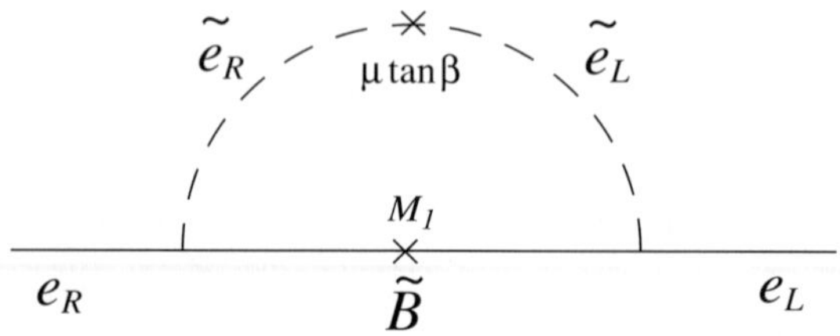

Fig. 6. Supersymmetric contribution to the electron's electric dipole moment. A photon attached to the charged internal line is implicit.

sion, $\mu \to e\gamma$, and other processes, cause similar difficulties. If supersymmetry is discovered, one immediate question will be how the supersymmetric flavor puzzle is resolved.

3.4. *CP Violation*

Weak-scale supersymmetry also generically violates bounds on CP violation. Bounds from ϵ_K are numerically the most stringent, but are typically satisfied in models that eliminate flavor violation in some natural way. However, CP-violating, but flavor-conserving, constraints remain a problem. A prime example of these constraints is the electron's electric dipole moment (EDM). The supersymmetric contribution to this observable is shown in Fig. 6. This contribution requires [26]

$$\left[\frac{2\ \text{TeV}}{m_{\tilde{e}}}\right]^2 \left[\frac{\mu M_1}{m_{\tilde{e}}^2}\right] \tan\beta \sin\phi_{CP} \lesssim 1 \ , \tag{32}$$

that is, uncomfortably heavy selectrons or small CP-violating phases $\phi_{CP} \ll 1$. As in the flavor-violating case, this problem is not confined to one sector — a similarly stringent constraint in the hadronic sector follows from the neutron EDM. Again, the discovery of supersymmetry will raise the question of how these supersymmetric CP problems are solved.

3.5. *Proton Decay*

In supersymmetric theories, proton decay through renormalizable operators may be eliminated by R-parity conservation. However, in grand unified theories (GUTs), proton decay is again allowed through non-renormalizable operators suppressed by the GUT scale.

The leading contribution for large $\tan\beta$ in supersymmetric SU(5) theories is given in Fig. 7 [27–29]. For minimal SU(5), this contribution, nor-

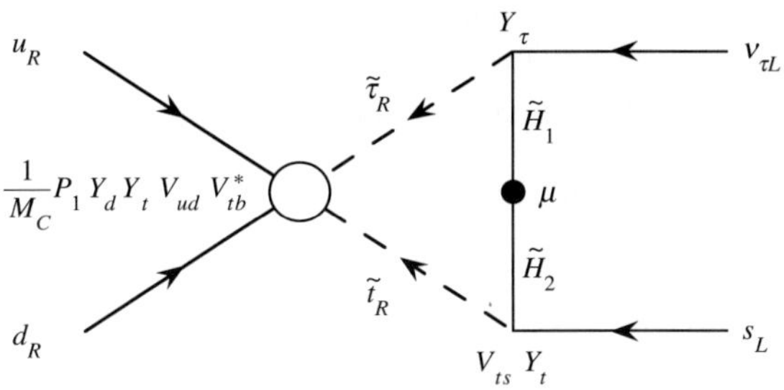

Fig. 7. Contribution to proton decay in supersymmetric grand unified theories [29].

malized to the current bounds from SuperKamiokande [30], implies

$$\left[\frac{\tau(p \to K^+\overline{\nu})}{1.6 \times 10^{33}\ \text{yr}}\right]\left[\frac{(5\ \text{TeV})^2\,\mu^2}{m_{\tilde{q}}^4}\right]\left[\frac{\tan\beta}{4}\right]^4\left[\frac{2 \times 10^{16}\ \text{GeV}}{M_C}\right]^2 \lesssim 1\,, \tag{33}$$

where M_C is the colored Higgs mass, typically expected to be of the order of the unification scale M_{GUT}. This operator relies only on standard model flavor violation, and so is present even in supersymmetric theories with degenerate scalars and CP conservation.

The absence of observed proton decay is becoming a severe constraint for simple supersymmetric GUTs. It motivates models with naturally heavy superpartners, extended GUT models, or even those with some mechanism to unify gauge coupling without GUT gauge groups. The discovery of light squarks will immediately exclude the simplest GUT models.

4. Models

If weak-scale supersymmetry is found in nature, the promise of linear colliders lies in the possibility of disentangling supersymmetry without model-dependent assumptions. However, without actual data, studies of the potential of colliders require some model framework to consider. The models studied are typically motivated by the desire to solve the problems without sacrificing the successes of Section 3. Here we describe several frameworks with distinct collider phenomenology.

For collider phenomenology, models are largely specified by their soft supersymmetry-breaking terms. What is the origin of the soft terms? It is typically expected that they are generated by some dynamical mechanism closely analogous to the mechanism of spontaneous gauge symmetry breaking. In electroweak symmetry breaking, for example, gauge symmetry is broken when the Higgs field gets a non-vanishing vev. For supersymmetry breaking, the F fields play the role of the Higgs field. To see this, note that supersymmetry transformations on chiral superfields may be defined as $\delta_\xi \Psi = [\xi Q + \overline{\xi Q}, \Psi]$. Expanding both sides in powers of θ, we find

$$\delta_\xi \phi = \sqrt{2}\xi\psi \tag{34}$$

$$\delta_\xi \psi = -i\sqrt{2}\sigma^\mu \bar{\xi}\partial_\mu \phi + \sqrt{2}\xi F \tag{35}$$

$$\delta_\xi F = -i\sqrt{2}\bar{\xi}\bar{\sigma}^\mu \partial_\mu \psi \; . \tag{36}$$

If F obtains a VEV,

$$\langle 0|F|0\rangle = \langle 0|[\xi Q + \overline{\xi Q}, \psi]|0\rangle \neq 0 \; , \tag{37}$$

implying that $Q|0\rangle \neq 0$, and supersymmetry is broken in the vacuum.

In most supersymmetric models, including those to be discussed below, the terms of the Lagrangian may be divided into three sectors: the supersymmetry breaking sector, containing only non-MSSM fields Z, the mediation sector coupling the Z and MSSM fields, and the MSSM sector, containing only MSSM fields. Supersymmetry is broken when dynamical effects generate a vev for the F component of a Z field. This vev is called the *scale of supersymmetry breaking.* Terms in the mediation sector then generate soft terms for MSSM fields. For example, when F_Z gets a vev, the soft scalar masses and trilinear A-terms of Eq. (17) will be generated by higher-dimensional operators suppressed by the mass scale M of the mediation sector:

$$\begin{aligned} \lambda_{ij}\frac{Z^\dagger Z}{M^2}\Psi_i^\dagger \Psi_j|_{\theta\theta} &\rightarrow \lambda_{ij}\frac{|\langle F_Z\rangle|^2}{M^2}\phi_i^*\phi_j \\ \lambda_{ijk}\frac{Z}{M}\Psi_i\Psi_j\Psi_k|_{\theta\theta} &\rightarrow \lambda_{ijk}\frac{\langle F_Z\rangle}{M}\phi_i\phi_j\phi_k \; . \end{aligned} \tag{38}$$

Gaugino masses and B-terms may also be generated in this way. The sparticle masses and other weak-scale parameters therefore contains an imprint of terms in the mediation sector which are determined by physics at energies much higher than the weak scale. Field condensation does not alter the high energy behavior of the theory, and terms generated from F condensation, such as those above, do not disturb the cancellation of quadratic divergences.

For most phenomenological applications, *how* supersymmetry is broken is largely irrelevant. Far more important is the supersymmetry-mediation mechanism, that is, the terms of the mediation sector, and the supersymmetry-breaking scale. Both are crucial in determining the signatures of supersymmetry. The mediation terms determine the pattern of soft terms and, consequently, the masses, mixings, and interactions of the standard model superpartners. The supersymmetry-breaking scale determines the properties of the gravitino. The gravitino's mass is

$$m_{\tilde{G}} = \frac{F_{\rm DSB}}{\sqrt{3}M_{\rm Pl}} \;, \tag{39}$$

and its interactions are given by

$$\mathcal{L} \supset \frac{1}{F_{\rm DSB}} \left[m_{\tilde{f}}^2 \overline{\tilde{f}} f + \frac{m_\lambda}{4\sqrt{2}} \overline{\lambda} \sigma^{\mu\nu} F_{\mu\nu} \right] \tilde{G} \;, \tag{40}$$

where $F_{\rm DSB} \equiv [\sum_i \langle F_i \rangle^2]^{1/2}$ is the total F vev generated by dynamical supersymmetry breaking. For simple cases where there is only one non-zero $\langle F \rangle$, $F_{\rm DSB}$ is equivalent to the $\langle F_Z \rangle$ appearing in Eq. (38), but in general, they need not be identical.

We will now describe several models, beginning with three supergravity theories. In supergravity, supersymmetry breaking is mediated through non-renormalizable gravitational interactions, and the large mass scale of Eq. (38) is the Planck mass $M_{\rm Pl}$. These interactions are generically present in all theories with supersymmetry and gravity, and supergravity theories are therefore natural models to consider. However, by their very nature, the coefficients of these non-renormalizable terms cannot be precisely determined without knowing the microscopic quantum theory of gravity from which they presumably derive. To proceed, various simple and phenomenologically attractive assumptions are made. These assumptions then fix the soft terms, allowing one to conduct concrete collider studies.

Another approach is to suppress the non-renormalizable supergravity terms and find new sources for soft terms. This requires new structure involving additional fields and interactions, but has the advantage that the soft terms are calculable in field theory. We describe models with gauge- and anomaly-mediated supersymmetry breaking, two prominent examples of this approach.

Finally, we conclude with comments on the implications of GUT and string models for soft supersymmetry-breaking parameters.

4.1. *Minimal Supergravity*

In supergravity, soft terms are generated by terms such as those in Eq. (38), where the large mass scale is $M = M_{\rm Pl}$. Soft scalar masses have the form $m_{ij}^2 = \lambda_{ij} m_0^2$, where $m_0 \equiv \langle F \rangle / M_{\rm Pl}$, and i, j are generational indices. These terms generically mediate flavor-changing transitions that violate the constraints discussed in Section 3.3.

In *minimal supergravity* [21–22], gaugino- one assumes that all of the soft scalar masses are given by a single parameter m_0 at some high scale, typically taken to be $M_{\rm GUT} \simeq 2 \times 10^{16}$ GeV. Similarly, one assumes a unified gaugino mass $M_{1/2}$, and a universal A-term parameter A_0. At $M_{\rm GUT}$, the model is completely specified by standard model parameters and m_0, $M_{1/2}$, A_0, B and μ.

The weak-scale MSSM Lagrangian is then determined by evolving all parameters to the weak-scale through renormalization group evolution. Renormalization group evolution of supersymmetry parameters is described by a complicated system of differential equations. Schematically, however, the 1-loop renormalization group equations are

$$\frac{dg}{dt} \sim \frac{1}{16\pi^2} g^3 \tag{41}$$

$$\frac{dy}{dt} \sim \frac{1}{16\pi^2} \left[g^2 y - y^3 \right] \tag{42}$$

$$\frac{dM}{dt} \sim \frac{1}{16\pi^2} g^2 M \tag{43}$$

$$\frac{dA}{dt} \sim \frac{1}{16\pi^2} \left[-g^2 M - y^2 A \right] \tag{44}$$

$$\frac{dm^2}{dt} \sim \frac{1}{16\pi^2} \left[g^2 M^2 - y^2 A^2 - y^2 m^2 \right] \ , \tag{45}$$

where $t \equiv \ln(Q_0/Q)$ with Q_0 some fixed renormalization scale, and *positive* numerical coefficients and gauge and flavor indices have been neglected. In evolving down from a high scale, gauge interactions raise m^2, while Yukawa interactions lower m^2. Unique among scalar mass parameters, $m_{H_u}^2$ is driven down by the large top quark Yukawa without a compensating positive contribution from the strong gauge interactions. It is typically driven negative, breaking electroweak symmetry. This feature of *radiative electroweak symmetry breaking* is often regarded as another success of supersymmetry.

At the weak scale, the supersymmetry parameters are constrained by the following tree-level relations:

$$\frac{1}{2} m_Z^2 = \frac{m_{H_d}^2 - m_{H_u}^2 \tan^2\beta}{\tan^2\beta - 1} - |\mu|^2 \tag{46}$$

$$2B = (m_{H_u}^2 + m_{H_d}^2 + 2|\mu|^2) \sin 2\beta \ . \tag{47}$$

Since m_Z is measured, these relations are used to exchange B and μ at the high scale for $\tan\beta$ and $\mathrm{Arg}(\mu)$. In minimal supergravity, μ is assumed real, and so the fundamental parameters of minimal supergravity are

$$m_0,\ M_{1/2},\ A_0,\ \tan\beta,\ \mathrm{sign}(\mu) \ . \tag{48}$$

The scale of supersymmetry breaking is set by the requirement that the soft terms are of the order of the weak scale, so $\langle F \rangle \sim m_W M_{\rm Pl} \sim 10^{20}$ GeV^2. By Eq. (39), the gravitino mass is also of the order of the weak scale. It is typically assumed to be heavier than the standard model superpartners. In this case, the gravitino is irrelevant for colliders, as its interactions are gravitational and so extremely weak.

The masses of the rest of the superpartners are determined by renormalization group evolution. At the weak scale, the gaugino mass parameters are given by

$$\frac{M_i}{\alpha_i} = \frac{M_{1/2}}{\alpha_{\rm unif}} \ , \tag{49}$$

in accord with Eq. (30). Numerically, this implies

$$M_1 \approx 0.4 M_{1/2} \ . \tag{50}$$

For the scalar masses, neglecting the effects of Yukawa couplings, we find the weak-scale values

$$\begin{aligned} m_Q^2 &= m_0^2 + 6.3 M_{1/2}^2 \\ m_U^2 &= m_0^2 + 5.8 M_{1/2}^2 \\ m_D^2 &= m_0^2 + 5.8 M_{1/2}^2 \\ m_L^2 &= m_0^2 + 0.5 M_{1/2}^2 \\ m_E^2 &= m_0^2 + 0.15 M_{1/2}^2 \ . \end{aligned} \tag{51}$$

If m_0 and $M_{1/2}$ are of the same order, the squark masses are dominated by $M_{1/2}$, while the slepton masses are dominated by m_0. Generically the renormalization group equations predict heavy squarks and light sleptons. The lightest scalars are $\tilde{\tau}_R$, followed by $\tilde{\mu}_R$, $\tilde{e}_R$, and the left-handed sleptons.

Minimal supergravity has a number of virtues. Grand unified boundary conditions are automatically incorporated. In addition, as is evident from Eqs. (50) and (51), in much of parameter space, the LSP is the lightest neutralino, and so is a possible dark matter candidate. The nature of the LSP in various regions of parameter space is presented in Fig. 8. For most of the region with $m_0 \lesssim 1$ TeV, the LSP is a Bino-like neutralino. Minimal supergravity also solves the supersymmetric flavor and CP problems, but only through *ad hoc* assumptions. From a universal scalar mass at $M_{\rm GUT}$, renormalization group evolution generates mass splittings and flavor mixing, but these are generally small. The assumption of a real μ parameter also guarantees the absence of supersymmetric CP violation, suppressing dangerous contributions to electric dipole moments.

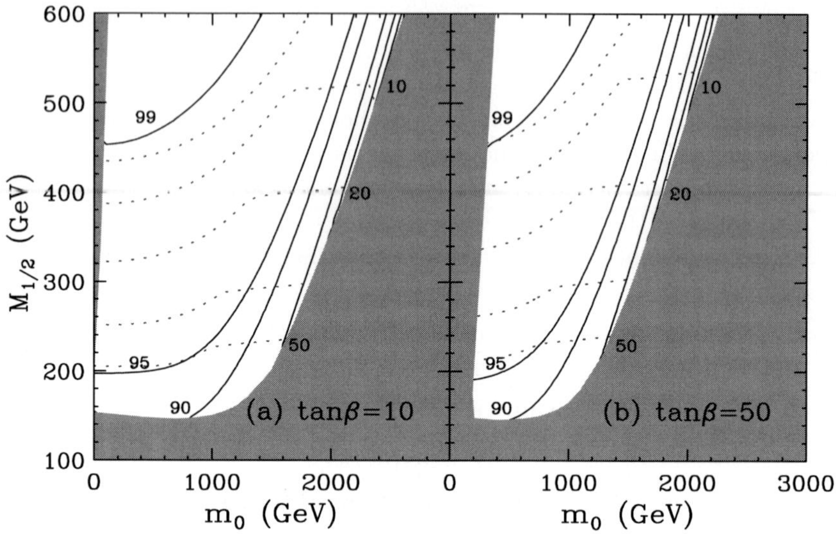

Fig. 8. The character of the LSP in minimal supergravity with $A_0 = 0$ and $\mu < 0$ [34]. In the unshaded regions, the LSP is the lightest neutralino, and contours of the gauginoness of the LSP $R_\chi \equiv |N_{11}|^2 + |N_{12}|^2$ are given (in percent). In the narrow upper left shaded regions, the LSP is the lighter stau. The lower right shaded regions are excluded by bounds on the chargino mass.

To summarize, a typical minimal supergravity scenario predicts

$$\begin{aligned} \text{NLSP candidates}: \ & \tilde{l}_R \text{ or } \tilde{\chi}_2^0, \tilde{\chi}_1^\pm \approx \tilde{W}^0, \tilde{W}^\pm \\ \text{LSP}: \ & \tilde{\chi}_1^0 \approx \tilde{B} \ , \end{aligned} \tag{52}$$

where NLSP denotes the next-to-lightest supersymmetric particle, which is the lightest detectable in colliders. One therefore generally expects neutralinos, charginos, and sleptons to be produced in the greatest numbers at linear colliders.

4.2. *Focus Point Supersymmetry*

Perhaps the most straightforward solution to the flavor and CP constraints of Sections 3.3 and 3.4 is to assume heavy superpartners. However, large supersymmetry parameters typically re-introduce the gauge hierarchy and so are considered unnatural. As an example, consider Eq. (46): if $|m_{H_u}^2| \gg m_Z^2$, the parameter μ must be highly fine-tuned to cancel most of $m_{H_u}^2$ to give the correct Z mass.

Naturalness does not require all supersymmetry-breaking parameters to be small, however. In focus point models [35, 36], large fundamental

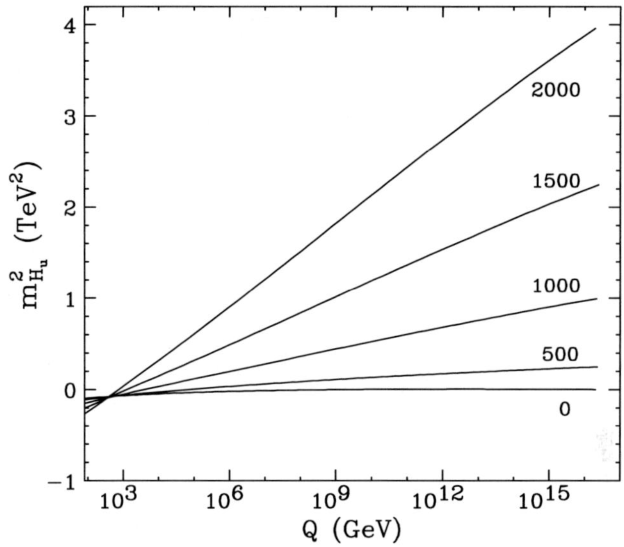

Fig. 9. Renormalization group trajectories of $m^2_{H_u}$ in minimal supergravity for various m_0 as labeled and $M_{1/2} = 300$ GeV, $A_0 = 0$, $\tan\beta = 10$ [36]. These focus to a point at $Q \sim m_W$.

supersymmetry parameters do not imply large cancellations in Eq. (46). A simple example is provided by the minimal supergravity model just discussed. Rewriting Eq. (46) in terms of the fundamental parameters, we find through numerical analysis

$$\frac{1}{2} m_Z^2 = -0.04\, m_0^2 + 8.8 M_1^2 - |\mu|^2 \ . \tag{53}$$

The small m_0^2 coefficient implies that even large values of m_0 do not result in large fine-tuning in the electroweak potential. The small coefficient may be understood in terms of a focus point in the $m^2_{H_u}$ renormalization group trajectories, shown in Fig. 9. For a large range of initial values m_0, these trajectories focus to a point with $m^2_{H_u} \approx 0$ at the weak scale. Naturalness bounds $m^2_{H_u}$ at the weak scale, but constrains the GUT-scale parameter m_0 only loosely. While $m^2_{H_u}$ is insensitive to m_0, the masses of all sleptons and squarks are of order m_0, and so very heavy scalars are consistent with naturalness in minimal supergravity, while at the same time alleviating flavor and CP constraints [37].

The focus point region in minimal supergravity is the region of $m_0 \gtrsim 1$ TeV in Fig. 8. From Eq. (53), we see that, for fixed $M_{1/2}$, large m_0 requires small $|\mu|$. This is also evident in Fig. 8, where, for large m_0 there is a region with $M_{1/2}$ and $|\mu|$ comparable, and therefore significant Higgsino content in the lightest neutralino. The prediction of focus point supersymmetry, then, is that, while sleptons and squarks are typically heavy, possibly even with multi-TeV masses beyond the LHC, the electroweak gauginos and Higgsinos are all light.

In fact, constraints on the dark matter relic density typically require significant gaugino–Higgsino mixing in focus point supersymmetry. The requirement that dark matter not overclose the universe implies that the processes of Fig. 3 be sufficiently efficient. In focus point supersymmetry, the processes mediated by sleptons and squarks are highly suppressed, and so dark matter annihilation relies on the processes with gauge boson final states. These processes are, however, absent for Bino-like LSPs, as they rely on gaugino–Higgsino mixing. As a result, dark matter constraints require gaugino–Higgsino mixing, and $|\mu|$ not too far above the gaugino mass parameters.

In focus point supersymmetry, then, all 6 charginos and neutralinos may be within reach of the linear collider, allowing detailed studies of these systems. A typical focus point spectrum is

$$\begin{aligned} \text{Heavy}: &\ \tilde{l}, \tilde{q} \\ \text{NLSP candidates}: &\ \tilde{\chi}_2^0, \tilde{\chi}_1^{\pm} \approx \text{ gaugino} - \text{Higgsino mixtures} \\ \text{LSP}: &\ \tilde{\chi}_1^0 \approx \tilde{B}, \tilde{H}^0 \text{ mixture}\,. \end{aligned} \qquad (54)$$

4.3. *Superheavy Supersymmetry*

Superheavy supersymmetry scenarios also alleviate the flavor and CP problems by decoupling. The soft Higgs masses $m_{H_u}^2$ and $m_{H_d}^2$ are constrained to the weak scale by naturalness. They are determined by the scalar mass renormalization group equation of Eq. (45). Note, however, that the other scalar mass parameters enter this renormalization group equation proportional to their Yukawa couplings.

This then naturally suggests a split spectrum with light third generation sfermions, and superheavy sfermions of the first two generations [38]. The light third generation sfermions preserve naturalness and do not induce violations of existing bounds, since low energy bounds on third generation observables are relatively weak. At the same time, superheavy sleptons and squarks of the first two generations, with masses of the order of 10 TeV, alleviate the most stringent flavor- and CP-violating constraints, such as those in from the kaon system, and μ-e transitions, but do not sacrifice naturalness, since their effect on the $m_{H_u}^2$ and $m_{H_d}^2$ renormalization group equations are suppressed by tiny Yukawa couplings. The split spectrum of superheavy supersymmetry may be generated immediately upon supersymmetry-breaking [39] or by renormalization group evolution effects [40].

A typical superheavy supersymmetry spectrum is

$$\begin{aligned} \text{Superheavy} : &\ \tilde{e}, \tilde{\mu}, \tilde{u}, \tilde{d}, \tilde{c}, \tilde{s} \\ \text{NLSP candidates} : &\ \tilde{\tau}, \tilde{\nu}_\tau, \tilde{b}, \tilde{t}, \tilde{\chi}_2^0, \tilde{\chi}_1^\pm \\ \text{LSP} : &\ \tilde{\chi}_1^0 \ . \end{aligned} \tag{55}$$

4.4. *Gauge Mediation*

In supergravity scenarios, the soft terms are assumed to arise from non-renormalizable gravitational interactions. As noted above, however, these are not calculable and *ad hoc* assumptions must be made to avoid constraints from flavor and CP bounds. In gauge-mediated supersymmetry breaking, one assumes that $\langle F \rangle \ll 10^{20}\ \text{GeV}^2$, so that these troublesome gravitational contributions to soft masses are highly suppressed.

Of course, weak-scale soft terms must still be generated. This is achieved by introducing N "messenger fields" with mass M_{mess} [41, 42]. These have standard model gauge quantum numbers and generate MSSM soft masses at the scale M_{mess} of the form

$$M_i = N \frac{g_i^2|_{Q=M_{\text{mess}}}}{16\pi^2} \frac{\langle F \rangle}{M_{\text{mess}}} \tag{56}$$

$$m_{\tilde{f}}^2 = 2N \sum_i C_{if} \left[\frac{g_i^2|_{Q=M_{\text{mess}}}}{16\pi^2} \right]^2 \left[\frac{\langle F \rangle}{M_{\text{mess}}} \right]^2 \tag{57}$$

through diagrams such as those of Fig. 10. Here C_{if} are coefficients determined by thegauge representation of sfermions. Yukawa effects are typically small and may be neglected. These soft masses are therefore dependent only on standard model gauge quantum numbers to an excellent approximation, and gauge mediation therefore elegantly predicts flavor-blind soft terms, which suppress contributions to flavor violation. Generically, CP-violating EDMs are still problematic, however.

The gaugino and sfermion masses of Eq. (57) are then evolved from the messenger scale to the weak scale. The gaugino masses are in the same proportion as in minimal supergravity, even though they are never unified, but the scalar masses are very different. The scalar spectrum is determined by only three parameters in the simplest models: $\langle F \rangle$, M_{mess}, and N. The possibility of distinguishing such spectra from supergravity predictions will be discussed in Section 9.

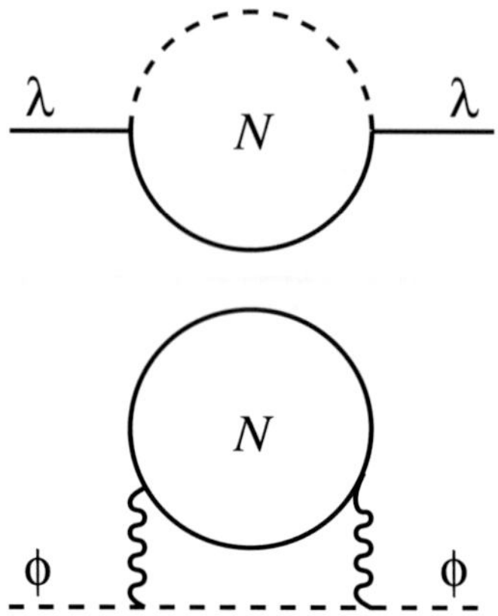

Fig. 10. Contributions to scalar and gaugino masses in gauge-mediated supersymmetry breaking.

For colliders, a key feature of gauge-mediated supersymmetry breaking is the prediction of $m_{\tilde{G}} \sim \langle F_{\rm DSB}\rangle / M_{\rm Pl} \ll m_{\rm Weak}$. The gravitino is therefore always the LSP. The NLSP is determined by a number of factors. For example, although the gaugino masses grow linearly with N, the scalar masses grow only as $\sqrt{N}$. For low N, the NLSP is typically a neutralino, but for larger N, the NLSP is typically a slepton [43]. This may be seen in Fig. 11, where the character of the NLSP in various regions of parameter space is shown. In general, then, we have the possibilities

$$\begin{aligned} \text{NLSP candidates}: &\ \tilde{\chi}_1^0\ ,\ \tilde{l} \\ \text{LSP}: &\ \tilde{G}\ . \end{aligned} \tag{58}$$

The NLSP decays to the gravitino with decay length

$$L \sim 0.1\ \text{mm} \times \left[\frac{E_{\rm NLSP}^2 - m_{\rm NLSP}^2}{m_{\rm NLSP}^2}\right]^{\frac{1}{2}} \left[\frac{\sqrt{F_{\rm DSB}}}{10^5\ \ \text{GeV}}\right]^4 \left[\frac{100\ \ \text{GeV}}{m_{\rm NLSP}}\right]^5 . \tag{59}$$

$\sqrt{F_{\rm DSB}}$ may range from 10^5 GeV to 10^9 GeV, where the lower limit is set by lower bounds on superpartner and messenger masses and the upper limit follows from requiring the supergravity contributions to be sub-dominant. For low values of $\sqrt{F_{\rm DSB}}$ in this range, the NLSP decays within collider detectors, producing signals with energetic photons and leptons, depending on what sparticle is the LSP [45]. However, for large $\sqrt{F_{\rm DSB}}$, the NLSP may travel macroscopic distances before decaying. In the case of neutralino NLSPs, this produces the conventional missing energy signals of supersym-

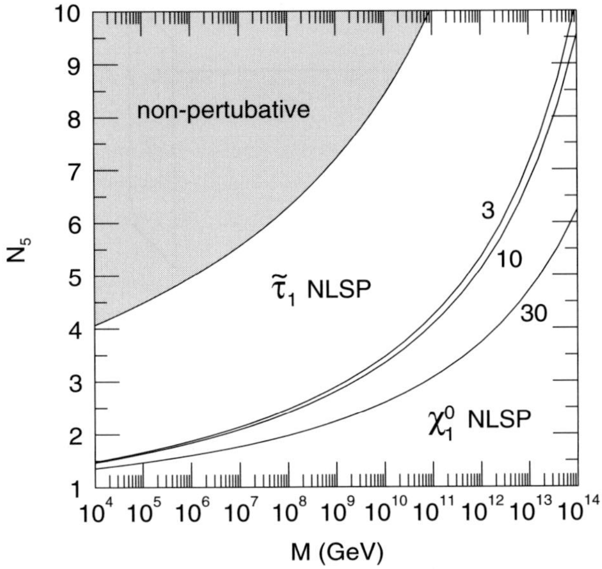

Fig. 11. The NLSP in various regions of minimal gauge-mediation model parameter space [44]. M is the messenger mass or scale, and N_5 is the number of messenger multiplets. In the region above the labeled contours, the NLSP is a stau; in the region below, the NLSP is a neutralino. The boundary between the two regions depends slightly on $\tan\beta$; the boundaries for $\tan\beta = 3$, 10, and 30 are shown. In the shaded region, gauge coupling constants become non-perturbative below the GUT scale under two-loop renormalization group evolution.

metry. For slepton NLSPs, however, non-relativistic sleptons may produce highly-ionizing charged tracks and tracks with unusual time-of-flight signatures.

4.5. *Anomaly Mediation*

In anomaly-mediated supersymmetry breaking [46, 47], the dangerous supergravity contributions are suppressed geometrically — supersymmetry-breaking fields are placed on one 3-brane, with MSSM fields on another, as shown in Fig. 12. The usual supergravity contributions are then suppressed because the supersymmetry-breaking and MSSM wave functions are separated in space.

In contrast to gauge mediation, however, in anomaly mediation, the soft terms are generated by loop-suppressed supergravity contributions of order $\sim (1/16\pi^2)\langle F\rangle/M_{\rm Pl}$. These arise from the Weyl rescaling anomaly. They are

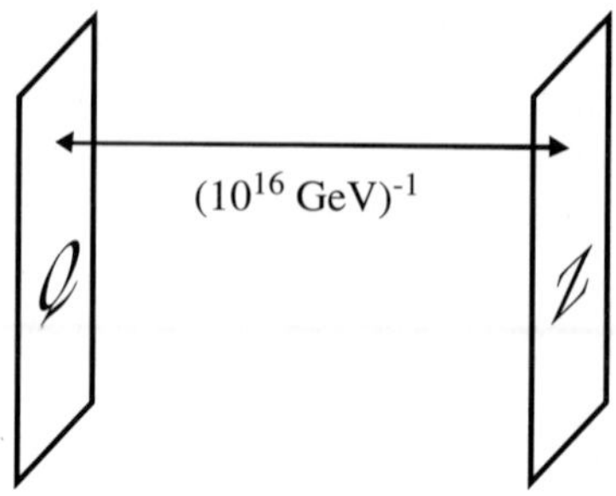

Fig. 12. Fields in anomaly-mediation: MSSM fields Q are separated from supersymmetry-breaking fields Z in extra dimensions.

always present, but are typically dominated by tree-level contributions in conventional four-dimensional supergravity. The required supersymmetry-breaking scale is therefore larger than in conventional supergravity, and the gravitino mass is of order 100 TeV. As with gauge mediation, the soft terms generated are essentially determined by gauge couplings, and so solve the supersymmetry flavor problem through degeneracy. The minimal models predict tachyonic sleptons — this catastrophe may be avoided either by assuming an additional contribution to scalar masses [48] or by postulating other additional structure [49]. Also, as in gauge-mediated models, additional structure is typically required to naturally satisfy EDM constraints.

The most striking phenomenological feature for colliders is the gaugino mass spectrum. The gaugino masses generated by the Weyl anomaly are $M_i \propto |b_i g_i^2|$, where b_i is the one-loop β-function coefficient for the corresponding gauge group in the MSSM. In the MSSM, $b_i = (-33/5, -1, 3)$, and the Wino mass is therefore predicted to be the smallest. The lightest states are then a triplet of Winos, and the typical spectrum is

$$\begin{aligned} \text{NLSP}: &\ \tilde{\chi}_1^{\pm} \approx \tilde{W}^{\pm} \\ \text{LSP}: &\ \tilde{\chi}_1^{0} \approx \tilde{W}^{0} \ . \end{aligned} \tag{60}$$

These Wino states are extraordinarily degenerate — at tree-level, their splitting is typically of order 1 MeV. One-loop contributions raise this to 150 MeV or more, allowing $\tilde{\chi}_1^{\pm} \to \tilde{\chi}_1^0 \pi^{\pm}$ [50]. However, the decay length may be of order 1 cm, and the resulting pion is very soft, leading to a peculiar and qualitatively new supersymmetry signature at colliders.

4.6. *GUT and Planck Scale Frameworks*

In the discussion above, we noted that the soft terms are in principle not predictable in supergravity. Properties such as scalar universality appeared as phenomenological assumptions, motivated by simplicity and experimental constraints. This is not the complete story, however. As also noted above, one strong motivation for supersymmetry is the unification of forces, including the three standard model forces and also gravity. The circle of ideas surrounding GUTs and string theory therefore naturally have implications for the GUT scale boundary conditions.

As an example, in string theory, the low energy effective theory is expected to be supersymmetric. In addition to the MSSM fields, this theory also contains many light scalar fields, including the dilaton and moduli, with gravitationally-suppressed couplings. The dilaton superfield's scalar component is the source of gauge and gravitational couplings, and its F term may break supersymmetry if it condenses. If this is the main source of supersymmetry breaking, the resulting models, called *dilaton-dominated supersymmetry breaking models* [51], predict mass spectra similar to minimal supergravity, but with specific relations among the average scalar mass, $M_{1/2}$, and A_0 at the Planck scale. Small splittings of scalar masses are introduced by moduli fields, and there are sum rules for scalar masses.

At slightly lower scales, consider GUTs motivated by the unification of couplings described in Section 3.1. In these theories, the MSSM gauge multiplets are merged into GUT multiplets, and have GUT interactions. For example, in SO(10) GUTs, the τ and t superfields are unified in one GUT representation. As a result, above the GUT scale, renormalization group evolution of the $\tilde{\tau}$ mass includes radiative corrections arising from the triplet Higgs boson with strength proportional to the top Yukawa coupling [52]. This effect is not present for selectrons and smuons, leading to a substantial splitting of slepton masses, even if they are initially degenerate. GUT symmetries may also lead to alternative parameterizations of the GUT scale boundary conditions. For example, in SO(10) theories, the Higgs and matter multiplets are members of separate multiplets, the **10** and **16** representations, respectively. Gauge symmetry does not require their scalar masses to be unified. Furthermore, even if unified, SO(10) breaking to SU(5) as an intermediate step will split these scalar masses by D-terms [53, 54]. As a result, the Higgs and matter scalar masses are more generally parameterized by two independent parameters at the GUT scale, opening up many new possibilities in the low energy MSSM spectrum.

While we cannot describe all models investigated to date here, we can extract important lesson from the above discussion. The soft terms of the MSSM may carry the imprint of interactions that are not amenable to direct experimental investigation. Measurements of soft mass relations may therefore shed light on physics at the highest energy scales.

5. Slepton Studies

We now turn to specific studies of supersymmetry at the linear collider. We begin with sleptons, which are the lightest observable superpartners in many models. As we will see, slepton studies may shed light on a large number of important questions. They also provide an arena for exploiting many of the most powerful techniques available to study new physics the linear collider.

5.1. *Signal and Background*

Slepton pair production proceeds through s-channel gauge bosons, $e^+e^- \to \gamma, Z \to \tilde{l}^+\tilde{l}^-$ and $e^+e^- \to Z \to \tilde{\nu}\tilde{\nu}^*$. In the case of 1st generation sleptons, there are also contributions from t-channel neutralino and chargino exchange. In the presence of supersymmetric LFV, t-channel contributions are also present for the 2nd and 3rd generations. We will assume lepton flavor is conserved here, deferring discussion of LFV to Section 5.4.

Let us focus on supergravity scenarios with R-parity conservation.[b] When sleptons are heavier than neutralinos or charginos, they decay through

$$\tilde{l} \to \tilde{\chi}_i^0 l,\ \tilde{\chi}_i^- \nu_l \tag{61}$$

$$\tilde{\nu} \to \tilde{\chi}_i^+ l,\ \tilde{\chi}_i^0 \nu_l \ . \tag{62}$$

Sleptons have spin 0 and so decay isotropically in their rest frames. Right-handed sleptons decay dominantly to the lighter neutralino with the largest $\tilde{B}$ component. Left-handed sleptons decay primarily to the lighter neutralinos and charginos with the largest $\tilde{W}$ components. When decay into heavy charginos or neutralinos is possible, sleptons may initiate cascade decays ending in the LSP with many visible particles produced along the way.

[b]In general, R-parity violating theories are not difficult to study at linear colliders, as the relevant backgrounds are not overwhelming and all superpartner decay products are visible, allowing full momentum and energy reconstruction [55].

We now focus on the simplest case in which a charged slepton is the NLSP, and the LSP is the lightest neutralino $\tilde{\chi}_1^0$. The signal is then

$$e^+e^- \to \tilde{l}^+\tilde{l}^- \to l^+l^-\tilde{\chi}_1^0\tilde{\chi}_1^0 \ . \tag{63}$$

The momenta of the two leptons may be measured precisely, but the two LSPs escape detection. The signal is same flavor, opposite sign leptons with missing transverse momentum $\not{p}_T$.

The major standard model background to this process is W pair production, leading to $e^+e^- \to W^+W^- \to l^+\nu_l l'^-\overline{\nu}_{l'}$. The total W^+W^- cross section is enormous compared to typical supersymmetry cross sections, but it proceeds largely through t-channel exchange of the light electron, and so is very forward-peaked. The total W^+W^- cross section may therefore be reduced from about 10 pb at $\sqrt{s} = 500$ GeV to less than 2 pb with the requirement $|\cos\theta| < 0.7$, where θ is the polar angle of the W relative to the beam axis. Of this, only 1% produces a given same flavor final state. For comparison, the typical $\tilde{\mu}_R^+\tilde{\mu}_R^-$ production cross section is 50 fb^{-1} not far above threshold, and such processes are well-represented in the central region.

Beam polarization may also greatly reduce the W^+W^- background [56, 57]. An electron's spin along its direction of motion is essentially determined by its chirality. We define polarizations for the electron and positron beams through

$$\begin{aligned} P_{e^-} &= \frac{N\left(h=\frac{1}{2}\right) - N\left(h=-\frac{1}{2}\right)}{N\left(h=\frac{1}{2}\right) + N\left(h=-\frac{1}{2}\right)} \\ P_{e^+} &= \frac{N\left(\overline{h}=-\frac{1}{2}\right) - N\left(\overline{h}=\frac{1}{2}\right)}{N\left(\overline{h}=\frac{1}{2}\right) + N\left(\overline{h}=-\frac{1}{2}\right)} \ , \end{aligned} \tag{64}$$

where h and $\overline{h}$ are the e^- and e^+ helicities, respectively. The cross section for pair production of any particles may then be expressed as

$$\begin{aligned} \frac{d}{d\cos\theta}\sigma(h,\overline{h}) &= \frac{1}{32\pi s}\beta_f \sum_{h,\overline{h}} |\mathcal{M}(h,\overline{h})|^2 \\ &\times \frac{1}{4}\left[1+(-)^{-1/2+h}P_{e^-}\right]\left[1+(-)^{1/2+\overline{h}}P_{e^+}\right] \ . \end{aligned} \tag{65}$$

Here $\beta_f = (1 - m_f^2/E_{\text{beam}}^2)^{1/2}$ is the velocity of the final state particles, where m_f is their mass, and $E_{\text{beam}} = \sqrt{s}/2$ is the beam energy. $\mathcal{M}(h,\overline{h})$ is the amplitude for incoming electrons and positrons with helicities h and $\overline{h}$, respectively, and depends on $\cos\theta$.

Much of W pair production proceeds through $\overline{\nu}\gamma_\mu P_L e W^\mu$ couplings. For right-polarized beams, then, the production cross section is reduced by the factor $(1-P_{e^-})(1-P_{e^+})$ relative to the unpolarized case. For $P_{e^-}=0.8$ and unpolarized positrons, the cross section is reduced by a factor of 5. In the limit where beam polarization is perfect ($P_{e^-}=P_{e^+}=1$) and $\sqrt{s}\gg m_W$, the production cross section of transverse W bosons goes to zero and the cross section is dominated by the process with longitudinal W bosons in the s-channel.[c]

Another important background is $e^+e^-\to e^+e^-l^+l^-$. This is a background to slepton pair production when the two photons radiated off from the initial-state e^+ and e^- collide to produce a l^+l^- pair, while the initial-state e^+ and e^- escape into the beam pipe. Note that beam polarization cannot reduce this background. In current linear collider detector designs, electrons and positrons traveling within $\theta_{\rm min}\sim 5$ mrad are undetected, as this region is not covered by luminosity monitors. The maximum $\not{p}_T$ carried by these electrons and positrons is then

$$\not{p}_T{}^{\rm max} = 2.5\ \ {\rm GeV}\frac{\theta_{min}}{5\ {\rm mrad}}\frac{E_{\rm beam}}{500\ \ {\rm GeV}}\ . \tag{66}$$

Supersymmetric events with $\not{p}_T \lesssim 2.5$ GeV are therefore obscured by this background. Of course, the two-photon background may be greatly reduced with $\not{p}_T$ and acoplanarity cuts, with little impact on most supersymmetry signals.

5.2. *Slepton Masses*

Kinematic Endpoints

At linear colliders, the parton energy is fixed, neglecting the effects of initial state radiation, beam energy spread, and beamstrahlung. To a reasonable approximation, then, sleptons are produced with energy equal to the beam energy $E_{\rm beam}=\sqrt{s}/2$, and velocity

$$\beta_{\tilde{l}} = \frac{p}{E_{\rm beam}} = \sqrt{1-\frac{m_{\tilde{l}}^2}{E_{\rm beam}^2}}\ . \tag{67}$$

For sleptons produced not too far above threshold, as will typically be the case at least initially, $\beta_{\tilde{l}}$ is substantially smaller than 1.

[c]In this limit, the cross section reduces to that of Goldstone boson production, as required by the equivalence theorem.

The slepton then decays to $l+\tilde{\chi}_1^0$. In the slepton's rest frame, the lepton's energy is

$$E_l^* = \frac{m_{\tilde{l}}^2 - m_\chi^2 + m_l^2}{2m_{\tilde{l}}} , \tag{68}$$

where the lepton's mass may be neglected. Boosting to the lab frame, the lepton's energy is

$$E_l(\cos\theta) = E_l^* \frac{1 + \beta_{\tilde{l}} \cos\theta}{\sqrt{1 - \beta_{\tilde{l}}^2}} , \tag{69}$$

where θ is the polar angle of l in the slepton rest frame measured relative to the slepton's boost direction. In the lab frame, the lepton's energy distribution is therefore flat.

From Eq. (69), one can see that, provided $\beta_{\tilde{l}}$ is significantly less than 1, the $\tilde{l}$ and $\tilde{\chi}_1^0$ masses may be well-determined by measuring the two endpoints of the lepton energy distribution [56, 57]. The relations between the masses and endpoints are

$$m_{\tilde{l}}^2 = \frac{sE_{\text{max}}E_{\text{min}}}{(E_{\text{max}} + E_{\text{min}})^2} \tag{70}$$

$$1 - \frac{m_{\tilde{\chi}_1^0}^2}{m_{\tilde{l}}^2} = 2\frac{E_{\text{max}} + E_{\text{min}}}{\sqrt{s}} . \tag{71}$$

In Fig. 13 we show the muon energy distribution from $\tilde{\mu}_R$ pair production, along with the dominant backgrounds [58]. The distribution is not perfectly flat: this distortion is caused by initial state radiation, beam energy spread, and beamstrahlung, along with acoplanarity cuts. Such effects smear the energy distribution near the edges and must be corrected through measurements of beam properties and Monte Carlo simulations.

Doing so, one finds that $m_{\tilde{\mu}_R}$ can be measured to within 0.4 GeV from kinematic endpoints, assuming an integrated luminosity of 160 fb^{-1} for this case. Of course, the same measurement can be done for selectrons, and the precision of the mass measurement is of the same order [59]. This provides an immediate and stringent test of scalar mass relations, such as the universality of scalar masses at the GUT scale discussed in Section 4.1.

Minimum Mass Method

In the kinematic endpoint method, the fact that leptons come in pairs in slepton events is never used. An analysis that incorporates this pairing information is the minimum mass method [12], where for each event, one

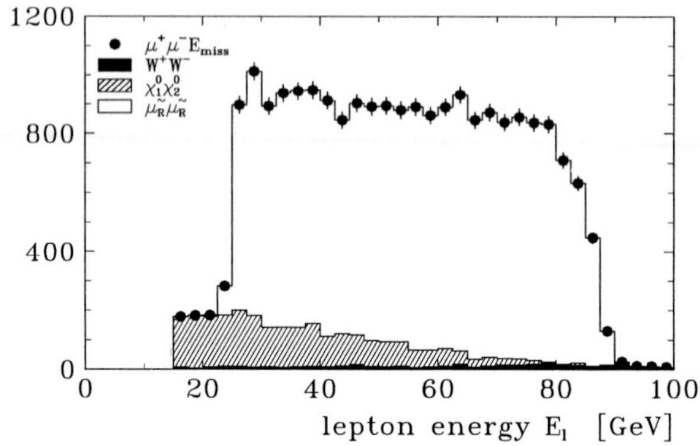

Fig. 13. Muon energy distribution from the signal $e_R^- e_L^+ \to \tilde{\mu}_R^+ \tilde{\mu}_R^- \to \mu^+ \tilde{\chi}_1^0 \mu^- \tilde{\chi}_1^0$ (open), and the dominant backgrounds (shaded), for $\sqrt{s} = 320$ GeV and an integrated luminosity of 160 fb^{-1} [58]. The underlying supersymmetric masses are $m_{\tilde{\mu}_R} = 132$ GeV and $m_{\tilde{\chi}_1^0} = 71.9$ GeV.

determines the minimum slepton mass consistent with the measured lepton momenta and a postulated $\tilde{\chi}_1^0$ mass. The minimum mass distribution peaks sharply at the actual mass [10, 12, 60, 61], as may be seen in Fig. 14, making possible mass measurements that are more precise than those from the kinematic endpoint method discussed above. For scenarios in which the $\tilde{\chi}_1^0$ mass is well-known from some other process, such as chargino production, the minimum mass method provides another avenue for precise slepton mass determination.

Threshold Scans

Slepton masses may also be determined through threshold scans [58, 62–67]. For pair-production of right-handed sleptons, the required initial state has angular momentum $J_z = 1$. Since the final state particles have spin 0, they must be produced in a P wave, and the production cross section grows as β^3 near threshold, in contrast to fermion pair production, where the cross section grows as β. The measurement of this behavior is a telling check that sleptons are, in fact, scalars. An example of this threshold behavior is given in Fig. 15. In view of the β^3 suppression, such measurements require high luminosity, and finite width effects, threshold corrections, and sub-dominant

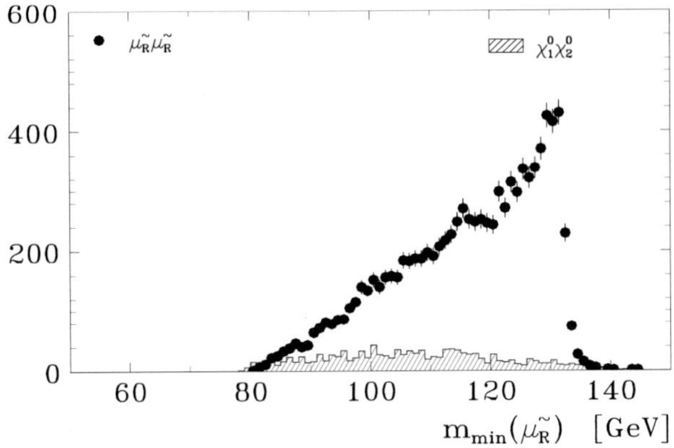

Fig. 14. Minimum mass distribution for smuons from the process $e_R^- e_L^+ \to \tilde{\mu}_R^+ \tilde{\mu}_R^- \to \mu^+ \tilde{\chi}_1^0 \mu^- \tilde{\chi}_1^0$, for $\sqrt{s} = 320$ GeV and an integrated luminosity of 160 fb^{-1} [10]. The underlying supersymmetry parameters are $m_{\tilde{\mu}_R} = 132$ GeV and $m_{\tilde{\chi}_1^0} = 71.9$ GeV.

diagrams must all be carefully controlled. However, e^+e^- threshold scans may allow precise mass measurements at the 100 MeV level, assuming luminosities of ~ 100 fb^{-1} [58, 64].

In e^-e^- collisions, the relevant initial state for $\tilde{l}_R^+ \tilde{l}_R^-$ production has angular momentum $J_z = 0$, and so slepton pair production has β threshold behavior [62, 64]. Cross sections for $\tilde{e}_R$ pair production in e^-e^- and e^+e^- modes are compared in Fig. 15. Mass measurements with 100 MeV precision can be achieved with a total luminosity of $\mathcal{O}(1)$ fb^{-1}, two orders of magnitude less than required in e^+e^- collisions for similar precision [64]. The full arsenal of linear collider modes may even allow one to extend this mass measurement to the rest of the first-generation sleptons through a series of β threshold scans: $e^-e^- \to \tilde{e}_R^- \tilde{e}_R^-$ yields $m_{\tilde{e}_R}$; $e^+e^- \to \tilde{e}_R^\pm \tilde{e}_L^\mp$ yields $m_{\tilde{e}_L}$; $e^+e^- \to \tilde{\chi}_1^+ \tilde{\chi}_1^-$ yields $m_{\tilde{\chi}_1^\pm}$; and $e^-\gamma \to \tilde{\nu}_e \tilde{\chi}_1^-$ yields $m_{\tilde{\nu}_e}$ [68]. If completed, this would yield a high precision test of Eq. (25), a robust prediction of supersymmetry and gauge invariance, and possibly also a high precision measurement of $\tan\beta$.

Mass Measurements at the LHC

The LHC may also measure superpartner masses with some precision. At hadron colliders, the parton-parton center-of-mass energy is unknown. In supersymmetric events, typically two LSPs are missing, but only the total missing momentum can be reconstructed, and there is a large uncertainty in the longitudinal momentum of the system. Nevertheless, it is still possible

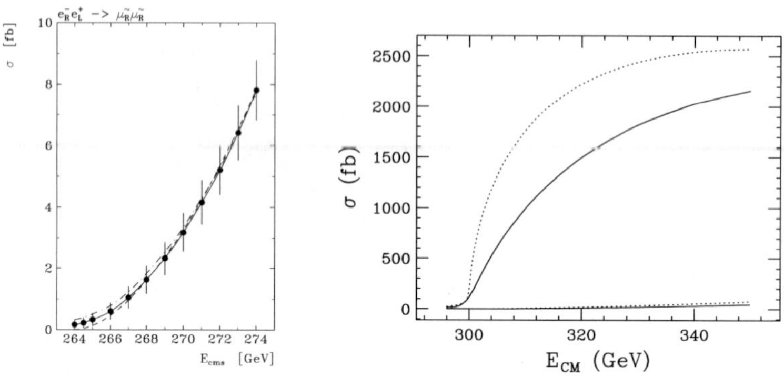

Fig. 15. Left: Threshold cross sections for $e^+e^- \to \tilde{\mu}_R^+\tilde{\mu}_R^-$, with $m_{\tilde{\mu}} = 132$ GeV and smuon width $\Gamma_{\tilde{\mu}_R} = 0.3$ (solid) and 0 and 0.6 GeV (dashed) [63]. The error bars are for a 10 point scan with total integrated luminosity $\mathcal{L} = 100\ \text{fb}^{-1}$. Right: Threshold behavior for $\sigma(e^-e^- \to \tilde{e}_R^-\tilde{e}_R^-)$ (upper two contours) and $\sigma(e^+e^- \to \tilde{e}_R^+\tilde{e}_R^-)$ (lower two contours) for $m_{\tilde{e}_R} = 150$ GeV and $M_1 = 100$ GeV and beam polarizations $P_{e^-} = 0.8$ and $P_{e^+} = 0$ [64]. In each pair, the dotted curve neglects all beam effects, and the solid curve includes the initial state radiation, beamstrahlung, and beam energy spread of flat beams.

to obtain reliable mass measurements from cascade decays if one can isolate an invariant mass distribution from a particular step in a cascade decay [69, 70]. For example, in the process $\tilde{\chi}_2^0 \to \tilde{l}_R l \to \tilde{\chi}_1^0 l^+ l^-$, the l^+l^- invariant mass distribution is predicted to be proportional to m_{ll}, with maximum

$$m_{ll}^{\max} = \sqrt{\frac{(m_{\tilde{\chi}_2^0}^2 - m_{\tilde{l}_R}^2)(m_{\tilde{l}_R}^2 - m_{\tilde{\chi}_1^0}^2)}{m_{\tilde{l}_R}^2}} \,. \tag{72}$$

The endpoint of the m_{ll} distribution is therefore determined by the three masses involved in the cascade decay. An example of this mass distribution is shown in Fig. 16. When the jets from the parent decay $\tilde{q} \to \tilde{\chi}_2^0 q$ are also identified, the number of identified end points increases to four: the upper endpoints of the m_{jl} and m_{jll} distributions, and the lower endpoint of the m_{jll} distribution for $m_{ll} > m_{ll}^{\max}/\sqrt{2}$. In principle, then, all 4 masses entering the cascade may be reconstructed. The resulting mass determinations for two sample points are also shown in Fig. 16. The mass difference $m_{\tilde{l}_R} - m_{\tilde{\chi}_1^0}$ is determined rather precisely. Note, however, that the absolute sparticle masses are poorly constrained, with uncertainties of the order of several tens of GeV.

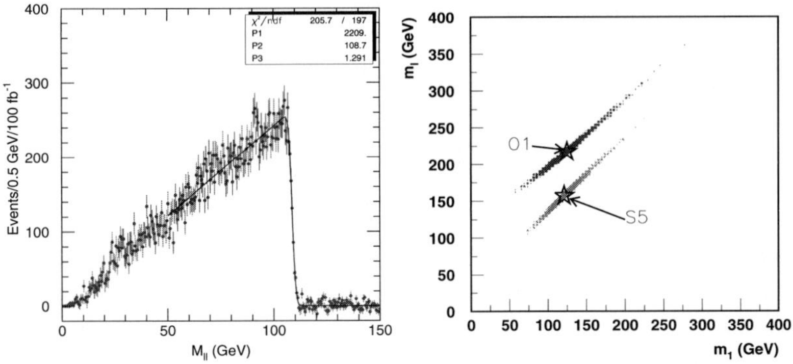

Fig. 16. Left: The $e^+e^- + \mu^+\mu^- - e^\pm e^\mp$ mass distribution for LHC minimal supergravity Point 5 with $\tilde{\chi}_2^0 \to \tilde{l}l \to \tilde{\chi}_1^0 l^+ l^-$ [71]. Right: Scatter plot of reconstructed values of the $\tilde{l}_R$ and $\tilde{\chi}_1^0$ masses for LHC point 5 (S5) and for a different model (O1) using the decay chain $\tilde{q}_L \to \tilde{\chi}_2^0 q \to \tilde{l}_R lq \to \tilde{\chi}_1^0 llq$ [72].

The example above illustrates the complementarity of the LHC and the linear collider. At the LHC, squarks and gluinos are produced with large cross sections, and their cascade decays typical include many charginos and neutralinos, as well as charged sleptons and sneutrinos. Distributions within these cascade decays may provide important kinematic information. The LHC therefore provides a "top view" of the superparticle spectrum. On the other hand, the linear collider produces the lighter sparticles first and determines their properties systematically from the bottom up. Remarkably, the uncertainties of masses and other parameters may be determined at each step with percent level uncertainty. Combining the data from the linear collider with that available at the LHC, all sparticle masses may be determined with a precision of the order of a few percent.

It is also noteworthy that, while the LHC determines decay patterns, detailed determinations of the spins and couplings of sleptons and other particles is difficult. Using polarized beams, the linear collider may determine these slepton properties, as we now describe.

5.3. *Polarized Cross Sections*

We now consider selectron pair production. As noted in Eq. (65), cross sections for various beam polarizations are the sum of four independent polarized cross sections determined by the amplitudes $\mathcal{M}(h, \overline{h})$. Each $\mathcal{M}(h, \overline{h})$ is a function of gauge-sfermion-sfermion and gaugino-sfermion-fermion couplings. For $\tilde{e}_R^+\tilde{e}_R^-$ production,

$$i\mathcal{M}(h,\overline{h}) = -i\lambda e^{i\lambda\phi}\sin\theta s\beta_f \left[\frac{g^2}{\cos^2\theta_W}\frac{A_h A_{\frac{1}{2}}}{s-m_Z^2+i\Gamma_Z} + \frac{e^2}{s} + \frac{(1\pm(-)^{\overline{h}+\frac{1}{2}})}{2}\sum_j \frac{1}{2}\frac{|A_{jR}^2|}{t-m_{\tilde{\chi}_j}^2}\right], \tag{73}$$

where $h=\pm 1/2$ is the helicity of the initial state electron, $\overline{h}=\pm 1/2$ is the helicity of the initial state positron, and $\lambda \equiv h-\overline{h}$. The angles θ and ϕ specify the $\tilde{e}_R^-$ production direction in polar coordinates, β_f is the selectron velocity, and $t=-\frac{s}{4}(1-2\cos\theta\beta_f+\beta_f^2)$. The couplings A_h and A_{jR} are given by

$$A_{\frac{1}{2}} = \sin^2\theta_W \ , \ A_{-\frac{1}{2}} = -\frac{1}{2}+\sin^2\theta_W \tag{74}$$

$$A_{jR} = -\sqrt{2}gN_{j1}\tan\theta_W \ . \tag{75}$$

The amplitude for $\tilde{e}_L^+\tilde{e}_L^-$ pair production is obtained from Eq. (73) by the replacements

$$A_h \to A_{-h} \tag{76}$$

$$1\pm(-)^{\overline{h}+1/2} \to 1\mp(-)^{\overline{h}+1/2} \tag{77}$$

$$A_{jR} \to A_{jL} = \frac{g}{\sqrt{2}}(N_{j2}+N_{j1}\tan\theta_W) \ . \tag{78}$$

The contribution of t-channel neutralino exchange to the amplitudes above is easy to understand. For $\tilde{e}_R$ pair production, the t-channel amplitude contributes only to $\mathcal{M}(1/2,-1/2)$; for $\tilde{e}_L$ pairs, it contributes only to $\mathcal{M}(-1/2,1/2)$. This follows from the fact that gaugino interactions preserve chirality, and so only the $\tilde{e}_R e_R\tilde{B}$ and $\tilde{e}_L e_L\tilde{B}$ couplings are non-vanishing.

The s-channel amplitude contributes to both amplitudes $\mathcal{M}(1/2,-1/2)$ and $\mathcal{M}(-1/2,1/2)$. In the limit $s \gg m_Z^2$, and for right-polarized e^- beams with $h=1/2$, one finds

$$e^+e_R^- \to \tilde{e}_R^+\tilde{e}_R^- \quad i\mathcal{M}(1/2,-1/2) = -iCg'^2 \tag{79}$$

$$e^+e_R^- \to \tilde{e}_L^+\tilde{e}_L^- \quad i\mathcal{M}(1/2,-1/2) = -iC\frac{1}{2}g'^2 \ , \tag{80}$$

where $C=\lambda e^{i\lambda\phi}\sin\theta\beta_f$. In this high-energy limit, the Z mass may be neglected, and the γ and Z mass eigenstates may be replaced by the gauge eigenstates B and W^0. Since e_R does not couple to W^0, only the hypercharge gauge boson contributes, and the amplitudes are proportional to the hypercharge coupling g', with hypercharges 1 and 1/2 for $\tilde{e}_R$ and $\tilde{e}_L$.

This example illustrates the power of beam polarization. By polarizing the beam, we can switch off some Feynman diagrams 'by hand.' One may then isolate particular amplitudes and determine fundamental parameters precisely [56, 57]. For example, if we use right-polarized e^- beams and observe $\tilde{e}_R$ and $\tilde{e}_L$ pair production with the rates predicted by Eqs. (79) and (80), we can establish the chirality-preserving nature of the supersymmetric gaugino interactions.

For $e^+e^- \to \tilde{e}_L^+\tilde{e}_R^-$ production, the s-channel amplitude is absent because gauge interactions preserve chirality. The same is true for $e^-e^- \to \tilde{e}_{R(L)}^-\tilde{e}_{R(L)}^-$, where the s-channel amplitude is forbidden by total lepton number conservation. In this case, the t-channel amplitude has a new feature: it is proportional to the mass of the neutralino exchanged in the t-channel, which is required to flip chirality. For example,

$$\begin{aligned} e^+e^- \to \tilde{e}_L^+\tilde{e}_R^- \quad i\mathcal{M}(h,\overline{h}) &= i\sqrt{s}\delta_{\lambda,0}\left[1+(-)^{\overline{h}-\frac{1}{2}}\right] \\ &\times \sum_j \frac{1}{2} m_{\tilde{\chi}_j^0} \frac{A_{jL}A_{jR}}{t-m_{\tilde{\chi}_j^0}^2} \,. \end{aligned} \tag{81}$$

The existence of the t-channel amplitude for $e^-e^- \to \tilde{e}^-\tilde{e}^-$ proves that the exchanged t-channel particle is a Majorana fermion. Note also that the amplitude is now not proportional to the velocity β, and so the cross section now grows like β near threshold. This behavior was discussed in Section 5.2 in the context of threshold mass measurements. Finally, the amplitude is proportional to M_1 in both of the limits $M_1 \ll M_2, |\mu|$ and $M_1 \gg M_2, |\mu|$. The total cross section is therefore directly related to fundamental mass parameters.

At high enough beam energies, selectrons will be produced through all mechanisms ($\tilde{e}_L^+\tilde{e}_L^-$, $\tilde{e}_R^+\tilde{e}_R^-$, and $\tilde{e}_L^\pm\tilde{e}_R^\mp$). All of these processes may result in the signal $e^+e^- + \not{p}_T$, as discussed above in Section 5.1. However, the various signals may often be disentangled, both kinematically, as the energy ranges of the electrons and positrons differ for the different processes, and by comparing the cross sections with various beam polarizations.

Slepton production cross sections have interesting dependences on the fundamental supersymmetry parameters. Figure 17 shows cross sections for $\tilde{e}_R$ pair production in the $(M_1, \tan\beta)$ plane. Here we fix the LSP $\tilde{\chi}_1^0$ mass. The GUT relation for M_1 and M_2 is also assumed. Not surprisingly, there is almost no dependence on $\tan\beta$. This dependence enters only through the neutralino mass matrix in the form of $\cos\beta$ and $\sin\beta$, which are both fairly constant for moderate and large $\tan\beta$. For a completely right-polarized

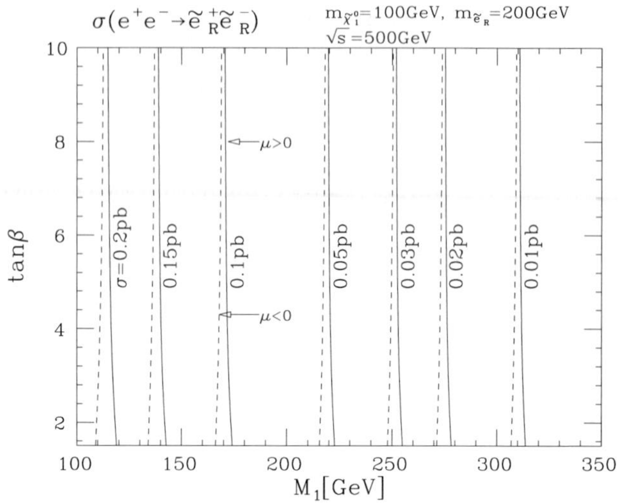

Fig. 17. Contours of $\sigma(e^+e^-_R \to \tilde{e}^+_R\tilde{e}^-_R)$ for $m_{\tilde{e}_R} = 200$ GeV and $\sqrt{s} = 500$ GeV in the $(M_1, \tan\beta)$ plane [73]. At each point in the plane, μ is chosen so that $m_{\tilde{\chi}^0_1} = 100$ GeV. Solid lines correspond to a $\mu > 0$ solution and dashed lines to $\mu < 0$.

beam, the only state exchanged in the t-channel is the Bino. This t-channel contribution dominates at the left-hand side of the plot, and then decouples for large M_1. At the left-hand side of the plot, the LSP is Bino-like, but becomes Higgsino-like on the right. However, the production proceeds through the neutralino with largest $\tilde{B}$ component. The $\tan\beta$ independence of the cross section is even more prominent for the process $e^-e^- \to \tilde{e}^-\tilde{e}^-$, where chirality requires a gaugino mass insertion, which is independent of neutralino mixings. This process also provides sensitivity to even very large M_1, and may allow for a measurement of the Bino mass even when the Bino is far from kinematically accessible [64, 67].

We have concentrated on selectrons here, as they have many interesting properties related to the existence of t-channel amplitudes. It should be stressed that smuon and stau studies are also interesting. In particular, stau studies may be very important at the linear collider. The lighter stau is often the lightest scalar superpartner as a result of GUT-scale interactions (see Section 4.6), renormalization group evolution between the GUT and weak scales (see Section 4.1), and left-right mixing (see Section 2.5). Note that staus may also be studied at the LHC, but such studies are significantly complicated by τ decay to low energy leptons or mesons.

The signal of $\tilde{\tau}$ production is very complicated and will be considered in more detail in Section 5.5. Here we merely note that the production amplitude for $e^+e^- \to \tilde{\tau}_1^+\tilde{\tau}_1^-$ in the limit $\sqrt{s} \gg m_Z$ for right-polarized electrons is [74]

$$i\mathcal{M}(1/2, -1/2) = -iCg'^2 \left(\frac{1}{2} \cos^2 \theta_{\tilde{\tau}} + \sin^2 \theta_{\tilde{\tau}} \right) , \tag{82}$$

where $\theta_{\tilde{\tau}}$ is the stau left-right mixing angle. The measurement of the $\tilde{\tau}$ production cross section for right-polarized electron beams will determine $\theta_{\tilde{\tau}}$. If one finds $\tilde{\tau}_2$ in addition to $\tilde{\tau}_1$, the full $\tilde{\tau}$ mass matrix may be reconstructed. The reconstruction of the $\tilde{\tau}$ mass matrix will provide yet another opportunities to test the universality of scalar masses at the high scale. In addition, the off-diagonal stau mass m^2_{LR} is proportional to $\tan\beta$ and μ as noted in Eq. (22), and the decay distribution also depends on $\tan\beta$, as we will see. Stau studies may therefore provide a rare opportunity to measure $\tan\beta$ for large $\tan\beta$.

5.4. *Lepton Flavor Violation*

As noted in Section 2.5, there is no reason for the fermion and sfermion mass matrices to be simultaneously diagonalizable. This implies that the discovery of superpartners will lead to a whole new sector of flavor physics to explore. In the standard model, after the initial discoveries and mass measurements, the focus has naturally turned to measurements of flavor violation, such as in the quark and neutrino sectors. A similar progression may be expected if superpartners are discovered.

At the moment, there is no standard explanation of fermion masses. Recently, our knowledge of neutrino mixing has expanded tremendously, but no compelling theory of flavor has emerged. One might wonder if supersymmetric flavor studies will also lead to a bewildering wealth of data without furthering our fundamental understanding of flavor. While possible, there are important differences in these two cases. In the case of neutrinos, the newly discovered mixings are in principle unrelated to those of the quarks. For example, in theories that attempt to explain flavor through horizontal flavor symmetries, neutrino mixing is dependent on a whole new set of flavor representation assignments. For superpartners, however, the new mixings are tied to the flavor properties of standard model fermions by supersymmetry. For example, standard model particles and their superpartners must be governed by the same flavor symmetries, since they are in the same supermultiplet. In these frameworks, then, superpartners do not introduce additional degrees of freedom, but rather provide new constraints on the same flavor physics governing the standard model particles. For this reason, the careful investigation of superpartner flavor symmetries provides a promising avenue for understanding not only superpartner properties, but also the masses and mixings observed in the standard model.

The full 6×6 slepton mass matrix of Eq. (18) is very complicated. As a first step, one may begin by neglecting left-right mixing, a reasonable approximation for selectrons and smuons. One can further specialize to two generation mixing and consider, for example, $\tilde{e}_R$-$\tilde{\mu}_R$ mixing. A convenient basis to consider is the mass eigenstate basis for both charged leptons and sleptons, in which all mixing is confined to gaugino vertices. Lepton flavor violation (LFV) then occurs in slepton pair production through t-channel neutralino exchange and in decay vertices [75, 76]. In simple scenarios where the sleptons decay directly to a neutralino LSP, the resulting signals of flavor-violating slepton pair production at the linear collider are $e^+e^- \to \tilde{l}^+\tilde{l}^- \to e^\pm\mu^\mp\tilde{\chi}_1^0\tilde{\chi}_1^0$ and $e^-e^- \to \tilde{l}^-\tilde{l}^- \to e^-\mu^-\tilde{\chi}_1^0\tilde{\chi}_1^0$.

The probability for flavor-violating decay is

$$P(\tilde{e}_R \to \mu) = \frac{1}{2}\sin^2 2\theta_R \frac{(\Delta m_R^2)^2}{4m_R^2\Gamma^2 + (\Delta m_R^2)^2} , \tag{83}$$

where $\Delta m_R^2 = m_1^2 - m_2^2$ and $m_R = (m_1 + m_2)/2$, with m_1 and m_2 the physical slepton masses, and θ_R is the flavor mixing angle. Γ is the slepton decay width. If $m_R\Gamma \ll \Delta m_R^2$, the mixing probability has its maximal value of $(\sin^2 2\theta_R)/2$; however, for $m_R\Gamma \gg \Delta m_R^2$, P vanishes, since sleptons decay before they have time to mix.

In the e^+e^- case, flavor violation requires a careful treatment of t-channel and s-channel interference, as flavor violation is present in t-channel diagrams, but absent in s-channel processes. However, as with neutrino studies, results are conveniently presented in the $(\sin 2\theta_R, \Delta m_R^2)$ plane. In Fig. 18, cross sections for the flavor-violating signal are given, along with the discovery reach of the linear collider. Beam polarization again provides a useful tool to reduce background from processes such as W^+W^- and $e^+\nu W^-$. Given an integrated luminosity of 50 fb^{-1}, e-μ flavor violation can be discovered for mixing angles of order $\theta_R \sim 0.05$ and mass degeneracies at the 1% level. For the same luminosity, the e^-e^- mode provides an even better discovery potential, as backgrounds such as W^+W^- are completely prohibited by total lepton number conservation.

Supersymmetric flavor violation is already constrained by low energy data, such as μ-e conversion and $\mu \to e\gamma$. In Fig. 18, these constraints

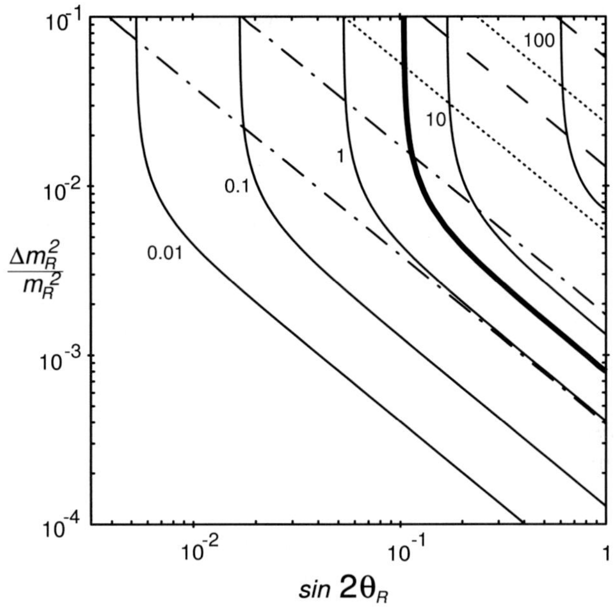

Fig. 18. The solid contours are for $\sigma(e^+e^-_R \to e^\pm\mu^\mp\tilde{\chi}^0_1\tilde{\chi}^0_1)$ in fb for $\sqrt{s} = 500$ GeV, $m_{\tilde{e}_R}, m_{\tilde{\mu}_R} \approx 200$ GeV, and $M_1 = 100$ GeV [75]. The thick contour represents the LFV discovery reach given 50 fb^{-1}. Constant contours of $B(\mu \to e\gamma) = 4.9 \times 10^{-11}$ and 2.5×10^{-12} are also plotted for degenerate left-handed sleptons with mass 350 GeV and $-(A + \mu \tan\beta)/m_R = 0$ (dotted), 2 (dashed), and 50 (dot-dashed).

are given by straight lines. These are highly dependent on supersymmetry parameters, notably $\tan\beta$, but are suppressed by $\Delta m^2_R/m^2_R$ through the supersymmetric analogue of the GIM mechanism. Note that the mass splitting below which LFV is suppressed is set by the slepton width Γ for the collider signal, but by the slepton mass m_R for low energy constraints. Since $\Gamma \ll m_R$, there is a large range of mass splittings in which LFV is suppressed in low energy experiments, but observable at colliders.

At the same time, because the high-energy and low-energy LFV rates have different functional dependences on the mass splitting and mixing angle, simultaneous measurements of LFV in both high- and low-energy experiments will provide complementary information. The combined results may even allow the extraction of mass splittings and mixing angles separately, providing valuable information for attempts to identify the fundamental origin of quark and lepton masses and mixings.

5.5. *Tau Polarization from Stau Decay*

The polarizations of all leptons resulting from slepton decay carry information about the underlying supersymmetry parameters. Typically, of course, these polarizations are unobservable. However, in the case of taus, polarization can be measured at colliders. Here we describe the parameter dependence of the polarization of taus from stau decay and show how this polarization can be reconstructed experimentally.

Tau leptons produced in stau decay have the simple energy distribution of Eq. (69). The τ decays into $A\nu_\tau$, where $A = e\nu_e, \mu\nu_\mu, \pi, \rho, a_1$. The heavy mesons further decay into pions through $\rho^\pm \to \pi^\pm\pi^0$ and $a_1^\pm \to \pi^\pm\pi^\mp\pi^\pm, \pi^\pm\pi^0\pi^0$. The signature of $\tilde{\tau}$ pair production is therefore two acoplanar jets with low multiplicity.

Because $E_\tau \gg m_\tau$, the decay products maintain the original direction of the parent τ. However, the total energy of the decay products is substantially reduced since the tau neutrinos escape detection. The $E_{\text{jet}(\tau)}$ distribution is therefore not flat, but decreases to zero at the endpoint, degrading the sharp edge expected in $\tilde{\mu}$ or $\tilde{e}$ pair production. Fits to the total E_{jet} distribution yield both $m_{\tilde{\tau}}$ and $m_{\tilde{\chi}_1^0}$, but the sensitivity to $m_{\tilde{\tau}}$ is reduced.

On the positive side, however, this jet energy distribution contains information about the τ polarization, which in turn depends on the stau and neutralino mass eigenstates. The chirality of a sfermion is preserved in the resulting fermion in gauge, and therefore, gaugino interactions. On the contrary, Higgs, and therefore Higgsino, interactions flip the chirality. This is depicted in Fig. 19, where the arrows indicate the flow of chirality. The two types of interaction also have different couplings, with the $\tau\tilde{\tau}\tilde{H}$ coupling proportional to tau Yukawa coupling $y_\tau \propto m_\tau/\cos\beta$ and the $\tau\tilde{\tau}\tilde{B}$ and $\tau\tilde{\tau}\tilde{W}^0$ couplings proportional to gauge couplings g_1 and g_2, respectively.

For a general stau mass eigenstate, in the limit where the lightest neutralino is a pure Bino state, the τ polarization is given by

$$P_\tau(\tilde{\tau}_1 \to \tilde{B}\tau) = \frac{4\sin^2\theta_\tau - \cos^2\theta_\tau}{4\sin^2\theta_\tau + \cos^2\theta_\tau} \,, \tag{84}$$

depending only on θ_τ, as expected. On the other hand, if the neutralino is a mixed object, there is a non-trivial dependence on both gauge and Yukawa couplings. For a general neutralino eigenstate, in the limit of a pure $\tilde{\tau}_R$ state, the resulting tau polarization is

$$P_\tau(\tilde{\tau}_R \to \tilde{\chi}_1^0\tau) = \frac{(g\sqrt{2}N_{11}\tan\theta_W)^2 - (y_\tau N_{13})^2}{(g\sqrt{2}N_{11}\tan\theta_W)^2 + (y_\tau N_{13})^2} \,. \tag{85}$$

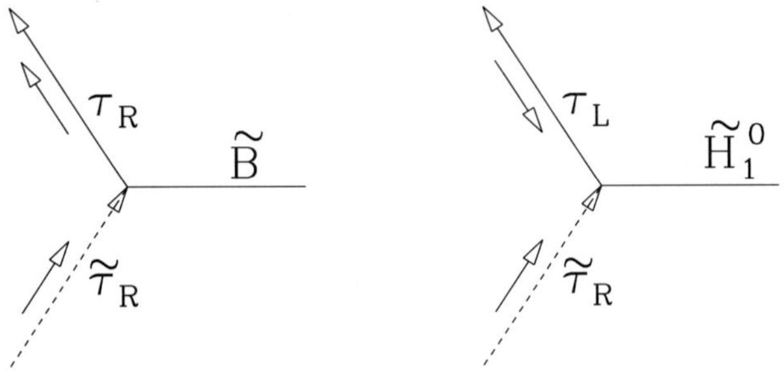

Fig. 19. The flow of chirality in interactions of $\tilde{\tau}_R$ with gauginos and Higgsinos.

Because $\theta_{\tilde{\tau}}$ can be measured through the polarization dependence of the stau production cross section, as noted in Section 5.3, a measurement of the τ polarization in this case will provide information on the tau Yukawa coupling and $\tan\beta$. As $y_\tau = gm_\tau/(\sqrt{2}m_W \cos\beta)$ is comparable to gauge couplings only for large $\tan\beta$, the sensitivity to $\tan\beta$ is high only for large $\tan\beta \gtrsim 10$ when the charginos and neutralinos involved have significant Higgsino components.

The branching fraction of taus to ρ mesons is large, with $B(\tau \to \rho) \approx 23\%$, and the ρ polarization is mostly longitudinal for τ_R and transverse for τ_L. One can thus determine P_τ by measuring P_ρ. The pion energy distributions in $\rho_{L(T)} \to \pi^\pm\pi^0$ decay are simple functions of $z_c = E_{\pi^\pm}/E_{\text{jet}}$, where E_{jet} is the total energy of the jet to which the $\pi^\pm$ belongs. These distributions are

$$\frac{d\Gamma(\rho_T \to 2\pi)}{dz_c} \sim 2z_c(1-z_c) - 2m_\pi^2/m_\rho^2$$
$$\frac{d\Gamma(\rho_L \to 2\pi)}{dz_c} \sim (2z_c - 1)^2 \ , \tag{86}$$

where $(1-\beta_\pi)/2 \le z_c \le (1+\beta_\pi)/2$ and $\beta_\pi = \sqrt{1 - 4m_\pi^2/m_\rho^2}$.

A detailed Monte-Carlo study of tau polarization is done in Ref. [73]. In this work, detector granularity is incorporated in the detector simulation, and tracking information is used to subtract the charged track fraction of the energy. The study finds reasonable π, ρ, and a_1 separation through the measurement of jet invariant mass, and ρ candidates are used to determine the τ polarization. Determination of P_τ by measuring E_{π^+}/E_{jet} with the error of $\delta P_\tau = 0.08$ appears possible given 10^4 stau pairs. (See Fig. 20.) However, conclusive results may require a full detector simulation.

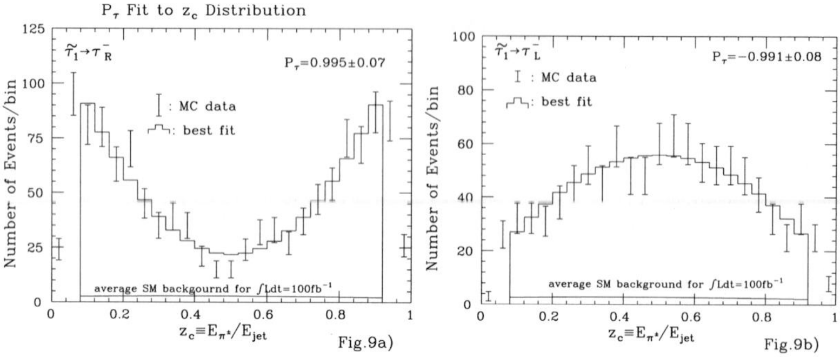

Fig. 20. Distributions in z_c for ρ candidates selected from 10^4 $\tilde{\tau}_1$ pairs decaying exclusively into τ_R (left) and τ_L (right), together with the best fit histograms [73].

6. Chargino and Neutralino Studies

6.1. *Signal and Background*

The gaugino mass relations $M_1 : M_2 : M_3 \simeq 1 : 2 : 7$ are predicted in scenarios with unified gaugino masses at the GUT scale, and also in gauge-mediated supersymmetry breaking scenarios. This hierarchy implies that Binos and Winos, or more generally, charginos and neutralinos, are among the lightest superpartners, and so are among the most likely to be produced at the linear collider.

The relevant production processes include

$$e^+e^- \to \tilde{\chi}_1^0\tilde{\chi}_1^0,\ \tilde{\chi}_1^0\tilde{\chi}_2^0,\ \tilde{\chi}_1^\pm\tilde{\chi}_1^\mp,\ \tilde{\chi}_2^0\tilde{\chi}_2^0\ . \tag{87}$$

In favorable scenarios where $|\mu| \sim M_1, M_2$, such as the focus point models discussed in Section 4.2, the $\tilde{\chi}_2^\pm$, $\tilde{\chi}_3^0$, and $\tilde{\chi}_4^0$ states may also be produced, allowing access to the entire neutralino and chargino systems. Neutralino production proceeds through s-channel Z boson exchange and t-channel $\tilde{e}$ exchange. Charginos are produced through s-channel exchange of γ and Z and t-channel $\tilde{\nu}_e$ exchange.

In many models, the lightest neutralino $\tilde{\chi}_1^0$ is stable and therefore invisible. In this case, the first process of Eq. (87) is unobservable. (It is a frustrating fact that thousands of neutralinos may already have been produced in colliders without our knowledge.) The heavier states typically decay visibly, however, with decay modes such as

$$\begin{aligned}\tilde{\chi}_1^+ &\to W^+\tilde{\chi}_1^0 \\ &\to \tilde{\tau}^+\nu,\ \tilde{l}^+\nu,\ \tilde{\nu}l^+ \\ &\to \overline{q}'q\tilde{\chi}_1^0,\ l^+\nu\tilde{\chi}_1^0\end{aligned}$$

$$\begin{aligned}\tilde{\chi}_2^0 &\to h\tilde{\chi}_1^0,\ Z\tilde{\chi}_1^0 \\ &\to \tau\tilde{\tau},\ l\tilde{l} \\ &\to ll\tilde{\chi}_1^0,\ \nu\overline{\nu}\tilde{\chi}_1^0,\ \overline{q}q\tilde{\chi}_1^0\ .\end{aligned} \tag{88}$$

The signal of chargino pair production is therefore missing momentum accompanied by $4j$, $2j + l$, or $l^+l'^-$. Neutralino pair signals are similar, but with each charged lepton replaced by a same flavor lepton pair. Note that if the lightest neutralino is stable, the second lightest neutralino is also invisible if it decays to $\nu\overline{\nu}\tilde{\chi}_1^0$.

As in the case of sleptons, the dominant backgrounds for chargino and neutralino events are again various standard model gauge boson processes, such as WW and ZZ. These may be removed with cuts similar to those used for slepton production. Note, however, that Wino-like states couple only to left-handed (s)electrons. In the case of pair production of Wino-like states, then, a right-polarized electron beam reduces both the signal and backgrounds.

6.2. *Masses and Polarized Cross Sections*

Chargino production is among the most well-studied supersymmetric processes, and we will concentrate on charginos here, although neutralinos also provide many insights. The fundamental supersymmetry parameters entering the chargino mass matrix of Eq. (27) are μ, M_2, and $\tan\beta$. These determine not only the chargino masses, but also the chargino mixing matrices U_{ij} and V_{ij} of Eq. (28). The production cross section also depends on the sneutrino mass that enters through the t-channel process. The chargino pair production cross section is therefore determined by four supersymmetry parameters:

$$\mu\,, \qquad M_2\,, \qquad \tan\beta\,, \qquad m_{\tilde{\nu}_e}\ . \tag{89}$$

In addition to these, the observables of chargino events depend on the parameters entering chargino decay. These include M_1 and other sparticle masses. Note that chargino production and decay are intertwined through spin correlations, and so cannot be treated separately in principle.

The flexibility of the linear collider provides the possibility of disentangling all of these parameters even if only one chargino state is kinematically accessible [57, 77, 78, 79]. First, for center-of-mass energies reasonably far above threshold, the impact of decay parameters on many observables may be reduced to low levels. Second, beam polarization allows a variety of cross section measurements with different dependences on the fundamental parameters. Among the most important quantities are

$$m_{\tilde{\chi}_1^+}\,, \qquad \sigma_R\,, \qquad A_R^{FB}\,, \qquad \sigma_L\,, \qquad A_L^{FB}\,, \tag{90}$$

where σ and A^{FB} are the total cross section and forward-backward asymmetry for chargino production, and the subscripts L and R denote left- and right polarized electron beams. Each of these quantities has an interesting and unique dependence on the fundamental parameters.

The chargino mass $m_{\tilde{\chi}_1^+}$ depends in principle on M_2, $|\mu|$, and $\tan\beta$. However, when the off-diagonal entries are small compared to the diagonal entries, as is often the case, $m_{\tilde{\chi}_1^+}$ provides essentially a direct measurement of M_2 or $|\mu|$, whichever is smaller. The chargino mass can be measured by kinematic endpoints through analyses similar to those described in Section 5.2. It may also be measured very precisely through threshold scans, since the cross section is proportional to $\beta_{\tilde{\chi}_1^+}$, as discussed in Section 5.2.

Although the chargino cross section has both s-channel and t-channel contributions in general, the t-channel sneutrino contribution is eliminated in the limit of purely right-polarized electron beams. The right-polarized cross section therefore contains only s-channel contributions mediated by γ and Z gauge bosons. As discussed in Section 5.3, in the high energy limit $\sqrt{s} \gg m_Z$ where the Z mass is negligible, the exchanged gauge bosons may be replaced by the gauge eigenstates B and W^0 to an excellent approximation. The W^0 contribution is absent for right-polarized electrons, while B exchange is absent for Wino-like charginos. The right-polarized cross section σ_R is therefore highly suppressed for Winos and is a sensitive measure of the Higgsino content of the chargino. This can be seen in Fig. 21. For $M_2 \ll |\mu|$ where the lighter chargino is Wino-like, σ_R is all but absent, while for $M_2 \gg |\mu|$, where the lighter chargino is Higgsino-like, σ_R is large. Note that σ_R becomes near maximal even very close to threshold (the hatched region) as a result of the β threshold behavior. A_R^{FB} is also sensitive to chargino mixing, with a dependence different from σ_R. The right-polarized quantities σ_R and A_R^{FB} therefore provide information on chargino mixing that is highly complementary to the information provided by $m_{\tilde{\chi}_1^+}$.

Finally, the left-polarized quantities σ_L and A_L^{FB} are sensitive to the mass of the exchanged sneutrino. The t-channel amplitude is proportional to $1/(t - m_{\tilde{\nu}_e}^2)$ where $t = m_{\tilde{\chi}_1^+} - (1 - \beta_{\tilde{\chi}_1^+} \cos\theta)s/2$ and θ is the angle between the $\tilde{\chi}_1^-$ momentum and the electron beam. The s- and t-channel contributions interfere destructively, and σ_L provides a measurement of $m_{\tilde{\nu}_e}$ even when the sneutrino mass is several hundreds of GeV and sneutrinos are too heavy to be produced directly. In addition, the forward-backward

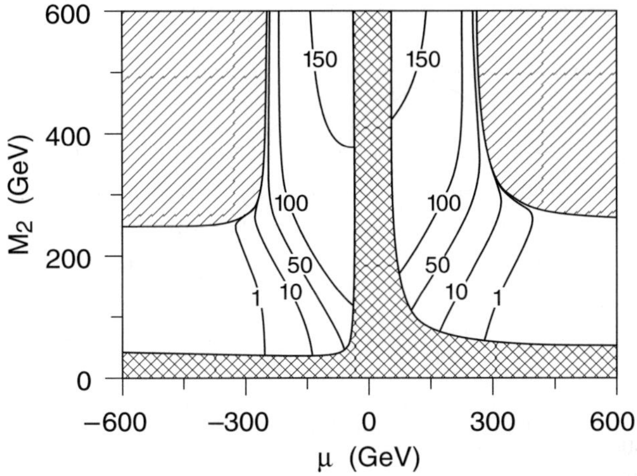

Fig. 21. Contours of $\sigma(e^+e^-_R \to \tilde{\chi}_1^+\tilde{\chi}_1^-)$ (in fb) in the (μ, M_2) plane for fixed $\tan\beta = 4$ and $\sqrt{s} = 500$ GeV. The cross-hatched region is excluded by current bounds, and charginos are kinematically inaccessible in the hatched region [77].

asymmetry is sensitive to the sneutrino mass. This asymmetry may be transferred to the decay products. For example, in Fig. 22, the forward-backward asymmetry of the final state e in chargino events is shown as a function of $m_{\tilde{\nu}_e}$ for various $m_{\tilde{e}_L}$, assuming a left-polarized electron beam. Assuming the MSSM $m_{\tilde{\nu}_e}$-$m_{\tilde{e}_L}$ splitting relation of Eq. (25), A_L^{FB} provides another measurement of $m_{\tilde{\nu}_e}$. Alternatively, one can check the validity of the slepton splitting relation using the combined measurements.

Chargino studies therefore provide a rich arena for supersymmetry studies. Many important implications follow from such studies. For example, if the charginos are determined to be Wino-like, the chargino and neutralino masses become highly correlated with M_2 and M_1, respectively, allowing a model-independent measurement of these parameters and providing a test of GUT or gauge-mediated predictions. On the other hand, if they are determined to have significant Wino–Higgsino mixing, upper bounds on the heavier chargino and neutralinos may be obtained. Chargino studies may also determine, or set upper bounds on, sneutrino masses even when sneutrinos are far beyond the kinematic reach of the collider, providing another target energy for future supersymmetry searches.

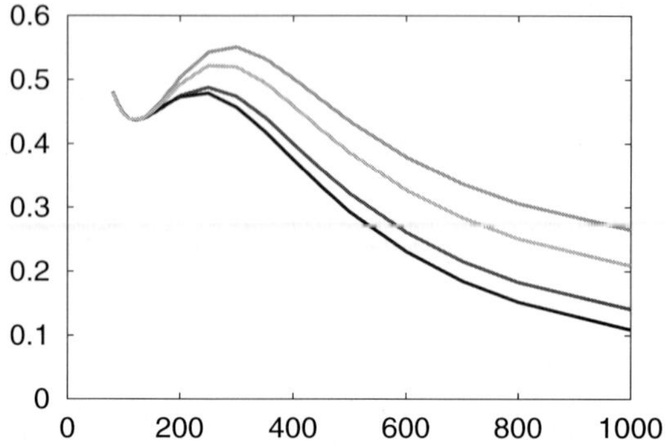

Fig. 22. The forward-backward asymmetry of the final state electron from chargino decay in chargino pair events as a function of the sneutrino mass (in GeV) [79]. The curves are for $m_{\tilde{e}_L} = 130$ GeV, 150 GeV, 200 GeV, and $m^2_{\tilde{e}_L} = m^2_{\tilde{\nu}_e} - m^2_W \cos 2\beta$, from above. The other underlying parameters are fixed to $M_2 = 152$ GeV, $\mu = 316$ GeV, $\tan\beta = 3$, and $M_1 = 79$ GeV, and the center-of-mass energy is $\sqrt{s} = 500$ GeV, with beam polarizations $P_{e^-} = -0.85$ and $P_{e^+} = -0.6$.

6.3. *CP Violation*

Because the linear collider allows one to overconstrain the chargino system, as discussed above, it also allows one to explore additional degrees of freedom assumed absent in the simplest models. For example, one may explore supersymmetric sources of CP violation, or even the basic supersymmetry relations between particle and sparticle couplings. We discuss the former here and the latter in Section 7. A general discussion of CP violation at linear colliders may also be found in Chapter 10.

Many supersymmetric parameters are in general complex. For example, the gaugino and Higgsino mass parameters are, in general,

$$\mu = |\mu| e^{i\phi_\mu} \; , \; M_i = |M_i| e^{i\phi_i} \; . \tag{91}$$

As we already discussed in Section 3.4, CP-violating phases of supersymmetry parameters are strongly constrained by neutron and electron EDMs. Large $\mathcal{O}(1)$ phases therefore require fine-tuning among parameters [80]. However, baryogenesis requires some source of CP violation beyond the standard model, and it is natural to ask how CP violation in supersymmetry may be discovered.[d]

The CP phases of supersymmetry parameters may be measured through their impact on CP-*conserving* quantities such as masses and decay distributions. For example, allowing the μ parameter to be complex, the two chargino masses are given by [78]

$$m^2_{\tilde{\chi}^{\pm}_{1,2}} = \frac{1}{2}\left[M_2^2 + |\mu|^2 2m_W^2 \mp \Delta_C\right] , \tag{92}$$

where

$$\begin{aligned}\Delta_C^2 = (M_2^2 - |\mu|^2)^2 + 4m_W^4 \cos^2 2\beta + 4m_W^2(M_2^2 + |\mu|^2)\\ +8m_W^2 M_2|\mu| \sin 2\beta \cos\phi_\mu \ .\end{aligned} \tag{93}$$

If the full chargino system is available and the two chargino masses and two mixing angles ϕ_+ and ϕ_-, suitably modified to diagonalize the complex chargino mass matrix [78], are constrained, deviations of ϕ_μ from zero may be observed. CP-violating phases also distort the neutralino mass spectrum [81].

The dependence of various CP-conserving observables on the phases of μ and M_1 are shown in Fig. 23. Significant variations are clearly possible. In these figures, however, the magnitudes of all parameters are held fixed. It is therefore not clear whether the effects of CP-violating phases may be mimicked simply by suitable adjustments of real parameters. To answer this question, a CP-violating scenario was studied in Ref. [82]. The predicted values for three cross sections, $\sigma(\tilde{\chi}^0_1\tilde{\chi}^0_2)$, $\sigma(\tilde{\chi}^0_2\tilde{\chi}^0_2)$, and $\sigma(\tilde{\chi}^{\pm}_1\tilde{\chi}^{\mp}_1)$, and three masses, $m_{\tilde{\chi}^0_1}, m_{\tilde{\chi}^0_2}$, and $m_{\tilde{\chi}^{\pm}_1}$ were then determined. The measured values of these observables were then assumed to lie in Gaussian distributions around these central values, where the Gaussian widths were chosen to simulate realistic experimental resolutions and statistical errors. 10,000 "pseudo-data sets" were formed by choosing values of these observables within these distributions, and for each pseudo-data set, the best fit underlying supersymmetry parameters were determined.

Figure 24 shows the results of this study. The input model parameters are shown as arrows in the figures. The left panel shows best fit values of M_1 and ϕ_1 for the 10,000 data sets, assuming $\tan\beta = 4$ fixed to its underlying value (dark points) and including $\tan\beta$ among the fitted parameters (light

[d]Large CP-violating phases are also motivated by string theory, where all couplings, including the standard model Yukawa couplings, are expected to result from the dynamical condensation of moduli fields. If soft supersymmetry-breaking terms arise from the same source, large phases in soft supersymmetry parameters are also expected.

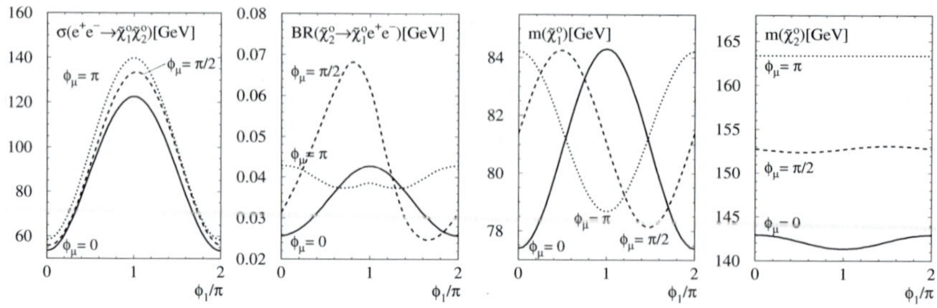

Fig. 23. The dependence of various observables on CP-violating phases for the μ parameter and Bino mass, assuming magnitudes of the supersymmetry parameters as given in minimal supergravity with $m_0 = 100$ GeV, $M_{1/2} = 200$ GeV, $A_0 = 0$, $\tan\beta = 4$ and $\mu > 0$ [82].

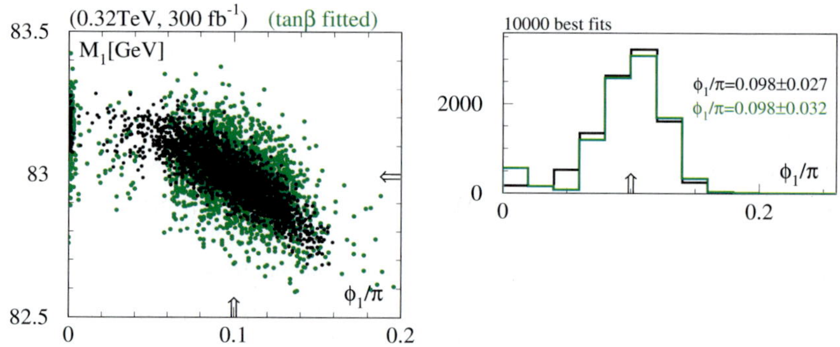

Fig. 24. The best fit values of M_1 and ϕ_1 given measurements of three cross sections, $\sigma(\tilde{\chi}_1^0\tilde{\chi}_2^0)$, $\sigma(\tilde{\chi}_2^0\tilde{\chi}_2^0)$, and $\sigma(\tilde{\chi}_1^\pm\tilde{\chi}_1^\mp)$, and three masses, $m_{\tilde{\chi}_1^0}, m_{\tilde{\chi}_2^0}$, and $m_{\tilde{\chi}_1^\pm}$ [82]. The effects of experimental resolution and finite statistics have been included (see text). The parameter $\tan\beta$ is either fixed to its underlying value (dark) or included among the fitted parameters (light green). The right-hand panel summarizes the distribution of best-fit ϕ_1 phases.

green points). The ϕ_1 distribution is summarized in the right-hand panel. As can be seen, the hypothesis of a vanishing phase is strongly disfavored. In these examples, the underlying value of $|\mu| = 310$ GeV is assumed known from chargino studies. However, uncertainties of 5 GeV in the μ parameter do not alter these conclusions.

CP violation may also be measured directly through manifestly CP-*violating* observables [78]. Such observables are required to be sensitive to the polarization of charginos and neutralinos relative to the plane of production. These observables typically require production of heavy charginos and neutralinos, as in $\tilde{\chi}_1^{\pm}\tilde{\chi}_2^{\mp}$ and $\tilde{\chi}_i^0\tilde{\chi}_j^0$ $(i \neq j)$ events, and are challenging, as the dominant CP-even contribution becomes a background for the extraction of a small CP-odd component. However, measurements of CP violation in this way are possible, and such observables could provide direct and unambiguous evidence for supersymmetric CP violation.

7. Testing Supersymmetry

Newly discovered particles need not conform to our preconceived expectations — the μ-π and c-τ puzzles are well-known cautionary tales. A doubling of the standard model spectrum is often thought to be the smoking gun signal of supersymmetry. However, there is no guarantee that all superpartners will be discovered together. In addition, even if all of the appropriate degrees of freedom are discovered, the new physics need not be supersymmetric.[e] Can new particles be identified as supersymmetric through incisive tests? Also, if some superpartners are missing, can we constrain their properties, much as precision electroweak data constrained the mass of the top quark previously and constrains the Higgs boson mass now?

It turns out that these two questions may both be answered by testing purely supersymmetric relations. We have already noted in Sections 2.2 and 2.3 that the gaugino-sfermion-fermion and Higgsino-sfermion-fermion couplings are determined by the gauge and Yukawa couplings of the standard model, respectively. Tests of the equivalence of these couplings are therefore model-independent tests of supersymmetry [77]. At the same time, measurements of small deviations from these identities are measurements of supersymmetry breaking, and so constrain soft parameters and the superparticle spectrum [73, 84–89].

7.1. *Verifying Supersymmetry*

Exact supersymmetry relates both dimensionful and dimensionless couplings. The dimensionful relations, such as the equivalence of the electron and selectron masses, must be broken, and the new contributions are the model-dependent soft supersymmetry-breaking terms of Section 2.4. In contrast, the relations between dimensionless couplings are preserved in all

[e]For example, see Ref. [83] for an extra-dimensional model where KK modes provide a new particle spectrum identical to that predicted by supersymmetry. In addition, the colored KK particles are heavier than the electroweak ones, and even the missing energy signals predicted in many supersymmetry frameworks are mimicked by the existence of a stable weakly-interacting lightest KK particle.

attractive models, as these relations are required if the gauge hierarchy is to be preserved. They therefore provide model-independent predictions that may be exploited to confirm that newly-discovered particles are indeed superpartners.

How well may the dimensionless supersymmetric identities be tested at the linear collider? To quantify this, we may treat the standard model and supersymmetric couplings as independent parameters, and then determine how well the supersymmetric couplings may be determined. These new couplings must be determined along with all of the usual unknown supersymmetry parameters, and the introduction of a yet another degree of freedom implies that many independent measurements of a given reaction are necessary. Testing supersymmetry in this model-independent way therefore makes full use of the flexibility and potential of linear colliders.

Tests of supersymmetry are possible with many different superpartners and many different coupling relations. As an example, consider $\tilde{e}_R$ pair production. The t-channel neutralino exchange process depends on the Bino coupling $g_{\tilde{B}\tilde{e}_R e_R}$. To investigate the level of sensitivity to this coupling, we define the parameter

$$Y_{\tilde{B}} \equiv \frac{g_{\tilde{B}\tilde{e}_R e_R}}{\sqrt{2}g'} \ . \tag{94}$$

Supersymmetry predicts $Y_{\tilde{B}} = 1$. In the limit $m_Z \ll M_1, |\mu|$ and assuming 100% right-handed electron beams, the amplitude for $\tilde{e}_R$ pair production is approximately

$$\mathcal{M} \propto \beta_{\tilde{e}} \left[1 - \frac{4Y_{\tilde{B}}^2}{1 - 2\cos\theta\beta_{\tilde{e}_R} + \beta_{\tilde{e}_R}^2 + 4M_1^2/s} \right] , \tag{95}$$

where the first and second terms come from s- and t-channel processes, respectively. We may therefore constrain both $Y_{\tilde{B}}$ and M_1 by measuring the differential cross section $d\sigma/d\cos\theta$.

To determine the differential cross section, we must know the slepton production angle. This is, of course, not directly observable. However, in any given event, we may determine it up to a two-fold ambiguity [57]. The angle between an observed lepton and its parent slepton is determined by the lepton energy through Eq. (69). The slepton direction then lies on a cone with known opening angle centered on the lepton direction. The direction of each slepton may be constrained in this way; because the slepton pair must be produced back-to-back, we find two possible solutions for the slepton production angle in each event, as shown in Fig. 25. The distribution of the wrong solution turns out to be more or less flat, and so the combined

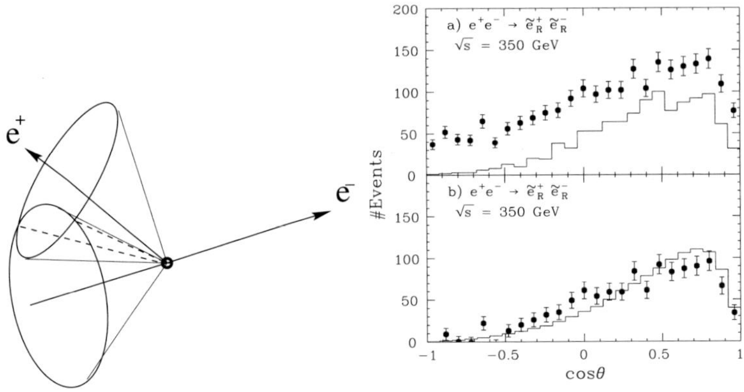

Fig. 25. Left: Reconstruction of the slepton production angle [57]. Each parent slepton's direction lies on a cone around the observed lepton momentum. Since the sleptons are produced back-to-back, the slepton directions are then determined up to a two-fold ambiguity to lie along the dashed lines. The cone for $\tilde{e}^-$ solution has been inverted in the figure. The dashed lines represent the solutions for the $\tilde{e}^+$ direction. Right: Angular distribution of $e^+e^- \to \tilde{e}_R^+\tilde{e}_R^-$ reconstructed from the final state e^+e^- for underlying parameters $\sqrt{s} = 350$ GeV, $\tilde{e}_R = 142$ GeV, and $\tilde{\chi}_1^0 = 118$ GeV [57]. The data points with error bars show angular distributions with (a) both right and wrong solutions and (b) with the wrong solution distribution, assumed flat, subtracted (see text). The histograms are the distribution of correct solutions (a) after and (b) before selection cuts.

distribution plotted in Fig. 25 allows one to determine the sparticle angular distribution in a statistical manner.

Given the differential cross section for selectron pair production, as well as kinematic endpoint information to constrain $m_{\tilde{e}_R}$ and $m_{\tilde{\chi}_1^0}$, one may then constrain $Y_{\tilde{B}}$. The results of such an analysis are given in Fig. 26, where we see that $Y_{\tilde{B}}$ may be determined at the 2% level for integrated luminosity 100 fb^{-1} [73], providing precise and model-independent evidence that the produced scalar particle is indeed a selectron. Constraints on $Y_{\tilde{B}}$ of even higher precision are possible from $e^-e^- \to \tilde{e}_R^-\tilde{e}_R^-$ production [85], and the equivalent parameters for the SU(2) and SU(3) couplings may also be constrained by chargino and squark studies [85] and also through studies of triple gauge vertices [89]. An exhaustive list of observables with the potential for testing supersymmetry is given in Ref. [84].

7.2. *Super-oblique Parameters*

So far we have been discussing the possibility of testing supersymmetry by verifying dimensionless coupling identities. These identities are exact at

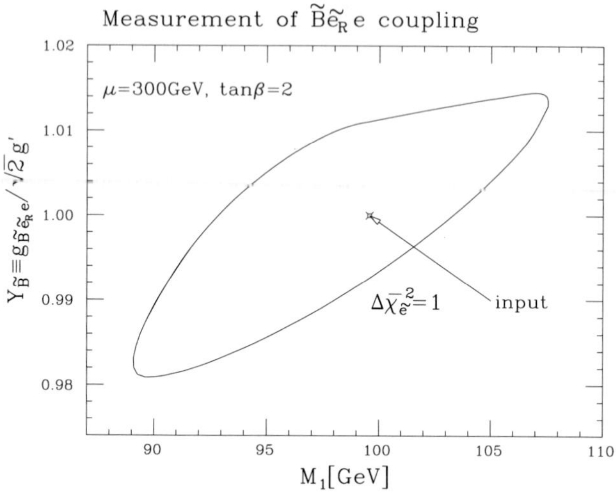

Fig. 26. $\Delta\overline{\chi}^2_{\tilde{e}} = 1$ contour in the M_1-$Y_{\tilde{B}}$ plane for integrated luminosity 100 fb^{-1} [73]. $\Delta\chi^2_{\tilde{e}} = 1$ is defined by constraints from electron energy and differential cross sections measurements and is roughly equivalent to a significance of $\Delta\chi^2 = 1$. Input values are $m_{\tilde{e}_R} = 200$ GeV, $\mu = 300$ GeV, $M_1 = 99.57$ GeV, and $\tan\beta = 2$.

tree-level. At the loop-level, however, even these relations receive corrections [90]. The breaking of dimensionless relations is called *hard supersymmetry breaking* [91].

These corrections may be understood by analogy to the oblique corrections [92] of the standard model. In the standard model, SU(2) multiplets with custodial SU(2)-breaking masses, such as the (t, b) multiplet, induce splittings in the couplings of the (W, Z) vector multiplet at the quantum level. Similarly, in supersymmetric models, supermultiplets with soft supersymmetry-breaking masses, such as the $(\tilde{f}, f)$ supermultiplets, induce splittings in the couplings of the (gauge boson, gaugino) vector supermultiplet at the quantum level. This analogy can be made very precise [84, 86, 87]. Corrections to hard supersymmetry relations are therefore called *super-oblique corrections*, and the splittings are typically written in terms of *super-oblique parameters*.

An enlightening viewpoint is to consider the evolution of dimensionless couplings from some high scale in these theories. Above the scale of superpartner masses, the theory is completely supersymmetric, and so the gaugino and gauge couplings remain identical. However, at scales below some sparticle mass, the effective theory is not supersymmetric, and so the supersymmetry relations get corrected. If there are still sparticles in this effective theory, they may be produced at colliders, and their couplings will show evidence of the heavy superpartners.

This viewpoint makes obvious an interesting fact: the super-oblique parameters are *non-decoupling*. The heavier the decoupled superparticles are, the longer the couplings evolve in the non-supersymmetric theory and the greater their splitting. Measurements of super-oblique parameters at the linear collider therefore provides a method for probing superpartner masses at arbitrarily high scales. Super-oblique parameter measurements are particularly relevant in models with large hierarchies in the supersymmetry spectrum. Such hierarchies are certainly possible, and possibly even favored, given the constraints discussed in Section 3, and may be found in models with heavy colored superpartners, focus point supersymmetry, or the superheavy supersymmetry scenarios of Section 4.3.

For concreteness, let us consider models with

$$M_3, m_{\tilde{q}} \gg m_{\tilde{l}}, M_2, M_1 \ . \tag{96}$$

Characterizing the superparticle mass scale with two parameters, $M_{\tilde{Q}}$ for all the heavy states, and $M_{\tilde{L}}$ for all the light states, we find [84]

$$\tilde{U}_1 \equiv Y_{\tilde{B}} - 1 = \frac{11}{5} \frac{g_1^2}{16\pi^2} \ln \frac{M_{\tilde{Q}}}{M_{\tilde{L}}} \tag{97}$$

$$\tilde{U}_2 \equiv Y_{\tilde{W}} - 1 = 3 \frac{g^2}{16\pi^2} \ln \frac{M_{\tilde{Q}}}{M_{\tilde{L}}} \ . \tag{98}$$

where $\tilde{U}_1$ and $\tilde{U}_2$ are the super-oblique parameters for the U(1) and SU(2) gauge groups. Both of these may be measured at linear colliders. For $M_{\tilde{Q}}/M_{\tilde{L}} \sim 10$, $\tilde{U}_2 \approx 2.0\%$ and $\tilde{U}_1 \approx 0.7\%$, leading to enhancements of t-channel contributions of about 8.0% and 2.8%, respectively. Given the results described above, such deviations are certainly observable at the linear collider. Non-vanishing super-oblique parameters are signals of supersymmetry breaking, and precise measurements could even constrain the mass scale of superpartners that are far beyond the reach of colliders.

The super-oblique corrections are, of course, just a subset of all radiative corrections. They may be particularly large, given the decoupling picture leading to the $\log M_Q$ corrections discussed above. However, additional radiative corrections appear in all supersymmetry processes. Radiative corrections to masses have been presented in Ref. [93]. For some production and decay processes, the full one-loop correction is also available [94]. If supersymmetry is discovered, the full radiative analysis of all available processes will become important in precision supersymmetric physics, just as it was for precision electroweak physics. These studies will be important also for global fits and extrapolation to higher mass scales, which we will discuss below in Section 9.

8. Determining the Scale of Supersymmetry Breaking

If supersymmetry is discovered, the mediation of supersymmetry breaking will be investigated in detail by measuring the soft supersymmetry-breaking terms, as we have discussed. Intimately connected to the mediation mechanism, however, is the mechanism of supersymmetry breaking. As a first step, one would like to know the scale of supersymmetry breaking $F_{\rm DSB}$ discussed in Section 4. This scale determines both the gravitino mass and its couplings. In frameworks such as supergravity, where the scale of supersymmetry breaking is high, the gravitino is decoupled from collider phenomenology, and so colliders cannot provide much information about the supersymmetry-breaking scale, other than to provide lower bounds.

In low-scale supersymmetry-breaking scenarios, including the gauge-mediated models discussed in Section 4.4, however, standard model superpartners may decay to gravitinos on observable scales. This decay provides a window on the gravitational sector of supersymmetry, allowing a quantitative measurement of the scale of supersymmetry breaking, and possibly providing information about the gravitino mass $m_{3/2}$, the messenger scale $M_{\rm mess}$, and so on.

The robust prediction of gauge-mediated models is the existence of a very light gravitino LSP with significant couplings to standard model particles. In these scenarios, standard model superpartners may decay into gravitinos in collider detectors. The decay length for this decay was given in Section 4.4. More precisely, for a neutralino NLSP, the decay length for $\tilde{\chi}_1^0 \to \tilde{G}\gamma$ is

$$L \simeq 0.10 \text{ mm} \frac{1}{\kappa_\gamma^2} \left[\frac{100 \text{ GeV}}{m_{\tilde{\chi}_0^1}} \right]^5 \left[\frac{\sqrt{F_{\rm DSB}}}{10^5 \text{ GeV}} \right]^4 \left[\frac{E^2_{\tilde{\chi}_0^1} - m^2_{\tilde{\chi}_0^1}}{m^2_{\tilde{\chi}_0^1}} \right]^{\frac{1}{2}} , \qquad (99)$$

where $\kappa_\gamma = |N_{11} \cos\theta_W + N_{12} \sin\theta_W|$. For typical parameters, this is a macroscopic distance. The NLSP's decay products therefore typically do not point back to the interaction point. Depending on what superpartner is the NLSP, we would then see the following unusual signals at the linear collider:

$$e^+e^- \to \tilde{l}^+\tilde{l}^- \to ll\tilde{G}\tilde{G} + \text{displaced vertices}$$
$$e^+e^- \to \chi_1^0\chi_1^0 \to \gamma\gamma\tilde{G}\tilde{G} + \text{displaced vertices}\ . \qquad (100)$$

These signals are spectacular, and may be used to differentiate supersymmetry from backgrounds.[f]

The signals of Eq. (100) have been studied carefully in Ref. [95]. For slepton NLSPs with decay lengths longer than $\sim 10\ \mu$m, the $\tilde{l}$ momentum can be measured by the inner tracking detector. The NLSP lifetime can then be reconstructed with relative ease. In conjunction with NLSP mass measurements, $F_{\rm DSB}$ will determined to within a few percent.

The case of neutralino NLSPs is more difficult, but still promising. The linear collider calorimeter has some angular resolution, and will measure decay lengths between 5 cm and 2 m with a statistical precision of a few percent. For decay lengths above $\sim 10\ \mu$m but below 10 cm, three-body decays $\tilde{\chi}_1^0 \to f\bar{f}\tilde{G}$ may be used to reconstruct the vertex. The branching fraction for such decays is typically a few percent. Finally, for very long decay lengths of the order of 10 m, one can compare the number of 2 photon events to the number of 1 photon events, where, in the latter case, one of the two NLSPs decays outside the detector. This may be sensitive to supersymmetry-breaking scales $\sqrt{F_{\rm DSB}}$ as large as 2000 TeV. In this last case, high luminosity is of great importance.

9. Extrapolation to the Planck Scale

In Sections 5 and 6, we saw numerous examples in which the weak-scale parameters of the MSSM can be determined in a model-independent manner. As described in Section 4, these soft supersymmetry-breaking parameters are expected to derive from soft masses at higher scales, and the weak-scale parameters are determined by the more fundamental high energy parameters by renormalization group evolution.

In practice, the renormalization group flow will be inverted. Measurements of superpartner properties at the linear collider and the LHC will provide determinations of weak-scale supersymmetry parameters. These can then be evolved to high scales to determine their fundamental, microscopic values.

---

[f]Note that displaced vertices are particularly helpful at the LHC. For example, given displaced vertices, one can reconstruct the masses of various sparticles, such as $\tilde{\chi}_3^0$ and $\tilde{\chi}_4^0$, which might not be accessible in supergravity scenarios [71].

An example of the power of this approach is given in Fig. 27 [96]. The assumed physical framework is a minimal supergravity model. The precision with which the weak-scale parameters will be measured depends, of course, on the collider and detector parameters. The authors assumed mass precisions of

$$\begin{aligned} \Delta m_{\tilde{\chi}^{\pm,0}} &\approx 0.3 \ \text{GeV} \\ \Delta m_{\tilde{l}} \approx \Delta m_{\tilde{\nu}} &\approx 0.1 \ \text{GeV} \\ \Delta m_{\tilde{\tau}} &\approx 0.6 \ \text{GeV} \\ \Delta m_{\tilde{t},\tilde{b}} &\approx 1 \ \text{GeV} \ . \end{aligned} \tag{101}$$

Along with analyses of cross sections and other quantities in slepton and chargino/neutralino events, these measurements would provide determinations of the gaugino parameters M_1 and M_2 at the percent level. The gluino mass M_3 is assumed to be measured by a combination of LHC and linear collider data. Note that the linear collider information improves the M_3 precision by over an order of magnitude through the complementarity described in Section 5.2.

Extrapolations of the supersymmetry parameters from the weak scale to the GUT scale are shown in Fig. 27. The unification of gaugino masses is striking, providing clear evidence for gauge group unification. The sfermion masses also unify at the same scale, with great precision for the sleptons. The quark and Higgs parameters are less well-determined, and the uncertainty in weak-scale values is noticeably magnified as one evolves to higher energies. Nevertheless, taken as a whole, the sfermion mass trajectories also show clear evidence for unification at the GUT scale.

The identical analysis has been performed for a gauge-mediated supersymmetry breaking model. The results are show in Fig. 28. Recall that the simplest gauge-mediated models predict gaugino mass relations identical to those in unified scenarios, but the scalar mass predictions of Eq. (57) are very different. The results of Fig. 28 clearly disfavor scalar mass unification at any scale, and would thereby exclude minimal supergravity and many GUTs. In addition, the renormalization group trajectories show unification of the the left-handed slepton L_1 and H_u masses at 10^8 GeV. In gauge-mediated scenarios this identifies the messenger scale.

10. Connections to Cosmology

The discovery of supersymmetry with a potentially stable LSP implies the discovery of dark matter candidates. These discoveries will open many avenues for dark matter studies, with a wealth of connections between particle

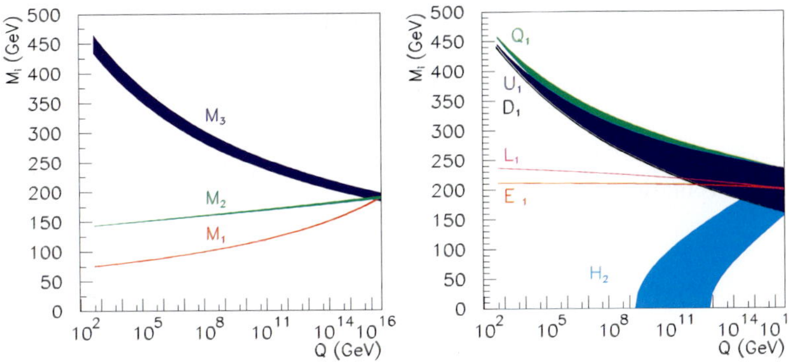

Fig. 27. Renormalization group evolution of gaugino (left) and sfermion (right) mass parameters from the weak scale to the GUT scale in a minimal supergravity model with $m_0 = 200$ GeV, $M_{1/2} = 190$ GeV, $A_0 = 500$ GeV, $\tan\beta = 30$, and $\mu < 0$ [96]. H_u denotes the m_{H_u} renormalization group trajectory. The weak-scale experimental inputs are discussed in the text, and the bands indicate 95% CL contours.

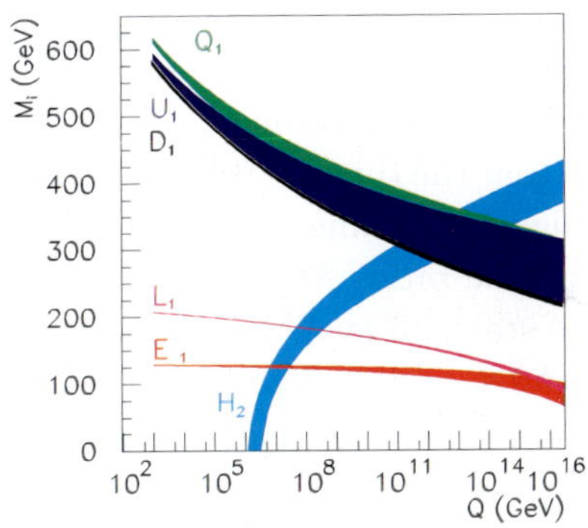

Fig. 28. Evolution of sfermion mass parameters in a gauge-mediated supersymmetry breaking model with $M_{\rm mess} = 2 \times 10^8$ GeV, $\Lambda = \langle F \rangle / M_{\rm mess} = 28$ TeV, $N = 3$, $\tan\beta = 30$ and $\mu < 0$ [96]. H_u denotes the m_{H_u} renormalization group trajectory.

physics and cosmology. A schematic picture of the resulting investigation of supersymmetric dark matter is given in Fig. 29.

In the MSSM, suitable dark matter candidates are the neutralino and the gravitino. In the case of neutralino dark matter, the thermal relic density is determined by the neutralino pair annihilation cross section as described in Section 3.2, and dark matter may be detected either directly through its interactions with ordinary matter or indirectly through its annihilation decay products. At linear colliders, supersymmetry parameters

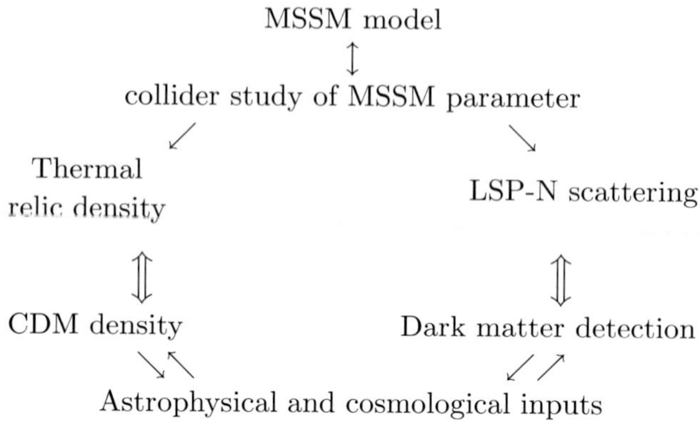

Fig. 29. The road to understanding supersymmetric dark matter.

may be determined at the percent level, and the thermal relic density and neutralino-nucleon scattering cross sections may be determined with similar precision. Such progress corresponds to the upper-half of Fig. 29, and the completion of this program will provide a great deal of information about the suitability of neutralinos as dark matter candidates.

As examples, recall that the thermal relic density and detection rates are sensitive to gaugino-Higgsino mixing, as explained in Sections 3.2 and 4.2. The mixing depends on the ratios $M_i/|\mu|$ and may be measured either by studying both chargino states, if available, or by studying a single chargino state with polarized beams, as discussed in Section 6.2. The relic density is also modified if there are other superpartners with masses within roughly 5% of the LSP mass [97]. In the case of minimal supergravity, for example, the lighter $\tilde{\tau}$ is lighter than all other charged superpartners and may be nearly degenerate with the neutralino LSP. In this case, the thermal relic density is reduced by coanihilation effects and is highly sensitive to the $\tilde{\tau}$-$\tilde{\chi}_1^0$ mass splitting. Precise mass measurements are therefore needed to pin down the relic density. The mass determination of highly degenerate sparticles is very challenging in general; in this case, it is complicated by the $ee\tau\tau$ background discussed in Section 5.1 and requires careful detector design to retain sensitivity to low momentum jets and leptons.

Even if the supersymmetry parameters are all precisely measured, this is not the whole story, however. The identification of the thermal relic density with the present day cold dark matter density is subject to cosmological assumptions. For example, the calculation of the thermal relic density assumes that the dominant source of dark matter is from dark matter particles falling out of thermal equilibrium. It is possible, however, that the bulk of the dark matter is created not through thermal equilibrium and freeze-out, but through the decay of a supermassive particle after freeze-out, but before $t = 1$ s. For example, in the anomaly-mediated supersymmetry breaking models of Section 4.5, the LSP thermal relic density is negligible, as Winos annihilate extremely efficiently in the early universe. However, anomaly-mediated scenarios also have supermassive gravitinos, with mass ~ 10 TeV. These may decay to LSPs after the usual freeze-out temperatures, producing Wino LSPs with a cosmologically interesting mass density [98].

The thermal relic density calculation also assumes that nothing unusual happens once the dark matter is produced at temperatures of $T \sim \mathcal{O}(10)$ GeV. At present, the thermal history of the universe is on sure footing only for times after Big Bang nucleosynthesis, that is, times $t \gtrsim 1$ s and temperatures $T \lesssim 1$ MeV. The thermal relic density calculation therefore requires an extrapolation of four orders of magnitude in temperature. Large entropy production by late-decaying particles may drastically alter calculated relic densities, reducing a seemingly too-large relic density to the ideal range, for example. The bottom line is that the cold dark matter density obtained following the path from the bottom of Fig. 29 need not coincide with the thermal relic density obtained by following the path from the top. Instead, discrepancies might provide new insights into the history of our universe.

In a similar vein, the neutralino-nucleon cross sections are not necessarily in one-to-one correspondence with dark matter detection rates. This correspondence requires information about the local density and velocities of dark matter. The uncertainties and problems associated with these issues have been discussed extensively [99–101]. If neutralinos are identified as the dark matter, future colliders will determine their properties and the neutralino-nucleon scattering cross sections. To the extent that these may be checked against actual detection rates, future colliders will also provide important information about dark matter halo densities and velocity distributions.

The second possible supersymmetric dark matter candidate is the gravitino. Gravitino dark matter is somewhat beyond the scope of this paper; for an excellent review, see Ref. [42]. Recall, however, that the NLSP lifetime is related to the gravitino mass, as both are determined by $F_{\rm DSB}$ as in Eqs. (39) and (40). The gravitino's relic density is proportional to the gravitino's mass, and so the observed dark matter density provides an upper limit on the gravitino's mass of $m_{3/2} \lesssim 0.2$ KeV typically. In this range, the gravitino is hot dark matter. Gravitinos with masses around 1 keV may be interesting warm dark matter candidates. In this range, the displaced vertices of NLSP decays to gravitinos may be accessible at linear colliders through the studies discussed in Section 8.

11. Conclusions

Supersymmetry and the linear collider are a near perfect fit — it is difficult to envision a richer linear collider program than that provided by the superpartner spectrum, and it is hard to imagine a more incisive tool for studying supersymmetry than the linear collider. Of course, the program of supersymmetry studies at the linear collider depends on what superpartners are kinematically accessible. However, as evident from the many examples discussed above, even if only one or a few superpartners are kinematically accessible, the linear collider will be able to provide model-independent measurements of a host of soft supersymmetry-breaking parameters. In conjunction with the information provided by hadron colliders, this information will likely provide precise information about sparticles beyond the linear collider's reach.

In this overview, we have highlighted many of the fundamental questions that will be addressed by such supersymmetry measurements. These range from issues specific to supersymmetry, such as the testing of supersymmetric identities, the resolution of the supersymmetric flavor and CP puzzles, and the determination of the scale of supersymmetry breaking, to grand universal questions, such as the nature of dark matter, the unification of forces, and the geometry of spacetime. As discussed in this review, if supersymmetry is within reach, the linear collider may shed light on all of these issues, and will, in some cases, provide definitive answers.

At the same time, there are many outstanding problems. To name but a few, the study of radiative corrections in supersymmetry and its impact on precision measurements is still in its early days. In addition, the importance of experimental uncertainties, such as in the luminosity spectrum, beam energy, and polarimetry have been considered in a few studies, but have not been systematically investigated. And of course, new ideas for physics beyond the standard model continually arise, and the potential of linear colliders to probe such new ideas and to differentiate these ideas from other new physics possibilities will continue to be of interest. To fulfill the potential of linear colliders to study new physics, dedicated experimental and theoretical efforts are still needed. We hope this chapter will be of use to people who are interested in continuing the exploration of supersymmetry at new colliders.

Acknowledgments

We thank our collaborators for their many insights through the years, Keisuke Fujii, David Miller, and Amarjit Soni for their support of this review, and Michael Peskin for sparking our initial interest in this subject.

References

1. J. Bagger and J. Wess, *Supersymmetry And Supergravity*, Princeton University Press, Princeton, NJ (1990).
2. H. E. Haber, "Introductory low-energy supersymmetry," hep-ph/9306207.
3. J. A. Bagger, "Weak-scale supersymmetry: Theory and practice," hep-ph/9604232.
4. M. Drees, "An introduction to supersymmetry," hep-ph/9611409.
5. X. Tata, "What is supersymmetry and how do we find it?," hep-ph/9706307.
6. S. P. Martin, "A supersymmetry primer," in *Perspectives on Supersymmetry*, ed. G. L. Kane, pp. 1-98 [hep-ph/9709356].
7. S. Weinberg, *The Quantum Theory Of Fields. Vol. 3: Supersymmetry*, Cambridge University Press, Cambridge, UK (2000).
8. N. Polonsky, *Supersymmetry: Structure and phenomena. Extensions of the standard model*, Lect. Notes Phys. **M68**, 1 (2001) [hep-ph/0108236].
9. T. Abe *et al.* [American Linear Collider Working Group Collaboration], hep-ex/0106055; hep-ex/0106056; hep-ex/0106057; hep-ex/0106058.
10. J. A. Aguilar-Saavedra *et al.* [ECFA/DESY LC Physics Working Group Collaboration], hep-ph/0106315.
11. K. Abe *et al.* [ACFA Linear Collider Working Group Collaboration], hep-ph/0109166.
12. J. L. Feng and D. E. Finnell, Phys. Rev. **D49**, 2369 (1994) [hep-ph/9310211].
13. H. Baer, R. Munroe and X. Tata, Phys. Rev. **D54**, 6735 (1996) [Erratum-ibid. **D56**, 4424 (1997)] [hep-ph/9606325].
14. A. Bartl, H. Eberl, S. Kraml, W. Majerotto, W. Porod and A. Sopczak, Z. Phys. **C76**, 549 (1997) [hep-ph/9701336].
15. M. Drees and O. J. Eboli, Eur. Phys. J. **C10**, 337 (1999) [hep-ph/9902391].
16. A. Bartl, H. Eberl, S. Kraml, W. Majerotto and W. Porod, Eur. Phys. J. **C6**, 1 (2000) [hep-ph/0002115].
17. R. Kitano, T. Moroi and S. Su, hep-ph/0208149.
18. K. R. Dienes, Phys. Rept. **287**, 447 (1997) [hep-th/9602045].
19. P. Horava and E. Witten, Nucl. Phys. **B475**, 94 (1996) [hep-th/9603142].
20. H. Goldberg, Phys. Rev. Lett. **50**, 1419 (1983).
21. L. M. Krauss, Nucl. Phys. **B227**, 556 (1983).

22. J. Ellis, J. S. Hagelin, D. V. Nanopoulos and M. Srednicki, Phys. Lett. **B127**, 233 (1983).
23. G. Jungman, M. Kamionkowski and K. Griest, Phys. Rept. **267**, 195 (1996) [hep-ph/9506380].
24. F. Gabbiani, E. Gabrielli, A. Masiero and L. Silvestrini, Nucl. Phys. **B477**, 321 (1996) [hep-ph/9604387].
25. R. Contino and I. Scimemi, Eur. Phys. J. **C10**, 347 (1999) [hep-ph/9809437].
26. T. Moroi, Phys. Rev. **D53**, 6565 (1996) [Erratum-ibid. **D56**, 4424 (1997)] [hep-ph/9512396].
27. T. Goto and T. Nihei, Phys. Rev. **D59**, 115009 (1999) [hep-ph/9808255].
28. K. S. Babu and M. J. Strassler, hep-ph/9808447.
29. T. Goto and T. Nihei, hep-ph/9909251.
30. Y. Suzuki *et al.* [TITAND Working Group Collaboration], hep-ex/0110005.
31. A. H. Chamseddine, R. Arnowitt and P. Nath, Phys. Rev. Lett. **49**, 970 (1982).
32. R. Barbieri, S. Ferrara and C. A. Savoy, Phys. Lett. **B119**, 343 (1982).
33. L. J. Hall, J. Lykken and S. Weinberg, Phys. Rev. **D27**, 2359 (1983).
34. J. L. Feng, K. T. Matchev and F. Wilczek, Phys. Lett. **B482**, 388 (2000) [hep-ph/0004043].
35. J. L. Feng, K. T. Matchev and T. Moroi, Phys. Rev. Lett. **84**, 2322 (2000) [hep-ph/9908309].
36. J. L. Feng, K. T. Matchev and T. Moroi, Phys. Rev. **D61**, 075005 (2000) [hep-ph/9909334].
37. J. L. Feng and K. T. Matchev, Phys. Rev. **D63**, 095003 (2001) [hep-ph/0011356].
38. M. Drees, Phys. Rev. **D33**, 1468 (1986); M. Dine, A. Kagan and S. Samuel, Phys. Lett. **B243**, 250 (1990); S. Dimopoulos and G. F. Giudice, Phys. Lett. **B357**, 573 (1995) [hep-ph/9507282]; A. Pomarol and D. Tommasini, Nucl. Phys. **B466**, 3 (1996) [hep-ph/9507462].
39. G. Dvali and A. Pomarol, Phys. Rev. Lett. **77**, 3728 (1996) [hep-ph/9607383]; A. G. Cohen, D. B. Kaplan and A. E. Nelson, Phys. Lett. **B388**, 588 (1996) [hep-ph/9607394].
40. J. L. Feng, C. Kolda and N. Polonsky, Nucl. Phys. **B546**, 3 (1999) [hep-ph/9810500]; J. Bagger, J. L. Feng and N. Polonsky, Nucl. Phys. **B563**, 3 (1999) [hep-ph/9905292]; J. A. Bagger, J. L. Feng, N. Polonsky and R. Zhang, Phys. Lett. **B473**, 264 (2000) [hep-ph/9911255]; H. Baer, C. Balazs, P. Mercadante, X. Tata and Y. Wang, Phys. Rev. **D63**, 015011 (2001) [hep-ph/0008061]; H. Baer, C. Balazs, M. Brhlik, P. Mercadante, X. Tata and Y. Wang, hep-ph/0102156; J. Hisano, K. Kurosawa and Y. Nomura, Phys. Lett. **B445**, 316 (1999) [hep-ph/9810411]; Nucl. Phys. **B584**, 3 (2000) [hep-ph/0002286].
41. M. Dine and A. E. Nelson, Phys. Rev. **D48**, 1277 (1993) [hep-ph/9303230]; M. Dine, A. E. Nelson and Y. Shirman, Phys. Rev. **D51**, 1362 (1995) [hep-ph/9408384]; M. Dine, A. E. Nelson, Y. Nir and Y. Shirman, Phys. Rev. **D53**, 2658 (1996) [hep-ph/9507378].

42. G. F. Giudice and R. Rattazzi, Phys. Rept. **322**, 419 (1999) [hep-ph/9801271].
43. S. Dimopoulos, G. R. Dvali, R. Rattazzi and G. F. Giudice, Nucl. Phys. **B510**, 12 (1998) [hep-ph/9705307].
44. J. L. Feng and T. Moroi, Phys. Rev. **D58**, 035001 (1998) [hep-ph/9712499].
45. D. R. Stump, M. Wiest and C. P. Yuan, Phys. Rev. **D54**, 1936 (1996) [hep-ph/9601362]; S. Dimopoulos, M. Dine, S. Raby and S. Thomas, Phys. Rev. Lett. **76**, 3494 (1996) [hep-ph/9601367].
46. L. Randall and R. Sundrum, Nucl. Phys. **B557**, 79 (1999) [hep-th/9810155].
47. G. F. Giudice, M. A. Luty, H. Murayama and R. Rattazzi, JHEP **9812**, 027 (1998) [hep-ph/9810442].
48. T. Gherghetta, G. F. Giudice and J. D. Wells, Nucl. Phys. **B559**, 27 (1999) [hep-ph/9904378]; J. L. Feng and T. Moroi, Phys. Rev. **D61**, 095004 (2000) [hep-ph/9907319].
49. See, *e.g.*, A. Pomarol and R. Rattazzi, JHEP **9905**, 013 (1999) [hep-ph/9903448]; Z. Chacko, M. A. Luty, I. Maksymyk and E. Ponton, JHEP **0004**, 001 (2000) [hep-ph/9905390]; E. Katz, Y. Shadmi and Y. Shirman, JHEP **9908**, 015 (1999) [hep-ph/9906296]; I. Jack and D. R. Jones, Phys. Lett. **B482**, 167 (2000) [hep-ph/0003081].
50. C. H. Chen, M. Drees and J. F. Gunion, Phys. Rev. Lett. **76**, 2002 (1996) [hep-ph/9512230]; Phys. Rev. **D55**, 330 (1997) [Erratum-ibid. **D60**, 039901 (1999)] [hep-ph/9607421]; S. Thomas and J. D. Wells, Phys. Rev. Lett. **81**, 34 (1998) [hep-ph/9804359]; H. C. Cheng, B. A. Dobrescu and K. T. Matchev, Nucl. Phys. **B543**, 47 (1999) [hep-ph/9811316]; J. L. Feng, T. Moroi, L. Randall, M. Strassler and S. f. Su, Phys. Rev. Lett. **83**, 1731 (1999) [hep-ph/9904250]; D. K. Ghosh, P. Roy and S. Roy, JHEP **0008**, 031 (2000) [hep-ph/0004127]; J. F. Gunion and S. Mrenna, Phys. Rev. **D64**, 075002 (2001) [hep-ph/0103167]; D. K. Ghosh, A. Kundu, P. Roy and S. Roy, Phys. Rev. **D64**, 115001 (2001) [hep-ph/0104217].
51. V. S. Kaplunovsky and J. Louis, Phys. Lett. **B306**, 269 (1993) [hep-th/9303040]; A. Brignole, L. E. Ibanez, C. Munoz and C. Scheich, Z. Phys. **C74**, 157 (1997) [hep-ph/9508258]; T. Kobayashi, D. Suematsu, K. Yamada and Y. Yamagishi, Phys. Lett. **B348**, 402 (1995) [hep-ph/9408322]; A. Brignole, L. E. Ibanez and C. Munoz, Nucl. Phys. **B422**, 125 (1994) [Erratum-ibid. **B436**, 747 (1995)] [hep-ph/9308271].
52. R. Barbieri and L. J. Hall, Phys. Lett. **B338**, 212 (1994) [hep-ph/9408406]; R. Barbieri, L. J. Hall and A. Strumia, Nucl. Phys. **B445**, 219 (1995) [hep-ph/9501334].
53. M. Drees, Phys. Lett. **B181**, 279 (1986).
54. Y. Kawamura, H. Murayama and M. Yamaguchi, Phys. Rev. **D51**, 1337 (1995) [hep-ph/9406245].
55. See, *e.g.*, K. Huitu, J. Maalampi and K. Puolamaki, Eur. Phys. J. **C6**, 159 (1999) [hep-ph/9705406]; S. Bar-Shalom, G. Eilam and A. Soni, Phys. Rev. Lett. **80**, 4629 (1998) [hep-ph/9802251]; D. K. Ghosh, R. M. Godbole and S. Raychaudhuri, hep-ph/9904233; M. Besancon and G. Moreau, hep-

ph/9909441; D. K. Ghosh and S. Roy, Phys. Rev. **D63**, 055005 (2001) [hep-ph/0003225].
56. JLC Group Collaboration, "JLC-I," KEK-92-16.
57. T. Tsukamoto, K. Fujii, H. Murayama, M. Yamaguchi and Y. Okada, Phys. Rev. **D51**, 3153 (1995).
58. H. U. Martyn and G. A. Blair, hep-ph/9910416.
59. M. Dima *et al.*, Phys. Rev. **D65**, 071701 (2002) [hep-ex/0112017].
60. David Wagner, http://hep-www.colorado.edu/SUSY, COLO-HEP-400, COLO-HEP-401.
61. J. D. Lykken, hep-ph/9803427.
62. J. L. Feng, Int. J. Mod. Phys. **A13**, 2319 (1998) [hep-ph/9803319].
63. H. U. Martyn, hep-ph/0002290.
64. J. L. Feng and M. E. Peskin, Phys. Rev. **D64**, 115002 (2001) [hep-ph/0105100].
65. A. Freitas, D. J. Miller and P. M. Zerwas, Eur. Phys. J. **C21**, 361 (2001) [hep-ph/0106198].
66. J. K. Mizukoshi, H. Baer, A. S. Belyaev and X. Tata, Phys. Rev. **D64**, 115017 (2001) [hep-ph/0107216].
67. C. Blochinger, H. Fraas, G. Moortgat-Pick and W. Porod, Eur. Phys. J. **C24**, 297 (2002) [hep-ph/0201282].
68. V. Barger, T. Han and J. Kelly, Phys. Lett. **B419**, 233 (1998) [hep-ph/9709366].
69. I. Hinchliffe, F. E. Paige, M. D. Shapiro, J. Soderqvist and W. Yao, Phys. Rev. **D55**, 5520 (1997) [hep-ph/9610544].
70. H. Bachacou, I. Hinchliffe and F. E. Paige, Phys. Rev. **D62**, 015009 (2000) [hep-ph/9907518].
71. ATLAS Collaboration, *ATLAS Detector and Physics Performance Technical Design Report*, CERN/LHCC/99-14, http://atlasinfo.cern.ch/Atlas/-GROUPS/PHYSICS/TDR/access.html.
72. B. C. Allanach, C. G. Lester, M. A. Parker and B. R. Webber, JHEP **0009**, 004 (2000) [hep-ph/0007009].
73. M. M. Nojiri, K. Fujii and T. Tsukamoto, Phys. Rev. **D54**, 6756 (1996) [hep-ph/9606370].
74. M. M. Nojiri, Phys. Rev. **D51**, 6281 (1995) [hep-ph/9412374].
75. N. Arkani-Hamed, H. C. Cheng, J. L. Feng and L. J. Hall, Phys. Rev. Lett. **77**, 1937 (1996) [hep-ph/9603431].
76. N. Arkani-Hamed, J. L. Feng, L. J. Hall and H. C. Cheng, Nucl. Phys. **B505**, 3 (1997) [hep-ph/9704205]; D. Bowser-Chao and W. Y. Keung, Phys. Rev. **D56**, 3924 (1997) [hep-ph/9704219]; M. Hirouchi and M. Tanaka, Phys. Rev. **D58**, 032004 (1998) [hep-ph/9712532]; J. Hisano, M. M. Nojiri, Y. Shimizu and M. Tanaka, Phys. Rev. **D60**, 055008 (1999) [hep-ph/9808410]; J. J. Cao, T. Han, X. Zhang and G. R. Lu, Phys. Rev. **D59**, 095001 (1999) [hep-ph/9808466]; D. Nomura, Phys. Rev. **D64**, 075001 (2001) [hep-ph/0004256]; M. Guchait, J. Kalinowski and P. Roy, Eur. Phys. J. **C21**, 163 (2001) [hep-ph/0103161]; M. Dine, Y. Grossman and S. Thomas,

hep-ph/0111154; W. Porod and W. Majerotto, Phys. Rev. **D66**, 015003 (2002) [hep-ph/0201284].
77. J. L. Feng, M. E. Peskin, H. Murayama and X. Tata, Phys. Rev. **D52**, 1418 (1995) [hep-ph/9502260].
78. S. Y. Choi, A. Djouadi, H. K. Dreiner, J. Kalinowski and P. M. Zerwas, Eur. Phys. J. **C7**, 123 (1999) [hep-ph/9806279]; S. Y. Choi, A. Djouadi, H. S. Song and P. M. Zerwas, Eur. Phys. J. **C8**, 669 (1999) [hep-ph/9812236]; S. Y. Choi, M. Guchait, J. Kalinowski and P. M. Zerwas, Phys. Lett. **B479**, 235 (2000) [hep-ph/0001175]; S. Y. Choi, A. Djouadi, M. Guchait, J. Kalinowski, H. S. Song and P. M. Zerwas, Eur. Phys. J. **C14**, 535 (2000) [hep-ph/0002033]; see also D. Atwood, S. Bar-Shalom, G. Eilam and A. Soni, Phys. Rept. **347**, 1 (2001) [hep-ph/0006032].
79. G. Moortgat-Pick, A. Bartl, H. Fraas and W. Majerotto, hep-ph/0004181; Eur. Phys. J. **C18**, 379 (2000) [hep-ph/0007222].
80. T. Ibrahim and P. Nath, Phys. Rev. **D58**, 111301 (1998) [Erratum-ibid. **D60**, 099902 (1999)] [hep-ph/9807501].
81. S. Y. Choi, J. Kalinowski, G. Moortgat-Pick and P. M. Zerwas, Eur. Phys. J. **C22**, 563 (2001) [hep-ph/0108117]; Addendum, Eur. Phys. J. **C23**, 769 (2001) [hep-ph/0202039].
82. V. D. Barger, T. Falk, T. Han, J. Jiang, T. Li and T. Plehn, Phys. Rev. **D64**, 056007 (2001) [hep-ph/0101106].
83. H. C. Cheng, K. T. Matchev and M. Schmaltz, Phys. Rev. **D66**, 056006 (2002) [hep-ph/0205314].
84. H. C. Cheng, J. L. Feng and N. Polonsky, Phys. Rev. **D56**, 6875 (1997) [hep-ph/9706438].
85. H. C. Cheng, J. L. Feng and N. Polonsky, Phys. Rev. **D57**, 152 (1998) [hep-ph/9706476].
86. M. M. Nojiri, D. M. Pierce and Y. Yamada, Phys. Rev. **D57**, 1539 (1998) [hep-ph/9707244].
87. E. Katz, L. Randall and S. Su, Nucl. Phys. **B536**, 3 (1998) [hep-ph/9801416].
88. S. Kiyoura, M. M. Nojiri, D. M. Pierce and Y. Yamada, Phys. Rev. **D58**, 075002 (1998) [hep-ph/9803210].
89. U. Mahanta, Phys. Rev. **D59**, 015017 (1999) [hep-ph/9810344].
90. P. H. Chankowski, Phys. Rev. **D41**, 2877 (1990).
91. K. Hikasa and Y. Nakamura, Z. Phys. **C70**, 139 (1996) [hep-ph/9501382]; Erratum, Z. Phys. **C71**, 356 (1996).
92. M. E. Peskin and T. Takeuchi, Phys. Rev. Lett. **65**, 964 (1990).
93. J. Bagger, K. T. Matchev and D. Pierce, Phys. Lett. **B348**, 443 (1995) [hep-ph/9501277]; D. M. Pierce, J. A. Bagger, K. T. Matchev and R. j. Zhang, Nucl. Phys. **B491**, 3 (1997) [hep-ph/9606211]; T. Fritzsche and W. Hollik, Eur. Phys. J. **C24**, 619 (2002) [hep-ph/0203159].
94. M. A. Diaz, S. F. King and D. A. Ross, Nucl. Phys. **B529**, 23 (1998) [hep-ph/9711307]; M. A. Diaz, S. F. King and D. A. Ross, Phys. Rev. **D64**, 017701 (2001) [hep-ph/0008117]; M. A. Diaz and D. A. Ross, hep-ph/0205257; J. Guasch, W. Hollik and J. Sola, hep-ph/0207364.

95. S. Ambrosanio and G. A. Blair, Eur. Phys. J. **C12**, 287 (2000) [hep-ph/9905403].
96. G. A. Blair, W. Porod and P. M. Zerwas, Phys. Rev. **D63**, 017703 (2001) [hep-ph/0007107]; G. A. Blair, W. Porod and P. M. Zerwas, hep-ph/0210058.
97. K. Griest and D. Seckel, Phys. Rev. **D43**, 3191 (1991).
98. T. Moroi and L. Randall, Nucl. Phys. **B570**, 455 (2000) [hep-ph/9906527].
99. P. Sikivie, I. I. Tkachev and Y. Wang, Phys. Rev. **D56**, 1863 (1997) [astro-ph/9609022].
100. G. Gelmini and P. Gondolo, Phys. Rev. **D64**, 023504 (2001) [hep-ph/0012315].
101. C. Calcaneo-Roldan and B. Moore, Phys. Rev. **D62**, 123005 (2000) [astro-ph/0010056].

CHAPTER 5

Dynamical Electroweak Symmetry Breaking

Wolfgang Kilian

Institut für Theoretische Teilchenphysik, University of Karlsruhe
76128 Karlsruhe, Germany
E-mail: kilian@particle.uni-karlsruhe.de

Dynamical symmetry breaking provides a possible solution to the electroweak hierarchy problem. It requires new strong interactions that are effective at some high-energy scale. If there is no light Higgs boson, this scale is constrained to be in the TeV range, and signals of the new interactions can be observed, directly or indirectly, in collider experiments. Even if no observable states in the Higgs sector are kinematically accessible, a Linear Collider will cover the low-energy parameter space that arises in a systematic model-independent analysis of dynamical electroweak symmetry breaking.

1. Introduction

1.1. *Particle Masses*

The most striking property of the particle spectrum is clearly the huge range in the fundamental scales. On a logarithmic mass scale, the known elementary particles (except for the neutrinos) populate a small region centered almost 20 orders of magnitude below the Planck mass, the fundamental scale of space-time geometry:

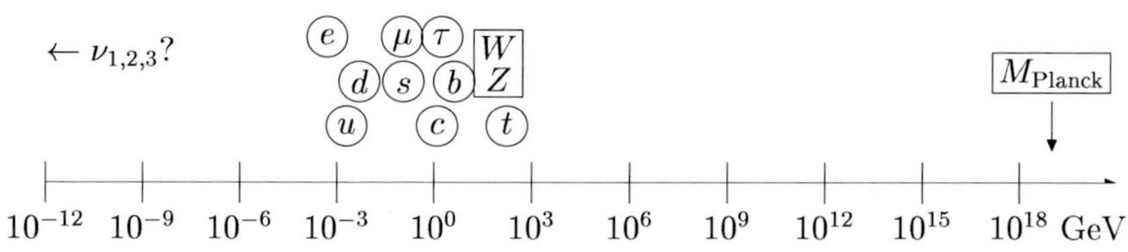

Their interactions, as far as we know, respect the gauge symmetry principle. Gauge invariance of the electroweak interactions in particular forbids mass terms for all known elementary particles (except for right-handed neutrinos). In reality, all particles but the photon are massive, so the electroweak $SU(2)_L \times U(1)_Y$ gauge symmetry is softly broken. It is manifest in the interactions, but hidden in the mass spectrum.

The masses of vector bosons are best understood as mixing terms. The electroweak gauge bosons which form a $SU(2)_L$ triplet $W^{\pm 0}$ and a $SU(2)_L$ singlet B, are coupled to a multiplet of *scalar* states in a different representation of the electroweak symmetry group. Diagonalizing the mass matrix, the resulting eigenstates consist of the massless photon and the three massive vector bosons $W^\pm$ and Z. Their longitudinal components can be identified with three components of the scalar multiplet. Similarly, the (Dirac) masses and mixings of the fermions originate from bilinear couplings of left-handed and right-handed two-component (Weyl) fermions, which are otherwise unrelated and belong to different $SU(2)_L \times U(1)_Y$ representations.

The three scalar states (i.e., the longitudinal vector bosons) in the observed spectrum do not make up a complete linear representation of $SU(2)_L$. Therefore, a field theory based just on the degrees of freedom that have been established experimentally is non-renormalizable. We can make use of such an effective theory to get a consistent low-energy expansion of the relevant physics [1]. This formalism will be described below in Section 2. However, at high energies a non-renormalizable theory has an inherent cutoff scale where it ceases to be predictive. For instance, the scalars may turn out as composite at the cutoff scale, or there may be additional scalars which have not yet been observed. In any case, the pattern of mass generation via mixing suggests a new interaction, and the complexity of the flavor couplings indicates a similar complexity of this new *Higgs* sector [2], of which we might just be scratching the surface. In the Standard Model and its extensions the problem is solved by *brute force*, postulating the existence of scalar multiplets with just the appropriate quantum numbers and couplings [3], but this need not be the solution chosen by Nature.

Since the actual properties of the Higgs sector are unknown, in a first approach one may identify a parameter v as the characteristic scale of electroweak symmetry breaking (EWSB), by convention taken as

$$v = (\sqrt{2}\,G_F)^{-1/2} \approx 246\ \text{GeV}. \tag{1}$$

v is an abstract quantity at this stage, and while its value is fixed by the measurement of a low-energy process (muon decay), its exact meaning depends on the underlying dynamics which is not yet accessible directly.

1.2. *Exponentials*

In searching for possible sources of the terms that softly break the electroweak symmetry in the fundamental Lagrangian, one may take either one of two different views:

(1) The electroweak gauge symmetry is an accidental approximate symmetry.
(2) The electroweak gauge symmetry is exact in the dynamics, but spontaneously broken in the low-energy states.

The first explanation is generally rejected since it implies that the smallness of the breaking terms ($v/M_{\text{Planck}} \sim 10^{-17}$) is pure coincidence. For a natural explanation of this ratio [4], we not only need to adopt the second scenario, but we also are led to postulate *dynamical* symmetry breaking. Such an extremely small number should be the result of solving fundamental dynamical equations *without* particularly small parameters. Ultimately, we are looking for a fundamental theory without adjustable parameters at all.

Fortunately, we know at least one solution to this problem. Strongly-coupled quantum field theories generate large scale ratios from the quantum effect of renormalization group running [5]. They allow the interpretation of the scale hierarchy as an exponential factor

$$M_{\text{Planck}}/v = e^{(\text{numberoforder10})} \tag{2}$$

within the context of quantum field theory.[a]

QCD-type condensation is the one example where this actually happens in particle physics. Neglecting all other interactions for simplicity, we could solve the renormalization group equation for the running strong coupling constant $g_s(\mu)$ with a "natural" initial condition at the Planck scale, say $g_s(M_{\text{Planck}}) = 1/2$. Identifying the QCD scale Λ_{QCD} with the scale where $g_s(\mu)$ becomes strong at low energies, we obtain a scale ratio $M_{\text{Planck}}/\Lambda_{\text{QCD}}$ similar to Eq. (2). The exponent is proportional to the inverse of the initial value of $\alpha_s = g_s^2/4\pi$ at the high scale. In the QCD case, the low-energy singularity in the perturbative evolution of the coupling is

[a] Alternatively, such an exponential could be a consequence of the space-time structure. This possibility is discussed in Chapter 7.

resolved by the condensation of gluons and of left-handed and right-handed fermions, breaking the chiral symmetry at the scale $\Lambda_{\text{QCD}} \lesssim 1$ GeV. Incidentally, this mechanism would also trigger EWSB in the GeV range, if the electroweak symmetry was not already broken at the higher scale $v = 246$ GeV.

In fact, dynamical symmetry breaking is the explanation for a wide variety of physical phenomena, from superconductivity to the laser effect. The overall description is simple. Out of the fundamental fields of the model one constructs scalar field multiplets Φ with nonvanishing quantum numbers under the symmetry in question. If the effective potential for such a field has a nontrivial solution which is energetically favorable, it will get a vacuum expectation value, breaking the symmetry and allowing for new effective couplings which exact symmetry would forbid. There is no need for Φ to correspond to independent observable degrees of freedom, since it can be a composite of other fields present in the theory (such as Cooper pairs in BCS superconductivity). Thus, while fundamental Higgs particles as Φ quanta are not excluded, they are not necessary for the Higgs mechanism of EWSB to work.

1.3. *Higgs or No Higgs?*

Just as in QCD, one can associate a scale Λ with dynamical symmetry breaking, the *compositeness scale.* Λ might be significantly higher than the EWSB scale v. Then, in the intermediate scale range $v \ldots \Lambda$ the model will reduce to a renormalizable effective theory. This can only be the case if the scalars which provide the longitudinal W and Z states are accompanied by extra scalar states such that a complete linear representation of $SU(2)_L \times U(1)_Y$ is formed. The extra states can be observed as particles, the Higgs bosons (cf. Chapter 2).

As will be discussed below, in models without an observable Higgs state the compositeness scale cannot be higher than [6]

$$\Lambda \lesssim 4\pi v \approx 3 \text{ TeV}. \tag{3}$$

If there is a single Higgs boson with mass m_H, the compositeness scale is constrained instead by

$$\Lambda' \lesssim m_H \exp \frac{\Lambda^2}{12 m_H^2} \qquad \text{with } \Lambda = 4\pi v, \tag{4}$$

the location of the Landau pole of the Higgs self-coupling. The relation (4) implies that the concept of a Higgs boson makes sense only if $m_H \lesssim 1$ TeV, such that $\Lambda' > m_H$ [7].

The ratio Λ'/m_H could be large, but dynamical symmetry breaking by itself provides no obvious mechanism for this. One would rather expect this ratio to be of order one, so the window for composite Higgs states is narrow. On the other hand, scalar states could actually be elementary degrees of freedom. This is consistent with a dynamical solution of the hierarchy problem only if their masses are protected by a symmetry. In that case, the scalar interactions which trigger EWSB could only be generated indirectly, which shifts dynamical symmetry breaking to a new (hidden) sector of the theory. This is the way supersymmetric models are constructed, which are considered in Chapter 4.

Another possibility which opens some window for composite Higgs bosons is the identification of the Higgs multiplet with pseudo-Goldstone scalars generated by spontaneous symmetry breaking at a higher scale [8]. Recently, realistic models with this structure have been proposed, the so-called *Little Higgs* models [9]. Like supersymmetric models, they leave room for the electroweak scale being generated dynamically through strong interactions, but in these scenarios the effective theory is weakly interacting up to energies significantly beyond the TeV scale.

By contrast, in the absence of light Higgs multiplets no weakly interacting effective theory can be constructed that is valid up to that energy range. Instead, the strong interactions which accompany dynamical scale generation may be directly coupled to Standard Model particles and show up in scattering processes once the available energies are sufficiently close to the new compositeness scale. Therefore, one can not just expect to observe the effects *associated* with EWSB at colliders (e.g., Higgs particles, superpartners, pseudo-Goldstone bosons), but there is hope to get a handle on those (strongly-interacting) new degrees of freedom which are *responsible* for EWSB.

1.4. *Models of Dynamical Symmetry Breaking*

Within the context of four-dimensional field theory, dynamical symmetry breaking must be assigned to new gauge interactions. This is a common feature of all dynamical models of EWSB [10]. The models differ in the role of fermions:

(1) Since, as in QCD, fermions which feel strong gauge interactions are confined, it is likely that the fermions (*technifermions*) directly associated with EWSB are unobservable at low energies. This is the original *technicolor* (TC) idea [11]. Technifermion condensation is able to account

for effective scalar states which provide the longitudinal components of W and Z bosons.

(2) Pure TC will not generate any left-right couplings for the observable fermions, just as QCD generates constituent masses for quarks, but not for leptons. However, such couplings could be due to additional dynamically broken gauge interactions at even higher energies which are felt both by ordinary and by technifermions. This mechanism is known as *extended technicolor* (ETC) [12]. Below the ETC scale, such interactions lead to four-fermion couplings. When technifermions condense at the compositeness scale, the desired bilinear couplings are generated.

(3) Exchanging the roles of TC and ETC, a dynamically broken gauge interaction might trigger EWSB by fermion condensation not too far above the electroweak scale. In that case, the affected fermions need not be confined. The heavy top quark is the prime candidate (*topcolor*) for such a strongly-interacting object [13]. Since for this mechanism to actually work the top quark is somewhat too light, more recent models that follow this pattern implement a combined *topcolor-assisted technicolor* scheme [14].

(4) Alternatively, the physical top mass could be suppressed by a two-state mixing effect (*top see-saw*) [15]. In such models, the electroweak scale v is typically suppressed compared to the compositeness scale Λ, and there is room for composite Higgs bosons in the intermediate range.

The complicated pattern of flavor physics stands as the main obstacle for constructing a simple theory of dynamical symmetry breaking. It is difficult to simultaneously accomodate (i) very light leptons and quarks, (ii) a heavy top quark, (iii) the smallness of flavor-changing neutral currents [10]. Turning the argument around, the fact that fermion mass generation cannot be separated from electroweak physics in strongly-interacting models, opens the possibility that some of the puzzles of flavor physics are resolved at future collider experiments.

2. Effective Theories of Electroweak Interactions

2.1. *The Bottom-up Approach*

While it is clearly worthwhile to search for signals of specific models in collider experiments, the limited energy range of a next-generation Linear Collider may not allow to access new states associated with the Higgs sector directly. In such a situation, a model-independent treatment of the dynamics is more appropriate. Fortunately, the formalism of effective (or phenomenological) Lagrangians [1] provides a generic framework for the bottom-up description of electroweak interactions in the absence of a complete renormalizable field theory.

At energies much below the W and Z (and Higgs) masses, the dominant interactions of leptons and quarks are QED and QCD interactions, governed by an effective Lagrangian of operator dimension four and less:

$$\begin{aligned}\mathcal{L} &= \mathcal{L}_3 + \mathcal{L}_4 \\ &= -(\overline{Q}_L M_Q Q_R + \overline{L}_L M_L L_R + \text{h.c.}) - (\overline{N}_L^c M_{N_L} N_L + \overline{N}_R^c M_{N_R} N_R) \\ &\quad + \overline{Q}_L i\not{D} Q_L + \overline{Q}_R i\not{D} Q_R + \overline{L}_L i\not{D} L_L + \overline{L}_R i\not{D} L_R - \tfrac{1}{4} A_{\mu\nu} A^{\mu\nu}\end{aligned} \tag{5}$$

For simplicity, we ignore QCD interactions here. The building blocks consist, first of all, of left- and right-handed quark and lepton fields,

$$Q_L = \begin{pmatrix} U_L \\ D_L \end{pmatrix}, \quad Q_R = \begin{pmatrix} U_R \\ D_R \end{pmatrix}, \quad L_L = \begin{pmatrix} N_L \\ E_L \end{pmatrix}, \quad L_R = \begin{pmatrix} N_R \\ E_R \end{pmatrix}. \tag{6}$$

We omit generation and color indices, so all coupling constants should be understood as matrices. While gauge couplings are diagonal by definition, the bilinear quark and lepton couplings M_Q and M_L and the Majorana mass matrices M_{N_L} and M_{N_R} are not.[b] Upon diagonalization, from this structure one obtains the fermion masses together with the 3×3 mixing matrices which determine the weak interactions of quarks and neutrinos.

QED gauge interaction are present in the dimension-four terms which contain the electromagnetic field strength and covariant derivative

$$A_{\mu\nu} = \partial_\mu A_\nu - \partial_\nu A_\mu, \qquad D_\mu = \partial_\mu + ieqA_\mu, \tag{7}$$

where e is the positron charge and q the multiplying factor for the charge of a given fermion species. In the doublet notation used here, the normalized charge q is a diagonal 2×2 matrix which reads

$$q_{Q,L} = \tfrac{1}{2}\left(y_{Q,L} + \tau^3\right) \quad \text{with} \quad y_Q = \tfrac{1}{3} \quad \text{and} \quad y_L = -1. \tag{8}$$

for the quarks and leptons, respectively.

At very low energies, the full electroweak symmetry is present only in the higher-dimensional operators which induce weak interactions. While magnetic-moment type operators (dimension five)

$$\mathcal{L}_5 = \overline{Q}_R \mu_Q \sigma_{\mu\nu} A^{\mu\nu} Q_L + \overline{L}_R \mu_L \sigma_{\mu\nu} A^{\mu\nu} L_L + \text{h.c.} \tag{9}$$

[b]Electroweak symmetry requires M_{N_L} to vanish, but M_{N_L} is re-introduced in the low-energy effective theory if M_{N_R} is very large and the right-handed neutrinos are integrated out.

are strongly suppressed, four-fermion operators (dimension six)

$$\mathcal{L}_6 = \sum_{f=Q_L,Q_R,L_L,L_R} s_{ijkl}(\overline{f}_i f_j)(\overline{f}_k f_l) + v_{ijkl}(\overline{f}_i \gamma^\mu f_j)(f_k \gamma_\mu f_l) \tag{10}$$

are more significant. Experiment has shown that their structure is consistent with the specific factorizable form of the Fermi model,

$$\mathcal{L}_6 = -4\sqrt{2}\, G_F \left(2 J^+_\mu J^{\mu -} + c_w^2 J^0_\mu J^{\mu 0}\right) \tag{11}$$

with the charged and neutral currents

$$J^\pm_\mu = \tfrac{1}{\sqrt{2}} \left[\overline{Q}_L \tau^\pm \gamma_\mu Q_L + \overline{L}_L \tau^\pm \gamma_\mu L_L\right] \tag{12}$$

$$\begin{aligned} J^0_\mu = \tfrac{1}{c_w} \Big[& \overline{Q}_L \left(-q_Q s_w^2 + \tfrac{\tau^3}{2}\right) \gamma_\mu Q_L + \overline{Q}_R(-q_Q s_w^2)\gamma_\mu Q_R \\ & + \overline{L}_L \left(-q_L s_w^2 + \tfrac{\tau^3}{2}\right) \gamma_\mu L_L + \overline{L}_R(-q_L s_w^2)\gamma_\mu L_R \Big] \end{aligned} \tag{13}$$

Here, $s_w \approx 0.48$ is the sine of the weak mixing angle, and $c_w = \sqrt{1 - s_w^2}$.

We do not know with certainty that the factorizable structure of four-fermion interactions is exact, but flavor-physics experiments have not yet revealed any deviations from this picture. Assuming that deviations are absent, the Fermi Lagrangian can be rewritten with the help of the vector fields $W^\pm$ and Z,

$$\mathcal{L}_6 = -g_W(W^{+\mu} J^+_\mu + W^{-\mu} J^-_\mu) - g_Z(Z^\mu J^0_\mu) + \mathcal{L}_{2(W)} \tag{14}$$

where

$$\mathcal{L}_{2(W)} = M_W^2 W^{+\mu} W^-_\mu + \tfrac{1}{2} M_Z^2 Z^\mu Z_\mu, \tag{15}$$

such that Eq. (14) becomes equivalent to Eq. (11) when the W and Z fields are integrated out. After rearranging the basis,

$$W^+ = \tfrac{1}{\sqrt{2}}(W^1 - iW^2), \qquad Z = c_w W^3 - s_w B, \tag{16}$$

$$W^- = \tfrac{1}{\sqrt{2}}(W^1 + iW^2), \qquad A = s_w W^3 + c_w B, \tag{17}$$

the Fermi Lagrangian assumes the form

$$\begin{aligned} \mathcal{L} = & \overline{Q}_L i\not{D}_L Q_L + \overline{Q}_R i\not{D}_R Q_R + \overline{L}_L i\not{D}_L L_L + \overline{L}_R i\not{D}_R L_R \\ & - \tfrac{1}{4} A_{\mu\nu} A^{\mu\nu} + \mathcal{L}_{2(W)} + \mathcal{L}_3 \end{aligned} \tag{18}$$

with a local $SU(2)_L \times U(1)_Y$ symmetry manifest in the terms of the first line. With respect to this symmetry, the vector fields have the usual gauge transformation properties, and the fermions couple via covariant derivatives

$$D_{L\mu} = \partial_\mu - ig'(q + \tfrac{\tau^3}{2})B_\mu + ig\tfrac{\tau^a}{2} W^a_\mu \tag{19}$$

$$D_{R\mu} = \partial_\mu - ig' q B_\mu. \tag{20}$$

To describe physics at and above the mass scale of the electroweak vector bosons $W^\pm$ and Z, we have to include kinetic terms

$$\mathcal{L}_{4(W)} = -\tfrac{1}{2}\,\mathrm{tr}\left[\mathbf{W}_{\mu\nu}\mathbf{W}^{\mu\nu}\right] - \tfrac{1}{2}\,\mathrm{tr}\left[\mathbf{B}_{\mu\nu}\mathbf{B}^{\mu\nu}\right], \qquad (21)$$

where the field strength tensors are defined in terms of vector fields W^a $(a = 1, 2, 3)$ and B

$$\mathbf{W}_{\mu\nu} = \partial_\mu \mathbf{W}_\nu - \partial_\nu \mathbf{W}_\mu + ig[\mathbf{W}_\mu, \mathbf{W}_\nu] \qquad (22)$$

$$\mathbf{B}_{\mu\nu} = \partial_\mu \mathbf{B}_\nu - \partial_\nu \mathbf{B}_\mu \qquad (23)$$

with $\mathbf{W}_\mu = W^a_\mu \frac{\tau^a}{2}$ and $\mathbf{B}_\mu = B_\mu \frac{\tau^3}{2}$. (This is not the most general form allowed by the spontaneously broken gauge invariance, but we postpone the discussion of anomalous couplings to Section 2.2.)

While the dimension-four part of the effective Lagrangian exhibits full electroweak gauge symmetry, this symmetry is not manifestly present in the dimension-two and dimension-three operators, the fermion and gauge boson mass terms. However, by introducing an extra field Σ with a suitable transformation law, this problem can formally be solved without losing the universality of the effective-theory formalism [16]–[18]. This field parameterizes our ignorance about the true nature of the Higgs sector. While the actual dynamics at high energies could be very complicated, at low energies a generic description is dictated by symmetry considerations only.

The field $\Sigma(x)$ is a 2×2 matrix which is defined to have the appropriate behavior under gauge transformations $U(x) \in SU(2)_L$ and $V(x) = \exp(i\beta\tau^3) \in U(1)_Y$:

$$\Sigma(x) \to U(x)\,\Sigma(x)\,V^\dagger(x). \qquad (24)$$

Spontaneous symmetry breaking is implemented by a nonzero vacuum expectation value

$$\langle \tfrac{1}{2}\,\mathrm{tr}\left[\Sigma^\dagger(x)\Sigma(x)\right]\rangle = 1 \qquad (25)$$

which for practical purposes can be replaced by

$$\langle \Sigma(x) \rangle \to 1 \quad \text{for } x \to \infty. \qquad (26)$$

For perturbative calculations, it is often useful to adopt the *unitary* or *unitarity* gauge where $\Sigma(x) \equiv 1$, but when discussing the electroweak symmetry structure, we should not impose this restriction.

A unitary matrix has three degrees of freedom, therefore the minimal number of degrees of freedom parameterizing Σ is three. A possible, but not unique, parameterization is given by [16]–[18]

$$\Sigma(x) = \exp\left(-\tfrac{i}{v}\mathbf{w}(x)\right) \quad \text{with } \mathbf{w}(x) = w^a(x)\,\tau^a; a = 1, 2, 3. \qquad (27)$$

where v is the electroweak scale (1). We do not make any attempt to further constrain the dynamics associated with Σ. This question has to be solved experimentally by measuring the free parameters, looking for new states related to Σ, and comparing this to any theoretical predictions in specific models.

The fermion mass term in the effective Lagrangian is replaced by

$$\mathcal{L}_3 = -(\overline{Q}_L\Sigma M_Q Q_R + \overline{L}_L\Sigma M_L L_R + \text{h.c.}) - \overline{L}^c_R M_{N_R}\tfrac{1+\tau^3}{2}L_R \tag{28}$$

which has the required $SU(2)_L \times U(1)_Y$ symmetry.

The boson mass term is replaced by a kinetic-energy term for the Σ field. We introduce further abbreviations

$$V_\mu = \Sigma(D_\mu\Sigma)^\dagger \quad \text{and} \quad T = \Sigma\tau^3\Sigma^\dagger \tag{29}$$

$$\text{where} \quad D_\mu\Sigma = \partial_\mu\Sigma + ig\mathbf{W}_\mu\Sigma - ig'\Sigma\mathbf{B}_\mu \tag{30}$$

to write this as

$$\mathcal{L}_{2(W)} = -\tfrac{v^2}{4}\,\text{tr}\,[V_\mu V^\mu] \tag{31}$$

In unitary gauge, the vector field V_μ corresponds to a particular combination of the W and B fields, namely

$$V_\mu = -ig\mathbf{W}_\mu\Sigma + ig'\Sigma\mathbf{B}_\mu \tag{32}$$

which defines the physical massive vector bosons, $W^\pm_\mu$ and Z_μ.

The dimension-four part of the Lagrangian does not involve the Σ field

$$\begin{aligned}\mathcal{L}_4 = {} & \overline{Q}_L i\not{D} Q_L + \overline{Q}_R i\not{D} Q_R + \overline{L}_L i\not{D} L_L + \overline{L}_R i\not{D} L_R \\ & - \tfrac{1}{2}\,\text{tr}\,[\mathbf{W}_{\mu\nu}\mathbf{W}^{\mu\nu}] - \tfrac{1}{2}\,\text{tr}\,[\mathbf{B}_{\mu\nu}\mathbf{B}^{\mu\nu}]\,,\end{aligned} \tag{33}$$

Allowing for Σ self-interactions $\mathcal{L}_\Sigma$, the complete effective Lagrangian

$$\mathcal{L} = \mathcal{L}_{2(W)} + \mathcal{L}_3 + \mathcal{L}_4 + \mathcal{L}_\Sigma \tag{34}$$

is invariant under the full electroweak symmetry group. In analogy with low-energy QCD, this effective theory is called the *chiral Lagrangian* of electroweak interactions. Its validity is not restricted to the particular scenario of dynamical symmetry breaking it is usually associated with. The minimal SM with a Higgs boson is just a special case of Eq. (34), where the field Σ is given a definite linear representation.

2.2. *Anomalous Couplings*

The guideline for constructing the effective Lagrangian (34) has been to start with the Fermi model Lagrangian and add the minimal set of fields that make the weak-interaction symmetries manifest. This requires the addition of kinetic terms for the new fields and the inclusion of Σ factors in the boson and fermion mass terms. However, if one does one-loop calculations with Eq. (34), proper gauge-fixing and ghost terms taken into account, one will observe that additional operators are needed to make the theory finite at next-to-leading order. This is natural since the Lagrangian (34) does not yet contain all possible operators of dimension four or less consistent with electroweak symmetry.

At dimension two, there is an additional operator not yet considered, namely

$$\mathcal{L}'_{2(W)} = -\beta' \tfrac{v^2}{8} \operatorname{tr}[TV_\mu] \operatorname{tr}[TV^\mu] . \tag{35}$$

Imposing CP-invariance on the effective Lagrangian[c], the complete list of dimension-four operators not contained in Eq. (34) reads [17]

$$\mathcal{L}_1 = \alpha_1 g g' \operatorname{tr}\left[\Sigma \mathbf{B}_{\mu\nu} \Sigma^\dagger \mathbf{W}^{\mu\nu}\right] \tag{36}$$
$$\mathcal{L}_2 = i\alpha_2 g' \operatorname{tr}\left[\Sigma \mathbf{B}_{\mu\nu} \Sigma^\dagger [V^\mu, V^\nu]\right] \tag{37}$$
$$\mathcal{L}_3 = i\alpha_3 g \operatorname{tr}\left[\mathbf{W}_{\mu\nu} [V^\mu, V^\nu]\right] \tag{38}$$
$$\mathcal{L}_4 = \alpha_4 (\operatorname{tr}[V_\mu V_\nu])^2 \tag{39}$$
$$\mathcal{L}_5 = \alpha_5 (\operatorname{tr}[V_\mu V^\mu])^2 \tag{40}$$
$$\mathcal{L}_6 = \alpha_6 \operatorname{tr}[V_\mu V_\nu] \operatorname{tr}[TV^\mu] \operatorname{tr} TV^\nu \tag{41}$$
$$\mathcal{L}_7 = \alpha_7 \operatorname{tr}[V_\mu V^\mu] \operatorname{tr}[TV_\nu] \operatorname{tr} TV^\nu \tag{42}$$
$$\mathcal{L}_8 = \tfrac{1}{4} \alpha_8 g^2 (\operatorname{tr}[T\mathbf{W}_{\mu\nu}])^2 \tag{43}$$
$$\mathcal{L}_9 = \tfrac{i}{2} \alpha_9 g \operatorname{tr}[T\mathbf{W}_{\mu\nu}] \operatorname{tr}[T[V^\mu, V^\nu]] \tag{44}$$
$$\mathcal{L}_{10} = \tfrac{1}{2} \alpha_{10} (\operatorname{tr}[TV_\mu] \operatorname{tr}[TV_\nu])^2 \tag{45}$$
$$\mathcal{L}_{11} = \alpha_{11} g \epsilon^{\mu\nu\rho\lambda} \operatorname{tr}[TV_\mu] \operatorname{tr}[V_\nu \mathbf{W}_{\rho\lambda}] \tag{46}$$

In the general case of a nonlinear symmetry representation the Lagrangian contains terms of arbitrarily high dimension. Therefore, this list is not sufficient to make the theory finite to *all* orders. In each order of perturbation theory new terms are introduced with the dimension of the Σ-dependent terms increased by two.

[c]A discussion of CP violation is beyond the scope of this review.

This fact does not make the effective-Lagrangian approach useless. It merely implies that at each order of the perturbative expansion one should be prepared for new contributions which are generically of the order $1/16\pi^2$ (since they are induced as loop corrections) with the operator dimension increased by two [6]. The two extra powers of fields or derivatives are compensated by two powers of $1/v$, the expansion parameter of Σ in Eq. (27). As long as the energy is small enough, one can truncate the perturbative series to obtain an approximation of the true amplitude. In matrix elements, the loop expansion therefore becomes a *low-energy* expansion in terms of

$$\frac{E^2}{(4\pi v)^2} = \frac{E^2}{\Lambda^2}. \tag{47}$$

where E is any linear combination of energies, masses and momenta assigned to the external particles. This sets the scale where perturbation theory breaks down in the absence of Higgs-like states:

$$\Lambda = 4\pi v \approx 3 \text{ TeV}. \tag{48}$$

Of course, there may be larger contributions of the operators (36–46) than predicted by the loop expansion. Since the effect of anomalous couplings increases with energy, generically this leads to a lower cutoff scale $\Lambda' < \Lambda$.

2.3. *Custodial Symmetry*

As we have seen in the previous section, there are free parameters which arise when massive vector bosons are introduced to regularize the Fermi model. One of those affects the M_W/M_Z ratio which is usually referred to as the ρ parameter [19]:

$$\frac{M_W^2}{M_Z^2 c_w^2} = \rho \quad \text{where} \quad \rho = \frac{1}{1+\beta'}. \tag{49}$$

Experimentally, ρ is approximately equal to unity. Hence, the coefficient β' of the operator (35) in the chiral Lagrangian vanishes to leading order in perturbation theory.

This can be attributed to an approximate symmetry [1, 19]. If we take the global symmetry of the Lagrangian not to be $SU(2)_L \times U(1)_Y$ but enlarge it to $SU(2)_L \times SU(2)_R$ (where $U(1)_Y \subset SU(2)_R$) and impose the transformation law on Σ

$$\Sigma \to U\Sigma V^\dagger \tag{50}$$

with $U \in SU(2)_L$ and $V \in SU(2)_R$, then the term (35) is forbidden. Simultaneously, this symmetry forbids all operators in the chiral Lagrangian which contain a T factor in the trace, namely $\mathcal{L}_6$ to $\mathcal{L}_{11}$ [Eqs. (41–46)].

It is natural to take the right-handed fermions to be doublets under $SU(2)_R$, just as the left-handed fermions are doublets under $SU(2)_L$. However, for the fermions this is not a good symmetry. It is violated by the up-down mass differences (and by right-handed neutrino Majorana masses).

The hypercharge part of weak interactions also violates this symmetry. The B vector field, in our notation, couples to Σ by a τ^3 matrix factor, which is not consistent with a right-handed $SU(2)_R$ symmetry. Nevertheless, this breaking is proportional to s_w^2 which is not a large parameter, and the fermion couplings affect vector and scalar interactions at the loop level only. Looking at bosonic interactions, $SU(2)_R$ invariance is a reasonable approximation.

Spontaneous symmetry breaking in this sector then takes the form

$$SU(2)_L \times SU(2)_R \to SU(2)_C \tag{51}$$

where $SU(2)_C$ is the diagonal subgroup, the *custodial* $SU(2)$ *symmetry* of electroweak interactions. Bosonic states should come as $SU(2)_C$ multiplets, and in fact, the Z and $W^\pm$ masses are degenerate up to the factor $1/c_w^2 \approx 1 + s_w^2$. Similarly, one expects new particles associated with EWSB also to be organized as $SU(2)_C$ multiplets.[d]

3. Goldstone Boson Scattering

In the chiral-Lagrangian approach described in this chapter, electroweak symmetry breaking and fermion mass generation are mediated by a matrix-valued field Σ which parameterizes the Higgs sector dynamics. The degrees of freedom that make up Σ are a probe for the mechanism of electroweak symmetry breaking. A complete theory would probably replace Σ by a multitude of new (elementary or composite) fields, but in any case the scalars w^a introduced in the minimal parameterization (27) must be present to serve as longitudinal W and Z bosons. This is the triplet of Goldstone bosons associated with spontaneous electroweak symmetry breaking [20]. Thus, the only experiment which *guarantees* information about electroweak symmetry breaking is a measurement of Goldstone boson interactions.

[d]This argument is independent of the origin of EWSB; in Higgs models, the spectrum also tends to follow this pattern.

In the high-energy limit (s, t, u all going to infinity) corresponding scattering amplitudes of $W_L^\pm, Z_L$ on the one hand and $w^\pm, w^0$ on the other hand become equal. This fact is known as the *Equivalence Theorem* [21]. We can make use of it by considering scattering amplitudes of Goldstone bosons in place of the electroweak gauge bosons that are observed in the detector (i.e., their decay products). Doing this, we should always keep in mind that for a quantitative analysis one has to compute the full electroweak amplitudes, since realistic collider energies are far from the asymptotic region.

3.1. *Quasielastic Scattering at Leading Order*

The lowest-order effective Lagrangian (34) provides a unique prediction for the quasielastic $2 \to 2$ scattering amplitudes of W and Z bosons. In the absence of $SU(2)_C$ violation, projecting onto longitudinal states and taking the high-energy limit one obtains the *Low-Energy Theorem* (LET) [22]

$$A(W_L^- W_L^- \to W_L^- W_L^-) = -s/v^2 \tag{52}$$
$$A(W_L^+ W_L^- \to W_L^+ W_L^-) = -u/v^2 \tag{53}$$
$$A(W_L^+ W_L^- \to Z_L Z_L) = s/v^2 \tag{54}$$
$$A(Z_L Z_L \to Z_L Z_L) = 0 \tag{55}$$

The cross sections for on-shell scattering are calculated by squaring the amplitudes, inserting phase space factors and dividing by a symmetry factor of two for like-sign W and for ZZ final states. This symmetry factor will not be included in the amplitude in any of the relations given here.

To be precise, the LET predicts the numerical coefficient of the lowest-order term of an expansion of the scattering amplitudes in terms of E/v, terms of order M_W/E neglected. Since $M_W = gv/2$, this is in fact the limit $g \to 0$ with v fixed. As an approximation to the exact amplitude, the LET is useful for energies larger than M_W and below the scale where either new states appear or partial-wave unitarity is saturated otherwise (see Sec. 3.4).

3.2. *Custodial Symmetry Relations*

The Goldstone bosons transform under $SU(2)_C$ transformations as a triplet. Therefore, if this symmetry is exact, all quasielastic scattering amplitudes are expressible in terms of a single function $A(s,t,u)$:

$$A(W_L^- W_L^- \to W_L^- W_L^-) = A(t,s,u) + A(u,t,s) \tag{56}$$

$$A(W_L^+ W_L^- \to W_L^+ W_L^-) = A(s,t,u) + A(t,s,u) \tag{57}$$

$$A(W_L^+ W_L^- \to Z_L Z_L) = A(s,t,u) \tag{58}$$

$$A(W_L^- Z_L \to W_L^- Z_L) = A(t,s,u) \tag{59}$$

$$A(Z_L Z_L \to Z_L Z_L) = A(s,t,u) + A(t,s,u) + A(u,t,s) \tag{60}$$

The function $A(s,t,u)$ satisfies

$$A(s,u,t) = A(s,t,u). \tag{61}$$

As we have seen, its Taylor expansion begins with

$$A(s,t,u) = s/v^2. \tag{62}$$

Note that these relations are strongly violated in forward scattering where photon exchange is important. There, elastic WW scattering becomes singular while $WW \to ZZ$ stays finite. This is outside the validity region of the Equivalence Theorem.

3.3. *Next-to-leading Order Contributions*

Even without additional knowledge about the high-energy behavior of the theory, the next-to-leading order corrections to Goldstone scattering can be computed. Only the logarithmic terms are scheme-independent and thus physically meaningful [6, 23, 24]:

$$\begin{aligned}\Re A(s,t,u) = \frac{s}{v^2} + \frac{1}{16\pi^2 v^4}\left\{-\frac{(t-u)}{6}\left[t\ln\frac{-t}{\mu^2} - u\ln\frac{-u}{\mu^2}\right] - \frac{s^2}{2}\ln\frac{s}{\mu^2}\right\} \\ + \alpha_4^0\frac{4(t^2+u^2)}{v^4} + \alpha_5^0\frac{8s^2}{v^4}.\end{aligned} \tag{63}$$

The result depends on a renormalization scale μ. This dependence can be absorbed in a redefinition of the coefficients of the operators $\mathcal{L}_4$ and $\mathcal{L}_5$:

$$\alpha_4(\mu) = \alpha_4^0 - \frac{1}{12}\frac{1}{16\pi^2}\ln\frac{\mu^2}{\mu_0^2}, \qquad \alpha_5(\mu) = \alpha_5^0 - \frac{1}{24}\frac{1}{16\pi^2}\ln\frac{\mu^2}{\mu_0^2} \tag{64}$$

Finite corrections depend on the calculational scheme, i.e., on the UV completion of the theory. They are contained in the constant coefficients $\alpha_{4,5}^0$, which therefore represent the relevant information. The same applies to the $SU(2)_C$-violating couplings $\alpha_{5,6,10}$ which are scale-independent to this order, if the coefficient β' (35) is indeed zero (or $\rho = 1$), as suggested by data.

3.4. *Unitarity Constraints*

The optical theorem states that the total cross section for any process is equal to the imaginary part of the elastic forward scattering amplitude. If there is only elastic $2 \to 2$ scattering, this can be translated into a relation for the scattering amplitude $A(s,t,u)$. Expanding it in partial waves

$$A(s,t,u) = 32\pi \sum_\ell a_\ell(s)\,(2\ell+1)\,P_\ell(1+2t/s), \tag{65}$$

each partial-wave amplitude a_ℓ has to satisfy

$$|a_\ell(s) - i/2| = 1/2, \tag{66}$$

i.e., as a curve in the complex plane parameterized by s it has to stay on the *Argand circle*, a circle with radius $\frac{1}{2}$ around the point $\frac{i}{2}$. In particular, the real part of the partial-wave amplitude can never exceed $1/2$.

In terms of the amplitude function $A(s,t,u)$, the $SU(2)_C$ eigenamplitudes are given by [25]

$$A(I=0) = 3A(s,t,u) + A(t,s,u) + A(u,t,s), \tag{67}$$

$$A(I=1) = A(t,s,u) - A(u,t,s), \tag{68}$$

$$A(I=2) = A(t,s,u) + A(u,t,s). \tag{69}$$

Inserting the LET expressions (52–55) and expanding in terms of partial-wave amplitudes, one obtains the nonvanishing terms

$$a^{I=0}_{J=0} = \frac{s}{16\pi v^2}, \qquad a^{I=1}_{J=1} = \frac{s}{96\pi v^2}, \qquad a^{I=2}_{J=0} = -\frac{s}{32\pi v^2}. \tag{70}$$

There is no higher spin involved if we remain with the LET amplitudes. The critical value $a = 1/2$ is reached at the energies

$$I=0: \qquad E = \sqrt{8\pi}\,v = 1.2\ \text{TeV}, \tag{71}$$

$$I=1: \qquad E = \sqrt{48\pi}\,v = 3.5\ \text{TeV}, \tag{72}$$

$$I=2: \qquad E = \sqrt{16\pi}\,v = 1.7\ \text{TeV}. \tag{73}$$

At these energies, perturbation theory ceases to be predictive. In order to have amplitude expressions that are at least in accord with unitarity beyond these scales, one can try to resum the perturbation series in a particular way. The result depends on the chosen resummation prescription and does not tell anything about the actual high-energy behavior. However, it can serve as a consistent implementation of particular models with distinct features at energies beyond the unitarity saturation threshold.

The idea of such unitarization models is to project each eigenamplitude function $a_\ell(s)$ onto the Argand circle. Doing this, one assumes implicitly that no new scattering channels are open, so that $2 \to 2$ quasielastic scattering dominates at all scales. Two particular models have become popular, representing extreme cases:

(1) The K-matrix unitarization model [26, 27] is not limited to the perturbative expansion. Assuming that $a(s)$ is a real-valued amplitude function one starts with, the unitarized amplitude is given by

$$a_K(s) = a(s)\,\frac{1 + ia(s)}{1 + a(s)^2}. \tag{74}$$

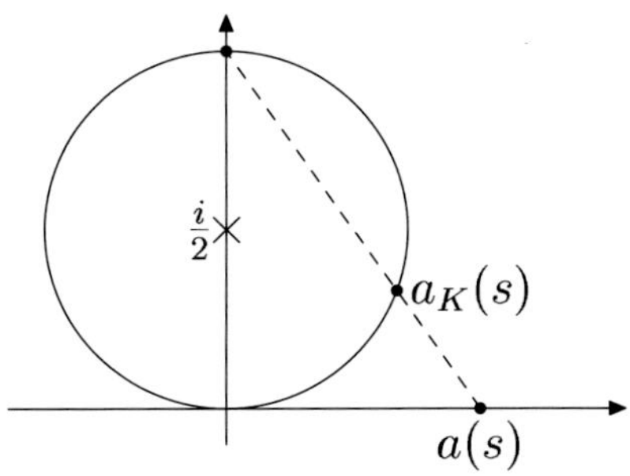

Fig. 1. K matrix construction for projecting a real scattering amplitude onto the Argand circle.

Geometrically, the value $a_K(s)$ corresponds to the projection of the point $a(s)$ onto the Argand circle along the straight line connecting $z = a(s)$ with $z = i$ (Fig. 1). By construction, the K-matrix prescription will never generate a resonance if there is none within the function $a(s)$. In this respect it can be regarded as a minimal unitarization model. In particular, if the LET expression $a(s) = s/v^2$ is inserted, the amplitude function $a(s)$ translates into

$$a_K(s) = a_0 s\,\frac{v^2 + ia_0 s}{v^4 + a_0^2 s^2}. \tag{75}$$

This unitarized amplitude will asymptotically approach the fixed point $a_K(s) = i$, a resonance at infinity.

(2) To obtain the Padé unitarization model (also known as the inverse amplitude method) [28], one separates the amplitude into two pieces.

Usually, one takes the leading term $a^{(0)}(s)$ and the real part of the next-to-leading order term $a^{(1)}(s)$ in the chiral expansion which are proportional to s and to s^2, respectively. Then, the unitarized amplitude reads

$$a_P(s) = \frac{a^{(0)}(s)^2}{a^{(0)}(s) - a^{(1)}(s) - ia^{(0)}(s)^2} \tag{76}$$

If $a^{(1)}(s)$ vanishes, this coincides with the K-matrix model. However, if $a(s)$ has the form

$$a(s) = a_0 \left(\frac{s}{v^2} + \alpha \frac{s^2}{v^4} \right) \tag{77}$$

the Padé-unitarized amplitude is

$$a_P(s) = \frac{-a_0 s/\alpha}{s - v^2/\alpha + ia_0 s/\alpha} \tag{78}$$

This is a resonance with mass $M = v/\sqrt{\alpha}$ and width $\Gamma = a_0 M/\alpha$. In other words, adopting the Padé unitarization method is equivalent to the assumption that a $2 \to 2$ scattering amplitude is entirely dominated by a single resonance.

Padé unitarization works remarkably well for pion-pion scattering in low-energy QCD, consistent with vector meson dominance. Unfortunately, this does not imply anything for the scattering of electroweak Goldstone bosons.

3.5. *Resonances and New Particles*

A striking signature of new physics is a resonance in some scattering channel. If such resonances appear in Goldstone boson scattering, this would almost certainly give a clue in the search for the origin of electroweak symmetry breaking. While resonances are more likely to be observed at the LHC where the available energy for Goldstone boson scattering is somewhat higher than for a first-stage Linear Collider, the strongest resonance in a particular scattering channel will have a low-energy tail that contributes to the parameters in the low-energy expansion. Resonances could be elementary particles (such as a Higgs boson) or bound states of more fundamental objects yet to be discovered.

As discussed before, one expects states associated with EWSB to be grouped in multiplets of the $SU(2)_C$ custodial symmetry. Given the $SU(2)_C$

quantum numbers of Goldstone bosons, this leaves the following possibilities:

$$\begin{aligned}
&\text{Scalar singlet } \sigma: && \mathcal{L}_\sigma = g_\sigma \sigma \tfrac{v}{2} \operatorname{tr}[V_\mu V^\mu] && (79)\\
&\text{Vector triplet } \rho_\mu^a: && \mathcal{L}_\rho = g_\rho \tfrac{v^2}{2} \operatorname{tr}\left[\rho_\mu^a \tau^a V^\mu\right] && (80)\\
&\text{Tensor singlet } \tau^{\mu\nu}: && \mathcal{L}_\tau = g_\tau \tfrac{v}{2} \tau^{\mu\nu} \operatorname{tr}[V_\mu V_\nu] && (81)\\
&\quad\vdots && \quad\vdots
\end{aligned}$$

The amplitude functions corresponding to these couplings for the scalar and vector cases take the form

$$A_\sigma(s,t,u) = g_\sigma^2 \frac{s^2}{v^4} \frac{1}{s-M^2}, \tag{82}$$

$$A_\rho(s,t,u) = g_\rho^2 \left(\frac{-u}{t-M^2} + \frac{s-t}{u-M^2} + 3\frac{s}{M^2} \right). \tag{83}$$

where the resonance width has to be inserted when a pole falls inside the physical region.

While scalar and tensor resonances are electrically neutral, a vector resonance multiplet ρ has neutral and charged components analogous to the $W^\pm, Z$ triplet. The interaction resulting from evaluating (80) is antisymmetric, forbidding the coupling of the ρ to identical particles: It can decay into W^+W^- but not into ZZ.

If $SU(2)_C$ is violated, resonances transforming as scalar triplets (a_0), vector singlets (ω) etc. are also accessible in Goldstone scattering, and the amplitude relations (56–60) between Z and W external states are lost.

A scalar resonance σ with arbitrary coupling g_σ is the generalization of a Higgs resonance. The width of such a state is given by

$$\Gamma_\sigma = \frac{3g_\sigma^2}{32\pi} M_\sigma. \tag{84}$$

if Goldstone bosons are the only decay channels. In the SM, the coupling g_σ itself is proportional to the mass, $g_\sigma = \sqrt{2}\, M/v$, and thus $\Gamma \propto M^3$. Such a state becomes very broad and loses its identity if $M \gtrsim 1$ TeV. Beyond the SM, the σ mass and coupling need not be related, and narrow scalar resonances may be present.

A vector resonance triplet is a characteristic feature of QCD-like technicolor models [11, 29]. If there are no other decay channels than Goldstone bosons (i.e., longitudinal vector bosons), the resonance width is given by

$$\Gamma_\rho = \frac{g_\rho^2}{48\pi} M_\rho \tag{85}$$

which is smaller than the width of a scalar with equal mass and coupling.

Vector resonances have the special property that they can mix with electroweak gauge bosons. This may be interpreted as a remnant of the electroweak interactions of their constituents. As a result, one expects a significant sensitivity to ρ properties not just in Goldstone scattering, but also in the $e^+e^- \to W_L^+W_L^-$ scattering amplitude.

A special case are neutral pseudoscalar resonances, π_T^0 and η_T^0. While they do not couple to Goldstone pairs, they can have a coupling to pairs of *transversal* vector bosons which is induced by the triangle anomaly: The coupling strength is much smaller than for σ and ρ states, and one expects the coupling to ZZ, W^+W^-, $\gamma\gamma$ and even gg to be similar in magnitude. This is to be contrasted with σ resonances for which the (longitudinal) ZZ and WW couplings are dominant. However, this picture may be complicated by CP violation in the strong dynamics, which would induce $\eta - \sigma$ mixing. Clearly, if any resonance appears in vector boson scattering, it is important to determine its spin and the polarization of vector bosons in the decay by angular correlation analysis.

Apart from Goldstone (vector boson) decays, all such states are likely to have a significant or even dominant fraction of heavy-quark decays: $t\bar{t}$, $t\bar{b}$, $b\bar{b}$. The channels $\tau^+\tau^-$ and $\tau^+\nu_\tau$ are also possible.

While resonant production of new states in Goldstone scattering is restricted by the symmetries of Goldstone pairs, any state associated with EWSB can in principle be pair-produced. Some particles can also be directly pair-produced in e^+e^- annihilation, but Goldstone scattering gives access to additional members of new multiplets. The coupling strength extracted from the production cross section of such a particle is an independent piece of information.

In models of dynamical symmetry breaking, low-lying pseudo-Goldstone boson scalars are a common feature [10, 11]. Classifying them according to their $SU(2)_C$ properties, multiplets analogous to the low-energy QCD spectrum can be expected, among them $SU(2)_C$ triplets (π), doublets (K) and singlets (η). All can in principle be produced in Goldstone scattering, and the cross sections may be sizable. As dominant decay modes one expects longitudinally polarized vector bosons and heavy quarks, possibly accompanied by transversally polarized vector bosons (radiative decays).

Pseudo-Goldstone bosons (technipions) share quantum numbers with the $H^\pm, A^0$ states that are present in models with more than one Higgs doublet, e.g., the MSSM. As a consequence, the detection of Higgs-like scalars is *not* sufficient to establish a weakly interacting scenario of EWSB.

Only by a careful analysis of the complete pattern of masses and couplings a particular model can be favored or excluded, and the measurement of the couplings to Goldstone bosons (longitudinal W, Z as opposed to transversal gauge bosons) is an important ingredient. Many of the corresponding measurements are difficult or impossible at a hadron collider even if production rates are large, but are straightforward in the low-background environment of a Linear Collider.

4. Measuring Higgs Sector Parameters at a Linear Collider

4.1. *Precision Observables*

The effective Lagrangian (34) and (35–46) encodes all known facts about the structure of electroweak interactions. Our current knowledge about the free parameters of this effective theory can be summarized as follows:

(1) The masses of charged fermions and vector bosons have been measured or derived from hadronic data with high accuracy.
(2) The neutrino mass matrices are much less certain. We know about three different eigenstates with very low mass and considerable mixing, but nothing about any other eigenstates.
(3) The current-current structure of weak interactions is well established for the first two generations of quark and leptons. For the third generation, there is still some room for contributions that do not fit in this picture.
(4) Dimension-five operators (magnetic moments, flavor-changing penguin operators, etc.) are suppressed, and all measurements and limits are consistent with loop effects of the known particles.

In short, the gauge symmetry structure is extremely well tested, at least for the first two fermion families. This makes the chiral-Lagrangian approach a meaningful parameterization. Any anomalous effect can consistently be parameterized by the coefficients of gauge-invariant (higher-dimensional) operators. In the bosonic sector in particular, a basis of the CP-conserving dimension-four operators is given by (35–46). These terms are sensitive to Higgs sector physics since they contain factors of the symmetry-breaking Σ field, and therefore carry the information about EWSB that is available at low energies until new degrees of freedom are observed directly.

Only three parameters in this list are significantly constrained by the electroweak precision data gathered during the last decade. In the analysis of Z pole observables, of the W mass and of the flavor-independent low-energy data, all deviations from the SM prediction can, to leading nontrivial order, be parameterized by the coefficients of the three operators that are bilinear in the vector fields. In our terminology, these are α_1, β' and α_8 [Eqs. (35, 36, 43)]. Another parameterization has been introduced by Peskin and Takeuchi [30], who considered the Taylor expansion of vector boson propagators. The leading deviations from the SM relations are given by three parameters ΔS, ΔT and ΔU, which are related to the chiral Lagrangian coefficients by

$$\Delta S = -16\pi\alpha_1, \qquad \alpha\Delta T = -\beta', \qquad \Delta U = -16\pi\alpha_8. \tag{86}$$

ΔT parameterizes custodial $SU(2)_C$ violation in the Higgs sector, analogous to the ρ parameter which is related to it (49) by

$$\Delta\rho = \alpha\,\Delta T. \tag{87}$$

ΔS describes anomalous mixing of weak and hypercharge bosons and thus affects the measured value of the weak mixing angle, while ΔU parameterizes $SU(2)_C$ violation in the left-handed gauge sector. In most models, there is little room for the latter effect, so ΔU is usually not considered.

It should be emphasized that, in the absence of a Higgs boson, the SM radiative corrections to the interactions parameterized by ΔS and ΔT are logarithmically divergent. To obtain numeric values for them, one has to introduce a cutoff in the low-energy effective theory. It is customary to introduce a (fictitious) Higgs boson for that purpose, so that values for ΔS and ΔT have to be understood in reference to some fixed Higgs boson mass.

The experimental constraints on S and T are summarized in Fig. 2, where the reference Higgs boson mass has been set to 100 GeV. For this value, the exclusion contour encloses the origin, and ΔS and ΔT are consistent with zero. Incidentally, this is the prediction of the minimal SM with a light Higgs boson, which is therefore consistent with the precision data. If the reference Higgs mass m_H is changed to higher values, the origin of the ST plane moves into the lower right direction, as indicated in the plot. Then, the Higgs sector has to contribute nonzero shifts ΔS and ΔT. In fact, the actual location of the exclusion contour requires a positive value of ΔT to make the data consistent with a high effective Higgs mass. In the absence of a light Higgs boson one should therefore expect new physics to provide a certain amount of custodial $SU(2)_C$ violation [31].

In models of dynamical symmetry breaking, S and T (or α_1 and β') receive contributions from the compositeness scale Λ which are of the order

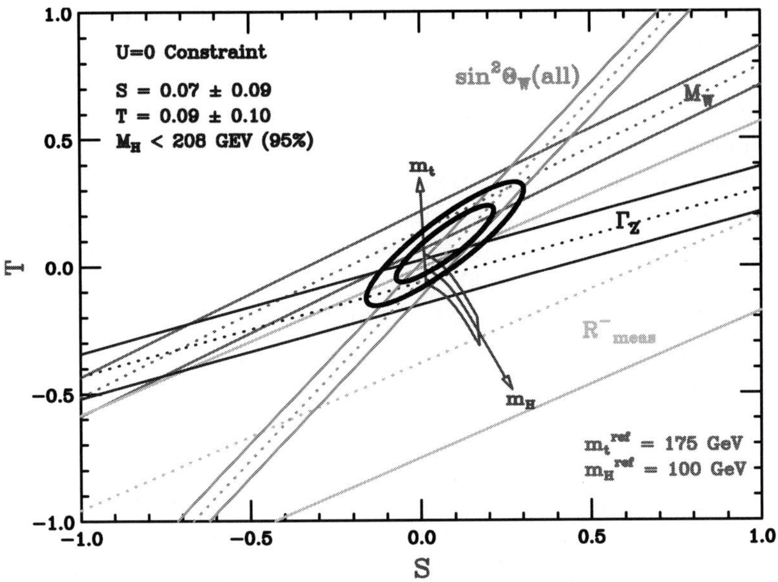

Fig. 2. Exclusion countour in the ST plane allowed by the electroweak precision data. The arrows denote the directions of increasing top mass and increasing Higgs mass [32].

$\Delta\alpha_1 \sim v^2/\Lambda^2$. If $\Lambda \sim 4\pi v$, the natural upper limit for Λ in the absence of a Higgs resonance, this correction is of the same order as the shift from a light to a heavy Higgs boson in Fig. 2. Of course, the sign and precise value of the new contributions is model-dependent. While QCD-like technicolor models are disfavored by the data since they typically have $\Delta S > 0$ and $\Delta T \approx 0$ (conserved $SU(2)_C$), topcolor models, for instance, predict custodial symmetry violation and positive ΔT, which makes them consistent with data for a larger range of physical Higgs masses.

The bands in Fig. 2 indicate the areas in the ST plane allowed by individual observables which depend on ΔS and ΔT. It is important that they all intersect in the same region, so that a meaningful exclusion contour can be drawn. If this were not the case, one would have to include operators of dimension six in the analysis. The magnitude of their contribution is parameterically of the same order as two-loop radiative corrections which, however, are strongly scheme-dependent in the absence of a physical Higgs boson.

The consistency of the individual bands in the present plot indicates that the inclusion of higher-order effects is not yet necessary, but significant improvements in the experimental sensitivity (as expected from a Giga-Z experiment, cf. Chapter 6) would provide such a level of precision that a two-parameter analysis in terms of S and T becomes obsolete. This is obviously desirable in the context of weakly interacting models where all observables are computable to higher order in perturbation theory, and information on additional parameters in the theory can be gained in this way. However, since it is unlikely that for strong interactions a reliable calculation of next-to-next-to-leading order effects is possible, in the context of dynamical EWSB the experimental coverage of a larger subset of the operator coefficients (36–46) will be more valuable.

4.2. *Triple Gauge Couplings*

In the operator basis (36–46), the operators $\mathcal{L}_2$, $\mathcal{L}_3$, $\mathcal{L}_9$ and $\mathcal{L}_{11}$ do not contribute to vector boson two-point functions, but modify the trilinear couplings of the photon and of the W and Z bosons. A standard parameterization of a CP-conserving triple gauge boson vertex has been introduced in Ref. [33],

$$\begin{aligned}\mathcal{L}_{WWV} = g_{WWV}\Big[& ig_1^V V_\mu (W^{-\nu}W^+_{\mu\nu} - W^-_{\mu\nu}W^{+\nu}) + i\kappa_V W^-_\mu W^+_\nu V^{\mu\nu} \\ & + i\frac{\lambda_V}{M_W^2} W_\mu^{-\nu} W_\nu^{+\rho} V_\rho{}^\mu + g_5^V \epsilon^{\mu\nu\rho\sigma}(W^-_\mu \partial_\rho W^+_\nu - \partial_\rho W^-_\mu W^+_\nu)V_\sigma\Big],\end{aligned} \tag{88}$$

where V denotes the photon ($V = \gamma$ or A_μ) and the Z boson interactions with prefactors $g_{WW\gamma} = e$ and $g_{WWZ} = gc_w$, respectively. If all anomalous operator coefficients vanish, we have

$$g_1^\gamma = g_1^Z = \kappa_\gamma = \kappa_Z = 1, \tag{89}$$
$$g_5^\gamma = g_5^Z = \lambda_\gamma = \lambda_Z = 0. \tag{90}$$

Nonzero coefficients of the dimension-four operators contribute the following shifts [34][e]

$$\Delta\kappa_\gamma = g^2\alpha_2 + g^2\alpha_3 + g^2\alpha_9, \tag{91}$$
$$\Delta\kappa_Z = -g'^2\alpha_2 + g^2\alpha_3 + g^2\alpha_9, \tag{92}$$
$$\Delta g_1^Z = \frac{1}{c_w^2} g^2\alpha_3, \tag{93}$$
$$\Delta g_5^Z = \frac{1}{c_w^2} g^2\alpha_{11}, \tag{94}$$
$$\Delta g_1^\gamma = \Delta g_5^\gamma = 0, \tag{95}$$
$$\Delta\lambda_\gamma = \Delta\lambda_Z = 0. \tag{96}$$

Due to electromagnetic gauge invariance, corrections to g_1^γ and g_5^γ have to vanish at zero momentum transfer. However, the absence of corrections to the λ couplings up to this order is a characteristic feature of the strongly-interacting scenario. Nonzero values for these coefficients are only introduced at higher order in the chiral expansion, i.e., by dimension-six operators. (In the weakly-interacting scenario where a light Higgs boson is present, all anomalous terms scale as dimension six.) The reason is that λ multiplies a term that involves transversal vector fields only and thus does not probe the Higgs sector directly. By contrast, the operators $\mathcal{L}_2$, $\mathcal{L}_3$, $\mathcal{L}_9$ and $\mathcal{L}_{11}$ (37, 38, 44, 46) involve Goldstone scalars, visible as the longitudinal components of vector bosons. Hence, the strongly interacting scenario predicts the transversal couplings λ_γ and λ_Z to be significantly suppressed compared to possible deviations in the g and κ parameters. Projecting onto longitudinal polarization states of the vector bosons by exploiting angular correlations of their decay products will enhance the relevant contributions.

With exact custodial $SU(2)_C$ symmetry we have $\alpha_9 = \alpha_{11} = 0$ and get the additional relations

$$\Delta\kappa_\gamma = -\frac{c_w^2}{s_w^2}(\Delta\kappa_Z - \Delta g_1^Z) \quad \text{and} \quad g_5^Z = 0, \tag{97}$$

If this symmetry assumption is valid, the leading anomalous effect on the couplings depends on just two parameters, α_2 and α_3, which could be measured by considering Z observables only, g_Z and κ_Z.

The measurement of W and Z pair production at LEP2 has provided the first meaningful bounds on anomalous triple gauge boson couplings.

[e]There are also shifts due to α_1, β' and α_8. They are constrained already now by the existing data as discussed in the previous section. These contributions, together with the one-loop radiative corrections, have to be included, but can be assumed to be known in a complete triple gauge boson coupling analysis.

The current precision is still low compared to that one already achieved for ΔS and ΔT. This situation will change when a Linear Collider is available. As discussed in Chapter 6, the experimental accuracy on the vector boson self-interactions will then become competitive, so the indirect sensitivity to the Higgs sector structure can be considerably improved by the complete coverage of the operator basis.

4.3. *W and Z Scattering Amplitudes*

The remaining parameters $\alpha_{4,5,6,7,10}$ [Eqs. (39–42, 45)] in the CP-conserving chiral Lagrangian do not contribute at tree-level to bilinear or trilinear vector boson couplings. Rather, they affect Goldstone boson scattering and thus show up as anomalous *quartic* couplings of (longitudinally polarized) W and Z bosons.

From the list of operators introducing quartic couplings, only $\mathcal{L}_4$ and $\mathcal{L}_5$ conserve custodial $SU(2)_C$, so in the symmetric case the quartic vector boson couplings depend on just two new parameters, α_4 and α_5. The other terms, $\mathcal{L}_6$, $\mathcal{L}_7$ and $\mathcal{L}_{10}$, describe new sources of $SU(2)_C$ violation. This counting does not include the effect of the parameters $\alpha_{3,8,9,11}$ which also contribute anomalous quartic vector-boson interactions. As discussed before, by combining low-energy and high-energy Linear Collider data the latter will be sufficiently constrained to be considered fixed in the analysis of quartic couplings, which must also include the (calculable) one-loop SM corrections.

Concentrating on the genuine quartic couplings, in the $SU(2)_C$ symmetric limit the contribution of α_4 and α_5 to the Goldstone scattering amplitude is given by (63). However, this amplitude is not accessible directly, but is embedded in multi-fermion processes involving intermediate gauge boson production and decay. There are three different classes of processes which probe Goldstone scattering at lepton colliders:

(1) The one-loop amplitude for the production of a final state VV' (where $V, V' = W^{\pm}, Z$) is affected by a *rescattering correction*:

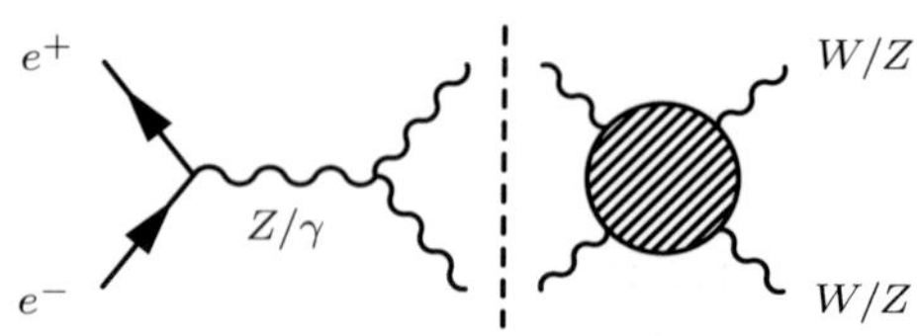

In this process rescattering of the vector bosons takes place at the full CM energy of the annihilating fermions. To make use of this fact, in a global fit of the parameters describing pair production their imaginary part has to be extracted. Only the channel with spin and isospin 1 can be accessed.

(2) The second process class which is sensitive to the symmetry-breaking sector is *triple vector boson production*:

$$e^- e^+ \to ZW^+W^- \quad (98)$$

$$e^- e^+ \to ZZZ \quad (99)$$

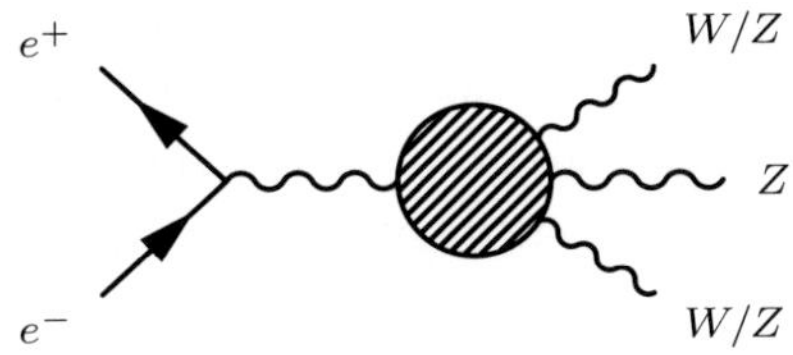

Gauge invariance makes the cross section for all processes of this type fall off with $1/s$. Therefore, the best measurements are not necessarily done at the highest energy but somewhat above threshold, and the sensitivity is limited by the available luminosity. The variable to project out resonances or to observe the effect of anomalous couplings is the invariant mass of vector boson pairs.

(3) The obvious place to look for Goldstone boson scattering amplitudes is *vector boson fusion*:

$$e^- e^+ \to e^- e^+ VV' \quad (100)$$

$$e^- e^+ \to \nu\bar{\nu}VV' \quad (101)$$

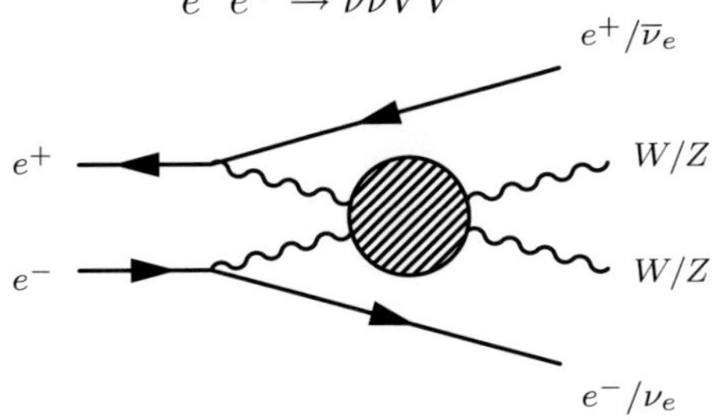

This class of processes has a cross section that rises logarithmically with energy, so increasing the energy as well as the luminosity will improve the experimental sensitivity. In the asymptotic limit where masses can be neglected, the intermediate vector bosons are essentially on-shell and the relevant Goldstone scattering amplitudes are probed directly, albeit at an effective CM energy which is significantly lower than the full collider energy.

Which process is actually most sensitive depends on machine parameters and experimental details (cf. Chapter 6). Concerning rescattering corrections, disentangling the imaginary parts of all form factors in vector boson pair production (i.e., four-fermion production) is a nontrivial task. Collider runs with different combinations of electron and positron polarization are needed for a clean separation of all contributions [35].

The other processes involve six-fermion production in e^+e^- collisions, where in the case of triple vector boson production all three fermion pairs originate from vector boson decays, such that the signal can be isolated by invariant mass constraints. In the case of vector boson fusion, the spectator neutrinos (electron/positron) go predominantly into the forward direction, hence the characteristic signature of this process is a large missing invariant mass (e^+e^- invariant mass).

For realistic Linear Collider parameters, triple vector boson production [36] appears to be less sensitive to the parameters of Goldstone scattering than vector boson fusion [37] once the collider energy is sufficient for the latter process to have a significant rate. In the asymptotic high-energy limit, this rate can be approximated by the cross section for the on-shell subprocess of $2 \to 2$ vector boson scattering, folded by the splitting probabilities $e \to W\nu$ (or $e \to Ze$). These structure functions are plotted in Fig. 3.

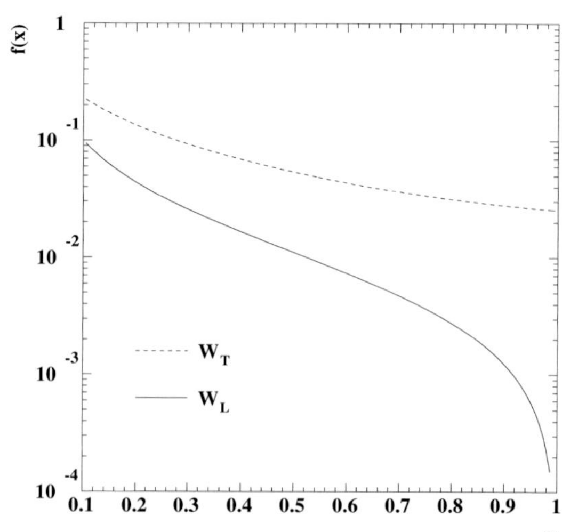

Fig. 3. Structure functions for the emission of a transversally and longitudinally polarized W boson from an electron or positron.

The figure shows that the emission of longitudinally polarized W bosons is significantly suppressed compared to transversally polarized ones, in particular towards the high-energy end of the spectrum. Only the former probe the anomalous quartic interactions we are interested in, so there is a significant W_T-induced background which has to be reduced by a suitable experimental strategy [37–39] Since the longitudinal spectrum drops sharply near $x = 1$, the energies that can be reached for Goldstone boson scattering are considerably lower than the collider energy. For that reason, a Linear Collider with an energy of at least 0.8 to TeV is necessary to achieve a reasonable precision in the determination of the quartic couplings. At these energies, unitary constraints on the amplitudes (Section 3.4) are not yet an issue, and the chiral Lagrangian parameterization describes the scattering processes in a model-independent way.

Anomalous quartic couplings involving photons will not be induced by the dimension-four operators $\mathcal{L}_4$ to $\mathcal{L}_{10}$. Therefore, in $e\gamma$ and $\gamma\gamma$ collisions vector boson pair production does not provide independent information on a strongly interacting Higgs sector at this level, and in e^+e^- collisions photon-induced processes should be considered as a background to vector boson fusion. Dimension-six operators, however, allow for anomalous quartic couplings involving photons and can be probed independently in these channels.

A complete coverage of the parameter space, which requires also the inclusion of the $SU(2)_C$ violating operators in the analysis, will be possible only by combining all available channels and including, in particular, results for the analogous processes at the LHC. Nevertheless, the results presented in Chapter 6 show that by considering only the dominant vector boson fusion channels at a Linear Collider, the experimental precision on the quartic couplings α_4 and α_5 will be in the percent range, not much worse that the expected accuracy in determining the bilinear and trilinear couplings discussed before.[f]

5. Conclusions

Dynamical symmetry breaking provides a natural explanation for the electroweak scale. If there is no weakly interacting effective theory which describes physics beyond the TeV range, as it might be the case if a light Higgs boson exists, the dynamics responsible for electroweak symmetry breaking

[f] In comparing numerical values, one should take into account that some authors extract a factor $16\pi^2$ from the α parameters, such that their natural values are $\mathcal{O}(1)$.

could be directly accessible at future colliders. The study of Goldstone-boson scattering amplitudes is then the key for accessing the (strongly-interacting) Higgs sector of the Standard Model.

The inherent scale of new strong interactions is likely beyond 1 TeV, therefore in this chapter we have considered mainly the indirect effects on precision observables at a Linear Collider where initially the energy may not be sufficient to produce new states. However, compared to existing precision data the set of observables that can be measured at a Linear Collider with at least percent accuracy is greatly enlarged, and essentially all interactions in the leading nontrivial order of the low-energy expansion are covered. Together with hadron collider data, this will allow to significantly constrain the possible scenarios and open the path towards a satisfactory theory of electroweak symmetry breaking and mass generation.

References

1. H. Georgi, *Weak interactions and modern particle theory*, Addison-Wesley, 1984; J.F. Donoghue, E. Golowich, and B.R. Holstein, *Dynamics of the Standard Model*, Cambridge University Press, 1992; A. Dobado, A. Gómez-Nicola, A.L. Maroto, and J.R. Pelaez, *Effective Lagrangians for the Standard Model* (Springer, 1997).
2. P.W. Higgs, Phys. Rev. Lett. **12**, 132 (1964); Phys. Rev. Lett. **13**, 508 (1964); Phys. Rev. **145**, 1156 (1966); F. Englert and R. Brout, Phys. Rev. Lett. **13**, 321 (1964); G.S. Guralnik, C.R. Hagen, and T.W. Kibble, Phys. Rev. Lett. **13**, 585 (1964); T.W. Kibble, Phys. Rev. **155**, 1554 (1967).
3. S.L. Glashow, Nucl. Phys. **22**, 579 (1961); A. Salam, in: N. Svartholm (ed.), *Elementary Particle Theory*, Almqvist and Wiksells, Stockholm, 1968; S. Weinberg, Phys. Rev. Lett. **19**, 1264 (1967).
4. L. Susskind, Phys. Rept. **104**, 181 (1984).
5. D. Gross and F. Wilczek, Phys. Rev. Lett. **30**, 1343 (1973); H.D. Politzer, Phys. Rev. Lett. **30**, 1346 (1973) .
6. S. Weinberg, Physica **96A**, 327 (1979).
7. For the current status of the Higgs boson mass limit, see: T. Hambye and K. Riesselmann, Phys. Rev. **D55**, 7255 (1997), and references therein.
8. G. 't Hooft, in: G. 't Hooft *et al.* (eds.), *Proc. Cargese Summer Institute* (NATO Advanced Study Institute, Series B: Physics, V. 59), 1979; reprinted in: E. Farhi and R. Jackiw (eds.), *Dynamical gauge symmetry breaking*, World Scientific, 1982; and in: G. 't Hooft (ed.), *Under the spell of the gauge principle* (World Scientific, 1994).
9. N. Arkani-Hamed, A. Cohen, and H. Georgi, Phys. Lett. **B513**, 232 (2001); N. Arkani-Hamed, A. Cohen, E. Katz, and A. Nelson, JHEP **0207**, 034 (2002); N. Arkani-Hamed, A. Cohen, T. Gregoire, and J. Wacker, JHEP **0208**, 020 (2002).

10. Models of dynamical symmetry breaking are reviewed in detail in: C. Hill and E. Simmons, hep-ph/0203079, submitted to Phys. Rept.
11. S. Weinberg, Phys. Rev. **D13**, 974 (1976); Phys. Rev. **D19**, 1277 (1979); L. Susskind, Phys. Rev. **D20**, 2619 (1979); E. Farhi and L. Susskind, Phys. Rev. **D20**, 3404 (1979).
12. S. Dimopoulos and L. Susskind, Nucl. Phys. **B155**, 237 (1979); E. Eichten and K. Lane, Phys. Lett. **B90**, 125 (1980).
13. W. Bardeen, C.T. Hill, and M. Lindner, Phys. Rev. **D41**, 1647 (1990); C.T. Hill, Phys. Lett. **B266**, 419 (1991).
14. C.T. Hill, Phys. Lett. **B345**, 483 (1995); K. Lane and E. Eichten, Phys. Lett. **B352**, 382 (1995).
15. B. Dobrescu and C.T. Hill, Phys. Rev. Lett. **81**, 2634 (1998); R. Sekhar Chivukula, B. Dobrescu, H. Georgi, and C.T. Hill, Phys. Rev. **D59**, 075003 (1999).
16. S. Weinberg, Phys. Rev. **166**, 1568 (1968); S.R. Coleman, J. Wess, and B. Zumino, Phys. Rev. **177**, 2239 (1969); C.G. Callan, S.R. Coleman, J. Wess, and B. Zumino, Phys. Rev. **177**, 2247 (1969).
17. T. Appelquist and C. Bernard, Phys. Rev. **D22**, 200 (1980); A. Longhitano, Phys. Rev. **D22**, 1166 (1980); Nucl. Phys. **B188**, 118 (1981);
18. J. Gasser and H. Leutwyler, Ann. Phys. (N.Y.) **158**, 142 (1984); Nucl. Phys. **B250**, 465 (1985).
19. M. Veltman, Act. Phys. Pol. **B8**, 475 (1977); Nucl. Phys. **B123**, 89 (1977); P. Sikivie, L. Susskind, M. Voloshin, and V. Zakharov, Nucl. Phys. **B173**, 189 (1980).
20. J. Goldstone, Nuovo Cim. **9**, 154 (1961); Y. Nambu, Phys. Rev. Lett. **4**, 380 (1960); J. Goldstone, A. Salam, and S. Weinberg, Phys. Rev. **127**, 965 (1962).
21. C.E. Vayonakis, Lett. Nuovo Cim. **17**, 383 (1976); M.S. Chanowitz and M.K. Gaillard, Nucl. Phys. **B261**, 379 (1985); G.J. Gounaris, R. Kögerler, and H. Neufeld, Phys. Rev. **D34**, 3257 (1986) ; Y.-P. Yao and C.-P. Yuan, Phys. Rev. **D38**, 2237 (1988); J. Bagger and C. Schmidt, Phys. Rev. **D34**, 264 (1990).
22. S. Weinberg, Phys. Rev. Lett. **17**, 616 (1966); M.S. Chanowitz, M. Golden, and H. Georgi, Phys. Rev. **D36**, 1490 (1987).
23. S. Dawson and S. Willenbrock, Phys. Rev. **40**, 2880 (1989).
24. O. Cheyette and M. Gaillard, Phys. Lett. **B197**, 205 (1987).
25. B. Lee, C. Quigg and H. Thacker, Phys. Rev. Lett. **38**, 883 (1977); Phys. Rev. **D16**, 1519 (1977); D. Dicus and V. Mathur, Phys. Rev. D7 (1973) 3111.
26. S.N. Gupta, *Quantum Electrodynamics*, Gordon and Breach, 1981.
27. M.S. Chanowitz, Phys. Rept. **320**, 139 (1999).
28. A. Dobado and J.R. Pelaez, Phys. Rev. **D56**, 3057 (1997).
29. R. Casalbuoni, S. De Curtis, D. Dominici, and R. Gatto, Phys. Lett. **B155**, 95 (1985); Nucl. Phys. **B282**, 235 (1987).
30. M. Peskin and T. Takeuchi, Phys. Rev. Lett. **65**, 964 (1990); Phys. Rev. **D46**, 381 (1992).
31. M.E. Peskin and J.D. Wells, SLAC-PUB-8763, hep-ph/0101342.
32. J. Bagger, talk at the 2001 Snowmass Summer Study, figure by M. Swartz.

33. K.J.F. Gaemers and G.J. Gounaris, Z. Phys. **C1**, 259 (1979); K. Hagiwara, K. Hikasa, R.D. Peccei, and D. Zeppenfeld, Nucl. Phys. **B282**, 253 (1987).
34. T. Appelquist and G.-H. Wu, Phys. Rev. **D48**, 3235 (1993).
35. M. Diehl, O. Nachtmann, and F. Nagel, HD-THEP-02-34, hep-ph/0209229.
36. V. Barger, T. Han, and R.J.N. Phillips, Phys. Rev. **D39**, 146 (1989); S. Dawson, A. Likhoded, G. Valencia, and O. Yushchenko, in: *Snowmass 1996, New directions for high-energy physics*, hep-ph/9610299; O. Eboli, M. Gonzalez-Garcia, and J. Mizukoshi Phys. Rev. **D58**, 034008 (1998).
37. E. Boos, H.-J. He, W. Kilian, A. Pukhov, C.-P. Yuan, and P. M. Zerwas, Phys. Rev. **D57**, 1553 (1998); Phys. Rev. **D61**, 077901 (2000).
38. V. Barger, K. Cheung, T. Han, and R.J.N. Phillips, Phys. Rev. **D52**, 3815 (1995).
39. R. Chierici, S. Rosati, and M. Kobel, LC-PHSM-2001-038, and in: J.A. Aguilar-Saavedra *et al.* (ECFA/DESY LC Physics Working Group), *TESLA: Technical design report. Part III: Physics at an* e^+e^- *Linear Collider*, DESY-2001-011, hep-ph/0106315.

CHAPTER 6

Physics of Electroweak Gauge Bosons

Klaus Mönig
DESY
Platanenalle 6, D-15738 Zeuthen, Germany
E-mail: Klaus.Moenigdesy.de

The study of gauge bosons is interesting in two respects. The properties of gauge bosons are modified by higher order effects that are sensitive to mass scales not directly accessible to experiment. On the other hand interactions amongst gauge bosons are sensitive to the symmetry structure of the theory where especially the interactions involving longitudinally polarised bosons teach us about the mechanism of electroweak symmetry breaking. This article explains how far a linear collider can increase our knowledge about the properties of and the interactions amongst gauge bosons.

1. Introduction

Within any local gauge theory the properties of the gauge bosons are uniquely defined by the structure of the gauge group. It is thus of special importance to study the physics of the gauge bosons. In the Standard Model of electroweak interactions there exist, for the unbroken $SU(2) \times U(1)$, a triplet (W^+, W^0, W^-), coupling to the weak isospin, and a singlet B, coupling to hypercharge. All bosons at this stage are massless vector-particles. Due to the mechanism of spontaneous symmetry breaking the two neutral particles mix with a mixing angle θ_W and the bosons acquire mass yielding a massive pair of charged gauge bosons, (W^+, W^-), a massive neutral boson, $(Z = \cos\theta_W W^0 - \sin\theta_W B)$, and the massless photon, $(\gamma = \sin\theta_W W^0 + \cos\theta_W B)$.

Once the gauge group is given the gauge sector is defined by three free parameters, the coupling constants for the $SU(2)$ and the $U(1)$, g, g' and the vacuum expectation value of the Higgs field, v.

The requirement that the photon is a massless particle with vector couplings to the fermions defines the mixing angle to be $g \sin\theta_W = g' \cos\theta_W = e$. Since the couplings of the Higgs boson are pure gauge couplings the ratio of the Z- and W-mass is given by $\cos\theta_W = m_W/m_Z$. This relation is actually valid as long as the symmetry breaking is only induced by doublet Higgs fields. As another consequence of the W-Higgs couplings being pure gauge couplings, the Fermi constant in muon decays depends only on the vacuum expectation value, $G_F = 1/\sqrt{2}v^2$.

All formulae given above are valid on Born level. Including higher orders in perturbation theory they receive process dependent corrections. In these loop corrections all parameters of the model enter and also new physics which might not be visible directly can produce measurable effects in the radiative corrections. To test the validity of a given model one has thus to measure a redundant set of parameters with the best possible precision. From measurements at lower energy the electromagnetic fine structure constant α and the Fermi constant G_F are known very accurately [1]. Since the Z can be produced singly in e^+e^--collisions and the energy of an e^+e^--storage ring like LEP can be calibrated with very good precision, the Z-mass (m_Z) is usually taken as the third parameter to fix the model. It turns out then that the quantities with the highest sensitivity to interesting radiative corrections are the vector and axial-vector couplings of the Z to fermions (g_{Vf}, g_{Af}) and the mass of the W.

All these quantities have been measured already with good precision at LEP, SLD, and the TEVATRON [2] and, as an example, the mass of the Higgs can be constrained to be less than about 200 GeV within the Standard Model of electroweak interactions. However, it turns out that all these quantities can be measured significantly more accurately at a linear collider allowing much more stringent tests of the Standard Model.

Another feature of a local gauge theory is that the gauge bosons have to interact amongst each other if the gauge group is non-Abelian. The structure of these interactions is completely given by the gauge group. The Standard Model predicts interactions between W^+W^-Z and $W^+W^-\gamma$ but no interactions amongst the neutral gauge bosons.

In case of a light Higgs boson electroweak interactions remain weak at high energies and the triple gauge couplings are only modified by loop effects. If no light Higgs boson exists electroweak interactions become strong at the TeV scale. One therefore expects new Born level contributions in the Lagrangian. In the latter case the process $VV \to V'V'$ ($V, V' = W, Z$) for longitudinally polarised vector bosons violates unitarity at $\sqrt{s} \sim 1.5$

TeV. In the Standard Model the cross section gets regularised by the Higgs contribution. At an e^+e^--linear collider this process is accessible with the sort of diagram shown in Fig. 1. However since the VV-centre of mass energy is usually much lower than the e^+e^--energy this process starts only to get interesting at $\sqrt{s} \sim 1$ TeV.

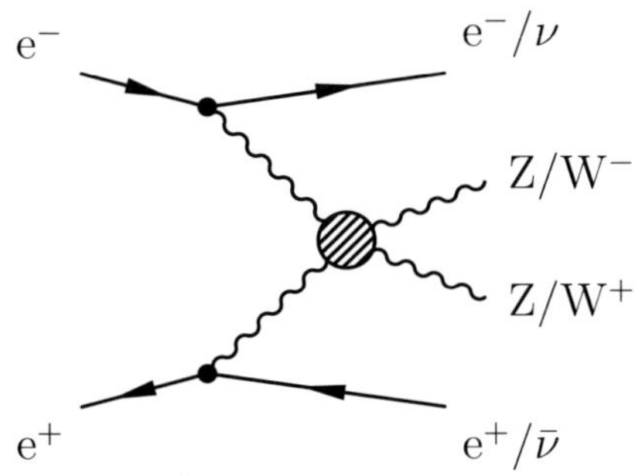

Fig. 1. Generic Feynman graph for vector-boson scattering in e^+e^--annihilation.

In a theory of strongly interacting gauge bosons the longitudinal degrees of freedom of the gauge bosons play the same role as the pions in QCD. In many theories also resonances corresponding to the ρ, visible in VV-scattering exist. In the same way, as for instance the ρ is seen in $e^+e^- \to \pi^+\pi^-$ also the pair production of longitudinal Ws is modified. Such an effect will be visible in a modification of the triple gauge couplings.

With a linear collider the physics outlined above can be tested in several ways [3, 4, 5]. In the e^+e^- mode, running on top of the Z-resonance the precision measurements done already at LEP and SLC can be repeated with much higher statistics and running close to the W-pair production threshold the W-mass can be measured with high precision. At high energies the triple gauge couplings can be tested as well as, at the highest energies, the strength of the vector boson scattering. In all cases the availability of polarised beams helps a lot and is completely indispensable for Z-pole running. In case indications for anomalous triple gauge-couplings are found, W-pair production in $\gamma\gamma$ collisions and single W production in $e\gamma$ collisions, which are both sensitive to the $WW\gamma$ vertex only help to understand the nature of the new couplings. Within the strongly interacting scenario the process $W^-W^- \to W^-W^-$ is accessible in e^-e^- collisions helping to disentangle the different models there.

2. Production of Gauge Bosons

Z-bosons can be produced singly in the s-channel (see Fig. 2). For $\sqrt{s} \approx m_Z$ this process is resonant and large samples can be produced. The total visible cross section is about 35nb so that billions of events can be recorded with high luminosity running at a linear collider.

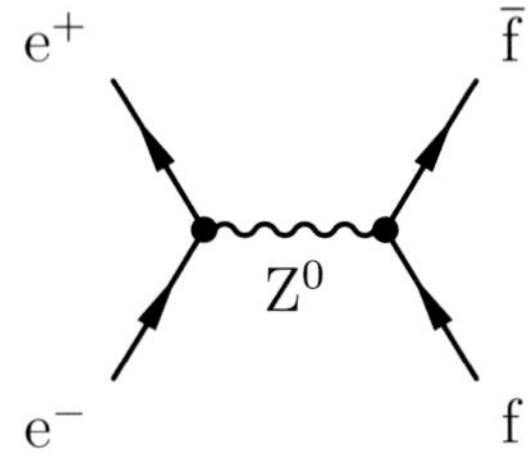

Fig. 2. Feynman graph for s-channel Z-production.

W-bosons are produced singly via γ-W fusion (see Fig. 3) or in pairs (see Fig. 4). Since both processes are non resonant the cross sections are only a few pb. Far above threshold, W-pair production falls like $1/s$ while single W production rises logarithmically. Figure 5 (a) shows the total cross section as a function of the centre of mass energy for both processes. At energies around $\sqrt{s} = 500$ GeV they are about equal.

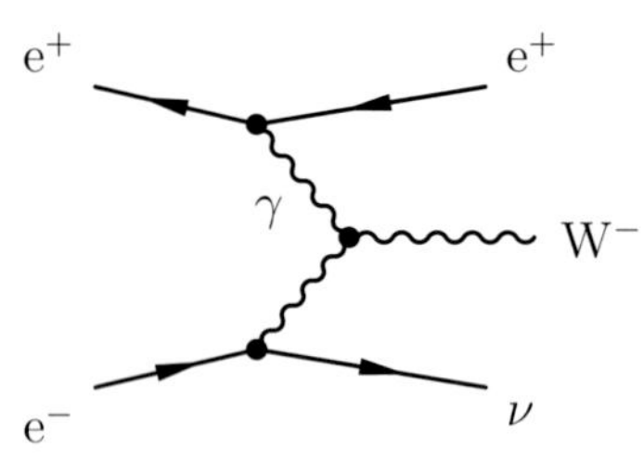

Fig. 3. Dominating Feynman graph for single W production in e^+e^--annihilation.

Since the width of the W is rather large, in principle it is not sufficient to compute the cross sections for stable Ws but instead the full four-fermion processes have to be calculated and the interferences between the graphs with resonant Ws and all the background diagrams have to be taken into account. These calculations exist and are used. At LEP-2 especially the

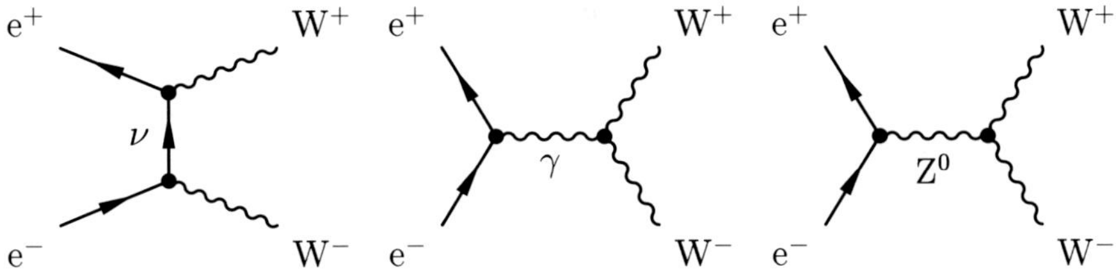

Fig. 4. Feynman graphs for the production of W-pairs in e^+e^--annihilation.

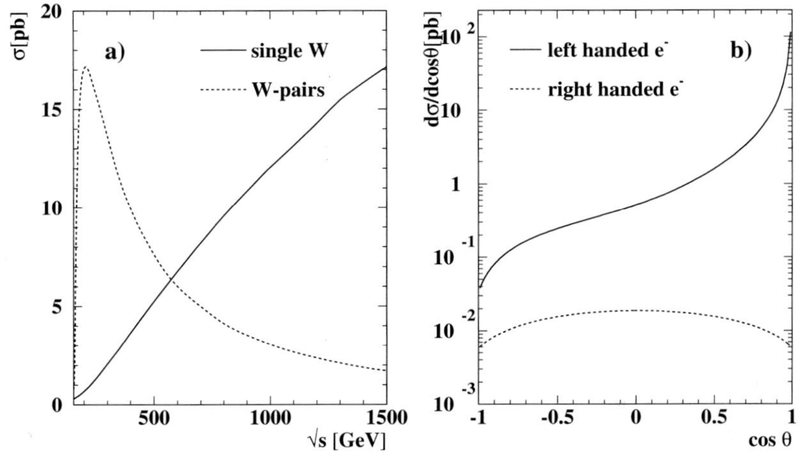

Fig. 5. (a): total cross section for single W [6] and W pair production [7] as a function of the centre of mass energy. (b): differential cross section for W-pair production for different beam polarisation.

interference between single W production and W-pair production with one W decaying into an electron and a neutrino turns out to be non negligible. At a linear collider, however the average momentum of the W and the available phase space are so large that with appropriate cuts the interferences and the contributions from the background diagrams are almost negligible.

The radiative corrections to W-pair production are large, so that good control of them is needed. A full one loop correction to the four fermion process is very complicated due to the huge number of diagrams involved and does not exist yet. However, above threshold it is sufficient to take the full one loop corrections for resonant Ws into account which is done in the double pole approximation [7, 8]. The present status of the radiative corrections is summarised in Ref. [9].

Due to the $(V-A)$ nature of the charged current couplings, only left-handed electrons and right-handed positrons couple to W-bosons. Single W^- production can thus be switched off or doubled by polarising the electron beam in the right way while single W^+ production can be modified accordingly by positron polarisation. For the same reason the t-channel diagram for W-pair production can be switched off by polarising one beam or enhanced by a factor of two or four polarising one or both beams. For energies that are much higher than the weak boson masses, the combined Z and γ exchange can be replaced by the neutral member of the W weak isospin triplet, because the orthogonal combination corresponding to the weak hypercharge boson does not couple to the $W^\pm$. Therefore the coupling to the electrons and positrons is also purely $(V-A)$ at high energies. Figure 5 b) shows the differential cross section for W-pair production for the two electron helicities at $\sqrt{s} = 500$ GeV. Already at linear collider energies, the cross section for right-handed electrons is suppressed by at least a factor of ten relative to left-handed electrons for all polar angles.

3. Properties of Gauge Bosons

3.1. *Standard Model Predictions*

As already said, the gauge sector of the Standard Model of electroweak interactions is determined by three parameters. Conveniently the three best measured parameters are used to define the theory and the other observables are then predicted as functions of those three. Typically the quantities used are [1]

- the electromagnetic fine structure constant at zero momentum transfer α, known to a precision of 5×10^{-8};
- the Fermi constant in μ-decays, known to 10^{-5};
- the mass of the Z-boson, known to 2×10^{-5}.

On Born level all couplings and the mass of the W can then be predicted in terms of these parameters and the partial and total widths of the W and Z can be derived from the couplings including small corrections due to fermion masses and due to the Kobayashi Maskawa Matrix. The consistency of the model on this level is known since long and the present interest is to test the theory at higher orders in perturbation theory. If loop corrections are included the predictions get sensitive to all parameters of the theory and also the properties of particles, that are too heavy to be produced directly, like the top-quark or the Higgs boson, modify the predictions. In a similar way also parameters of a theory beyond the Standard Model can alter these predictions, especially if the structure of the theory is such that the parameters connected to the new heavy particles do not decouple. All observables that can be measured with high precision can be expressed with three types of parameters, the mass of the W (m_W) and the vector- and axial-vector-coupling of a fermion, f, to the Z (v_f, a_f). On Born level one has $\cos\theta_W = m_W/m_Z$, $a_f = 1$ and $v_f/a_f = 1 - 4Q_f \sin^2\theta_W$.

The partial width of the Z decaying into $f\overline{f}$ is given by

$$\Gamma_{ff} = \frac{N_C^f G_F m_Z^3}{6\pi\sqrt{2}} \left(v_f^2 + a_f^2\right),$$

where the colour-factor N_C^f is 1 for leptons and 3 for quarks. Apart from corrections due to photon exchange, γ-Z-interference and initial state radiation, the total cross section close to the Z-pole can then be written as

$$\sigma_{\mathrm{f}}(s) = \frac{12\pi\Gamma_{ee}\Gamma_{ff}}{m_Z^2} \frac{s}{\left(s - m_Z^2\right)^2 + \frac{s^2}{m_Z^2}\Gamma_Z^2}$$

so that experimentally the Z-mass and total and partial widths can be obtained from a scan around the Z resonance. Ratios of partial widths can be measured from cross section ratios on the peak.

$\sin^2\theta_W$ can be measured with several asymmetries, all sensitive to the combination $\mathcal{A}_f = \frac{2v_f a_f}{v_f^2 + a_f^2}$: • the forward-backward-asymmetry with unpolarised beams,[a]

$$A_{\mathrm{FB}}^f = \frac{\sigma(\cos\theta > 0) - \sigma(\cos\theta < 0)}{\sigma_{\mathrm{tot}}} = \frac{3}{4}\mathcal{A}_e\mathcal{A}_f;$$

• the τ -polarisation and its forward-backward asymmetry for unpolarised beams

$$< \mathcal{P}_\tau > = \mathcal{A}_\tau,\ \mathcal{P}_\tau^{FB} = \mathcal{A}_e;$$

• the left-right-asymmetry

$$A_{\mathrm{LR}} = \frac{1}{\mathcal{P}} \frac{\sigma_L - \sigma_R}{\sigma_L + \sigma_R} = \mathcal{A}_e,$$

independent of the final state ($\sigma_{L,R}$ denotes the cross section for left/right-handed polarised electrons and $\mathcal{P}$ the beam polarisation); • the left-right-forward-backward-asymmetry

[a] All SM predictions given here are corrected for photon exchange, γ-Z-interference and initial state radiation.

$$A^f_{\rm LR,FB} = \frac{1}{\mathcal{P}} \frac{\left[\sigma^f_L(\cos\theta > 0) - \sigma^f_L(\cos\theta < 0)\right] - \left[\sigma^f_R(\cos\theta > 0) - \sigma^f_R(\cos\theta < 0)\right]}{\sigma^f_L + \sigma^f_R} = \mathcal{A}_f.$$

The W-mass is measured with several reconstruction techniques in $e^+e^- \to W^+W^-$ in the continuum above threshold and in $p\overline{p} \to WX$ and from the cross section in $e^+e^- \to W^+W^-$ near threshold.

Usually the deviations from the Born level prediction are written as

$$\begin{aligned} m_W^2 &= \frac{1}{2} m_Z^2 \left(1 + \sqrt{1 - \frac{4\pi\alpha}{\sqrt{2} G_F m_Z^2} \frac{1}{1 - \Delta r}}\right) \\ a_f &= \sqrt{1 + \Delta\rho_{\rm f}} \\ v_f/a_f &= 1 - 4Q_f \sin^2\theta^f_{\rm eff}, \end{aligned}$$

where Δr and $\Delta\rho_{\rm f}$ are small corrections and $\theta^f_{\rm eff}$ is the effective weak mixing angle.

In the Standard Model and in most extensions, where new physics appears only in loops all Z-couplings apart from the b-quark can be expressed in terms of the Z-couplings to charged leptons with small corrections depending only on the light fermion masses. Since the b-quark is the isospin partner of the very heavy top-quark some additional vertex corrections appear that are naturally enhanced by the top mass especially if the couplings are proportional to the fermion mass, like Higgs-couplings. In some models where new physics modifies already the Born-level predictions, like in models where the Z and a heavy Z' mix this need not be true.

In general there are two sorts of interesting corrections from heavy particles. The first class of corrections stems from the mass splitting within a weak isospin doublet. These corrections are proportional to the squared mass difference within the doublet and led already to the successful prediction of the top mass by LEP before its discovery. The second class are logarithmic corrections which are non-zero also for exact isospin symmetry. These corrections are responsible for the present predictions on the Higgs mass. The numerically largest corrections in most processes, however, are due to the running of the electromagnetic coupling to the Z-scale. α gets modified by about 10% due to light fermion loops introducing significant uncertainties in the predictions. The leptonic loops can be calculated reliably, however the hadronic loops are largely affected by QCD and bear a significant uncertainty.

In addition all observables with hadronic final states have to be corrected for virtual and real gluon radiation. As an example the hadronic decay width of the Z has the same QCD correction as the cross section $e^+e^- \rightarrow$ hadrons, $\Gamma_{\rm had} = \Gamma_{\rm had}^{(\rm no\ QCD)}(1 + \alpha_s(m_Z^2)/\pi + \cdots)$.

Frequently reparameterisations of the radiative correction parameters are used where the large isospin-breaking corrections are absorbed into one parameter, so that the others depend only on the logarithmic ones. One example are the so called ε parameters [10]

$$\begin{aligned}
\Delta\rho_\ell &= \varepsilon_1 \\
\sin^2\theta_{\rm eff}^\ell &= \frac{1}{2}\left(1 - \sqrt{1 - \frac{4\pi\alpha(m_Z^2)}{\sqrt{2}G_F m_Z^2}}\right)(1 - 1.43\varepsilon_1 + 1.86\varepsilon_3) \\
\frac{m_W^2}{m_Z^2} &= \frac{1}{2}\left(1 + \sqrt{1 - \frac{4\pi\alpha(m_Z^2)}{\sqrt{2}G_F m_Z^2}}\right)(1 + 1.43\varepsilon_1 - 1.00\varepsilon_2 - 0.86\varepsilon_3) \\
\Delta\rho_b - \Delta\rho_d &= 2\varepsilon_b
\end{aligned}$$

In this parameterisation ε_1 absorbs the large isospin-splitting corrections, ε_3 contains the logarithmic m_H dependence while ε_2 is almost constant in the Standard Model and most extensions. ε_b parameterises the additional corrections to the Zbb vertex.

The other set of parameters that is often used are the STU parameters [11]. The two parameterisations are basically equivalent [1]. In the STU-parameters the Standard Model expectations are subtracted and they are multiplied by factors of order $1/\alpha$.

3.2. *Status at Present Colliders*

During the first phase of LEP each of the four experiments has collected about four million Z-decays at energies close to the Z-resonance and SLD has recorded half a million Z-events with an average e^--polarisation of about 75%. In a second phase the LEP-experiments have collected about 10000 W-pairs each, from which the W-mass has been measured and at the TEVATRON about 80000 leptonic W-decays in $p\overline{p}$-collisions have been seen. In addition the TEVATRON has measured the top-mass to about 5 GeV precision. A summary of all results can be found in [2]. Table 1 summarises the present electroweak measurements and compares them to the Standard Model prediction after a fit to these data. The overall agreement is good. The two largest deviations are $A_{\rm LR}$ from SLD and $A_{\rm FB}^b$ from LEP. If interpreted in terms of $\sin^2\theta_{\rm eff}^\ell$, fixing $\mathcal{A}_b$ to its Standard Model prediction the discrepancy between the two measurements is 2.9σ. If on the contrary $\mathcal{A}_b$ is calculated from these two measurements plus the other measurements of $\sin^2\theta^\ell_{\rm eff}$ at LEP and the left-right-forward-backward asymmetry for b-quarks at SLD $\mathcal{A}_b$ deviates 2.6σ from the prediction. Since this is the largest deviation one can construct from the data[b] it can well be a statistical fluctuation.

Table 1. Results of the precision tests at LEP, SLD and the TEVATRON.

	Measurement with Total Error	Systematic Error	Standard Model fit	Pull
(a) LEP line-shape and lepton asymmetries:				
m_Z [GeV]	91.1875 ± 0.0021	0.0017	91.1875	0.0
Γ_Z [GeV]	2.4952 ± 0.0023	0.0012	2.4961	−0.4
$\sigma_0^{\rm had}$ [nb]	41.540 ± 0.037	0.028	41.480	1.6
R_ℓ	20.767 ± 0.025	0.007	20.741	1.0
$A_{\rm FB}^{0,\ell}$	0.0171 ± 0.0010	0.0003	0.0165	0.7
τ polarisation:				
$\mathcal{A}_\ell$	0.1465 ± 0.0033	0.0010	0.1483	−0.6
$q\bar{q}$ charge asym.:				
$\sin^2\theta^\ell_{\rm eff}$ ($\langle Q_{\rm FB}\rangle$)	0.2324 ± 0.0012	0.0010	0.2314	0.9
(b) SLD				
$\mathcal{A}_\ell$	0.1513 ± 0.00021	0.00010	0.1483	1.5
(c) LEP and SLD Heavy Flavour				
R_b^0	0.21644 ± 0.00065	0.00053	0.21578	1.0
R_c^0	0.1718 ± 0.0031	0.0022	0.1723	−0.1
$A_{\rm FB}^{b,0}$	0.0995 ± 0.0017	0.0009	0.1040	−2.6
$A_{\rm FB}^{c,0}$	0.0713 ± 0.0036	0.0017	0.0743	−0.8
$\mathcal{A}_b$	0.922 ± 0.020	0.016	0.935	−0.6
$\mathcal{A}_c$	0.670 ± 0.026	0.016	0.668	0.1
(d) $p\bar{p}$ and LEP II				
m_W [GeV]	80.449 ± 0.034		80.394	1.6
m_t [GeV] ($p\bar{p}$)	174.3 ± 5.1	4.0	174.3	0.0

As two examples, these data can be used to constrain the Higgs mass to be less than 190 GeV within the Standard Model or to exclude certain technicolour models (Fig. 6).

[b] A 3σ discrepancy between a $\sin^2\theta$ in νN scattering and the Standard Model prediction will be ignored here, since it is not relevant for linear collider physics and possible theoretical uncertainties are currently under discussion.

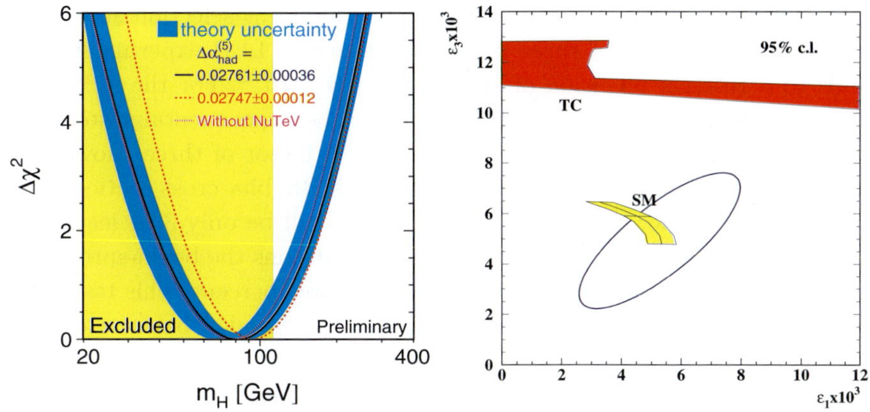

Fig. 6. left: $\Delta\chi^2$ of the electroweak fit as a function of the Higgs mass. right: ε-parameters obtained from the LEP and SLD data compared to the Standard Model and to a technicolour model.

3.3. *Prospects for the Linear Collider*

A linear collider will be able to run with a luminosity of $\mathcal{L} \sim 5 \times 10^{33}\,\mathrm{cm}^{-2}\mathrm{s}^{-1}$ at energies close to the Z-pole [12]. With this luminosity it is possible to collect 10^9 Z-decays in about 70 days of running. A similar luminosity should be possible close to the WW-threshold. In addition both beams have to be polarised with polarisations of $\mathcal{P}_{e^-} = \pm 80\%$ and $\mathcal{P}_{e^+} = \pm(40-60)\%$. This corresponds to an effective polarisation of

$$\mathcal{P}_{\mathrm{eff}} = \frac{\mathcal{P}_{e^+} + \mathcal{P}_{e^-}}{1 + \mathcal{P}_{e^+}\mathcal{P}_{e^-}} \approx 91 - 95\%.$$

Additionally, the polarisation should be switchable randomly from train to train.

The parameters extracted from the Z-lineshape are already today largely systematics limited. For that reason only a modest improvement can be expected. Expressed in the minimally correlated observables m_Z, Γ_Z, $\sigma_0^{\mathrm{had}} = \frac{12\pi}{m_Z^2}\frac{\Gamma_e\Gamma_{\mathrm{had}}}{\Gamma_Z^2}$, $R_\ell = \frac{\Gamma_{\mathrm{had}}}{\Gamma_l}$ the precision of all scan observables is about 0.1%, apart from m_Z which is known to 2×10^{-5}. At a linear collider it will probably not be possible to get an absolute calibration of the beam energy to the same level as it was possible at LEP. So m_Z will serve as calibration of the energy scale and thus cannot be improved.

However a relative calibration of the energy scale to 10^{-5} seems realistic, so that $\Delta\Gamma_Z = 1$ MeV is within reach. With the much better detector planned in the linear collider designs [13] and the higher statistics to perform cross checks the systematics on the event selection should get improved by a factor of three relative to the best LEP experiment. This leads to a relative precision of $\Delta R_\ell/R_\ell = 0.3 \times 10^{-3}$. For the $\sigma_0^{\rm had}$ measurement the absolute luminosity is needed. The experimental systematics on this measurement might also decrease by a factor of three. However if the theoretical uncertainty of the small angle Bhabha cross section stays at its present value, the improvement in $\sigma_0^{\rm had}$ will be only 30% leading to $\Delta\sigma_0^{\rm had}/\sigma_0^{\rm had} = 0.6 \times 10^{-3}$. To achieve these precisions the beam-spread and the beamstrahlung need to be understood to a few percent. This task looks difficult but possible, however further studies are needed before a definite statement can be made.

The possible improvements in $\alpha_s(m_Z^2)$, $\Delta\rho_\ell$ and the number of light neutrino species (N_ν), which can be seen as a general parameter for extra contributions to the invisible Z width are summarised in Table 2. Typical improvements are a factor of two to three.

Table 2. Possible improvements in the physics quantities derived from the Z-lineshape after high luminosity Z-running at TESLA. For α_s and $\Delta\rho$, $N_\nu = 3$ is assumed.

	LEP [2]	LC
$\alpha_s(m_Z^2)$	0.1181 ± 0.0027	± 0.0009
$\Delta\rho_\ell$	$(0.50 \pm 0.10) \cdot 10^{-2}$	$\pm 0.05 \cdot 10^{-2}$
N_ν	2.984 ± 0.008	± 0.004

Large improvements are expected for the asymmetries which in general have low systematic errors, especially when polarised beams are involved. From the gain in statistics and effective polarisation relative to SLD [14] a factor of 50 smaller statistical error can be expected. For $A_{\rm LR}$ this corresponds to a relative error of 2×10^{-4}. At present the relative error due to polarimetry is 0.5% and it cannot be expected that a polarimeter at a linear collider will be much better than the one at SLD. The error decreases immediately by a factor of three when positron polarisation is available with the assumed value due to the error propagation from the single beam polarisations to the effective polarisation, but this is still by far not enough. However, the required precision can be obtained with the so called Blondel scheme [15].

In general the cross section $e^+e^- \rightarrow Z \rightarrow f\overline{f}$ for polarised beams can be written as

$$\sigma = \sigma_u \left[1 - \mathcal{P}_{e^+}\mathcal{P}_{e^-} + A_{\rm LR}(\mathcal{P}_{e^+} - \mathcal{P}_{e^-})\right]. \tag{1}$$

Measuring all four possible helicity combinations $A_{\rm LR}$ can be obtained without explicit knowledge of the polarisation:

$$A_{\rm LR} = \sqrt{\frac{(\sigma_{++} + \sigma_{-+} - \sigma_{+-} - \sigma_{--})(-\sigma_{++} + \sigma_{-+} - \sigma_{+-} + \sigma_{--})}{(\sigma_{++} + \sigma_{-+} + \sigma_{+-} + \sigma_{--})(-\sigma_{++} + \sigma_{-+} + \sigma_{+-} - \sigma_{--})}}, \tag{2}$$

where in σ_{ij} i denotes the sign of the positron- and j the sign of the electron-polarisation. From an error analysis it turns out that only 10% of the luminosity needs to be spent on the small cross sections so that little luminosity is lost for other measurements and the statistical error increases only slightly due to the Blondel scheme.

In Eq. (2) it is assumed that the absolute values of the polarisation for the positive and negative helicity states are equal. To verify this, or to get the relevant corrections, polarimeters like the one used at SLD are still needed. However the absolute calibration of the polarimeter analysing power, which is the dominant error in the polarimetry, drops out in the measurement. One possible source of uncertainty in this difference measurement could be a difference in luminosity of the electron-laser interaction for the two laser polarisations. This can be overcome by using two polarimeter channels with different analysing power.

The statistical error on $A_{\rm LR}$ with 10^9 Z-decays will be $\Delta A_{\rm LR} = 3 \times 10^{-5}$. The variation of $A_{\rm LR}$ with the centre of mass energy is about

$$dA_{\rm LR}/d\sqrt{s} = 2 \times 10^{-2}/\ \mathrm{GeV},$$

so the beam energy relative to the Z-mass needs to be controlled to a precision of about 1 MeV. The amount of beamstrahlung expected in the high luminosity Z running changes $A_{\rm LR}$ by $\Delta A_{\rm LR} = 9 \times 10^{-4}$. It thus needs to be known to a precision of a few percent. If, however, the beamstrahlung in the Z-scan to calibrate the beam energy is the same as in the peak running it gets absorbed into an apparent shift of the calibration constants and basically no beamstrahl corrections for $A_{\rm LR}$ are needed.

It will be thus assumed that the linear collider can measure $A_{\rm LR}$ with a final precision of 10^{-4} corresponding to $\Delta \sin^2\theta^{\ell}_{\rm eff} = 0.000013$.

$\mathcal{A}_b$ can be measured from the left-right-forward-backward asymmetry with similar methods as the ones used by LEP and SLD [16]. If the lepton and the jetcharge methods from LEP are taken as a reference a statistical error of $\Delta\mathcal{A}_b \sim 5 \times 10^{-4}$ is possible for both methods. Not to be completely dominated by light quark background the excellent b-tagging capabilities of the LC-detector need to be exploited. For the jetcharge analysis a b-tagging with 99% purity is needed, which should be feasible with an efficiency of 75%. For the lepton-method a charm rejection of a factor of 50 is needed, which can be reached with 85 % efficiency.

For the lepton analysis the dominating error is then the statistical error from $b\bar{b}$-mixing ($\Delta\mathcal{A}_b(\mathrm{mix}) \approx 9 \times 10^{-4}$). This error is of purely statistical nature and cannot be reduced. For the jetcharge method light quark background and hemisphere correlations will both contribute to the systematic uncertainty with $\Delta\mathcal{A}_b \sim 10^{-3}$, so that, combining the two methods, a total error of $\Delta\mathcal{A}_b = 10^{-3}$ seems within reach.

Another Z-observable where significant progress can be achieved is $R_b^0 = \Gamma_{b\bar{b}}/\Gamma_{\mathrm{had}}$ which can be obtained with very small corrections from the cross section ratio $R_b = \sigma_{b\bar{b}}/\sigma_{\mathrm{had}}$. At LEP this quantity is mainly limited by the understanding of the charm background, the background from gluon splitting into $b\bar{b}$ and hemisphere correlations due to QCD effects. These correlations arise from the energy dependence of the b-tag, because a gluon, emitted at a large angle, takes energy from both jets. With the LC-detector the charm background can be suppressed by a factor of four relative to the LEP analyses, simultaneously reducing the statistical error by a factor of 20. The better b-tagging and increased statistics should also allow for a improvement in the measurement of the gluon splitting. The energy dependence in the b-tag comes from the large multiple scattering contribution to the impact parameter resolution in the LEP detectors. At the LC detector the losses are mainly due to the cut on the invariant mass of the reconstructed particles at the secondary vertex. Since the invariant mass is a Lorenz-invariant quantity the energy dependence and thus the hemisphere correlations should be much smaller. For R_b a total improvement by about a factor of five can be expected.

The W-mass is presently known with a precision of 34 MeV. However LHC should be able to improve its precision to about 15 MeV [17]. The linear collider in principle has two possibilities to measure the mass of the W, with a scan of the W-pair-production threshold or reconstructing Ws at higher energy similar to LEP. The phase space suppression close to threshold is $\propto \beta$ for the neutrino t-channel exchange while it is $\propto \beta^3$ for the s-channel photon and Z-exchange. For that reason mainly the well known $We\nu$-coupling enters in the prediction of the threshold cross section and the sensitivity to anomalous triple gauge couplings is very small.

t-channel neutrino exchange is only present for left-handed electrons and right-handed positrons. With 100% beam polarisation for both beams the cross section can thus be enhanced by a factor of four for the right helicity combination and already with one beam being polarised switched off completely. The left-right asymmetry for the background is much smaller, so that polarisation can be used to enhance the signal/background ratio for one polarisation state and basically switch off the signal to measure the background for the opposite one.

A possible W-threshold scan has been simulated for TESLA spending 100 fb^{-1} close to $\sqrt{s} = 161$ GeV corresponding to one year of running [18]. Efficiencies and backgrounds have been assumed to be the same as at LEP. With a total error of 0.25% on the luminosity and on the selection efficiencies m_W can be measured with a total precision of 6 MeV. If the efficiencies are left free in the fit, the error increases to only 7 MeV independent on the luminosity error. The achievable errors at the scan points are compared with the sensitivity to the W-mass in Fig. 7.

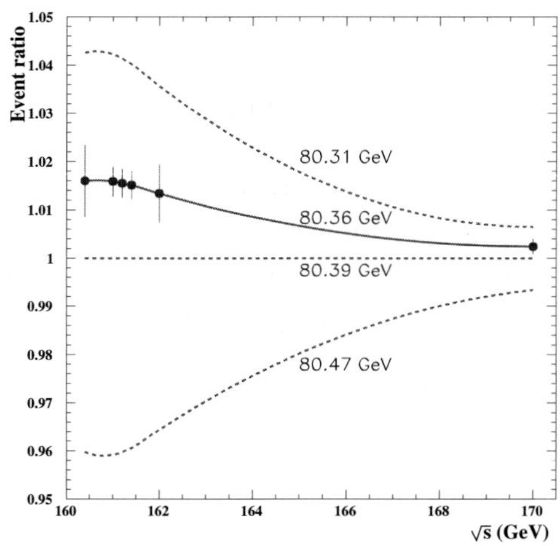

Fig. 7. Sensitivity of the W-pair threshold scan to the W-mass. The vertical axis shows the ratio of the cross section to the predicted cross section for $m_W = 80.39$ GeV. The error bars represent the expected errors for the scan described in the text.

With mass-reconstruction techniques using W-pair- and single-W-production at higher energies also a statistical error of a few MeV can be reached without dedicated luminosity. However it is not clear yet, if the systematics can be brought under control.

3.4. *Interpretation of the Precision Measurements*

To interpret the precision measurements one has to compare them to the model predictions at the loop level. In the theoretical prediction the uncertainties in all input parameters enter in the uncertainty of the prediction. The important parameters are:

• the running of α to the Z-mass. $\alpha(s)$ can be expressed as

$$\alpha(s)^{-1} = \Big(1 - \Delta\alpha_{\text{lep}}(s) - \Delta\alpha^{(5)}_{\text{had}}(s) - \Delta\alpha_{\text{top}}(s)\Big) \times \alpha^{-1}.$$

$\Delta\alpha_{\text{lep}}(s)$ represents the leptonic loops which can be calculated reliably, while the hadronic loops, represented by $\Delta\alpha^{(5)}_{\text{had}}(s)$ cause a large uncertainty. In principle they can be calculated from the cross section $e^+e^- \to$ hadrons for energies up to the Z-mass using the optical theorem. However this leads to large uncertainties and everybody agrees to use perturbative QCD starting at some energy. However the energy from which on QCD can be used is under debate at the moment [19]. If data are used until well above the resonance region one obtains from the low energy data [20, 21]

$$\Delta\alpha^{(5)}_{\text{had}}(m^2_Z) = (279.0 \pm 4.0) \times 10^{-4}$$

corresponding to an uncertainty in the Standard Model prediction of $\sin^2\theta^\ell_{\text{eff}}$ of 0.00014 and 7 MeV in m_W. However, the sensitivity to the details of the resonance region can be reduced significantly, if the low energy data is used to fit the coefficients of a QCD operator product expansion instead of integrating the total cross section. If the hadronic cross section is known to 1% up to the Υ-resonances the uncertainties are $\Delta\sin^2\theta^\ell_{\text{eff}} = 0.000017$ and $\Delta m_W = 1$ MeV [21].

• A top mass error of 1 GeV contributes an uncertainty of 0.000032 to the $\sin^2\theta^\ell_{\text{eff}}$ prediction and 6 MeV to m_W. At TESLA it should be possible to measure m_t to ~ 100 MeV, so that the top contribution to both observables will be negligible.

• A Z-mass error of 2 MeV results in a 0.000014 uncertainty of the $\sin^2\theta^\ell_{\text{eff}}$ prediction, about the same size as the experimental error and the uncertainty from $\alpha(m_Z)$. For m_W the direct uncertainty is 2.5 MeV. However, the linear collider calibrates for all methods the W-mass with the Z-mass, so that the relevant observable is m_W/m_Z. In this case the error is smaller by a factor of three. A 10% change in the Higgs mass will modify $\sin^2\theta^\ell_{\text{eff}}$ by 0.000031 and m_W by 6 MeV. When the Higgs is found by the time the measurements are done its mass will be known to better than 1% and thus not contribute to the uncertainty in the prediction. If the Higgs will not be found the data can be used to predict its mass to 5% accuracy. As an illustration Fig. 8 compares the Higgs mass fit of the LEP electroweak working group with the projected one after the LC precision measurements.

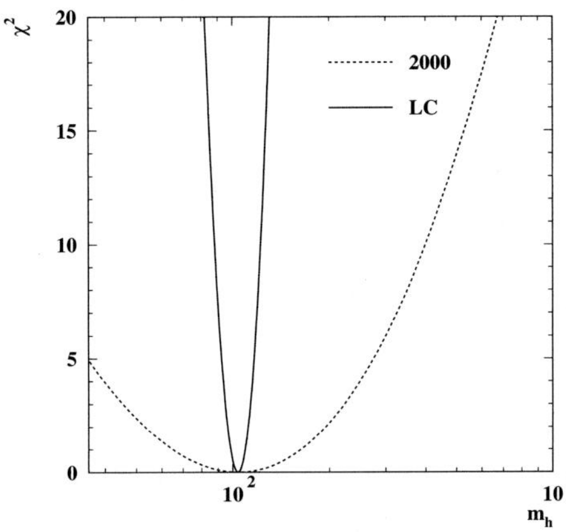

Fig. 8. $\Delta\chi^2$ as a function of the Higgs mass for the electroweak precision data now and after the linear collider Z-factory running.

The precision measurements can also be used to constrain extensions of the Standard Model. As an example from the MSSM Fig. 9 [22] shows the area in $m_A - \tan\beta$ that can be obtained from these data after certain assumptions on other SUSY-parameters, mainly affecting the stop-sector.

Figure 10 shows the precision data in the $\varepsilon_1 - \varepsilon_3$ plane for the present data and the LC-expectation with and without m_W. The data are compared to the SM expectation with $m_t = 174$ GeV and 100 GeV $< m_H <$ 1 TeV. Within the Standard Model the Higgs is already now significantly constrained because its trajectory is almost orthogonal to the long axis of the ellipse. Within many extensions of the SM, however, it is easy to generate an arbitrary ε_1 so that it is possible for basically any m_H to bring the prediction back into the ellipse. With the precision of the linear collider, especially including m_W, the Higgs mass will be tightly constrained without any ambiguity from ε_1.

As an example of an application of the model independent analysis, Fig. 11 shows in the S-T plane the predictions from the 2-Higgs-Doublet Model (2HDM) for cases, where a light Higgs exists but cannot be seen,

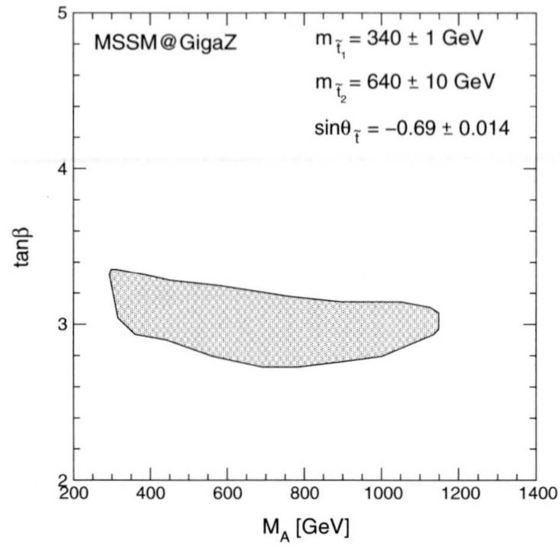

Fig. 9. The region in the $m_A - \tan\beta$ plane, allowed by $1\,\sigma$ errors by the measurements of m_W and $\sin^2\theta^\ell_{\rm eff}$: $m_W = 80.40$ GeV, $\sin^2\theta^\ell_{\rm eff} = 0.23138$, and by an LC measurement of m_H: $m_H = 110$ GeV. The other SUSY parameters including their uncertainties are given by $A_b = -640 \pm 60$ GeV, $\mu = 316 \pm 1$ GeV, $M_2 = 152 \pm 2$ GeV and $m_{\tilde{g}} = 496 \pm 10$ GeV.

compared to the present data and the expectations from a linear collider [23]. Only a linear collider can distinguish between the Standard Model with a light Higgs and the 2HDM. One can also see, that for these sort of analyses the good precision in the W-mass is needed.

In summary, the properties of gauge bosons, namely the mass of the W and the fermion couplings of the Z will be highly sensitive to loop corrections. They will thus tightly constrain the SM, in case no new physics is found until then or help to measure the parameters of the new theory, if new phenomena are seen.

4. Measurements of the CKM Matrix

The CKM matrix connects the weak eigenstates of down type quarks (D'_1) to their mass eigenstates (D_i) [1]:

$$D'_i = \sum_j V_{ij} D_j$$

The matrix has to be unitary to avoid flavour changing neutral currents on tree level, which have not been seen experimentally. For three flavours it

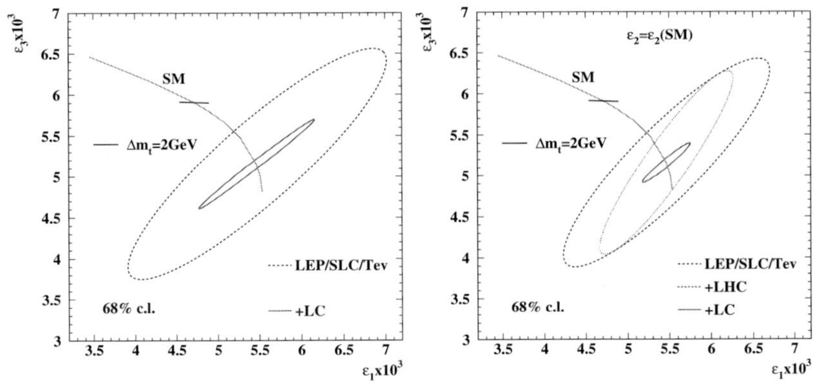

Fig. 10. Sensitivity $\varepsilon_1 - \varepsilon_3$ for the present data and the expectation at a linear collider. In the left plot ε_2 has been left free in the fit, equivalent to ignoring m_W in the $\varepsilon_1 - \varepsilon_3$ fit. In the right plot ε_2 has been fixed to its SM-value. The line marked "SM" shows the SM prediction with $m_t = 174$ GeV and varying m_H from 70 GeV (lower end) to 1 TeV (upper end). The effect of an uncertainty in m_t of 2 GeV is indicated. $\Delta m_t = 100$ MeV, expected from TESLA is inside the SM line width. For better comparability all central values gave been assumed to be on the SM prediction for $m_H = 100$ GeV.

can be parameterised by three angles and one complex phase, introducing CP violation. The coupling of an up-type quark U_i and a down-type quark D_j to the W-boson, normalised to the leptonic coupling of the W can thus be written as

$$g(U_i, D_j, W) = V_{ij} g(\ell, \nu, W).$$

Traditionally the CKM matrix elements are measured in weak meson decays. The absolute values of the elements can be measured from corresponding decay rates. Normally large event samples are available so that statistical errors are very small. However the hadronic matrix elements for the weak transitions are quite uncertain and the errors from this source are totally dominant. The phases of the elements only show up in interference patterns and are much harder to measure. A linear collider can contribute in three ways to the measurement of the CKM matrix elements:

- The elements V_{ti} can be accessed in top decays. This is discussed in chapter 3.
- The CP violating phases can be studied in B-decays in the Giga-Z running. The methods are essentially the same as at the e^+e^- B-factories and the B-physics experiments at hadron machines, namely BTeV and LHCb and the expected precision is typically in between the lepton- and

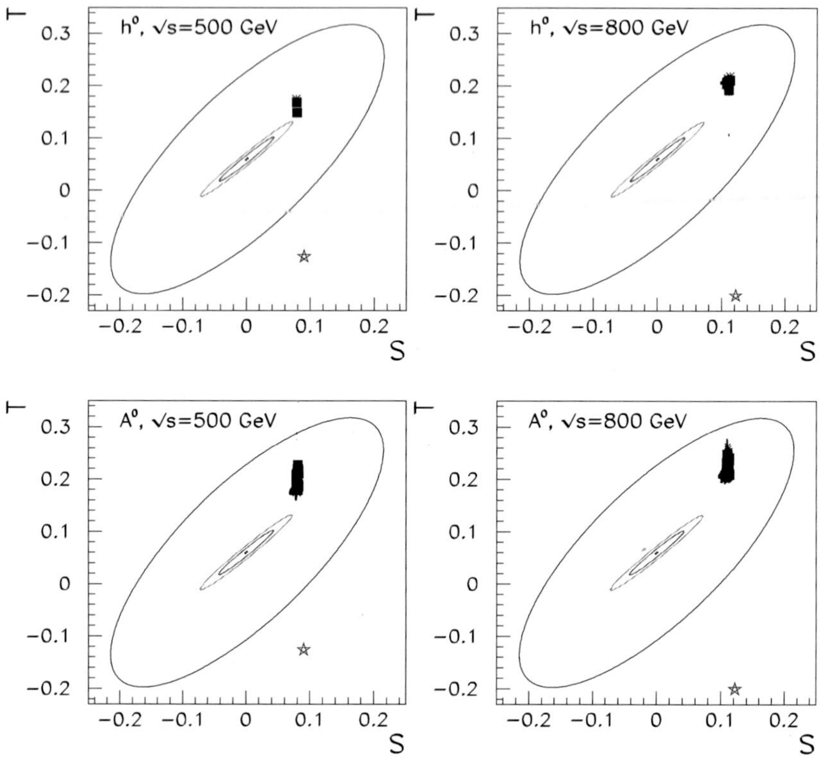

Fig. 11. Prediction for S and T from the 2 Higgs doublet model with a light Higgs for the cases where no Higgs is found compared to the current electroweak data and the projection for Giga-Z. The star denotes the Standard Model prediction if the Higgs mass is $\sqrt{s} - 100$ GeV.

the hadron machines. A detailed discussion can be found in Refs. [24, 25, 26].

• The decay width of a W into a quark pair depends on the corresponding matrix element:

$$\Gamma(W^+ \to U_i\overline{D_j}) = |V_{ij}|^2\Gamma(W^+ \to \ell\overline{\nu})(1 + \mathcal{C}_{\rm QCD}),$$

where $\mathcal{C}_{\rm QCD}$ parameterises QCD corrections. This method will be described in some detail in the following [27]. The experimental problem in the measurement of the CKM matrix elements from W decays is the tagging of the quark flavours. For b- and c-quarks there exist well understood methods, tagging the decays of long lived b- and c-flavoured hadrons with the microvertex detector. Already at LEP and SLD these algorithms have been used successfully in many analyses and they are essential in Higgs physics at a linear collider. For light quarks no high efficiency, high purity methods exist. However it is known that the leading hadron in a jet often contains the primary quark so that it can be used for flavour tagging. The Z decay widths to b- and c-quarks and to the sum of all quarks are known with very good precision from LEP and SLD and they agree well with the Standard Model prediction. It is thus reasonable to assume that also the partial width to u-, d- and s-quarks agree with the prediction to the same precision. The tagging purities and efficiencies for all quarks in Z-decays using the microvertex methods for b- and c-quarks and leading hadrons for light quarks can thus be measured accurately in Z-running at the linear collider. Fragmentation properties depend only logarithmically on the energy scale. The extrapolation from the Z-mass scale to the W-mass scale can thus be done safely using the usual parton shower fragmentation models. Typically the efficiencies are 2% larger for the Ws. The only complication comes from the fact that, due to isospin invariance, the probability for a primary u- or d-quark to end up in a charged pion is about the same, making the fit to the Z-data unstable. However, because of charge conservation the W always has to decay in one up- and one down-type (anti-)quark. If, as an example, one quark from a W is tagged as charm the other one can be a down-quark, but not an up-quark. So the problem can be solved if the Z- and the W-data are fitted simultaneously with the tagging efficiencies and the W-branching ratios as free parameters, without introducing additional model dependence. Table 3 shows the obtainable efficiencies for light quarks using $\pi^\pm$, $K^\pm$, K^0_s, p and Λ.

Table 3. Light quark tagging efficiencies of leading hadrons with $x_p > 0.2$.

	d	u	s
$\pi^\pm$	0.209	0.209	0.130
$K^\pm$	0.056	0.074	0.122
p	0.015	0.025	0.020
K^0_S	0.007	0.006	0.019
Λ	0.003	0.003	0.008

For the analysis of the CKM-matrix all events containing W bosons can be used. Especially attractive, however, are single W events. The W in these events has on average a very small momentum so that the typical jet energies are close to those in Z-pole events. So in these events the extrapolation of the detector efficiencies, which vary strongly with energy, especially for the dE/dx measurement, is very small. Table 4 shows the sensitivity of a linear collider with 1500 fb^{-1} integrated luminosity compared to present and expected accuracy.

Table 4. Current and projected uncertainty of the CKM matrix elements compared to the LC measurement from W-decays.

	current uncertainty	projected other	LC
$\|V_{ud}\|$	$\pm$ 0.0008		$\pm$0.0028
$\|V_{us}\|$	$\pm$0.0023		$\pm$0.0124
$\|V_{ub}\|$	$\sim \pm$0.008	$\pm$0.0004	$\pm$0.011
$\|V_{cd}\|$	$\pm$0.0016		$\pm$0.0072
$\|V_{cs}\|$	$\pm$0.01		$\pm$0.0017
$\|V_{cb}\|$	$\pm$0.0019	$\pm$0.0012	$\pm$0.0011

A large gain can be obtained in V_{cs}. However, if the unitarity of the CKM matrix is imposed, the present error in this element shrinks by more than one order of magnitude. Also in V_{cb} some progress is possible. This element is important for the interpretation of the CP-violation results in B-decays. If it is determined in B-decays the error is totally dominated by the theoretical uncertainty on the hadronic matrix elements, while in the determination from W-decays the error is statistical. The independent cross check is thus highly desirable.

5. Interactions amongst Gauge Bosons

The Standard Model predicts trilinear interactions between WWZ and $WW\gamma$, however there should be no interactions between neutral gauge bosons.

Single W production and W-pair production are sensitive to the triple gauge-boson couplings (TGC). In W-pair production the cross sections for neutrino t-channel exchange and γ,Z-s-channel exchange would rise individually with energy violating unitarity at some point. As demonstrated in Fig. 12 this rise is cancelled by the interference term, so that the resulting cross section shows the usual $1/s$ behaviour. For this reason W-pair production is extremely sensitive to the TGCs. To preserve the gauge cancellations, however, any anomalous coupling has to vanish for $\sqrt{s} \to \infty$. For the linear collider studies using W-pairs in principle this is not a problem, since the couplings are measured at a well defined scale. In the hadron collider studies the running of the couplings with the energy is often parameterised by a form factor $x' = \frac{x}{(1+s/\Lambda^2)^n}$ with $n > 0.5$ or $n > 1$ depending on the type of coupling [17]. Λ can be interpreted as the scale where the new physics, responsible for the anomalous couplings sets in, so that the effective parameterisation becomes meaningless above this scale.

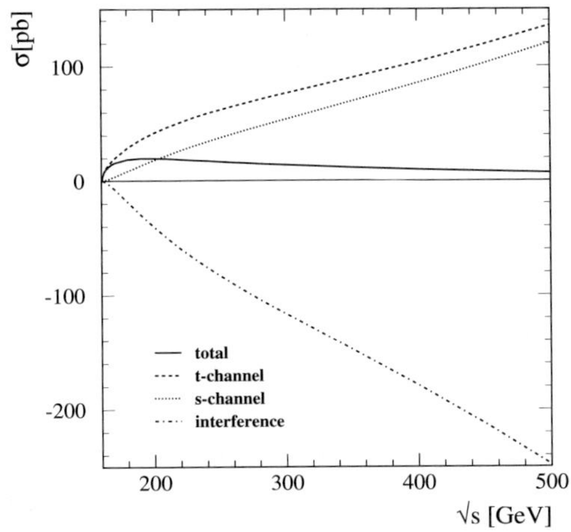

Fig. 12. W-pair production cross section for the different production processes separately, the interference and for the sum of all.

Contrary to W-pair production the effective scale of single W production is always much lower than the centre of mass energy. The flux functions for radiating a W or a photon off the initial state electrons, peak at low energy so that the produced W is on average almost at rest.

In models where the gauge interactions remain weak they can be parameterised with a linear Lagrangian. Keeping terms with dimension up to six the most general Lagrangian for WWV $(V = Z, \gamma)$ is

$$\begin{aligned} i\mathcal{L}_{eff}^{WWV} &= g_{WWV} \cdot [g_1^V V^\mu \left(W_{\mu\nu}^- W^{+\nu} - W_{\mu\nu}^+ W^{-\nu}\right) + \kappa_V W_\mu^+ W_\nu^- V^{\mu\nu} \\ &+ \frac{\lambda_V}{m_W^2} V^{\mu\nu} W_\nu^{+\rho} W_{\rho\mu}^- + i g_5^V \epsilon_{\mu\nu\rho\sigma} \left(\left(\partial^\rho W^{-\mu}\right) W^{+\nu}\right. \\ &\left. - W^{-\mu} \left(\partial^\rho W^{+\nu}\right)\right) V^\sigma + i g_4^V W_\mu^- W_\nu^+ \left(\partial^\mu V^\nu + \partial^\nu V^\mu\right) \\ &- \frac{\tilde{\kappa}_V}{2} W_\mu^- W_\nu^+ \epsilon^{\mu\nu\rho\sigma} V_{\rho\sigma} - \frac{\tilde{\lambda}_V}{2m_W^2} W_{\rho\mu}^- W^{+\mu}{}_\nu \epsilon^{\nu\rho\alpha\beta} V_{\alpha\beta}] \end{aligned}$$

with $V = \gamma$, Z, $g_{WW\gamma} = e$, $g_{WWZ} = e \cot\theta_W$ and $V_{\mu\nu} = \partial_\mu V_\nu - \partial_\nu V_\mu$.

Electromagnetic gauge invariance requires $g_1^\gamma(q^2 = 0) = 1$ and $g_5^\gamma(q^2 = 0) = 0$. In the Standard Model one has $g_1^V = \kappa_V = 1$. All other couplings are equal to zero.

In term of these parameters the magnetic dipole moment of the W is given by $\mu_W = \frac{e}{2m_W}(1 + \kappa_\gamma + \lambda_\gamma)$ and the electric quadrupole-moment by $q_W = -\frac{e}{m_W^2}(\kappa_\gamma - \lambda_\gamma)$.

From the terms in the effective Lagrangian the ones multiplied by g_1, κ and λ are C- and P-conserving, the one proportional to g_5 violates C and P, however is CP-conserving and the ones with g_4, $\tilde{\kappa}$ and $\tilde{\lambda}$ are CP-violating.

It is expected that deviations from the Standard Model should show up first in the C,P conserving couplings. At an e^+e^--collider the process that is by far most sensitive to triple gauge couplings is W-pair production. However, if no beam polarisation is available, it is not possible to separate $WW\gamma$ and WWZ couplings with this process. If one requires that the anomalous couplings are C- and P-symmetric, preserve $SU(2) \times U(1)$ symmetry without adding operators of higher dimension and are not excluded by the LEP1 precision data one is left with the three couplings [28]:

$$\begin{aligned} &\Delta g_1^Z \\ &\Delta \kappa_\gamma = -\frac{\cos^2\theta_W}{\sin^2\theta_W}\left(\Delta\kappa_Z - \Delta g_1^Z\right) \\ &\lambda_\gamma = \lambda_Z \end{aligned} \tag{3}$$

Since the initial state couplings $e^+e^-\gamma$ and e^+e^-Z depend differently on the beam polarisation, polarised beams can be used to separate $WW\gamma$ and WWZ couplings, as demonstrated in Fig. 13. With enough statistics it is thus possible to measure the five C,P-conserving couplings simultaneously. The C,P-violating couplings can be separated easily by constructing C,P-odd observables.

If the triple gauge couplings are modified by 1-loop corrections one expects deviations of the order $g^2/16\pi^2 \approx 2.7 \times 10^{-3}$. As an example Fig. 14 shows the expected effects on κ_γ from loop corrections in the MSSM [29].

5.1. *Experimental Procedures*

In $e^+e^- \to W^+W^-$ anomalous gauge couplings show up in a modification of the differential cross section with respect to the scattering angle, where the modification depends on the W polarisation state. The W-polarisation is accessible from the decay angles of the W-decay products, so that in total five observables per event are available (see Fig. 15):

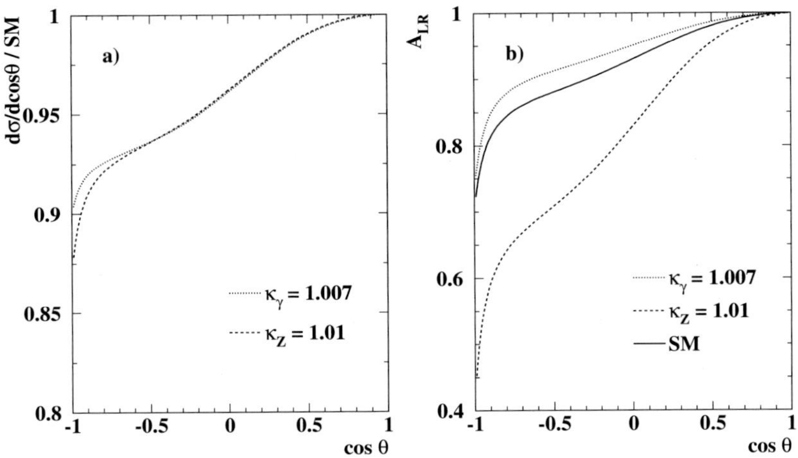

Fig. 13. Ratio of the differential cross section for W-pair production to the Standard Model prediction (a) and left-right asymmetry for this process (b) as a function of the W-production angle for anomalous κ_γ or κ_Z.

- the W production angle Θ;
- the decay angles θ^* of the fermions with respect to the W flight direction in the W rest frame, sensitive to the longitudinal polarisation;
- the decay angles ϕ^* of the fermions in the W-beam plane, sensitive to transverse polarisations.

About 46% of the W-pairs decay into four quarks, 43% into two quarks a charged lepton and a neutrino and 11% into two charged leptons and two neutrinos. Neglecting initial state radiation and beamstrahlung the Ws are exactly back to back. In this approximation the axis of the W production in the mixed decays is measured from the decay products of the hadronically decaying W and the sign ambiguity is resolved by the charge of the lepton. For the leptonically decaying W both decay angles can be measured uniquely. For the hadronically decaying W the quark and the antiquark cannot be distinguished in general, so that the ambiguity $(\theta^*, \phi^*) \rightarrow (\pi - \theta^*, \phi^* + \pi)$ remains. If the leptonically decaying W decays into $\tau\nu_\tau$ the resolution is worse, due to the additional neutrinos, but the measurement is still possible. For the fully hadronic W-pairs, even if the jet-pairing is done correctly, the decay angle ambiguities remain for both Ws and in addition the W^+ and the W^- cannot be separated, so that these events are only of very limited use. The ambiguities can be solved partly by jetcharge techniques or c-tagging with some c-charge determination, but

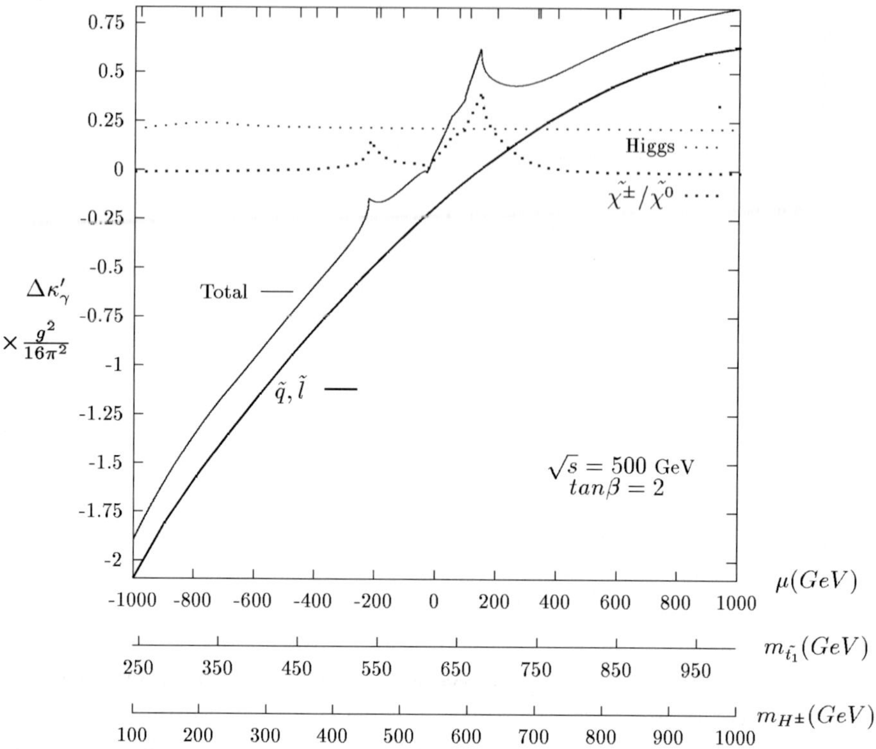

Fig. 14. Expected effects on κ_γ from MSSM loop corrections.

these methods have only a limited efficiency. If both W-bosons decay leptonically the full information can be reconstructed with a twofold ambiguity, if for the lepton-neutrino pairs the W mass is imposed. This works, however, only if no τ is involved, which is in about half of the events.

To analyse the data a binned analysis in a five-dimensional space is quite impractical. To circumvent this problem several methods are in use instead [28]:

Unbinned maximum likelihood fit: In general an unbinned maximum likelihood fit is the statistically most powerful method. However it turns out that it is very difficult to apply corrections for detector effects, backgrounds etc., so that this fit has not been used for gauge coupling analyses on real data up to now.

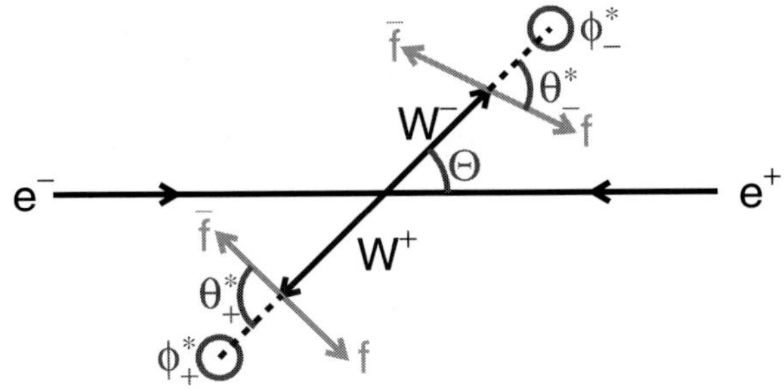

Fig. 15. Observables available in a W-pair event.

Optimal observable methods: If the differential cross section $d\sigma(q_1, ...q_m)/d\Omega$ $(\Omega = (\theta_1, ...\theta_n))$ is linear in the m parameters, q_j, to be determined, i.e. the differential cross section can be written as

$$\frac{d\sigma}{d\Omega} = S_0(\Omega) + \sum_j S_{1,j}(\Omega) q_j$$

one can construct an optimal observable $\mathcal{O}_j(\Omega) = S_{1,j}(\Omega)/S_0(\Omega)$ for which it can be shown that the q_j can be extracted in a statistically optimal way from the $\langle \mathcal{O}_j \rangle$ [30, 31]. If instead the distributions of the $\mathcal{O}_j(\Omega)$ are fitted, effects due to non linearity are accounted for automatically and for detector effects one can correct in the usual way. The $\mathcal{O}_j(\Omega)$ is just no longer exactly optimal, but when the non-linearity and detector corrections are small this is usually negligible. For $m < n$ one can thus reduce the number of parameters without a loss in statistical precision.

Spin density matrix: The full information of the W pair cross section can be expressed as the product of the differential cross section $\partial\sigma/\partial\cos\Theta$ and the spin density matrix

$$\rho_{\tau_-\tau'_-\tau_+\tau'_+}(\cos\Theta) = \frac{\mathcal{F}^{(\lambda)}_{\tau_-\tau_+}\left(\mathcal{F}^{(\lambda)}_{\tau'_-\tau'_+}\right)^*}{\sum_{\tau_+\tau_-}\left|\mathcal{F}^{(\lambda)}_{\tau_-\tau_+}\right|^2}, \tag{4}$$

where $\mathcal{F}^{(\lambda)}_{\tau_-\tau_+}$ is the amplitude for W^-W^+-pairs with W polarisations τ_- and τ_+, respectively, and an initial electron polarisations of λ.[c] In total this

[c]For unpolarised beams one has to sum over λ.

system has 81 density matrix elements where one is given by the normalisation

$$\sum_{\tau_+\tau_-} \rho_{\tau_-\tau_-\tau_+\tau_+} = 1.$$

This number reduces to 35 elements in a CP conserving theory, as is described in Ref. [32]. Only little information is lost if the spin density matrix for single Ws is used in the analysis, neglecting spin correlations:

$$\rho_{\tau_-\tau'_-}(\cos\Theta) = \sum_{\tau_+} \rho_{\tau_-\tau'_-\tau_+\tau_+}.$$

$\rho_{\tau\tau'}$ is hermitian, thus having 6 independent matrix elements. The diagonal elements, which are real, can be interpreted as the probability to find a W with helicity τ. The imaginary parts of the off-diagonal elements vanish if CP is conserved.

Experimentally $\rho_{\tau\tau'}$ can be obtained from

$$\rho_{\tau\tau'}\frac{\partial\sigma}{\partial\cos\Theta} = \int \frac{\partial^3\sigma}{\partial\cos\Theta\partial\cos\theta^*\partial\phi^*}\Lambda_{\tau\tau'}(\theta^*,\phi^*)d\cos\theta^*\,d\phi^*$$

The $\Lambda_{\tau\tau'}$ are projection operators for the different helicity states, for example $\Lambda_{00} = 2 - 5\cos^2\theta^*$. In this way the initial 5-dimensional parameter space can be reduced to 6 one-dimensional distributions.

5.2. *Results at LEP and the Tevatron*

Triple gauge couplings have already been measured at LEP2 and at the Tevatron. At LEP2 the measurements use the same processes as a linear collider will use. Since there one is closer to threshold the scales for W-pair and single W production are not so much different. However the single W production cross section is much smaller, so that also at LEP the dominant information comes from pair production. In $p\overline{p}$-collisions the most sensitive process is $W\gamma$-pair production, where the high p_t photon is radiated off a W. Also WZ- and WW-pairs are analysed, however with a lower sensitivity than the $W\gamma$-pairs. $p\overline{p}$-collisions are thus mainly sensitive to $WW\gamma$ couplings. In general all measurements agree well with the Standard Model prediction with a precision of a few percent, so that in no realistic model a deviation is expected. As an example figure 16 shows the results of the three parameter fit in the $(\Delta g_1^Z, -\Delta\kappa_\gamma)$ and $(\Delta g_1^Z - \lambda_\gamma)$ plane[2].

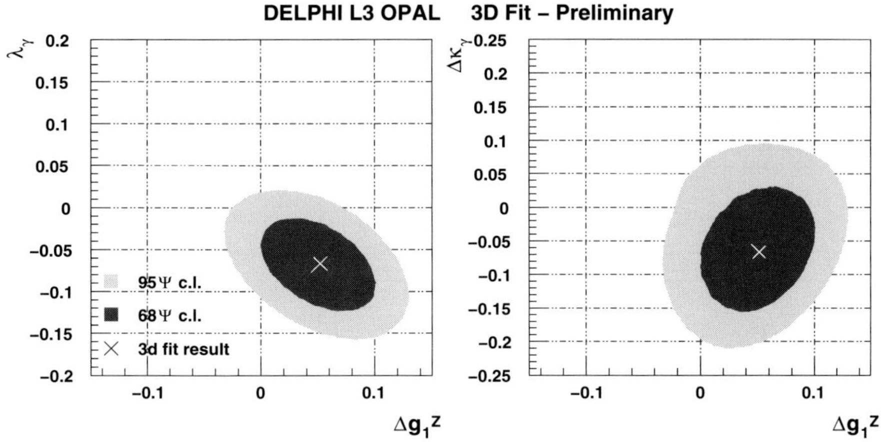

Fig. 16. Result of the 3-parameter fit to the triple gauge couplings at DELPHI, L3 and OPAL.

5.3. *Expectation from the Linear Collider*

For the linear collider studies the spin density matrix formalism has been used which produces close to optimal results [33]. Fits have been performed for several numbers of free parameters with and without beam polarisation. As already said the symmetries of the couplings manifest themselves in the spin density matrix elements that are non-zero, so that the sets of couplings with different symmetries are basically measured without correlations between the sets. Table 5 shows the results of the different single parameter fits including the C or P violating couplings for 500 fb^{-1} at $\sqrt{s} = 500$ GeV and 1000 fb^{-1} at $\sqrt{s} = 800$ GeV. For both cases an electron polarisation of $\pm 80\%$ and a positron polarisation of $\pm 60\%$ is assumed. Figure 17 shows the results of the five-parameter fit for $\sqrt{s} = 800$ GeV. Only the combinations with large correlations are shown.

Due to the large boost the resolution of the W-production angle is much better than at LEP. For the same reason the resolution in the decay angles is somewhat worse, however only sin and cos functions have to be distinguished here, so that the resolution is largely sufficient. In total the corrections due to detector effects are almost negligible and all experimental systematics are small. The radiative corrections however need to be known to significantly better than 1%. The beam polarisations can be measured from an extension of the Blondel scheme. The large forward peak of the cross section is dominated by neutrino t-channel exchange, so that its left-

Table 5. Expected uncertainties of the single parameter fits to the different triple gauge couplings. For $\sqrt{s} = 500$ GeV $\mathcal{L} = 500$ fb^{-1} and for $\sqrt{s} = 800$ GeV $\mathcal{L} = 1000$ fb^{-1} has been assumed. For both energies $\mathcal{P}_{e^-} = 80\%$ and $\mathcal{P}_{e^+} = 60\%$ has been used.

coupling	$\sqrt{s} = 500$ GeV	$\sqrt{s} = 800$ GeV
C,P-conserving, $SU(2) \times U(1)$ relations:		
Δg_1^Z	2.8	1.8
$\Delta\kappa_\gamma$	3.1	1.9
λ_γ	4.3	2.6
C,P-conserving, no relations:		
Δg_1^Z	15.5	12.6
$\Delta\kappa_\gamma$	3.3	1.9
λ_γ	5.9	3.3
$\Delta\kappa_Z$	3.2	1.9
λ_Z	6.7	3.0
not C or P conserving:		
g_5^Z	16.5	14.4
g_4^Z	45.9	18.3
$\tilde{\kappa}_Z$	39.0	14.3
$\tilde{\lambda}_Z$	7.5	3.0

right asymmetry is very close to one, independent of the triple couplings. This region can be used to measure the electron polarisation directly from data, even if no positron polarisation is available.

Without beam polarisation only the fits without separating $WW\gamma$ and WWZ couplings are possible. For these fits the errors increase by about a factor of two.

The precision of κ_γ for $\sqrt{s} = 500$ GeV for example is $0.12 \times (g^2/16\pi^2)$. If this is compared to the expected loop corrections from the MSSM, shown in Fig. 14 one sees, that the gauge couplings can probe extensions of the Standard Model with sensitivities similar to the Z-fermion couplings.

Figure 18 compares the expected errors on κ_γ and λ_γ at the linear collider with those from other machines. Especially for κ_γ the precision is much better than at the LHC. Since the dimension of the corresponding operator is lower for κ than for λ the suppression due to the high scale of new physics should be smaller. This parameter should thus be more sensitive to new effects for example from strong electroweak symmetry breaking.

Also the possibility to search for the neutral triple gauge couplings $Z\gamma\gamma$ and $ZZ\gamma$ has been studied [34]. However the possible limits are about an order of magnitude worse than predictions from loop effects in the Standard

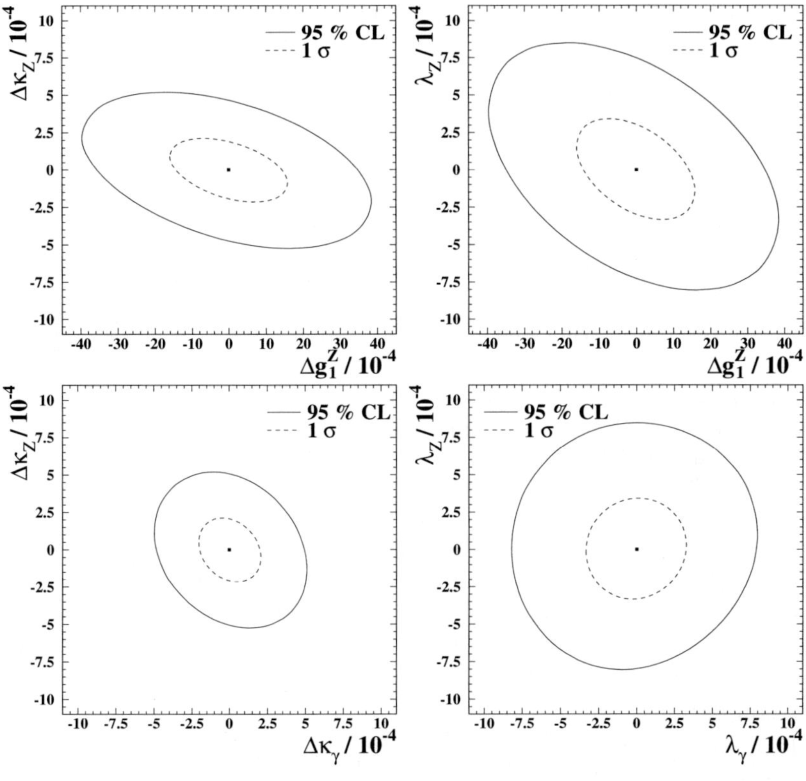

Fig. 17. 1σ and 95% c.l. (2D) contours for Δg_1^Z-$\Delta\kappa_Z$, Δg_1^Z-λ_Z, $\Delta\kappa_\gamma$-$\Delta\kappa_Z$ and λ_γ-λ_Z in the 5-parameter fit ($\sqrt{s} = 800$ GeV $\mathcal{L} = 1000$ fb^{-1}, $\mathcal{P}_{e^-} = 0.8$, $\mathcal{P}_{e^+} = 0.6$). For the combinations not shown the correlations are small.

Model and expectations from the LHC. For this reason they will not be discussed further.

6. Strong Electroweak Symmetry Breaking

If no Higgs boson exists electroweak interactions become strong at high energy and WW scattering violates unitarity at $\sqrt{s_{WW}} = 1.2$ TeV. The latest at this energy new physics has thus to set in, but with precision measurements it should be visible already at lower energies.

Strong electroweak symmetry breaking has already been discussed in detail in Chapter 5, so only the features relevant to gauge boson physics will be repeated here. Without a Higgs the longitudinal degrees of freedom of the

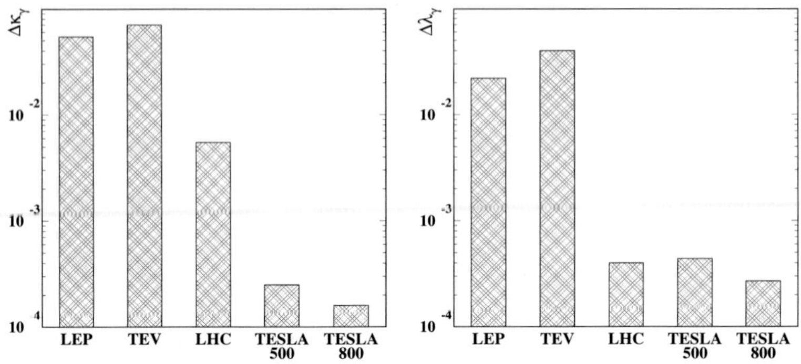

Fig. 18. Comparison of $\Delta\kappa_\gamma$ and $\Delta\lambda_\gamma$ at different machines. For LHC and TESLA three years of running are assumed (LHC: 300 fb^{-1}, TESLA $\sqrt{s} = 500$ GeV: 900 fb^{-1}, TESLA $\sqrt{s} = 800$ GeV: 1500 fb^{-1}).

W and the Z become the Goldstone bosons of a strongly interacting theory and the low energy equivalence theorem (LET) [35, 36] relates longitudinal vector boson scattering to pion scattering at low energy. Concrete models typically predict resonances like the ρ or the ω in QCD. However, as shown in Fig. 6 an exact copy of QCD is already excluded by the LEP/SLD precision data.

In a similar way as the ρ-resonance shows up in $e^+e^- \to \pi^+\pi^-$ also W-pair production should be sensitive to the strongly interacting scenario. In a model dependent analysis the amplitude for the production of two longitudinal Ws has been multiplied by a QCD like form factor assuming a resonance mass and width [37]. The LET is then the asymptotic value for high resonance masses. Figure 19 shows the expected sensitivity with $\sqrt{s} = 800$ GeV and $\mathcal{L} = 500$ fb^{-1}. Within the assumed model the LET and the SM model can clearly be distinguished and one is sensitive to resonance masses up to around 3 TeV.

To analyse the data in a model independent way effects from strong electroweak symmetry breaking can be parameterised in an effective Lagrangian. From LEP1/SLD it is known that the ρ-parameter is approximately one and the small deviations can fully be described by Standard Model loop corrections, mainly from the top quark. It is therefore reasonable to keep only terms that conserve the custodial $SU(2)_c$ symmetry. Under this assumption the effective Lagrangian for the triple gauge couplings,

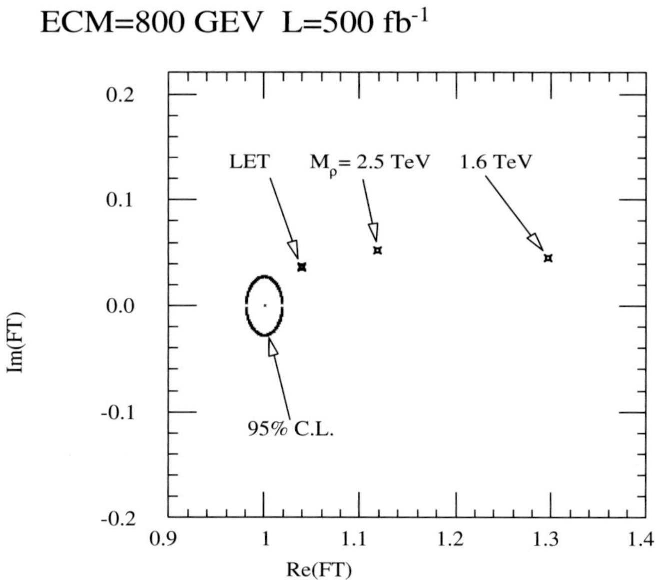

Fig. 19. Sensitivity to a ρ like resonance in W-pair production at $\sqrt{s} = 800$ GeV and $\mathcal{L} = 500\ \text{fb}^{-1}$.

keeping only terms of the lowest dimension, reads:

$$L_{TGC} = \frac{\alpha_1}{16\pi^2}\frac{gg'}{2}B_{\mu\nu}\,\mathrm{tr}\Big(\sigma_3 W^{\mu\nu}\Big) + \frac{\alpha_2}{16\pi^2}\mathrm{i}g'B_{\mu\nu}\,\mathrm{tr}\Big(\sigma_3 V^\mu V^\nu\Big) + \frac{\alpha_3}{16\pi^2}2\mathrm{i}g\,\mathrm{tr}\Big(W_{\mu\nu}V^\mu V^\nu\Big)$$

with $V_\mu = -\mathrm{i}g\frac{\sigma^i}{2}W_\mu^i + \mathrm{i}g'\frac{\sigma^3}{2}B_\mu$. From a naive dimensional analysis one expects

$$\frac{\alpha_i}{16\pi^2} = \left(\frac{v}{\Lambda_i^*}\right)^2$$

with $v = 246$ GeV and Λ_i^* being the scale of the new physics. From unitarity arguments one needs $\Lambda^* \sim 3$ TeV so that the α_i should be of $\mathcal{O}(1)$. The α_i

can be expressed in terms of g_1^Z, κ_γ, κ_Z as:

$$\begin{aligned}
g_1^Z &= 1 + \frac{e^2}{\cos^2\theta_W(\cos^2\theta_W - \sin^2\theta_W)}\frac{\alpha_1}{16\pi^2} + \frac{e^2}{\sin^2\theta_W\cos^2\theta_W}\frac{\alpha_3}{16\pi^2} \\
\kappa_\gamma &= 1 - \frac{e^2}{\sin^2\theta_W}\frac{\alpha_1}{16\pi^2} + \frac{e^2}{\sin^2\theta_W}\frac{\alpha_2}{16\pi^2} + \frac{e^2}{\sin^2\theta_W}\frac{\alpha_3}{16\pi^2} \\
\kappa_Z &= 1 + \frac{2e^2}{\cos^2\theta_W - \sin^2\theta_W}\frac{\alpha_1}{16\pi^2} - \frac{e^2}{\cos^2\theta_W}\frac{\alpha_2}{16\pi^2} + \frac{e^2}{\sin^2\theta_W}\frac{\alpha_3}{16\pi^2}.
\end{aligned} \tag{5}$$

Unfortunately Eq. 5 is singular in the blind direction

$$(\alpha_1,\ \alpha_2,\ \alpha_3)_{\rm blind} \propto (\cos^2\theta_W - \sin^2\theta_W,\ \cos^2\theta_W,\ -\sin^2\theta_W)$$

so that not all the α_i can be extracted from triple gauge couplings. However, α_1 can also be constrained from the Z-pole data. Figure 20 shows the expected errors in the $(\alpha_1, \alpha_2, \alpha_3)$ space for $\sqrt{s} = 800$ GeV $\mathcal{L} = 100\ \mathrm{fb}^{-1}, \mathcal{P}_{e^-} = 80\%$ and $\mathcal{P}_{e^+} = 60\%$ and from the Z-pole expectation at Giga-Z. Also the errors in (α_2, α_3) for $\alpha_1 = 0$ and for the combination of the high- and low-energy data are shown. The errors translate into mass scale limits of $\Lambda_i^* \approx 10$ TeV which are well above the unitarity limit of 3 TeV.

In the description of the quartic couplings with an effective Lagrangian there are two operators that conserve $SU(2)_c$ and that do not contribute already to the triple couplings:

$$\begin{aligned}
L_{QGC} = &\frac{\alpha_4}{16\pi^2}\,\mathrm{tr}\Big(V_\mu V_\nu\Big)\,\mathrm{tr}\Big(V^\mu V^\nu\Big) \\
&+ \frac{\alpha_5}{16\pi^2}\,\mathrm{tr}\Big(V_\mu V^\mu\Big)\,\mathrm{tr}\Big(V_\nu V^\nu\Big).
\end{aligned}$$

Experimentally the quartic couplings are measured through processes as sketched in Fig. 1. Since the electron coupling to the W is about a factor of two larger than the one to the Z, the sensitive processes have two neutrinos in the final state. This means that the gauge bosons in the final state have to be fully reconstructed so that decay modes involving neutrinos cannot be used.

In principle there are three processes sensitive to α_4 and α_5:

$$\begin{aligned}
e^+e^- &\to \nu\nu W^+W^- \\
e^+e^- &\to \nu\nu ZZ \\
e^-e^- &\to \nu\nu W^-W^-.
\end{aligned}$$

The sensitivity for the three processes is similar with different dependence on the two couplings. However, since the luminosity for e^-e^- is about an

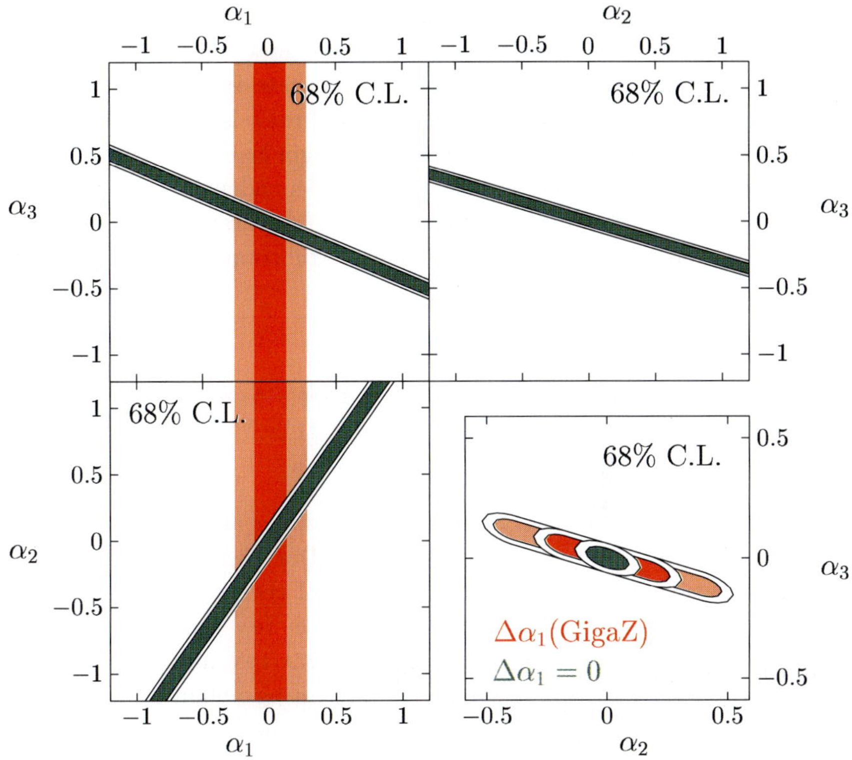

Fig. 20. Sensitivity to the effective couplings $\alpha_{1,2,3}$ from the measurement of the triple gauge couplings at a linear collider and from the Z-pole precision data. For the Z-pole constraint the outer region is without and the inner region with the accurate m_W measurement. In the lower right plot the $\alpha_2 - \alpha_3$ plane with $\alpha_1 = 0$ or with the Z-pole constraint is shown.

order of magnitude less than for e^+e^- the sensitivity per unit of running time in this channel is significantly smaller. A large potential background to these processes is $e^+e^- \rightarrow e^+e^-W^+W^-$ proceeding via intermediate photons. To reject this process it is important that electrons can be vetoed in the detector down to the lowest possible angles. If the electrons escape in the beampipe the total transverse momentum of the two-W system has to be very small. The $e^+e^- \rightarrow e^+e^-W^+W^-$ events can then be rejected by a cut on the total transverse momentum. Unfortunately such a cut suppresses also the longitudinal W-pairs more than the transverse ones. Another important background process is $e^+e^- \rightarrow e\nu WZ$. This background is in principle smaller than $e^+e^- \rightarrow e^+e^-W^+W^-$, but because of the missing neutrino the transverse momentum cut doesn't work. The process can only be separated from the signal if the energy flow resolution of the detector for hadronic jets is good enough to separate Ws and Zs. Another possibility to enhance the signal to background ratio is beam polarisation. Ws couple only to left-handed electrons and right-handed positrons while the electron-photon coupling is polarisation independent.

In a possible analysis signal events are first separated from the background and then analysed in terms of the quartic couplings. Variables sensitive to these couplings are:

- the VV ($V = W, Z$) centre of mass energy;
- the V production angle in the VV rest frame, to select hard VV-scattering;
- the V decay angles to select longitudinal Vs.

Figure 21 shows the sensitivity of the linear collider at $\sqrt{s} = 800$ GeV for $e^+e^- \to \nu\nu W^+W^-$ and $e^+e^- \to \nu\nu ZZ$ [38]. Due to the different dependence on the two parameters the combination of the two processes greatly reduces the error. The reach in Λ^* is $\Lambda_4^* \approx 1.8$ TeV and $\Lambda_5^* \approx 2.5$ TeV for the two parameter fit increasing to $\Lambda_4^* \approx 2.9$ TeV and $\Lambda_5^* \approx 4.9$ TeV for one-parameter fits, just reaching the interesting region of 3 TeV. The sensitivity increases strongly with the centre of mass energy. A factor of two in energy reduces the errors on $\alpha_{4,5}$ by almost an order of magnitude.

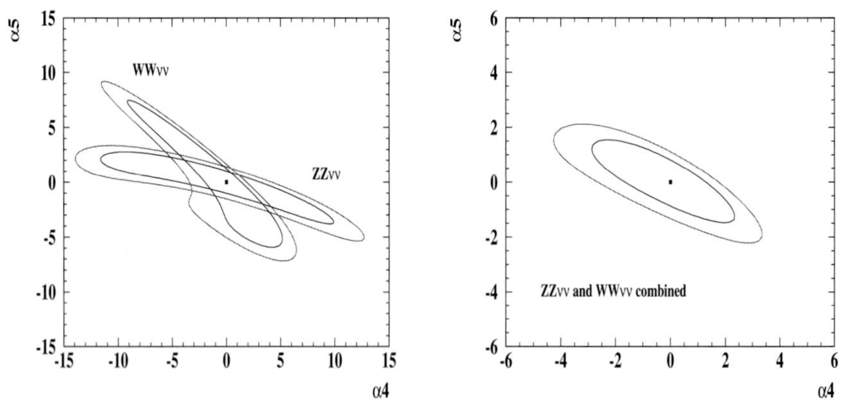

Fig. 21. Sensitivity to α_4 and α_5 at $\sqrt{s} = 800$ GeV from $e^+e^- \to \nu\nu W^+W^-$ and $e^+e^- \to \nu\nu ZZ$. The inner and outer contours represent 68% and 90% c.l. ($\mathcal{L} = 1000$ fb^{-1}, $\mathcal{P}_{e^-} = 0.8$, $\mathcal{P}_{e^+} = 0.4$).

The LHC can see resonances coupling to W-pairs up to masses of about 2 TeV [17]. If these resonances are vector-like a linear collider can easily measure their couplings in W-pair production (see *e.g.* Fig. 19). If they are scalar or tensor particles some information on their couplings can be obtained from the measurement of α_4 and α_5. However if the electroweak symmetry is really broken in a strongly interacting scenario finally a linear collider with few TeV centre of mass energy will be needed.

7. Conclusions

With a linear collider our knowledge of the gauge boson properties can be increased largely in many respects. Running on the Z-pole and the W-threshold allows significant progress in the precision of the observables that are most sensitive to electroweak loop corrections: the fermionic couplings of the Z and the mass of the W. These measurements test the consistency of the then-Standard Model on the 10^{-4} level or allow to estimate still unknown parameters for example in Supersymmetry. Also some progress in the measurement of the CKM matrix in W-decays is possible, helping in the understanding of CP-violation.

The most important progress, however, can be reached in the measurement of interactions amongst gauge bosons. The gauge interactions are a necessary feature of a non Abelian gauge group and the structure of triple gauge couplings can be checked with a precision of few$\times 10^{-4}$. If a light Higgs exists a completely new class of precision observables sensitive to loop corrections is accessible. The gauge coupling measurements, however, are especially interesting if no light Higgs exists. With the triple boson couplings one is sensitive to new physics that has to be present at the TeV scale to avoid violations of unitarity at these energies. In addition one can measure for the first time the scattering of gauge bosons. Since this process is the one that violates unitarity the earliest it is the most obvious place to look for strong electroweak symmetry breaking effects. The disadvantage of this process is that the effective WW centre of mass energy is significantly lower than the e^+e^- energy, but at $\sqrt{s} = 800$ GeV one should already be able to see effects from new physics if no light Higgs exists, although with smaller sensitivity than with the triple boson couplings. At higher energies this process will become ideal to see new resonances.

In summary, gauge boson physics is an essential part of the linear collider physics program. In a scenario with a light Higgs it will be indispensable to consolidate the model and to change its status from a working hypotheses to a physics theory. If no Higgs is seen the measurement of three and four boson interactions will open up a window to the physics of electroweak symmetry breaking which might not be directly visible at any next generation collider.

References

1. Particle Data Group, K. Hagiwara *et.al.*, Phys. Rev. **D66**, 010001 (2002).
2. The LEP collaborations, *A Combination of Preliminary Electroweak Measurements and Constraints on the Standard Model*, CERN-EP/2002-091, hep-ex/0212036.
3. J. A. Aguilar-Saavedra *et al.*, *TESLA Technical Design Report Part III: Physics at an* e^+e^- *Linear Collider*, DESY-01-011C.
4. T. Abe *et al.*, *Linear collider physics resource book for Snowmass 2001*, hep-ex/0106055 (part 1), hep-ex/0106056 (part 2), hep-ex/0106057 (part 3), and hep-ex/0106058 (part 4).
5. ACFA Linear Collider Working Group Collaboration, K. Abe *et al.*, *Particle physics experiments at JLC*, hep-ph/0109166.
6. A. Pukhov *et al.*, *CompHEP - a package for evaluation of Feynman diagrams and integration over multi-particle phase space. User's manual for version 33.*, hep-ph/9908288; E.Boos and M.Dubinin, *Single W-boson Production at Linear Colliders*, hep-ph/9909214.
7. S. Jadach *et al.*, Phys. Rev. **D61**, 113010 (2000).
8. A. Denner *et al.*, Phys. Lett. **B475**, 127 (2000); EPJdirect Vol. 2 **C4**, 1 (2000); Nucl. Phys. **B587**, 67 (2000).
9. M. Grünewald *et al.*, *Four Fermion Production in Electron Positron Collisions*, report of the *Four-fermion working group* of the *LEP2 Monte Carlo workshops*, CERN, 1999/2000, hep-ph/0005309.
10. G. Altarelli *et al.*, Phys. Lett. **B349**, 145 (1995).
11. M.E. Peskin, T. Takeuchi, Phys. Rev. **D46**, 381 (1992),.
12. N. Walker *TESLA luminosity at low centre of mass,* talk presented at the DESY/ECFA workshop Frascati, 8–10 November, 1998 `http://wwwsis.lnf.infn.it/ecfadesy/index.htm`..
13. G. Alexander *et al.*, *TESLA: The superconducting electron positron linear collider with an integrated X-ray laser laboratory. Technical design report. Part IV: A detector for TESLA*, DESY-01-011D.
14. P. C. Rowson, D. Su, and S. Willocq, Ann. Rev. Nucl. Part. Sci. **51**, 345 (2001).
15. A. Blondel, Phys. Lett. **B202**, 145 (1988).
16. K. Mönig, Rep. on Prog. in Phys. **61**, 8 (1998).
17. S. Haywood *et al.*, *Electroweak Physics*, CERN-2000-004, 117.
18. G. Wilson, *Precision Measurement of the W Mass with a Polarised Threshold Scan at a Linear Collider,* LC-PHSM-2001-009.
19. J. Erler, *Determinations of* $\alpha(m_Z)$*: Comparison and prospects*, hep-ph/0111005.
20. S. Eidelmann and F. Jegerlehner, Zeit. Phys. **C67**, 585 (1995).

21. F. Jegerlehner, *The effective fine structure constant at TESLA energies*, hep-ph/0105283.
22. J. Erler *et al.*, Phys. Lett. **B486**, 125 (2000).
23. J. Gunion, *Do Precision Electroweak Constraints Guarantee* e^+e^- *collider Discovery of at least one Higgs Boson of a Type II two Higgs Doublet Model?* talk given at LCWS2000, FNAL, October 24–28, 2000, hep-ph/0012199; P. Chankowski *et al.*, hep-ph/0009271.
24. P.F Harrison, H.R. Quinn (eds.) *The BaBar Physics Book*, SLAC-R-504.
25. P. Ball *et al.*, *B Decays*, CERN-2000-004, 305.
26. R. Hawkings, K. Mönig, EPJdirect **C8**, 1 (1999).
27. J. Letts, P. Mättig, *Direct Determination of the CKM Matrix from Decays of W Bosons and Top Quarks at High Energy* e^+e^- *Colliders*, LC-PHSM-2001-008.
28. G. Gounaris *et al.*, *Triple Gauge Boson Couplings* CERN-96-01, Vol. 1, 525.
29. A. Arhrib *et al.*, *Radiative Contributions to TGC in the MSSM* hep-ph/9603268.
30. D. Atwood and A. Soni, Phys. Rev. **D45**, 223 (1992).
31. M. Diehl, O. Nachtmann, Zeit. Phys. **C62**, 397 (1994).
32. M. Bilenky et al. Nucl. Phys. **B409**, 22 (1993).
33. W. Menges, *A Study of Charged Current Triple Gauge Couplings at TESLA*, LC-PHSM-2001-022.
34. J. Alcaraz, *Neutral Gauge Boson Pair Production at* e^+e^- *Linear Colliders*, Proceedings of the Worldwide Study on Physics and Experiments with Future Linear e^+e^--colliders, Sitges, April 1999, Vol. 1, 431.
35. M. Chanowitz and M. Gaillard, Nucl. Phys. **B261**, 379 (1985).
36. H. Veltman, Phys. Rev. **D43**, 2236 (1991).
37. T. L. Barklow, *Strong symmetry breaking at* e^+e^- *linear colliders*, in Proc. of the APS/DPF/DPB Summer Study on the Future of Particle Physics (Snowmass 2001) ed. R. Davidson and C. Quigg, hep-ph/0112286.
38. R. Chierici, S. Rosati, M. Kobel. *Strong electroweak symmetry breaking signals in WW scattering at TESLA*, LC-PHSM-2000-038.

CHAPTER 7

New Physics at the TeV Scale and Beyond

JoAnne L. Hewett
Stanford Linear Accelerator Center
2575 Sand Hill Rd
Menlo Park, CA 94025, U.S.A.

1. Introduction

Nature has presented us with many questions that have arisen from our past discoveries. What is the origin of mass? Do the forces unify? What is the Dark Matter and Dark Energy that permeates our universe? What is the origin of space and time? These questionscp appear to be related and point to the appearance of physics beyond the Standard Model at the TeV energy scale.

The discoveries of the last decade have both sparked these questions and provided evidence for the theoretical foundation which allows us to probe higher energy scales. The building of the Standard Model of the strong, electromagnetic, and weak interactions has been a remarkable achievement. The formulation of the theory as a renormalizable quantum theory preserves its predictive power beyond tree-level computations and allows for the probing of quantum effects. An array of experimental data confirm every feature of the theory to a high degree of precision, at the level of 0.1% in some cases, and thus tests the Standard Model at the level of higher order perturbation theory. In fact, at present, there are no compelling pieces of evidence that are in conflict with the Standard Model and it provides an accurate description of nature up to energies of order 100 GeV, corresponding to distances down to 10^{-16} cm.

Nonetheless, the fact that the Standard Model does not address several outstanding questions points to the existence of a more fundamental underlying theory of which the Standard Model is a low-energy manifestation.

Indeed, there are powerful theoretical arguments and questions which point to the existence of physics beyond the Standard Model: (i) gravity cannot be incorporated within the Standard Model, (ii) the origin of electroweak symmetry breaking is unproven, (iii) the mechanism which generates and stabilizes the hierarchy, *i.e.*, the sixteen orders of magnitude disparity between the electroweak scale at a TeV and the traditional scale of gravity defined by the Planck scale (10^{19} GeV), is unknown, (iv) the physics which generates the fermion masses and mixings remains a mystery, (v) neutrino masses and oscillations are not accommodated within the Standard Model, (vi) unification of the forces is incomplete, (vii) new CP, as well as baryon or lepton number, violating interactions are needed to explain the observed matter anti-matter asymmetry of the universe, and (viii) a viable dark matter candidate whose properties are in agreement with cosmological observations is not present in the Standard Model. These deficiencies of the Standard Model point towards the existence of a more fundamental underlying theory and a principal goal of particle physics during the next decade is to discover and elucidate this new theory.

The precise form that this new phenomena may take or the exact scale at which it will be unveiled cannot be predicted. However, there are strong indications that new physics will appear at the TeV scale. As discussed in a previous chapter, the nature of the dynamics of electroweak symmetry breaking must be revealed at the TeV scale. The full theory which gives rise to these dynamics needs to contain associated new interactions which explain why electroweak symmetry is broken at a TeV and how it is stablized. Standard Model loop contributions to the mass of the Higgs boson are quadratically divergent and drive the Higgs mass to unacceptably large values far above the TeV scale. In the absence of new interactions, a highly unnatural fine-tuning of the parameters must occur in order to cancel the large quantum corrections and retain a TeV scale Higgs boson. The presence of new physics at the TeV scale can also cancel, or possibly eliminate all together, these large corrections and avoid this fine-tuning. In addition, observations of the cosmic microwave background show that roughly 25% of the density of the universe is comprised of dark matter. If the Dark Matter has a component consisting of weakly interacting massive particles, then these particles must exist at the TeV scale in order to explain the observed density. Together these two arguments focus our attention to the TeV scale and demonstrate the necessity of a thorough exploration of the physics which takes place there.

Numerous ideas have been proposed to address the outstanding issues listed above and embed the Standard Model in a larger framework. An attractive and highly studied new theory is that of Supersymmetry, which stabilizes the electroweak scale, provides a natural dark matter candidate, and affords a mechanism for unification of the forces. It is the subject of another chapter. Here, we discuss the other possibilities which may resolve these questions. The ideas we examine are:

• Compositeness: This is a more historical approach in the notion that quarks and leptons are composite states built of more fundamental constituents.

• Extended Gauge Sectors: Many theories which unify the known forces contain new forces which reveal themselves at high energy scales with the existence of extra gauge bosons.

• Extra Spacetime Dimensions: The existence of additional spatial dimensions is required by string theory. Like Supersymmetry, theories of extra dimensions can also generate and stabilize the electroweak scale, unify the forces, and contain a Dark Matter particle candidate. However, they accomplish this via the geometry of a higher dimensional spacetime.

• Unconventional theories and possible surprises: New distinctive ideas about physics beyond the Standard Model are likely to appear in the future. We attempt to explore this unchartered territory by discussing collider tests of non-conventional possibilities arising from string theory in Section 7.

More generally, our limited imagination cannot span the full range of alternatives for new physics that might appear at the TeV scale and which is allowed by the current data set. We must prepare to discover the unexpected!

In this chapter, we demonstrate the ability of a Linear Collider operating at $0.5 - 1.0$ TeV to make fundamental progress in the illumination of new phenomena. The essential roles played by e^+e^- machines in this endeavor have a strong history. Lepton colliders have a good discovery potential and are complementary to hadron colliders operating at similar energy regions. The discovery of the gluon, charm, and tau sustain this assertion. Here, we show in what ways the Linear Collider adds to the search capability of the LHC in many scenarios. While most investigations of new physics effects at colliders focus on defining a search reach, the unmasking of the characteristics of the new phenomena is an equally important discovery. The precision of e^+e^- collisions offer the ideal tools with which to intensively study new phenomena, accurately determine its properties, and unravel the underlying theory. Here, we detail the diagnostic tests provided by a 0.5 – 1.0 TeV Linear Collider (LC) in various new physics scenarios. Together, the LC and LHC, will be crucial in the discovery of new physics, in the determination of its nature and how it relates to higher energy scales.

2. Contact Interactions and Compositeness

It is possible that the Standard Model particles are not elementary, but are composite structures built from smaller particles at a higher energy scale. This scenario has a strong historical basis, which is presently mirrored in the repetition of fermion generations or the large number of arbitrary parameters in the Standard Model. A composite nature for the boson fields would be revealed in non-standard values for the tri-linear gauge couplings discussed in Chapter 6. If the fermions have substructure, then their constituents are bound by a confining force at the scale Λ, which also characterizes the radius of the bound states. At energies above Λ, the composite nature of fermions would be revealed by the break-up of the bound states in hard $2 \to 2$ fermion scattering processes. At lower energies, deviations from the Standard Model may be observed in four fermion interactions via form factors or residual effective interactions induced by the binding force.

These composite remnants are usually parameterized by the introduction of contact terms in the low-energy Lagrangian. More generally, four fermion contact interactions represent a useful parameterization of many types of new interactions originating at high energy scales (above the direct kinematic reach of the collider), and specific cases will be discussed throughout this chapter. The lowest order four fermion contact terms are of dimension-6 and hence have dimensionful couplings proportional to g^2_{eff}/Λ^2. A general helicity conserving, flavor diagonal, invariant parameterization can be written as [1]

$$\mathcal{L} = \sum_{i,j=L,R} \frac{g^2_{eff}\eta_{ij}}{\Lambda_{ij}} \left(\overline{q}_i\gamma^\mu q_i + \mathcal{F}_\ell \overline{\ell}_i \gamma^\mu \ell_i\right) \left(\overline{q}_j \gamma_\mu q_j + \mathcal{F}_\ell \overline{\ell}_j \gamma_\mu \ell_j\right) , \qquad (1)$$

where the generation and color indices have been suppressed, the helicity coefficients $\eta_{ij} = \pm 1$, and $\mathcal{F}_\ell$ is inserted to allow for different quark and lepton couplings but is anticipated to be $\mathcal{O}(1)$. Since the binding force is expected to be strong when the momentum transfer Q^2 approaches Λ^2, it is conventional to define $g^2_{eff} = 4\pi$. L, R represent the usual chiral projections $(1 \mp \gamma_5)/2$, and the various choices for the chiral structure in Eq. (1) lead to different predictions for the angular distributions of the reactions where the contact terms contribute.

Since these contact interactions preserve chiral symmetry, the Standard Model matrix element for four fermion interactions, including fermion pair production at high energy colliders, is extended by an additional contact term. Interference between the contact terms and the usual gauge interactions can lead to observable deviations from Standard Model predictions at energies lower than Λ. Currents limits from various $2 \rightarrow 2$ fermion scattering processes at the Tevatron, HERA, and LEP II constrain Λ to be $\gtrsim 2 - 3$ TeV [2]. At the LHC, compositeness can be probed in $q\bar{q} \rightarrow \ell^+\ell^-$ scattering up to scales of order $\Lambda \sim 20 - 30$ TeV [3], where the range allows for various helicity structures of the contact term, for integrated luminosities of $10 - 100$ fb^{-1}. The case of four-quark interactions at the LHC is more problematic due to uncertainties in the parton distributions and the extrapolation of the calorimeter energy calibration to very high values of the jet transverse momentum. These problems may be overcome by the use of dijet angular distributions which may provide sensitivity to quark compositeness up to the scale $\Lambda \sim 25$ TeV for 30 fb^{-1} of integrated luminosity [4].

A LC may search for contact terms in $eeqq$ and $ee\ell\ell$ interactions, with $\ell = e, \mu, \tau$, in fermion pair production $e^+e^- \rightarrow f\bar{f}$. The use of polarized beams, combined with angular distributions, allows for a clear determination of the helicity of the contact term. An examination of contact effects for μ, charm-, and bottom-quark pair production at LC energies was performed in [5]. This study concentrated on tagged final states since contact effects are diluted when all quark flavors are summed due to cancellations between the up- and down-type quark contributions. In this study, both polarized and unpolarized angular distributions, assuming 90% electron beam polarization, were examined with tagging efficiencies of 60% and 35% for b- and c-quarks, respectively, and the detector acceptance was taken to be $|\cos\theta| < 0.985$. The resulting 95% CL sensitivity (after scaling to $\mathcal{L} = 500$ fb^{-1}) to Λ is listed for the various reactions in Table 1. The results of a study [6], including experimental systematic uncertainties, for light quark and muon pair production in e^+e^- annihilation is presented in Fig. 1. Here it is assumed that both the initial electron and positron beams are polarized and 1 ab^{-1} of integrated luminosity is used. For muon pair production, the dependence on the sign of η is negiglible and the average sensitivity for $\eta = +1$ and -1 is shown.

Compositeness searches in four-electron interactions in Møller and Bhabha scattering [7] summarized in Fig. 2. Here, $\Lambda_{\rm LL}$ and $\eta = +1$ is assumed. For equal luminosities, the sensitivity from Møller scattering is

Table 1. 95% CL search reach in TeV for contact interaction scales with various helicities in numerous processes, assuming 90% electron beam polarization and the final state tagging efficiencies discussed in the text. From [5].

	Λ_{LL}	Λ_{LR}	Λ_{RL}	Λ_{RR}
$\sqrt{s} = 0.5$ TeV				
$e_L^- e^+ \to \mu^+\mu^-$	57	52	18	18
$e_R^- e^+ \to \mu^+\mu^-$	20	18	52	55
$e_L^- e^+ \to c\bar{c}$	59	50	9	15
$e_R^- e^+ \to c\bar{c}$	21	20	43	57
$e_L^- e^+ \to b\bar{b}$	68	53	9	16
$e_R^- e^+ \to b\bar{b}$	30	21	59	59
$\sqrt{s} = 1.0$ TeV				
$e_L^- e^+ \to \mu^+\mu^-$	79	72	25	26
$e_R^- e^+ \to \mu^+\mu^-$	28	25	73	78
$e_L^- e^+ \to c\bar{c}$	82	72	12	21
$e_R^- e^+ \to c\bar{c}$	30	28	62	78
$e_L^- e^+ \to b\bar{b}$	94	77	14	23
$e_R^- e^+ \to b\bar{b}$	43	30	82	84

significantly better than that from Bhabha scattering. This is due to the availability of polarization of both electron beams in Møller scattering, as well as the Møller/Bhabha crossing relation in the central region of the detector. Limits on $\Lambda^+_{\rm LL}$ for different energies and luminosities can be calculated under the assumption that the compositeness limit scales as $\mathcal{L}^{1/4}s^{1/2}$.

These studies show that the LC is a powerful tool in the search for fermion compositeness.

Contact terms describe the effect of new high scale contributions to $e^+e^- \to \overline{f}f$ in a model independent manner. If such effects are discovered, their source could be the exchange of a new heavy particle rather than the existence of fermion substructure. In this case, the mass of the new particle would be of order $M_X \sim \Lambda \gg \sqrt{s}$, with Λ being the quantity measured by experiment. Examples of such heavy particle exchange are given in theories with extra gauge bosons, gravitons, leptoquarks, string resonances, and supersymmetry; most of these will be discussed in detail throughout this chapter. A strength of the Linear Collider is its ability to distinguish amongst these possibilities and determine the properties and the underlying theory of the new contributions. Such indirect determinations are possible due to available distributions and asymmetries and the precision to which the measurements can be performed.

A simple illustration of this point is given by the determination of the spin of a heavy new particle being exchanged in fermion pair production.

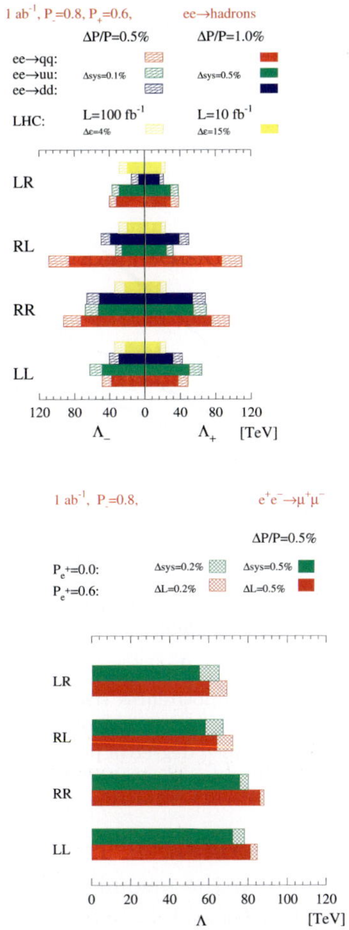

Fig. 1. 95% CL sensitivity to contact interaction scales Λ of different helicities for light quark and muon pair production at a 500 GeV Linear Collider. 1 ab^{-1} of integrated luminosity, 80% polarization of the electron beam, and 60% polarization of the positron beam are assumed. The systematic errors used in the study are as indicated. For the hadronic final states, the corresponding results for the LHC are shown. From [6].

The angular distributions and polarization asymmetries associated with $e^+e^- \to \overline{f}f$ are sensitive probes of the spin of the new particle. The discrimination power between the spin-1 exchange of a new neutral vector boson and the spin-0 exchange of a sneutrino in supersymmetric theories with R-parity violation is shown in Fig. 3. In this study [8], a two-parameter

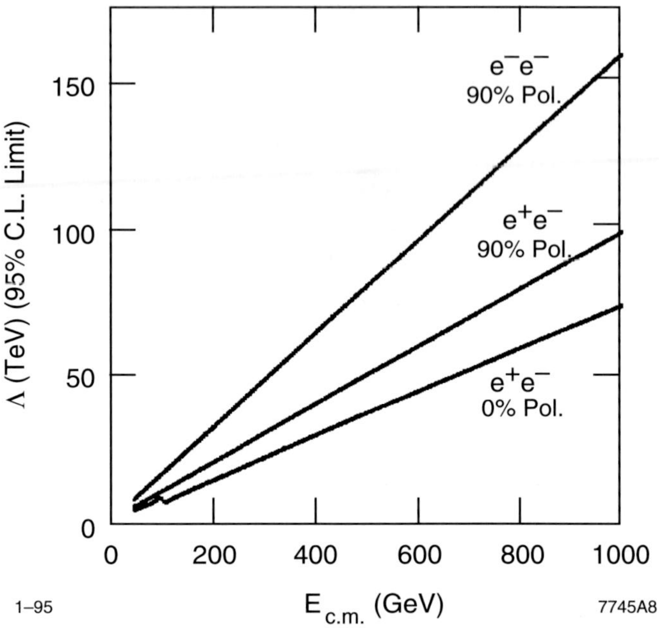

Fig. 2. The 95% CL limits for the compositeness scale $\Lambda^+_{\rm LL}$ from Møller and Bhabha scattering as a function of the e^-e^-or e^+e^- center-of-mass energy. The luminosity is given by $\mathcal{L} = 680{\rm pb}^{-1} \cdot s/M_Z^2$. The assumed degree of polarization of the electron beam(s) is indicated in the figure. From [7].

fit of a trial distribution of the form $\sim A(1+\cos\theta)^2 + B(1-\cos\theta)^2$, where θ is the scattering angle between the incoming electron and the outgoing fermion, was performed to the observables, with A, B being parameters determined by the fit. The mass of the new particle was taken to be 3 TeV, four different new gauge couplings were considered for the spin-1 exchange, and the sneutrino was allowed to mediate the reaction in both the s- and t-channels. As can be seen from the figure, the five regions are statistically well separated from each other and clearly distant from the Standard Model solution. More examples of this type will be presented throughout this chapter.

3. Extended Gauge Sectors

Extended gauge symmetries are a characteristic feature of many theories where the Standard Model is an effective low-energy representation of some

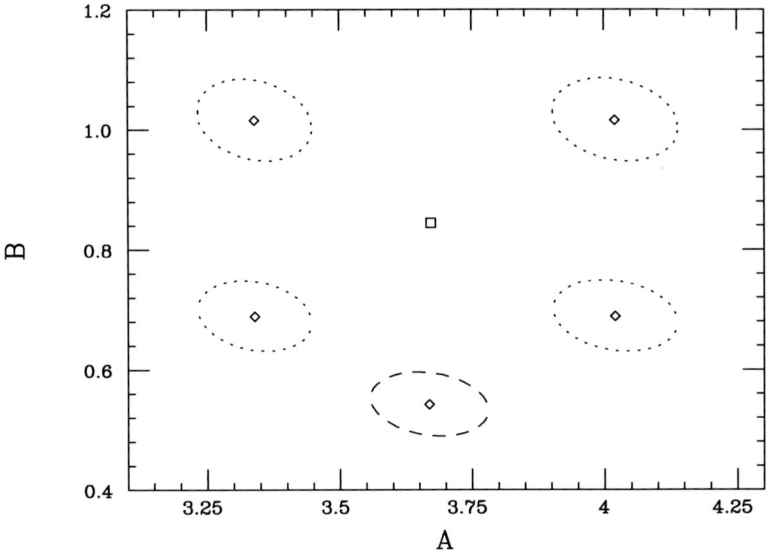

Fig. 3. Results of a 2-parameter fit to determine the spin of a new particle exchanged in $e^+e^- \to \overline{f}f$. The 95% CL contours circle the fitted values. The box in the center corresponds to the Standard Model, the dotted ellipses represent the fit to the four spin-1 cases considered, and the dashed ellipse is for the case of sneutrino exchange. The fit was performed with $\sqrt{s} = 1$ TeV with 150 fb^{-1}.

more fundamental theory at a high scale. Examples include Grand Unified Theories, Little Higgs models, Left-Right Symmetric models, Topcolor, and Superstring theories. Associated with the additional symmetries are new gauge bosons with masses related to the scale of the additional symmetry breaking; in some cases this is predicted to be near the TeV scale. These heavy gauge bosons are neutral and/or charged, denoted as Z' or W' respectively, depending on the structure of the new gauge group(s). The discovery of these new bosons at high energy accelerators is the hallmark signature for the existence of extended gauge sectors and would indicate that the Standard Model is embedded into a larger framework at higher energy scales. After discovery, the extraction of the properties, particularly the fermionic couplings, of the extra gauge bosons is of crucial importance. Such determinations would identify the origin of the new gauge fields, *i.e.*, the additional symmetries from which they arise, and hence elucidate the underlying physics at the high scale.

3.1. Z' Discovery Limits and Identification

The search capabilities for new neutral gauge bosons at existing and future colliders has been discussed by many authors [9, 10, 11]. Reviews of various models with extended gauge sectors, and the couplings of the new gauge fields within the theories, are also given in these references.

Hadron colliders observe the direct production of a Z' as a resonance in the Drell-Yan invariant mass spectrum via $\sigma(pp \to Z')B(Z' \to \ell^+\ell^-)$ with $\ell = e, \mu$. Depending on the underlying model, new particles can influence the total Z' width and hence its leptonic branching fraction. It is usually assumed that all Z' decay modes are known, but that may not necessarily be the case as outlined in the example given below. Assuming that only Standard Model decay channels are open, the 95% CL search reach at the LHC for a Z' from various models is displayed in Fig. 4 for 100 fb^{-1} and for the high luminosity option of 1 ab^{-1}. The figure shows that the search reach is essentially model independent and is in the range of 5 TeV for the Z' mass for the design luminosity of 100 fb^{-1}. The present bound on Z' production at the Tevatron Run II with 200 pb^{-1} is of order $M_{Z'} \gtrsim 800$ GeV. We note that an accurate mass determination of the Z' at the LHC/Tevatron is feasible once it is discovered.

At this point, it is worth noting a historical reference. As is well-known, the τ lepton was discovered in e^+e^- collisions at SPEAR in 1976. Due to its decays and weak production cross section, the tau is difficult to observe in hadronic collisions, and was first seen via $p\bar{p} \to Z \to \tau^+\tau^-$ after the Standard Model Z was discovered at the $Sp\bar{p}S$ collider. However, knowledge of the τ is crucial for predicting the correct Standard Model rate for Z and W production in Drell-Yan processes. Without the existence of e^+e^- colliders such as SPEAR, we would not have been able to apply this knowledge to Z and W production at the $Sp\bar{p}S$. Our analysis and interpretation of the Z/W discovery data would have been incorrect.

The Tevatron Run II bounds on Z' production preclude the direct production of new neutral gauge bosons at a 500 GeV LC for many (but not all) extended gauge models. However, a strong signal at the LC for the existence of a new Z' boson is given by its indirect effects through s-channel exchange. Through its interference with the Standard Model γ and Z exchange in $e^+e^- \to f\bar{f}$, deviations from Standard Model predictions can occur for $M_{Z'} > \sqrt{s}$. The combination of many observables where many different final states can be identified, such as the cross sections for $f\bar{f}$ final states, forward-backward asymmetries, A^f_{FB}, and left-right polarization asymmetries, A_{LR}^f, yields a large statistical sample with which to measure such deviations. The sensitivity of the LC to Z' bosons from various models is summarized in Fig. 4, summing over the final states $f = \mu, \tau, c, b$, including electron and positron beam polarization as well as detector systematic effects. Here, the sensitivity is shown at both the 2σ and 5σ level to facilitate comparison with the LHC. It is clear that overall evidence for Z' bosons with masses many times that of $\sqrt{s}$ may be observed, with a variation of the reach depending on the model. The sensitivity limits at the LC would improve by approximately 12% if the systematic effects were not taken into account. The variation of the LC sensitivity limits as a function of energy, luminosity, and polarization is shown in Fig. 5

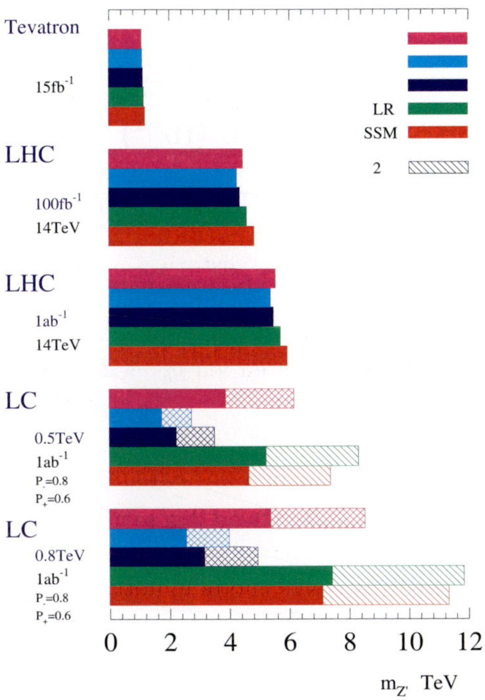

Fig. 4. 95% CL search limits for extra neutral gauge bosons at the Tevatron and LHC, and the 2σ and 5σ sensitivity bounds at the Linear Collider for various extended gauge models. From [11].

If the new neutral gauge boson is very weakly coupled, it will form a narrow resonance which will be difficult to observe directly at either the

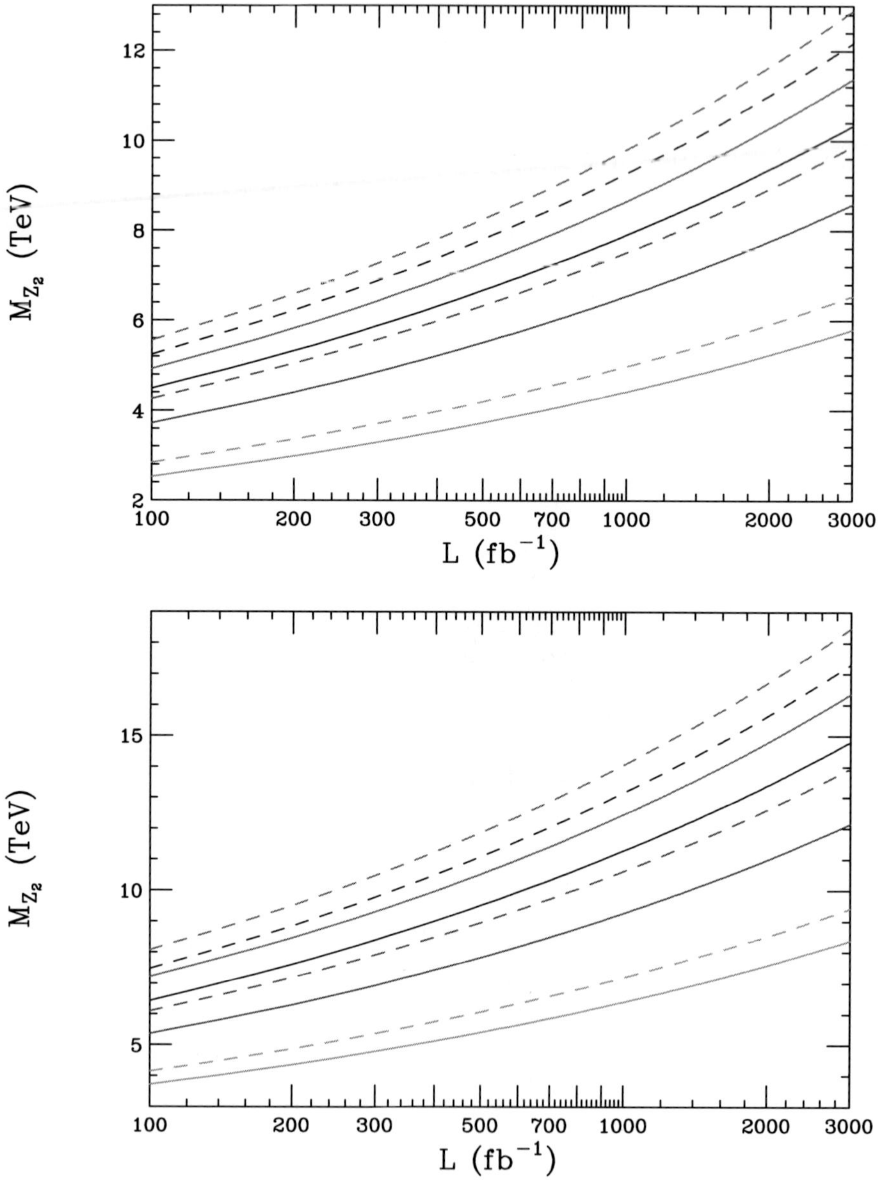

Fig. 5. 95% CL (or 2σ) sensitivity to Z' exchange at a 500 GeV (top panel) and 1 TeV (bottom panel) Linear Collider as a function of luminosity with $P_+ = 0$ (solid) and $P_+ = 0.6$ (dashed). From top to bottom, the model correspond to: E_6 Grand Unified theories ψ (green) and χ (red), the Left-Right Symmetric Model (blue) and the Sequential Standard Model (magenta). From [12].

LHC or the LC. However, if the mass of the gauge boson is below the center of mass energy for the Linear Collider, it can be produced through processes where the effective energy is reduced due to initial state radiation and beamstrahlung. This provides a sensitivity to such narrow particles which extends that of the LHC as shown in Fig. 6 [13].

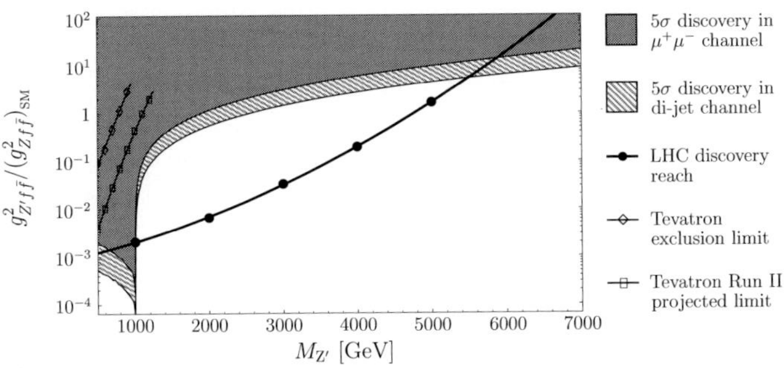

Fig. 6. 95% CL discovery reach at a 1 TeV LC with 1000 fb^{-1} for a sequential Z' boson in the $\mu^+\mu^-$ and dijet channels. Also shown are the discovery reach of the LHC from Drell-Yan production, the present limit from searches at the Tevatron and the expected sensitivity from the Tevatron Run II. From [13].

Once a new gauge boson is discovered, attention will be directed towards measuring its properties and deciphering the underlying symmetry. To accomplish this, it is essential to determine the couplings of the new boson. This is difficult at hadron colliders [9] alone due to both large backgrounds and low statistics. Some degree of model discrimination will be available at the LHC through the use of the forward-backward asymmetry and rapidity distribution of the final state lepton pair in Drell-Yan production [14]. However, these techniques are only viable for low mass Z' bosons, $M_{Z'} \lesssim 2$ TeV, and introduce some ambiguities.

Precision measurements at the LC can be used to measure the properties of the Z' and shed light on the underlying theory. This analysis works best when the LHC has first discovered a Z' and measured its mass, thus demonstrating the complementarity and interplay of the two machines. Fermion pair production at the LC is a process with high statistics and clear topologies. Coupled with an expected integrated luminosity of $500 - 1000$ fb^{-1}, the polarization of at least one initial beam, and a Z' mass measurement at the LHC, the Z' model can be resolved at the LC to a very good accuracy. Figure 7 displays the resolving power between Z' models assuming that the mass of the Z' was measured previously at the LHC. This study only considers leptonic final states and assumes lepton universality. If $M_{Z'}$ were beyond the LHC discovery reach or if the Z' does not couple to quarks then no prior knowledge of it would be obtained before the LC turns on. However, in this case, the LC can still yield some information on the Z' couplings and mass. Instead of extracting Z' couplings directly, "normalized" couplings, defined by

$$a_f^N = a'_f \sqrt{\frac{s}{M_{Z'}^2 - s}}; \quad v_f^N = v'_f \sqrt{\frac{s}{M_{Z'}^2 - s}} \tag{2}$$

could be measured. For demonstration of this case, the diagnostic power of a 1 TeV LC for a Z' with couplings of the E_6 model χ and mass $M_{Z'} = 5$ TeV is displayed in Fig. 7 for leptonic final states. An additional determination of the Z' mass and couplings could be performed at the LC [15] in this case from cross section and asymmetry measurements at several different values of $\sqrt{s}$.

A recent study of the process $e^+e^- \to \nu\bar{\nu}\gamma$ has demonstrated that the process can also be used to obtain information on $Z' - \nu\bar{\nu}$ couplings [16].

3.2. *W′* Discovery Limits and Identification

While considerable effort has been devoted to the study of Z' bosons at e^+e^- colliders, a corresponding endeavor for the W' sector has only recently been undertaken [16, 17, 18]. The processes under consideration are $e^+e^- \to \nu\bar{\nu}\gamma$ and $e\gamma \to \nu q + X$. We note that both charged and neutral extra gauge bosons contribute to the first reaction, while only W' bosons contribute to the latter.

In the W' analysis [16] based on $e^+e^- \to \nu\bar{\nu}\gamma$, the photon energy and angle with respect to the beam axis are restricted to $E_\gamma \geq 10$ GeV and $10^o \leq \theta_\gamma \leq 170^o$, to take into account detector acceptance. The most serious background, radiative Bhabha-scattering where the scattered e^+ and e^- go undetected down the beam pipe, is suppressed by restricting the photon's transverse momentum to $p_T^\gamma > \sqrt{s}\sin\theta_\gamma \sin\theta_v/(\sin\theta_\gamma + \sin\theta_v)$, where θ_v is the minimum angle at which the veto detectors may observe electrons or positrons; here, $\theta_v = 25$ mrad. The observable $d\sigma/dE_\gamma$ was found to provide the most statistically significant search reach. In the analysis based on $e\gamma \to \nu q + X$, the W' contribution can be isolated by imposing a kinematic cut requiring either the q or $\bar{q}$ to be collinear to the beam axis. In order to take into account detector acceptance, the angle θ_q of the detected quark relative to the beam axis is restricted to $10^0 \leq \theta_q \leq 170^0$. The kinematic

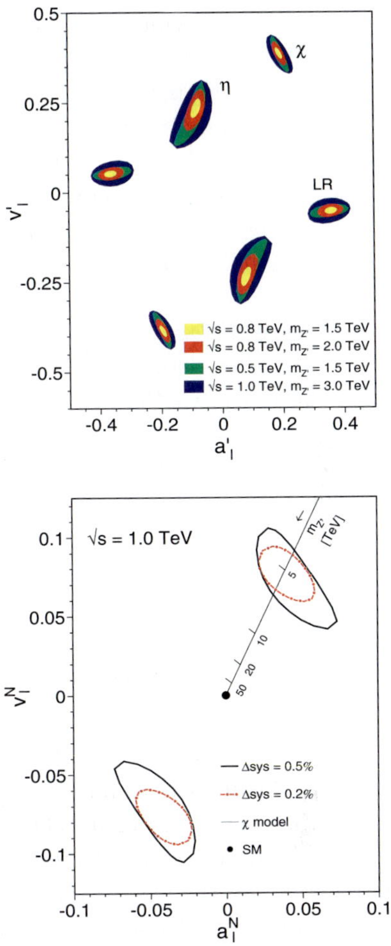

Fig. 7. Left Panel: Resolution power (95% C.L.) for different $M_{Z'}$ based on measurements of leptonic observables at $\sqrt{s} = 500$ GeV, 800 GeV, 1 TeV with a luminosity of $L_{\rm int} = 1\ {\rm ab}^{-1}$. The leptonic couplings of the Z' correspond to the χ, η, or LR model [6]. Right Panel: Resolution power (95% C.L.) for different $M_{Z'_\chi}$ based on measurements of leptonic observables at $\sqrt{s} = 1$ TeV with a luminosity of $L_{\rm int} = 1\ {\rm ab}^{-1}$ [6].

variable which is most sensitive to a W' is the p_{T_q} distribution. The quark's transverse momentum relative to the beam is restricted to $p_T^q > 40(75)$ GeV for $\sqrt{s} = 0.5(1.0)$ TeV, to suppress various Standard Model backgrounds. The 95% CL sensitivity for both processes are displayed graphically in Fig. 8. The corresponding W' search reach at the LHC is in the range $5-6$ TeV [9].

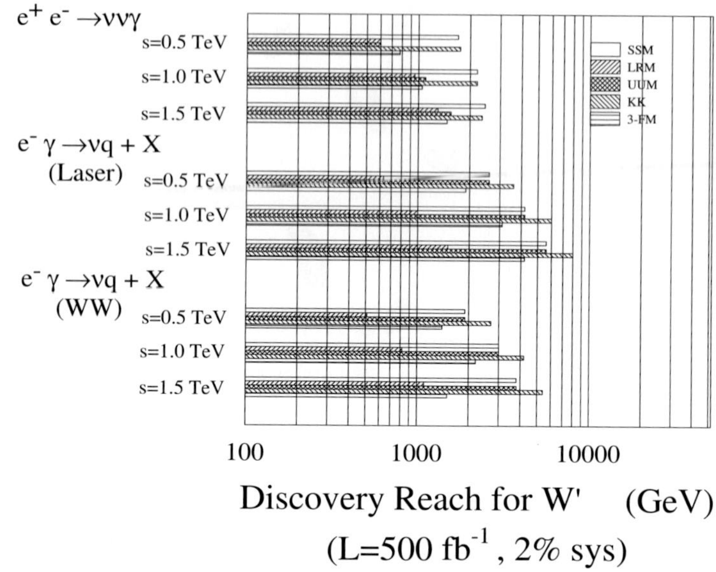

Fig. 8. 95% CL sensitivity reach for W' bosons at the LC. The $e\gamma$ reactions are computed for a back-scattered laser configuration and with the Weizsäcker Williams approximation. From [16, 18].

More important than as a discovery channel, is that these processes may be used to determine the couplings of a new W' boson. The 95% CL constraints which can be placed on the right- and left-handed couplings of a W' to fermions, assuming that the W' has Standard Model-like couplings, and that there is no corresponding Z' contribution to $e^+e^- \to \nu\bar{\nu}\gamma$, are shown in Fig. 9. Here, the total cross section σ and the left-right asymmetry A_{LR} are used as observables, with the systematic errors for $\sigma(A_{LR})$ taken as 2%(1%) and 80% electron and 60% positron polarization are assumed. The axes in this figure correspond to couplings normalized as $L(R)_f = C^{W'}_{L(R)} g/(2\sqrt{2})$. It is found that 2% systematic errors dominate the coupling determination. In addition, we note that the W' couplings can only be constrained up to a two-fold ambiguity, which could be resolved by reactions where the W' couples to a triple gauge vertex. Note that the polarization of the initial electron beam can be used to trivially determine whether the W' has a right-handed coupling to fermions.

If a W' were discovered at the LHC, measurements of its couplings in the processes $e^+e^- \to \nu\bar{\nu}\gamma$ and $e\gamma \to \nu q + X$ could provide valuable information

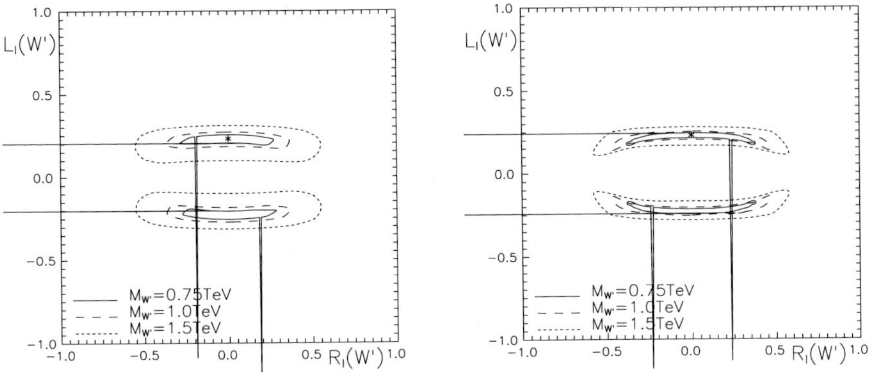

Fig. 9. Left Panel: 95% CL constraints from $e^+e^- \to \nu\bar{\nu}\gamma$ on couplings of a W' from the Sequential Standard Model indicated by a star for $\sqrt{s} = 0.5$ TeV and $L_{\rm int} = 1000$ fb^{-1} with a systematic error of 0.5% (0.25%) for $\sigma(A_{LR})$ for different W' masses. Right Panel: 95% C.L. constraints from $e\gamma \to \bar{\nu}q + X$ on couplings of a W' from the Sequential Standard Model for $\sqrt{s} = 0.5$ TeV and $L_{\rm int} = 1000\, fb^{-1}$ with a 2% systematic error for different W' masses.

regarding the underlying model, with the latter process serving to isolate the W' couplings from those of a Z'.

4. Extra Spatial Dimensions

The possible existence of additional dimensions beyond our usual 3 spatial +1 time dimensional world captivates our imagination and would dramatically alter our view of the universe. Extra spatial dimensions at very short distances are a prediction of string theory and theorists have studied the effects of higher dimensional spacetimes for decades. New theories, developed to address the hierarchy problem, propose that the effects of extra dimensions may be visible at larger distances, comparable to the TeV energy scale. These theories utilize the geometry of additional spatial dimensions to resolve the hierarchy, *i.e.*, the disparity between the electroweak scale ($\sim$ TeV) where electroweak symmetry breaking takes place and the traditional scale of gravity defined by the Planck scale (10^{19} GeV). These ideas make use of the fact that gravity has yet to be probed at energy scales much above 10^{-3} eV in laboratory experiments, admitting for the possibility that gravity behaves differently than expected at higher energies. If new dimensions are indeed related to the source of the hierarchy, then they should provide detectable signatures in experiments at the electroweak scale.

Theoretical frameworks with extra dimensions have some general features. In most scenarios, our observed 3-dimensional space is a 3-brane (sometimes called a 'wall'), where the terminology is derived from a generalization of a 2-dimensional membrane. This 3-brane is embedded in a higher D-dimensional spacetime, $D = 3 + \delta + 1$, which is known as the 'bulk.' The δ extra spatial dimensions are orthogonal to our 3-brane. If the additional dimensions are small enough, the Standard Model gauge and matter fields are phenomenologically allowed to propagate in the bulk; otherwise they are stuck to the 3-brane. Gravity, however, propagates throughout the full higher dimensional volume. Conventional wisdom dictates that if the additional dimensions are too large then deviations from Newtonian gravity would result and hence the extra dimensional space must be compactified.

As a result of compactification, fields propagating in the bulk expand into a series of states known as a Kaluza-Klein (KK) tower, with the individual KK excitations being labeled by mode numbers. Similar to the quantum mechanical behavior of a particle in a box, the momentum of the bulk field is quantized in the compactified dimensions. For an observer on the 3-brane, each quanta of momentum in the compactified volume appears as a KK excited state with mass $m^2 = \vec{p}_\delta^{\,2}$. This builds the KK tower of states, where each state carries identical spin and gauge quantum numbers. The collider signature for the existence of additional dimensions is the observation of a KK tower of states. The detailed properties of the KK states are determined by the geometry of the compactified space and their measurement would reveal the underlying geometry of the bulk. A review of extra dimensional models and their experimental signatures can be found in Ref. [19].

4.1. *Large Extra Dimensions*

In this scenario[20] the apparent gauge hierarchy is generated by a large volume of the extra dimensions. The Standard Model gauge and matter fields are confined to the 3-brane, while gravity propagates in the bulk. It is postulated that the higher dimensional gravitational interactions become strong near the electroweak scale. The Planck scale of the effective 4-dimensional theory is related to the Fundamental scale where gravity becomes strong in the full higher dimensional spacetime, M_D, via Gauss' Law,

$$M_{\rm Pl}^2 = V_\delta M_D^{2+\delta} \,, \tag{3}$$

where $V_\delta \sim (2\pi R_c)^\delta$ is the volume of the compactified space and R_c represents the radius of the compactified dimensions. Setting $M_D \sim 1$ TeV eliminates the gauge hierarchy and results in extra dimensions which range in size from a sub-millimeter to a few fermi for $\delta = 2$ to 6, assuming all radii are of equal size. The compactification scale ($M_c = 1/R_c$) associated with these parameters then ranges from $\sim 10^{-4}$ eV to a few MeV.

The case of $\delta = 1$ (which yields $R_c \approx 10^{11}$ m) is immediately excluded by astronomical data. Cavendish-type experiments, which search for departures from the inverse-square law gravitational force, exclude [21] $R_c > 130$ microns for $\delta = 2$, which translates to the bound $M_D > 1.7$ TeV. In addition, astrophysical and cosmological considerations [19] disfavor a value of M_D near the TeV scale for $\delta = 2$. Our knowledge of the electroweak and strong forces extends with great precision down to distances of order 10^{-15} mm, and thus the Standard Model fields do not feel the effects of the large extra dimensions present in this scenario and are confined to the 3-brane.

The Feynman rules for this scenario are obtained by considering a linearized theory of gravity in the bulk. The bulk graviton field strength tensor can be decomposed into spin-0, 1, and 2 states, each of which expands into a KK tower upon compactification. These KK states are evenly spaced with masses $m_{\vec{n}} = \sqrt{\vec{n}^2/R_c^2}$, where $\vec{n}$ represents the KK level number. The interactions of the KK gravitons with the Standard Model fields on the wall are governed by the conserved stress-energy tensor. The spin-1 states do not interact with fields on the 3-brane, the spin-0 states couple to the trace of the stress-energy tensor and will not be considered further here. Each state in the spin-2 KK tower, G_n, couples identically to the Standard Model fields on the 3-brane with inverse Planck strength, M_{Pl}^{-1}. The details of the KK decomposition for the bulk gravitons and the derivation of their Feynman rules can be found in [22, 23]. It is important to note that this description is an effective 4-dimensional theory, valid only for energies below M_D. The full theory above the fundamental scale is unknown.

Two classes of collider signatures arise in this model. The first is that of real emission of graviton states [22, 24] in scattering processes, with the relevant reaction at a Linear Collider being $e^+e^- \to \gamma(Z) + G_n$. The produced graviton, G_n, appears as missing energy in the detector, behaving as if it were a massive, non-interacting, stable particle. The cross section is computed for the production of a single massive graviton excitation and then summed over the full tower of KK states. Since the mass splittings of the KK excitations are so small, the sum can be replaced by an integral weighted by the density of KK states. The specific process kinematics cut off this integral rendering a finite and model independent result. This removes the expected suppression from the M_{Pl}^{-1} strength of the KK graviton couplings, and the cross section scales as simple powers of $\sqrt{s}/M_D$,

$$\sigma KK \sim \frac{1}{M_{Pl}^2}(\sqrt{s}R_c)^\delta \sim \frac{1}{M_D^2}\left(\frac{\sqrt{s}}{M_D}\right)^\delta . \tag{4}$$

It is important to note that due to integrating over the effective density of states, the emitted graviton appears to have a continuous mass distribution; this corresponds to the probability of emitting gravitons with different extra dimensional momenta. The observables for graviton production, such as the γ/Z angular and energy distributions, are thus distinct from those of other new physics processes involving fixed masses for the undetectable particles. In addition, the Standard Model background from $e^+e^- \to \nu\bar{\nu}\gamma$ also has different characteristics since it is a 3-body process.

The cross section for $e^+e^- \to \gamma G_n$ as a function of the fundamental Planck scale is presented in Fig. 10 for $\sqrt{s} = 1$ TeV. The level of Standard Model background is also shown, with and without electron beam polarization set at 90%. We note that the signal (background) increases (decreases) with increasing $\sqrt{s}$. The expected discovery reach from this process has been computed in [25] at a 800 GeV LC with 1000 ft^{-1} of integrated luminosity and various configurations for the beam polarization. These results are displayed in Table 2 and include kinematic acceptance cuts, initial state radiation, and beamstrahlung.

The associated emission process at hadron colliders, $q\bar{q} \to g + G_n$, results in a mono-jet signal. In this case, the effective low-energy theory breaks down for some regions of the parameter space, as the parton-level center of mass energy can exceed the value of M_D. The experiment is then sensitive to the new physics appearing above the fundamental scale that is associated with the low-scale UV theory of quantum gravity. An ATLAS simulation [26] of the missing transverse energy in signal and background events at the LHC with 100 fb^{-1} results in the discovery range for the effective theory displayed in Table 2. The lower end of the range corresponds to where the ultraviolet physics sets in and the effective 4-d theory fails, while the upper end represents the boundary where the signal is observable above background. Note that the search regions for a polarized LC are comparable to that of the LHC.

If an emission signal is observed, one would like to determine the values of the fundamental parameters, M_D and δ. The evolution of the emission cross section with center of mass energy in e^+e^- annihilation depends quite strongly on the values of these parameters. This is displayed in Fig. 11, where the cross section is normalized at 500 GeV for $M_D = 5$ TeV and

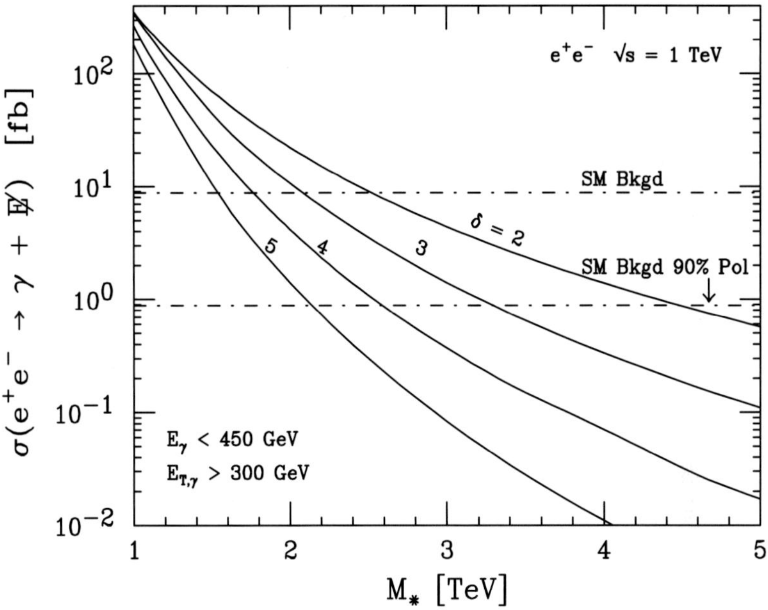

Fig. 10. The cross section for $e^+e^- \to \gamma G_n$ for $\sqrt{s} = 1$ TeV as a function of the fundamental Planck scale for various values of δ as indicated. The cross section for the Standard Model background, with and without 90% beam polarization correspond to the horizontal lines as labeled. The signal and background are computed with the requirement $E_\gamma < 450$ GeV in order to eliminate the $\gamma Z \to \nu\bar{\nu}\gamma$ contribution to the background. $M_* = M_D$. From [22].

$\delta = 2$. The measurement of the cross section at a Linear Collider at two different values of $\sqrt{s}$ can be used to determine these parameters and test the consistency of the data with the large extra dimensions hypothesis.

Lastly, we note that the cross section for the emission process can be reduced somewhat if the 3-brane is flexible, or soft, instead of being rigid [27]. In this case, the brane is allowed to recoil when the KK graviton is radiated; this can be parameterized as an exponential suppression of the cross section, with the exponential being a function of the brane tension. For reasonable values of the brane tension ($\sim M_D$), this suppression not large, as indicated by the dashed curves in the bottom panel of Fig. 11.

The second class of collider signals for large extra dimensions is that of virtual exchange of KK graviton states [22, 23, 28] in $2 \to 2$ scattering. This leads to deviations in cross sections and asymmetries in Standard Model processes, such as $e^+e^- \to f\overline{f}$, and gives rise to new reactions which are not present at tree-level in the SM, such as $e^+e^- \to hh$, or $\tilde{g}\tilde{g}$. The signature is

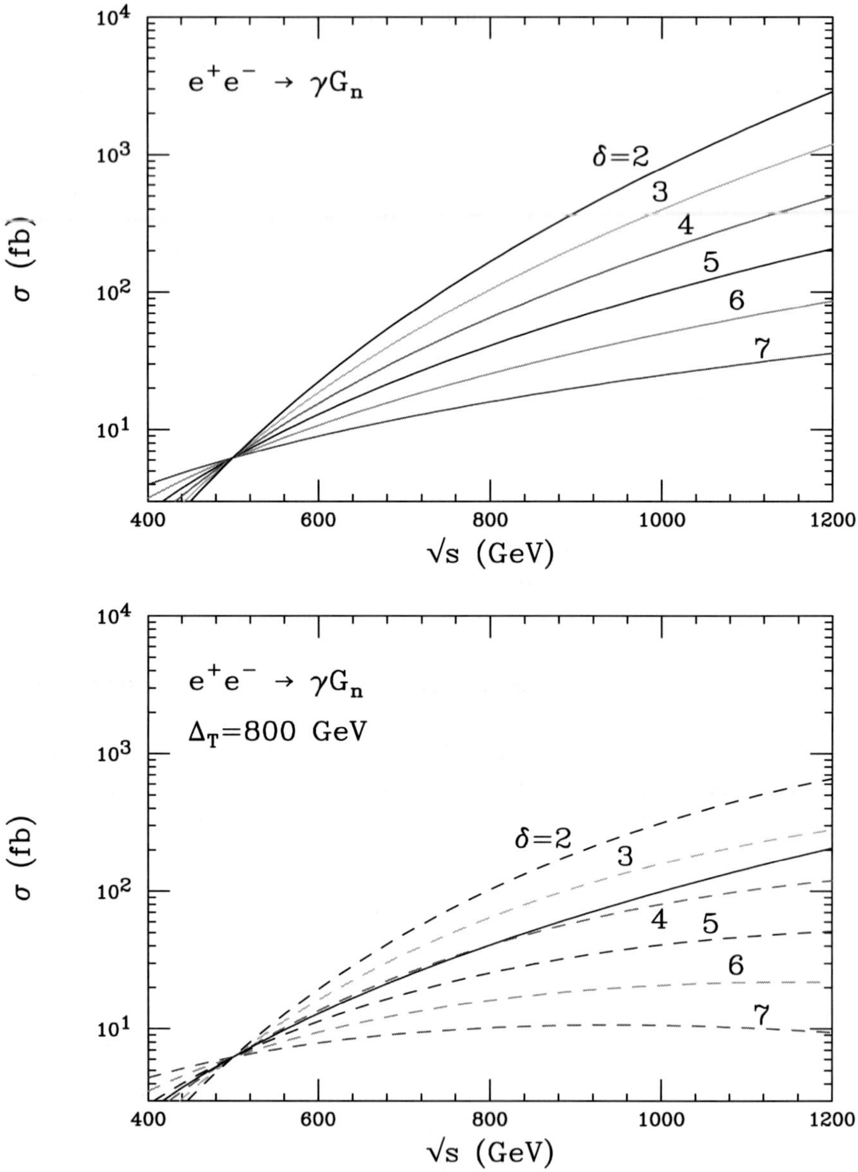

Fig. 11. Emission cross section in e^+e^- annihilation as a function of $\sqrt{s}$ for $\delta = 2 - 7$ from top to bottom on the right-hand side. The cross sections are normalized to $M_D = 5$ TeV and $\delta = 2$ at $\sqrt{s} = 500$ GeV. Top: Brane terms are not included. Bottom: The effects of finite brane tension are included, taking the relevant tension parameter to be $\Delta = 800$ GeV. Here, the lone solid curve represents the case without brane term effects for $\delta = 5$.

Table 2. 95% CL sensitivity to the fundamental Planck scale M_* in TeV for different values of δ, from the emission process for various polarization configurations and different colliders as discussed in the text.

$e^+e^- \to \gamma + G_n$		2	4	6
LC	$P_{-,+} = 0$	5.9	3.5	2.5
LC	$P_- = 0.8$	8.3	4.4	2.9
LC	$P_- = 0.8,\ P_+ = 0.6$	10.4	5.1	3.3
$pp \to g + G_n$		2	3	4
LHC		$4 - 7.5$	$4.5 - 5.9$	$5.0 - 5.3$

similar to that discussed above for composite theories and provides a good tool in the search for extra dimensions in the case $\sqrt{s} < M_D$.

The exchange process is governed by the effective Lagrangian

$$\mathcal{L} = i\frac{4\lambda}{\Lambda_H^2} T^{\mu\nu} T_{\mu\nu}\,, \tag{5}$$

employing the notation of [28]. This is a dimension-8 operator and $T^{\mu\nu}$ represents the conserved stress-energy tensor. The amplitude is proportional to the sum over the propagators for the graviton KK tower which, as discussed above, may be converted to an integral over the density of KK states. However, in this case, there is no specific cut-off associated with the process kinematics and the integral is divergent for $\delta > 1$. This introduces a sensitivity to the unknown ultraviolet physics which appears at the fundamental scale. Several approaches for regulating this sum may be employed. The most model independent is the introduction of a naive cut-off, with the cut-off being set to Λ_H which is of order the fundamental scale M_D. In addition, the parameter $\lambda = \pm 1$ is usually incorporated. The substitution

$$\mathcal{M} \sim \frac{i^2\pi}{M_{Pl}^2} \sum_{\vec{n}=1}^{\infty} \frac{1}{s - m_{\vec{n}}^2} \to \frac{\lambda}{\Lambda_H^4} \tag{6}$$

is then performed in the matrix element for s-channel graviton exchange with corresponding substitutions for t- and u-channel scattering. This substitution is universal for any $2 \to 2$ process.

The resulting angular distributions for fermion pair production are quartic in $\cos\theta$ and provide a smoking gun signal for spin-2 exchange. An illustration of this is given in Fig. 12 from [28] which displays the angular dependence of the polarized Left-Right asymmetry in $e^+e^- \to b\bar{b}$. Here, a 60% b-tagging efficiency, 90% electron beam polarization, 10° angular cut, and ISR have been included. Summing over all fermion final states (e, μ, τ, c, b, and t) and examining the deviations in the total cross section, forward-backward asymmetry, and left-right asymmetry in each case,

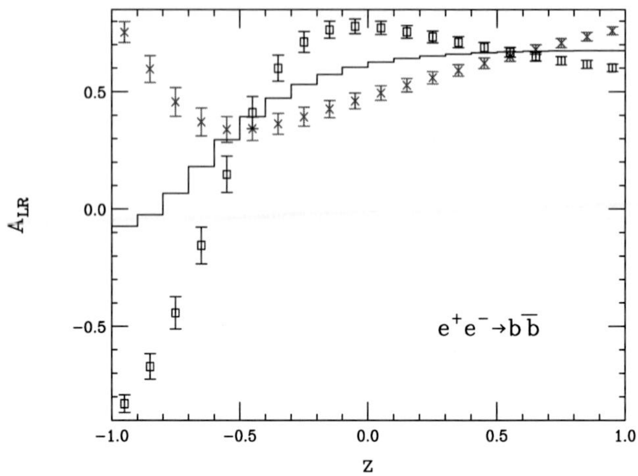

Fig. 12. Bin integrated angular dependence of the Left-Right asymmetry for $e^+e^- \to b\bar{b}$ at $\sqrt{s} = 500$ GeV. The solid histogram represents the Standard Model while the 'data' points are for $M_D = 1.5$ with $\lambda = \pm 1$. The error bars indicate the statistics in each bin for an integrated luminosity of 75 fb^{-1}. From [28]

as well as the tau polarization asymmetry, yields the search reach [29] for Λ_H shown in Fig. 13 as a function of integrated luminosity. We see that for a 1 TeV LC, scales of order $\Lambda_H \sim 8.4$ TeV can be observed with 500 fb^{-1}. The sensitivities for virtual graviton exchange at the LHC have been evaluated in [30] and are found to be $\Lambda_H \leq 7.5, 7.1$ TeV for Drell-Yan and di-photon production, respectively. We see that this search region is comparable to that of a Linear Collider.

If deviations are observed at the LC or LHC from the virtual exchange of new particles, then it will be necessary to have techniques available to differentiate between the possible scenarios giving rise to the effect. One such method is the use of transverse polarization at the LC. If longitudinal positron polarization is available at the LC, then spin rotators may be used to convert these to transversely polarized beams. This allows for new asymmetries to be constructed which are associated with the azimuthal angle formed by the directions of the $e^\pm$ polarization and the plane of the momenta of the outgoing fermions in $e^+e^- \to f\bar{f}$. The spin-averaged matrix element for this process can be written as

$$
\begin{aligned}
|\mathcal{M}|^2 = \frac{1}{4}(1 - P_L^- P_L^+)(|T_+|^2 + |T_-|^2) + (P_L^- - P_L^+)(|T_+|^2 - |T_-|^2) \\
+(2P_T^- P_T^+)[\cos 2\phi \Re(T_+ T_-^*) - \sin 2\phi \Im(T_+ T_-^*)] \,, \qquad (7)
\end{aligned}
$$

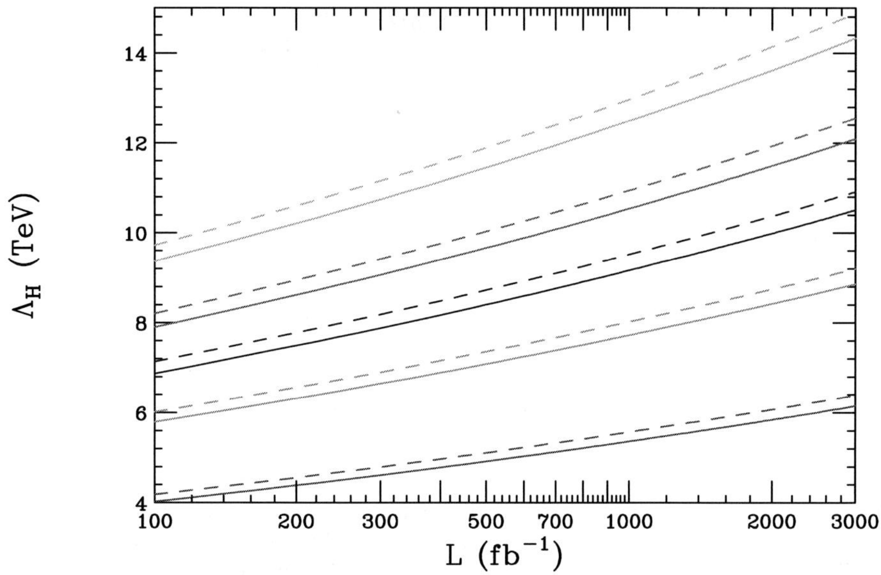

Fig. 13. 95% CL search reach for the scale associated with the exchange of graviton KK states in $e^+e^- \to f\bar{f}$. The solid (dashed) curves correspond to a initial state positron polarization of 0 (60)%. Each pair of curves is for $\sqrt{s} = 0.5\,, 0.8\,, 1.0\,, 1.2\,,$ and 1.5 TeV, from bottom to top, respectively. From [29].

where ϕ is the azimuthal angle defined on an event-by-event basis described above, $P^-_{L,T}(P^+_{L,T})$ represent the polarizations of the electron(positron), and $T_{+,-}$ correspond to the relevant helicity amplitudes. Note that the ϕ dependence in this expression is only accessible if both beams are transversely polarized. Spin-2 exchange does not introduce any new helicity amplitudes over those present for spin-1 exchange, but does yield an asymmetric distribution in $\cos\theta$ unlike the case of spin-1 exchange. Asymmetries in the differential cross section $d\sigma/d\cos\theta d\phi$ extend the search reach for graviton KK exchange by more than a factor of two, and provide an additional tool for isolating the signatures for spin-2 exchange up to mass scale in excess of $10\sqrt{s}$. These results [31] are shown in Table 3. If deviations due to virtual graviton exchange were observed in, *e.g.*, Drell-Yan production at the LHC, we see that a LC with positron polarization and $\sqrt{s} \geq 800$ GeV can identify the spin-2 nature of the exchange for the entire LHC search region.

If positron polarization is not available at the LC, then the spin-2 nature of virtual graviton exchange can be identified by expanding the fermion pair production cross section into multipole moments. Studies indicate [32] that this technique can uniquely identify spin-2 exchange at the 5σ level for fundamental scales M_D up to $6\sqrt{s}$.

Table 3. 95% CL search reach and identification of spin-2 exchange from the azimuthal asymmetries discussed in the text. From [31].

$\sqrt{s}$ (GeV)	Search Reach Λ_H (TeV)	ID Reach (TeV)
500	10.2	5.4
800	17.0	8.8
1000	21.5	11.1
1200	26.0	13.3
1500	32.7	16.7

Although this scenario resolves the hierarchy problem without invoking supersymmetry, if it is embedded in a string theory, then supersymmetry will also be present at the TeV scale. The full graviton supersymmetric multiplet would then reside in the bulk, resulting in a KK tower of gravitinos after compactification in addition to the KK gravitons. In supersymmetric models which expect a light zero-mode gravitino, such as gauge mediated supersymmetry breaking, the gravitino KK tower can yield interesting phenomenological effects. An example of this is in the process $e^+e^- \to \tilde{e}^+\tilde{e}^-$, which would now also receive contributions from t-channel KK gravitino exchange and s-channel KK graviton exchange, in addition to the usual Standard Model and Supersymmetric contributions. This has been computed in [33], which considered an $N = 2$ supersymmetry in the bulk, and after compactifying the gravitino sector, derived the KK gravitino couplings to $N = 1$ supersymmetric matter on the brane. The resulting dramatic effect on selectron pair production is highlighted by the ability to select various production channels via the use of electron beam polarization. This is displayed in Fig. 14 which shows the effects of the various contributions to $e_R^- e^+ \to \tilde{e}_R^\mp e_R^\pm$ for $M_D = 3\sqrt{s}$. The sensitivity of this process at $\sqrt{s} = 500$ GeV with 80% beam polarization and 500 fb^{-1} of integrated luminosity reaches $M_D = 12$ TeV for the case $\delta = 6$.

4.2. *TeV-scale Extra Dimensions*

The possibility of TeV^{-1}-sized extra dimensions naturally arises in braneworld theories [34]. By themselves, they do not allow for a reformulation of the hierarchy problem, but they may be incorporated into a larger structure in which this problem is solved. In these scenarios, the SM fields are phenomenologically allowed to propagate in the bulk. This presents a

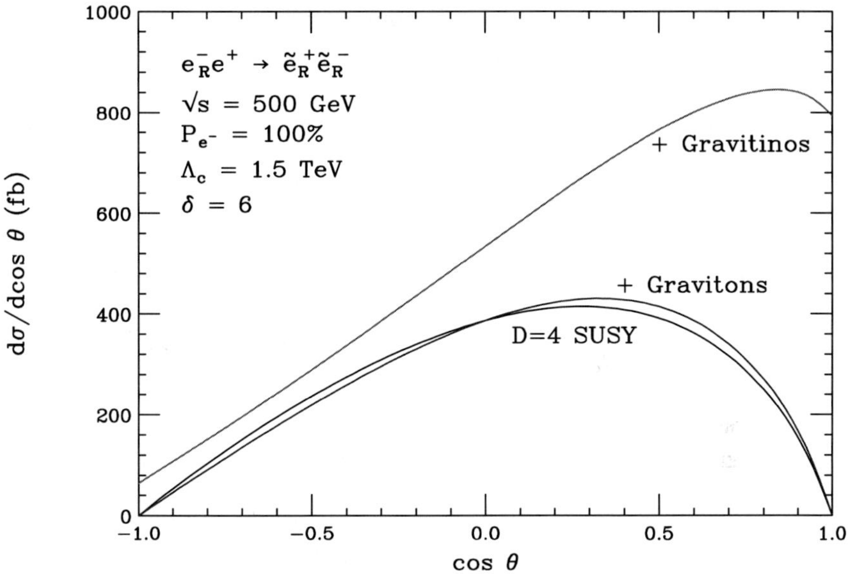

Fig. 14. The angular distribution for $e_R^- e^+ \rightarrow \tilde{e}_R^{\mp} e_R^{\pm}$ in standard $D = 4$ supersymmetry, plus the addition of bulk graviton KK tower exchange, and with bulk gravitino KK tower exchange, corresponding to the bottom, middle, and top curves, respectively. From [33]

wide variety of choices for model building: (i) all, or only some, of the Standard Model gauge fields exist in the bulk; (ii) the Higgs field may lie on the brane or in the bulk; (iii) the Standard Model fermions may be confined to the brane or to specific locales in the extra dimensions, or they may freely reside in the bulk. The phenomenological consequences of this scenario strongly depend on the location of the fermion fields. Unless otherwise noted, our discussion assumes that all of the Standard Model gauge fields reside in the bulk.

The masses of the excitation states in the electroweak gauge boson KK towers depend on where the Higgs boson is located. If the Higgs field propagates in the bulk, the zero-mode state of the Higgs KK tower received a vacuum expectation value (vev) which is responsible for the spontaneous breaking of the electroweak gauge symmetry. In this case, the mass matrix for the states in the gauge boson KK towers is diagonal and the excitation masses are shifted by the mass of the gauge zero-mode, which corresponds to the Standard Model gauge field, giving

$$m_{\vec{n}} = (m_0^2 + \vec{n} \cdot \vec{n}/R_c^2)^{1/2} \,. \tag{8}$$

However, if the Higgs is confined to the 3-brane, its vev induces off-diagonal elements in the mass matrix, generating a mixing amongst the electroweak gauge KK states of order $(M_0 r_c)^2$. The KK mass matrix must then be diagonalized to determine the excitation masses.

• *Fermions on the 3-brane.* We first discuss the case where the Standard Model fermions are rigidly fixed to the 3-brane and do not feel the effects of the additional dimensions. The simplest model of this class is the case of only one extra TeV^{-1}-sized dimension, compactified on a circle, with the orbifold S^1/Z_2 that maps $-y \rightarrow y$, where y represents the 5th dimensional coordinate. The fermions are constrained to lie at one of the two orbifold fixed points, $y = 0, \pi R_c$. In this scenario, the coupling strength of the gauge KK states to the Standard Model matter fields on the brane is $\sqrt{2}g$, where g is the corresponding Standard Model gauge coupling.

For models in this class, global fits to the precision electroweak data place strong constraints [35] on the mass of the first gauge KK excitation. Contributions to electroweak observables arise from the virtual exchange of gauge KK states and a summation over the entire KK tower must be performed. For $D > 5$, this sum is divergent and in the full higher dimensional theory, some new, as of yet unknown, physics would regularize this sum and render it finite. Since the $D = 5$ theory is finite, it is the scenario most often discussed [36] and is sometimes referred to as the 5-dimensional Standard Model (5DSM). In this case, an updated analysis of precision electroweak data constrains $M_c = 1/R_c \gtrsim 4 - 5$ TeV, independently (*i*) of whether the Higgs field vev is mostly in the bulk or on the brane, or (*ii*) of which orbifold fixed points the Standard Model fermions are located.

Such a large mass for the first electroweak gauge KK state is beyond the direct reach of a Linear Collider, but they can be produced as resonances at the LHC in the Drell-Yan channel provided $M_c \lesssim 7$ TeV [37]. (In addition, gluon KK states may be observable in the dijet channel at the LHC for $M_c \lesssim 15$ TeV [38].) However, the LC can indirectly observe the existence of heavy gauge KK states in the contact interaction limit via their virtual s-channel exchanges in fermion pair production, as discussed above in Section 3. In this case, the contribution from the entire KK tower must be summed. Combining the fermion final states and examining the deviations in the total cross section, forward-backward asymmetry, and left-right asymmetry for each final state as well as the tau polarization asymmetry, yields the search reach [29] for M_c in the 5DSM shown in Fig. 13. We see that for a 1 TeV LC, first KK excitation masses of order $M_c \sim 20$ TeV can be observed with 500 fb^{-1}. It has been shown [39] that the LC can distinguish the virtual exchange of gauge KK states from that of conventional heavy gauge states from Grand Unified scenarios, provided that the LHC measures M_c.

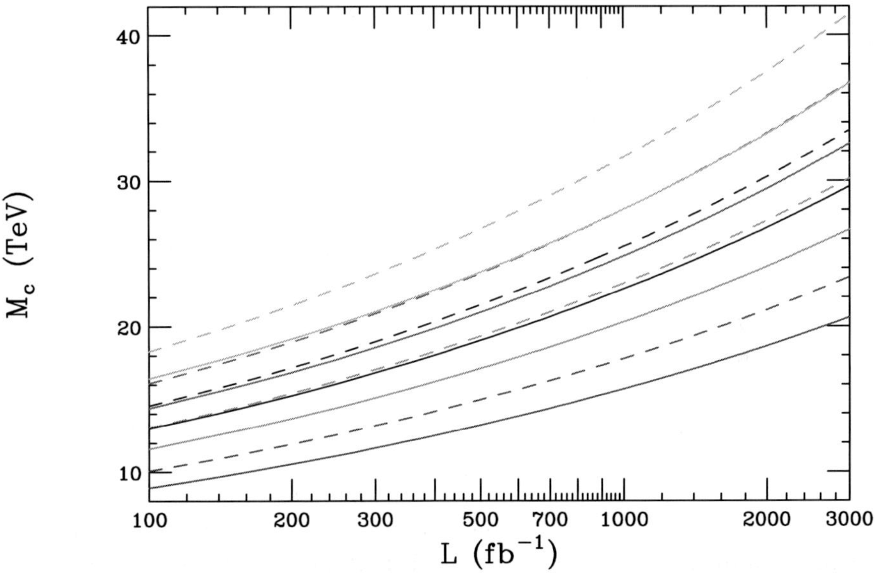

Fig. 15. 95% CL bound on the scale M_c from the reaction $e^+e^- \rightarrow f\bar{f}$, where $f = \mu, \tau, c, b, t$ have been summed over, as a function of the LC integrated luminosity . The solid (dashed) curves assume a positron polarization $P_+ = 0(0.6)$; an electron polarization of 80% has been assumed in all cases. From bottom to top the center of mass energy of the LC is taken to be 0.5, 0.8, 1, 1.2 and 1.5 TeV, respectively. From [29].

• *Localized fermions.* In this scenario, the Standard Model fermions are localized at specific points, y_i, in the extra TeV^{-1}-sized dimension, which are not necessarily at the orbifold fixed points. Here, the fermions have narrow gaussian-like wavefunctions in the extra dimension with the width of their wavefunction being much smaller than R_c^{-1}. The placement of the different fermions at distinct locations in the additional dimension, along with the narrowness of their wavefunctions, can then naturally suppress [40] operators mediating dangerous processes such as proton decay and also provide a mechanism for generating the fermion mass hierarchy [41]. An analysis of precision electroweak data in this case [35] shows that M_c is typically found to be $\geq 3 - 4$ TeV.

The exchange of gauge KK states in $2 \rightarrow 2$ scattering processes involving initial and final state fermions is sensitive to the placement of the fermions.

At a Linear Collider, this can be used to perform a cartography of the localized fermions [42], *i.e.*, measure the wavefunctions and locations of the fermions. For example, in this case Bhabha scattering can probe the relative locations of the left- and right-handed electrons. This is illustrated in Fig. 16 for a sample placement of $e_{L,R}$ in the 5th dimension with $M_c = 4$ TeV. The capability of beam polarization is particularly important for this analysis. In fact, at very large energies, the cross section for the polarized version of this process will tend rapidly to zero since the two particles completely miss each other in the extra dimension [43].

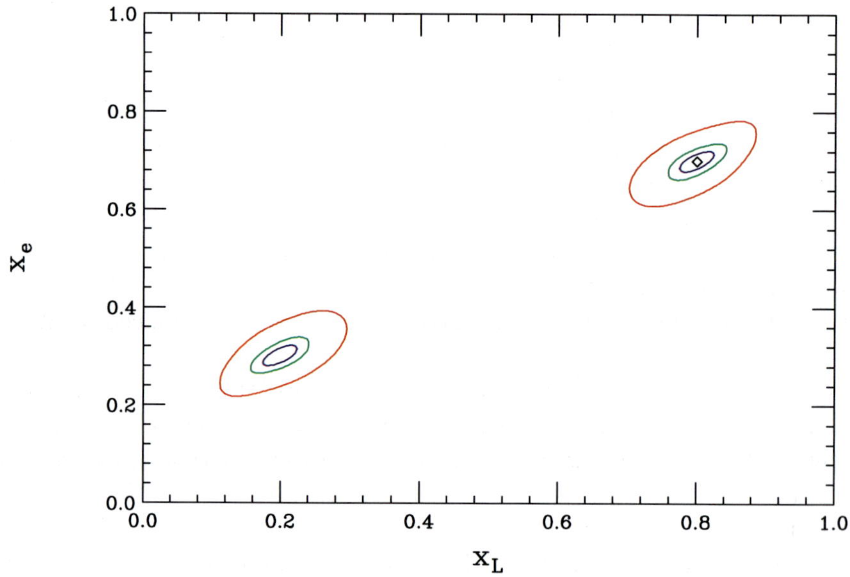

Fig. 16. 95% CL determination of the placement of the left- and right-handed electrons from polarized Bhabha scattering, assuming $M_c = 4$ TeV. From outermost to innermost, the concentric ellipses are for $\sqrt{s} = 0.5\,, 1.0\,, 1.5$ TeV. $x_{e(L)}$ represents the position of the right(left)-handed electron. The size of the extra dimension, πR_c has been normalized to unity. From [42].

• *Fermions in the bulk.* In the scenario known as universal extra dimensions [44], all Standard Model fields, including the fermions, propagate in the bulk. The simplest model of this type contains a single extra dimension compactified on an S_1/Z_2 orbifold. This case is different in that branes are not present; translational invariance in the higher dimensional space is thus preserved. This leads to the tree-level conservation of the δ-

dimensional momentum of the bulk fields, which in turn implies that KK number is conserved at tree-level while KK parity, $(-1)^n$, is conserved to all orders. Two immediate consequences of KK number and parity conservation are that (*i*) at tree-level, KK excitations can no longer be produced as s-channel resonances and can only be produced in pairs [45], and (*ii*) the lightest KK particle (LKP) is stable and is thus a dark matter candidate [46]. The former results in a substantial reduction of the sensitivity to such states in precision electroweak and collider data with the present bound on the mass of the first KK excitation being of order a few hundred GeV for $\delta = 1$ [44, 45].

Due to the conservation of KK parity, the phenomenology of universal extra dimensions resembles that of supersymmetric theories with R-parity conservation. Every Standard Model field has KK partners. These KK partners carry a conserved quantum number, KK parity, which guarantees that the LKP is stable. Heavier KK modes decay via cascades to the LKP, which escapes detection resulting in a missing energy signature. The Lagrangian of the model includes both bulk terms and interactions which are localized at the orbifold fixed points. The bulk interactions yield masses for the KK states as given in Eq. (4.2); these are highly degenerate within each KK level. The boundary interactions take the form of loop induced localized kinetic terms [47] and are not universal for the different Standard Model fields. In addition, non-local radiative corrections [48] also affect the KK mass spectrum. Combined, this has the effect of removing the degeneracy between the modes in a given KK level and also affects their decay chains. A typical KK spectrum [49] is displayed in Fig. 17. We see that the KK modes with strong interactions receive larger corrections than those with only electroweak interactions. This spectrum resembles that of the superpartners in a Minimal Supersymmetric Standard Model where the sparticles all lie relatively close together.

Once the KK mass degeneracy is lifted, the states decay promptly. The resulting collider phenomenology is strikingly similar to that of supersymmetry [49]. Indeed, if such a scenario were observed at the LHC, it is reasonable to expect that it would be mis-identified as the discovery of supersymmetry! Superpartner status would most likely be assigned to the various KK states. It is thus essential to develop techniques to distinguish the two scenarios.

The cleanest way to identify Universal Extra Dimensions would be the observation of the second KK level. However, due to the conservation of KK parity, second level KK states must be pair produced with each other (or singly at loop-level) and may be out of kinematic reach of the LHC. The distinguishing factor for this scenario is that the KK states have the same spin as their Standard Model partners, unlike the case of supersymmetry. It is thus imperative to measure the spin of the particles in the spectrum. In principle, particle spins can be determined at the LHC from examination of their decay distributions. However, given the cascade decay chains and the size of the pair production cross sections, this is expected to be problematic at the LHC.

The measurement of the threshold cross section for the production of new particles can determine their spin; in particular, scalar particle production is suppressed and has a slower turn on. This difference between the S-wave versus P-wave behavior of the threshold cross section can be cleanly determined at the LC as illustrated in Fig. 18. Further studies [51] have shown that a combined analysis at the LC of the angular distributions and energy spectrum of the final decay particles, as well as the total cross section measurement can easily distinguish Universal Extra Dimensions from Supersymmetry.

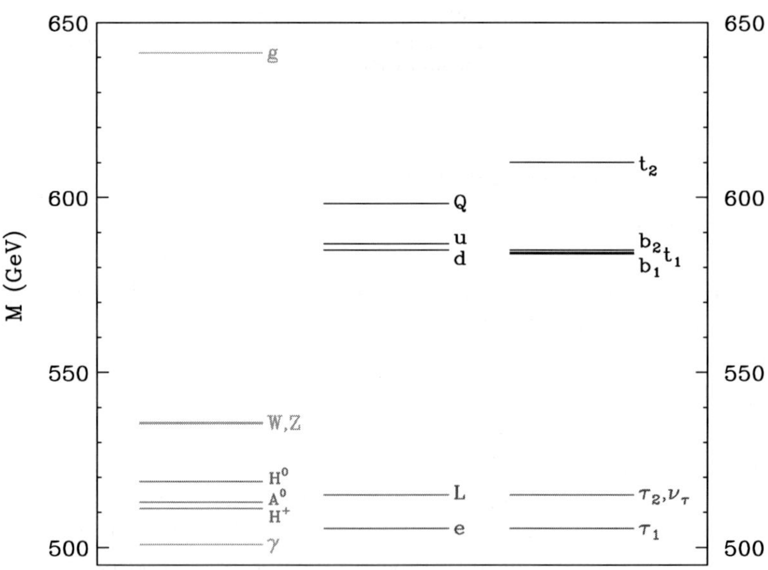

Fig. 17. Mass spectrum, including radiative corrections, of the first level KK states in the Universal Extra Dimensions scenario, taking $R^{-1} = 500$ GeV. From [49].

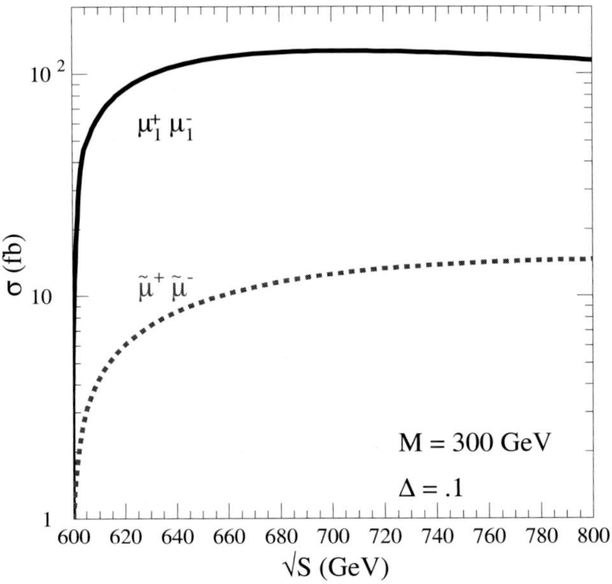

Fig. 18. Pair production of 300 GeV smuons and muon KK states at an e^+e^- collider as a function of center of mass energy. Δ represents the mass splitting between the smuon and LSP as well as of the muon KK and LKP. From [50].

4.3. *Warped Extra Dimensions with Localized Gravity*

In this scenario, known as the Randall-Sundrum (RS) model, the hierarchy between the electroweak and Planck scales is generated by a large curvature of the extra dimension [52]. The simplest such framework is comprised of just one additional spatial dimension of finite size in which gravity propagates. The geometry is that of a 5-dimensional Anti-de-Sitter space (AdS_5), which is a space of constant negative curvature. The extent of the 5th dimension is $y = \pi R_c$, and every slice of the 5th dimension corresponds to a 4-d Minkowski metric. Two 3-branes, with equal and opposite tension reside rigidly at orbifold fixed points at the boundaries of the AdS_5 slice. The Standard Model fields are conventionally constrained to the 3-brane located at $y = \pi R_c$, known as the TeV-brane, while gravity is localized about the opposite brane at the other boundary $y = 0$. This is referred to as the Planck brane.

The metric for this scenario is

$$ds^2 = e^{-2ky}\eta_{\mu\nu}dx^\mu dx^\nu - dy^2\,, \tag{9}$$

where the exponential function, or warp factor, multiplying the usual 4-d Minkowski term indicates a non-factorizable geometry. Here, the parameter k governs the degree of curvature of the AdS_5 space. Consistency of the low-energy description requires that the curvature be bounded by $k/\overline{M}_{Pl} \lesssim 0.1$ with $\overline{M}_{Pl} = M_{Pl}/\sqrt{8\pi} = 2.4 \times 10^{18}$. The fundamental scale M_5 is also of order $\overline{M}_{Pl}$.

If the gravitational wavefunction is localized on the Planck brane, then TeV scales can naturally be attained [52] on the TeV-brane providing $kR_c \simeq 10$. The scale $\Lambda_\pi \equiv \overline{M}_{Pl}e^{-kR_c\pi} \sim 1$ TeV, where $\overline{M}_{Pl} = M_{Pl}/\sqrt{8\pi}$ is the reduced Planck scale, then describes the scale of all physical processes on the TeV-brane. It has been demonstrated [53] that this value of kR_c can be stabilized without the fine tuning of parameters. The hierarchy is thus naturally established by the warp factor.

The 4-d phenomenology of this model is governed by two parameters, Λ_π and the ratio $k/\overline{M}_{Pl}$. The Feynman rules are obtained by a linear expansion of the flat metric, including the warp factor. After compactification, a KK tower of gravitons appears on the TeV-brane and have masses $m_n = x_n k e^{-kR_c\pi} = x_n\Lambda_\pi k/\overline{M}_{Pl}$ with the x_n being the roots of the first-order Bessel function, *i.e.*, $J_1(x_n) = 0$. Note that the first excitation is naturally of order a few hundred GeV and that the KK states are not evenly spaced. The interactions of the graviton KK tower with the Standard Model fields on the TeV-brane are [54]

$$\mathcal{L} = \frac{-1}{\overline{M}_{Pl}}T^{\mu\nu}(x)h^{(0)}_{\mu\nu}(x) - \frac{1}{\Lambda_\pi}T^{\mu\nu}(x)\sum_{n=1}^{\infty} h^{(n)}_{\mu\nu}(x)\,, \tag{10}$$

where $T^{\mu\nu}$ is the stress energy tensor. Note that the zero-mode decouples and that the couplings of the excitation states are inverse TeV strength. This results in a strikingly different phenomenology than in the case of large extra dimensions. Here, the graviton KK tower states are directly produced as single resonances. To exhibit how this may appear at a Linear collider, Fig. 19 displays the cross section for $e^+e^- \to \mu^+\mu^-$ as a function of $\sqrt{s}$, assuming $m_1 = 500$ GeV and taking $k/\overline{M}_{Pl} = 0.01 - 0.05$. Note that the curvature parameter controls the width of the resonance. The height of the third resonance is somewhat reduced as the higher KK excitations decay to the lighter graviton states, once it is kinematically allowed [55]. In this case, one can study graviton self-couplings and higher energy e^+e^- colliders may become graviton factories!

The LHC can discover these resonances in the Drell-Yan channel if $\Lambda_\pi < 10$ TeV [54, 56], provided that the resonance width is not too narrow, and determine their spin-2 nature via the angular distributions of the final-state lepton pairs [57] if enough statistics are available. An accurate determination of the branching fractions for the graviton KK decays to various final states could be performed at a LC and would probe the universal $T^{\mu\nu}$ structure of the couplings and verify the production of gravity.

If the KK gravitons are too massive to be produced directly, their contributions to fermion pair production may still be felt via virtual exchange. In this case, the uncertainties associated with a cut-off (as present in the large extra dimensions scenario) are avoided, since there is only one additional dimension and thus the KK states may be neatly summed. The resulting sensitivity to the scale Λ_π at the LHC and LC is displayed in Table 4. We see that the reach of the 500 GeV LC is complementary to that of the LHC and that a 1 TeV LC extends the discovery reach of the LHC. This degree of sensitivity to virtual graviton KK exchange at the LC implies that the KK coupling measurements discussed above should be viable.

Table 4. 95% CL search reach for Λ_π in TeV in the contact interaction regime taking 500 and 100 fb^{-1} of integrated luminosity at the LC and LHC, respectively. From [54].

	$k/\overline{M}_{Pl}$		
	0.01	0.1	1.0
LC $\sqrt{s} = 0.5$ TeV	20.0	5.0	1.5
LC $\sqrt{s} = 1.0$ TeV	40.0	10.0	3.0
LHC	20.0	7.0	3.0

• *Extensions of the RS Model: Standard Model bulk fields.* If the Standard Model fields are also allowed to propagate in the bulk [58], the phenomenology can be markedly different, and is highly dependent upon the value of the 5-dimensional fermion mass. In this scenario, the masses of the gauge KK states are generically constrained to be large by precision electroweak data. In fact the first gauge KK excitation usually lies above the LHC search reach of ~ 7 TeV. In this case, the precision available at the Giga-Z option of the Linear Collider may be able to discern the virtual effects of such heavy gauge KK states as illustrated in Fig. 20.

• *Extensions of the RS Model: Extended Manifolds.*

From a theoretical perspective, the RS model may be viewed as an ef-

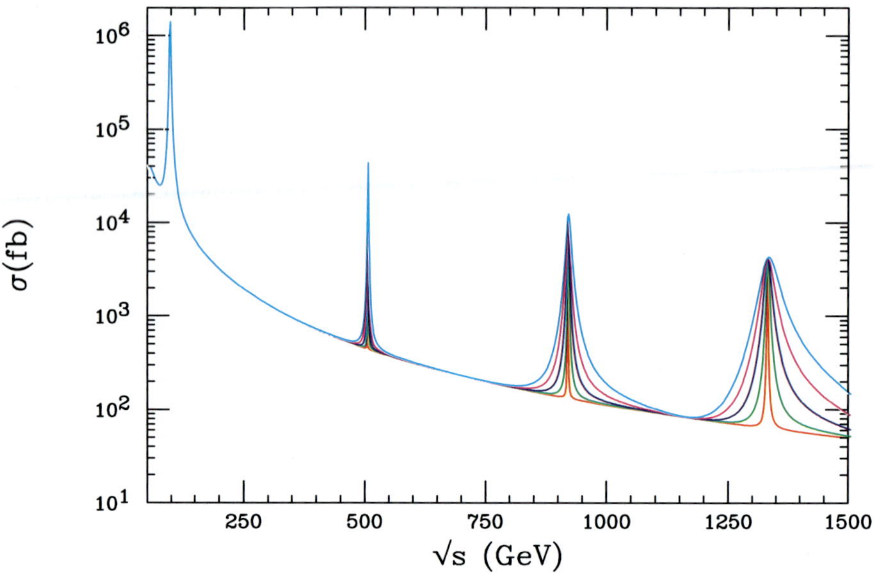

Fig. 19. The cross section for $e^+e^- \to \mu^+\mu^-$ including the exchange of a KK tower of gravitons with $m_1 = 500$ GeV. The curves correspond to $k/\overline{m}_{Pl} =$ in the range 0.01–0.05. From [54].

fective theory whose low energy features originate from a full theory of quantum gravity, such as string theory. One may thus expect that a more complete version of this scenario admits the presence of additional dimensions compactified on a manifold.

The existence of an extra manifold modifies the conventional RS phenomenology and collider signatures. As a first step, the additional manifolds S^δ with $\delta \geq 1$ have been considered [60]. In this case, additional graviton KK states appear, corresponding to orbital excitations of the S^δ manifold. For the simplest scenario of S^1, the RS metric is expanded to

$$ds^2 = e^{-2kr_c\phi}\eta_{\mu\nu}dx^\mu dx^\nu + r_c^2 d\phi^2 + R^2 d\theta^2 \,, \tag{11}$$

where θ parameterizes the S^1, and R represents its radius. The masses of the KK states are now given by $m_{n\ell} = x_{n\ell}\Lambda_\pi k/\overline{M}_{Pl}$, where the $x_{n\ell}$ are solutions of the equation

$$2J_\nu(x_{n\ell}) + x_{n\ell}J'_\nu(x_{n\ell}) = 0 \,, \tag{12}$$

with $\nu \equiv \sqrt{4 + (\ell/kR)^2}$. The KK mode number ℓ corresponds to the orbital excitations, while n denotes the usual RS AdS_5 mode levels. The couplings

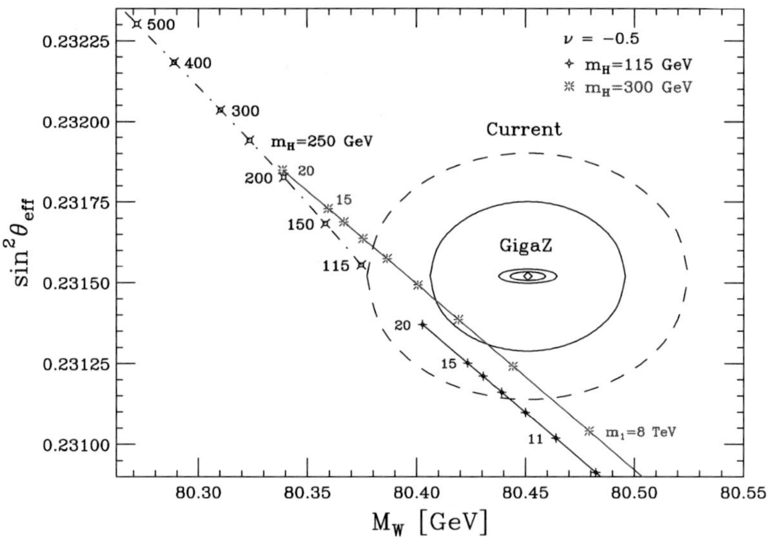

Fig. 20. The plane $\sin^2\theta_{eff}$ versus M_W showing current and future sensitivities, Standard Model and RS model predictions. The diamond show the current measured values of the observables. The large solid and dashed ellipses represent respectively the 68% and 95% CL regions from current data, while the smaller solid ellipses anticipate the same after operation of GigaZ. The black dashdot line shows the Standard Model predictions for different Higgs boson masses as labeled, while the solid colored lines show the RS model results for varying values of the first gauge KK excitation mass for two different Higgs masses. The value of ν denotes the location of the fermions in the 5th dimension. From [59].

of the $m_{n\ell}$ graviton KK states are then given by

$$\mathcal{L} = -\frac{1}{\overline{M}_{Pl}} T^{\mu\nu}(x) h^0_{\mu\nu}(x) - \frac{1}{\Lambda_\pi} T^{\mu\nu}(x) \sum_{n=1}^{\infty} \xi(n\ell) h^{(n,\ell)}_{\mu\nu}(x) , \qquad (13)$$

where $\xi(n\ell)$ depends on k , R, and $x_{n\ell}$ [60]. The KK spectrum and couplings are thus modified due to the presence of the new manifold.

In particular, the addition of the S^δ background to the RS setup results in the emergence of a forest of graviton KK resonances. These originate from the orbital excitations on the S^δ and occur in between the original RS resonances, and the separation between the states changes as the model parameters are varied. This is displayed in Fig. 21 which shows the cross section in $e^+e^- \to \mu^+\mu^-$ for $kR = 2.0$ and $m_{10} = 600$ GeV. Note that the density of the KK spectrum is substantially increased; in particular, both n and ℓ KK excitations are present in the same kinematic region and the peaks are not well separated. In the case shown here, the spacing between the $n = 1, \ell = 0$ and $n = 1, \ell = 1$ states is only 425 MeV, which is comparable to the individual KK widths of $\simeq$ 525 MeV. The μ pair resolution at a LC should be sufficient to resolve these two states if they are within kinematic reach.

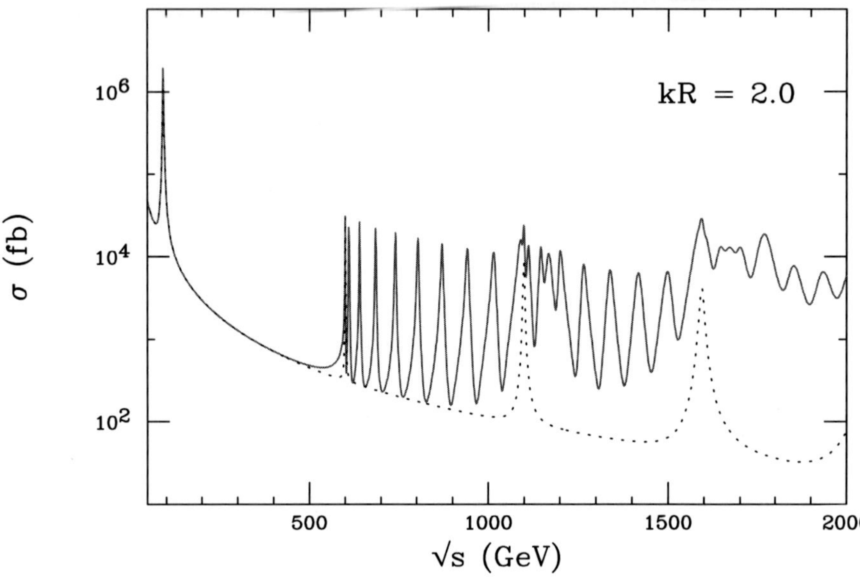

Fig. 21. The solid (red) curve corresponds to the cross section for $e^+e^- \to \mu^+\mu^-$ when the additional dimension is orbifolded, *i.e.*, for S^1/Z_2, with $m_{10} = 600$ GeV, $k/\overline{M}_{Pl} = 0.03$ and $kR = 2.0$. The result for the conventional RS model is also displayed, corresponding to the dotted curve. From [60].

5. Direct Production of New Particles

New particles are present in most theories containing physics beyond the Standard Model. Several classes of particles are generic features of many extensions of the Standard Model. Here, we consider two such examples, new fermions which carry lepton number, and leptoquarks. It is notoriously difficult [61] to observe the production of new heavy leptons above background at the LHC, and in this case a 0.5 – 1 TeV Linear Collider can compete with or extend the search reach of the LHC. Leptoquarks provide a good example of how the detailed properties of new particles can be measured at the Linear Collider, even if they are first discovered at the LHC.

5.1. *New Leptons*

Additional fermions beyond those contained in the Standard Model appear in many extensions of the Standard Model, notably in several grand unified theories [62]. They are usually called exotic fermions if they do not have the usual $SU(2)_L \times U(1)_Y$ quantum numbers of the Standard Model matter fields. Classes of new fermions include: (*i*) The sequential repetition of a Standard Model generation, (*ii*) Mirror fermions, which have chiral properties opposite to those of their Standard Model counterparts [63]. The restoration of left-right symmetry is a motivating factor for this possibility. (*iii*) Vector fermions, where the left- and right-handed fields have the same weak isospin. For instance, in E_6 grand unified theories, there are two additional isodoublets of leptons, one LH and one RH. (*iv*) Additional isosinglet fermions have also been much studied in general, particularly the right-handed neutrino in SO(10) and the singlet neutrinos of E_6.

Exotic fermions can mix with the ordinary Standard Model fermions; in principle, the mixing pattern may be complicated and is model independent. One simplifying factor is that intergenerational mixing is severely limited by the constraints on flavor changing neutral currents [10]. Global fits of low energy electroweak data and the high precision measurements of the properties of the Z boson provide upper limits for the remaining mixing angles of the order of $\sin^2\theta_{\rm mix} \leq 10^{-2} - 10^{-3}$ [64].

Exotic fermions may be produced in e^+e^- collisions either in pairs or singly in association with their Standard Model partner as a result of mixing. The cross section for pair production of exotic quarks via gluon fusion at the LHC is large enough that the reach of the LC is unlikely to be competitive [10]. On the other hand, the backgrounds to heavy lepton production are large in pp collisions, and production in e^+e^- collisions provides a promising alternative. Generally, the search reach for exotic leptons at the Linear Collider is up to the kinematic limit, given by the beam energy for pair production and the center of mass energy for single exotic lepton production [65]. The experimental signature requires knowledge of the $L^\pm$ decay mode, which is model dependent and also depends on the mass difference of the charged and neutral exotic leptons. Studies indicate that the signals for exotic lepton production are clear and easy to separate from Standard Model backgrounds [62, 65, 66], and the use of polarized beams is important in determining the electroweak quantum numbers [66].

Neutral heavy lepton production at the LC has been studied extensively [67]. Single heavy neutrino production via the process $e^+e^- \rightarrow \nu N \rightarrow$

$\nu e^{\pm} W^{\mp}$, including on-shell and off-shell heavy neutrinos N, occurs at a larger rate that pair production, and the Linear Collider can probe the existence of heavy Dirac and Majorana neutrinos up to $\sqrt{s}$ in the $\nu e^{\pm}+$ hadrons channel. Single heavy neutrino production can be clearly separated from Standard Model backgrounds, particularly on application of angular cuts on the final state particle distributions. Figure 22 displays the heavy neutrino production cross sections for various pair and single production processes, with all fermion mixing angles set to $\sin^2\theta_{\rm mix} = 0.0052$ [64]

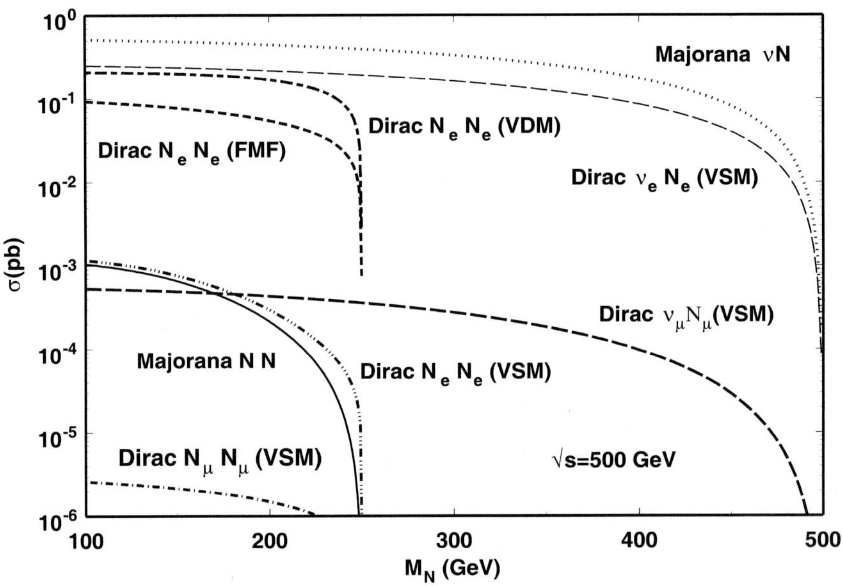

Fig. 22. Single and pair production cross sections of on-shell heavy Dirac and Majorana neutrinos at $\sqrt{s} = 500$ GeV for e^+e^- colliders. VDM(VSM) refers to whether the heavy neutrino resides in a SU(2) vector doublet(singlet). From [67].

5.2. *Leptoquarks*

Leptoquarks are a natural product of theories which relate leptons and quarks at a more fundamental level. These spin-0 or -1 particles carry both baryon and lepton number and are color triplets under $SU(3)_c$. They can be present at the electroweak scale in models where baryon and lepton number are separately conserved thus avoiding conflicts with rapid proton decay. Their remaining properties depend on the model in which they appear, and would need to be determined in order to ascertain the framework of the underlying theory. There are 14 different possible types of leptoquarks that couple to the Standard Model fermions; their classification can be found in [68]. Their fermionic couplings are given by a Yukawa interaction of unknown strength, while their gauge couplings are specified for a specific class of leptoquark. Low-energy data place tight constraints on inter-generational leptoquark Yukawa couplings and also require that these couplings be chiral.

At a Linear Collider, leptoquarks may be directly produced in pairs or as single particle, while virtual leptoquark exchange may be present in $e^+e^- \to$ hadrons. Pair production receives a t-channel quark-exchange contribution whose magnitude depends on the size of the Yukawa coupling. This only competes with the usual s-channel exchange, which depends on the leptoquark's gauge couplings, if the Yukawa coupling is of order electromagnetic strength. The possible final state signatures after the leptoquark decays are $\ell^+\ell^-$, $\ell^\pm\ \not\!\!E_T$, or $\not\!\!E_T + 2$ jets, where $\ell = e\,,\mu\,,\tau$, and observation is straightforward essentially up to the kinematic limit. A thorough study of the background and resulting search reach for each type of leptoquark can be found in [69]. Single leptoquark production proceeds via the reaction $e\gamma \to LQ+$ jet and is most easily studied in terms of the quark content of the photon [70]. In this case a lepton fuses with a quark from the Weiszacker-Williams (in e^+e^- mode) or laser backscattered (for $e\gamma$ LC option) photon to produce a leptoquark. The cross section is a convolution of the parton level process with distribution functions for the photon in the electron and the quark in the photon and is directly proportional to the $eqLQ$ Yukawa coupling. The kinematic advantage of single production is thus lost if the Yukawa coupling is too small. For Yukawa couplings of electromagnetic strength, leptoquarks with mass up to about 90% of $\sqrt{s}$ can be discovered at a LC [70]. If the Yukawa couplings are sizable enough, then virtual leptoquark exchange [71] will lead to observable deviations in the hadronic production cross section for leptoquark masses in excess of $\sqrt{s}$. A summary of the search reach from these three processes is shown in Fig. 23 in the leptoquark mass-coupling plane. In comparison, leptoquarks are produced strongly at the LHC, with search reaches in the 1.5 TeV range [72] independent of the Yukawa couplings.

The real strength of the LC is in the determination of the leptoquark's quantum numbers and the strength of its Yukawa couplings once it is discovered. Together, the production rate and polarized left-right asymmetry

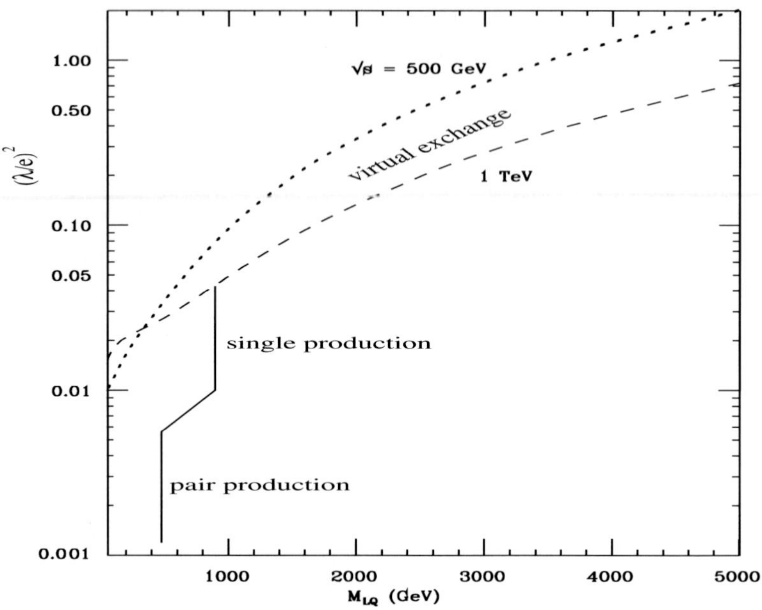

Fig. 23. Leptoquark search limits at a LC from the three processes discussed in the text. The Yukawa coupling is scaled to e. The pair and single production reaches are shown for $\sqrt{s} = 1$ TeV, while the indirect reach is displayed for both $\sqrt{s} = 0.5, 1$ TeV. From [69].

can completely determine the leptoquark's electroweak properties and identify its type [73] in both the pair and single production channels, up to the kinematic limit. In addition, the Yukawa coupling strength can be measured via the forward-backward asymmetry in leptoquark pair production (which is non-vanishing for significant Yukawa couplings), deviations in the hadronic cross sections, and the comparison of pair and single production rates.

6. Unconventional Theories and Possible Surprises

So far in this chapter, we have delineated the potential of a Linear Collider to explore the new physics which is present in set classes of established models. However, as likely as not, when nature finally reveals her mysteries it will be full of surprises containing phenomena outside the realm of our presently limited imaginations. Hence we must try to prepare to discover and elucidate the unexpected.

It is along these lines that we note that some of the most surprising recent developments in our field have been in string theory. While it is presently difficult to relate these theories to experiment, some of their ingredients, when considered in their own right, have interesting phenomenological consequences. Here, we consider two such examples of this approach, as a demonstration of the potential of the LC to uncover the unforseen.

6.1. *String Resonances*

If the scenario with large extra dimensions discussed in Section 4.1 is embedded in a string theory, then stringy effects must also appear at the TeV scale. If the string theory is weakly coupled, not only the gravitons, but also the Standard Model fields will exhibit an extended structure. The exchange of string Regge excitations of Standard Model particles in $2 \to 2$ scattering may appear as contact-like interactions with a strength that overwhelms the corresponding graviton exchange. This is deduced from simple coupling counting arguments; Yang-Mills bosons live at the end of open strings, while gravitons correspond to closed string states and require an additional coupling constant factor at the amplitude level. Hence the exchange of KK graviton states is suppressed by a factor of g^2 compared to the exchange of string Regge excitations.

This has been examined in [74, 75], where an illustrative string model was employed. Here, a string theory version of QED is modeled after scattering amplitudes on the 3-brane from weakly-coupled type IIB string theory. Electrons and photons then correspond to massless states of open strings ending on the 3-brane and are characterized by quantum fluctuations of an open string with specified boundary conditions. Within the context of this model, Bhabha scattering ($e^+e^- \to e^+e^-$) and pair annihilation ($e^+e^- \to \gamma\gamma$) receive contributions from string Regge exchanges. The differential cross section for these processes is modified by a form factor,

$$\frac{d\sigma}{d\cos\theta} = \left(\frac{d\sigma}{d\cos\theta}\right)_{SM} \left| \frac{\Gamma(1 - s/M_{str}^2)\Gamma(1 - t/M_{str}^2)}{\Gamma(1 - s/M_{str}^2 - t/M_{str}^2)} \right|^2 , \qquad (14)$$

which is of the form in the original work by Veneziano [76]. Here, M_{str} represents the string scale and can be related to the Fundamental scale in the large extra dimension scenario via $M_D/M_{str} = \pi^{-1/8}\alpha^{-1/4}$. Figure 24 displays the deviation from Standard Model expectations for Bhabha scattering from these string exchanges and compares their effect to those arising from other types of contact interactions. The 95% CL sensitivity for $\sqrt{s} = 1$ TeV and 100 fb^{-1} is $M_{str} < 3.1$ TeV, which corresponds to $M_D/\lambda^{1/4} < 9.3$ TeV.

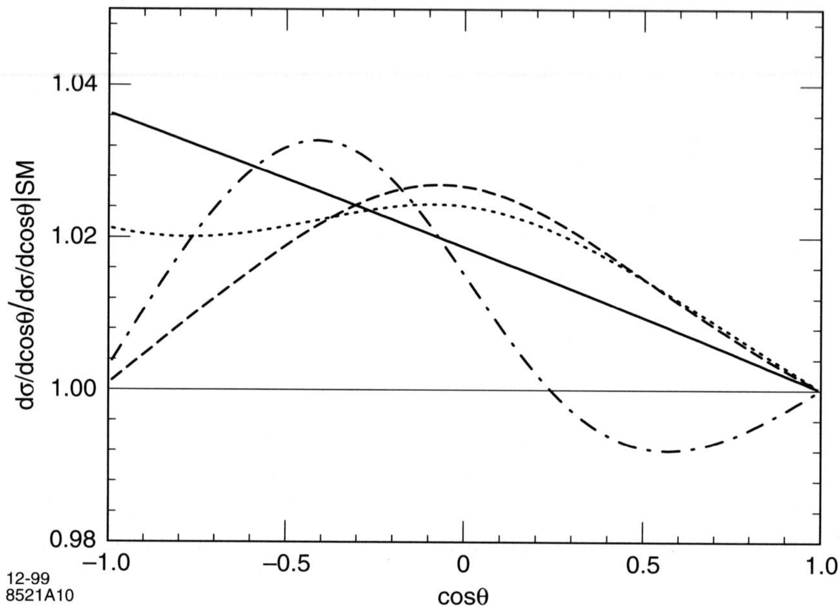

Fig. 24. Comparison of deviations from the Standard Model prediction for Bhabha scattering at 1 TeV due to corrections from higher dimension operators. The curves correspond to: string model with $M_{str} = 3.1$ TeV (solid), KK graviton exchange with $M_D/\lambda^{1/4} = 6.2$ TeV (dotted), VV contact interactions with $\Lambda = 88$ TeV (dashed), and AA contact interactions with $\Lambda = 62$ TeV (dot-dashed). From [74].

6.2. *Non-Commutative Field Theories*

Recent theoretical results have demonstrated that non-commutative quantum field theories (NCQFT) naturally appear within the context of string or M-theory in the presence of background electromagnetic fields [77]. In this scenario, the usual D-dimensional space associated with commuting space-time coordinates is generalized to one which is non-commuting. In such a space, the conventional coordinates are represented by operators which no longer commute,

$$[\hat{X}_\mu, \hat{X}_\nu] = i\theta_{\mu\nu} \equiv \frac{i}{\Lambda_{NC}^2} c_{\mu\nu} \,. \tag{15}$$

Here, the effect has been parameterized in terms of an overall scale Λ_{NC}, which characterizes the threshold where non-commutative (NC) effects become important, and a real antisymmetric matrix $c_{\mu\nu}$, whose dimensionless elements are presumably of order unity. The most likely value of Λ_{NC} is near the string or Planck scale, which could be a TeV. The matrix $c_{\mu\nu}$ is related to the Maxwell field strength tensor $F_{\mu\nu}$ in a straight-forward fashion. $c_{\mu\nu}$ is identical in all reference frames, defining a preferred NC direction in space; Lorentz invariance is thus violated at energies of order Λ_{NC}. This is affected by the rotation of the Earth and its revolution about the Sun, and thus the violation of Lorentz invariance is only observable in processes which are quadratic (or higher order even powers) in $\theta_{\mu\nu}$ and experiment must keep track of sidreal time.

Caution must be exercised to preserve orderings of the products of fields when formulating NCQFT. This can be accomplished with the introduction of the star product, $\phi(\hat{X})\phi(\hat{X}) = \phi(x) * \phi(x) = \phi(x)e^{[i\theta^{\mu\nu}\partial_\mu\partial_\nu/2]}\phi(x)$, which absorbs the effect of the commutation relation via a series of Fourier transforms. The NC action for a quantum field theory is thus obtained from the conventional one by replacing the products of fields by star products. A striking consequence of this is that the NC version of QED takes on a non-abelian nature in that both 3-point and 4-point photon couplings are generated. In addition, all QED vertices pick up additional phase factors which are dependent upon the momenta flowing through the vertex [78]. We note that propagators, however, are not modified since quadratic forms remain unchanged under the properties of the star product. NC effects thus have striking signatures in QED processes at a Linear Collider. The modifications to pair annihilation, Bhabha and Møller scattering, as well as $\gamma\gamma \to \gamma\gamma$ have been studied in Ref. [79]. Pair annihilation and $\gamma\gamma$ scattering both receive new diagrammatic contributions due to the non-abelian couplings, and all four processes acquire a phase dependence due to the relative interference of the vertex kinematic phases. The lowest order correction to the Standard Model in these processes occurs at dimension 8. The most dramatic result is that an azimuthal dependence is induced in $2 \to 2$ scattering processes due to the existence of the NC preferred direction in space-time. This azimuthal dependence in pair annihilation is illustrated in Fig. 25 for the case where the NC direction is perpendicular to the beam axis. The 95% CL sensitivity to Λ_{NC} is summarized in Table 5 for these four reactions. We see that these processes are complementary in their ability to probe different structures of non-commuting effects between space and time coordinates as well as between the space coordinates.

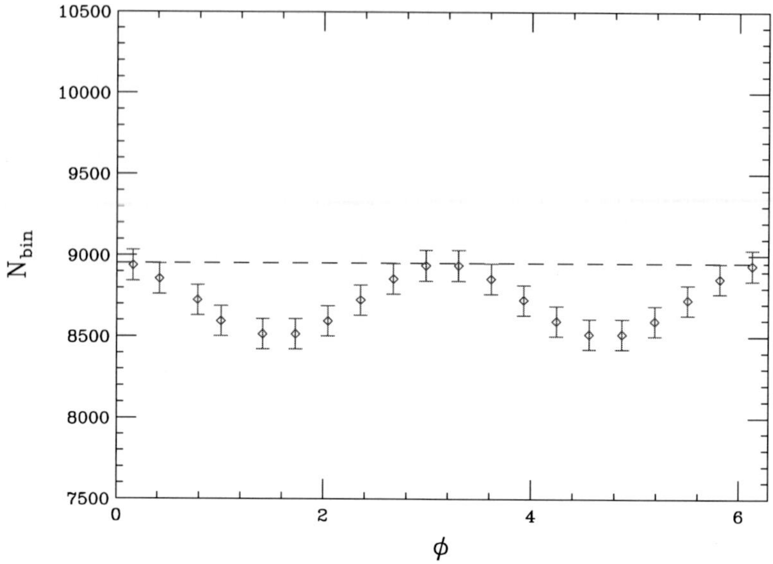

Fig. 25. ϕ dependence of the $e^+e^- \to \gamma\gamma$ cross section, taking $\Lambda_{NC} = \sqrt{s} = 500$ GeV and a luminosity of $500\,\mathrm{fb}^{-1}$. A cut of $|\cos\theta| < 0.5$ has been employed. The dashed line corresponds to the SM expectations and the 'data' points represent the NCQED results with statistic error bars. From [79].

Table 5. Summary of the 95% CL search limits on the NC scale Λ_{NC} from various processes at a 500 GeV Linear Collider with an integrated luminosity of 500 fb^{-1}. From [79].

Process	Structure Probed	Bound on Λ_{NC}
$e^+e^- \to \gamma\gamma$	Space-Time	740 – 840 GeV
Møller Scattering	Space-Space	1700 GeV
Bhabha Scattering	Space-Time	1050 GeV
$\gamma\gamma \to \gamma\gamma$	Space-Time	700 – 800 GeV
	Space-Space	500 GeV

7. Conclusions

In this chapter, we have discussed several classes of motivated models that contain new phenomena, and we have delineated the ability of a Linear Collider to explore them. We have seen that in numerous scenarios the LHC and the Linear Collider have a comparable and complementary discovery potential. In most of these cases, a signal for new physics will first be observed at the LHC, and the Linear Collider will then precisely determine its properties and reconstruct the fundamental underlying theory.

The human imagination is limited and does not span the full range of alternatives that might be present in nature and are allowed by present data. We must be prepared to discover and elucidate the unexpected. This is best accomplished by the exploration of the energy frontier by both e^+e^- and hadron colliders.

References

1. E. Eichten, K. Lane, and M. Peskin, Phys. Rev. Lett. **50**, 811 (1983).
2. S. Eidelman *et al.* [Particle Data Group Collaboration], Phys. Lett. **B592**, 1 (2004).
3. ATLAS Detector and Physics Performance Technical Design Report, LHCC 99-14/15 (1999).
4. Z.U. Usubov and O.B. Abdinov, Phys. Atom. Nucl. **64**, 1288 (2001) [Yad. Fiz. **64**, 1364 (2001)].
5. K. Cheung, S. Godfrey, and J.L. Hewett, in *DPF/DPB Summer Study on New Directions for High-Energy Physics*, Snowmass, CO, 1996, ed. D. Cassel *et al.*, hep-ph/9612257.
6. S. Riemann, "Indications of new physics in fermion pair production," Proceedings of the *5th International Linear Collider Workshop (LCWS 2000), Fermilab, Batavia, Illinois, 24–28 Oct 2000*, LC-TH-2001-007.
7. T. L. Barklow, Int. J. Mod. Phys. **A11**. 1579 (1996).
8. T. G. Rizzo, Phys. Rev. **D59**, 113004 (1999) [arXiv:hep-ph/9811440].
9. For a set of reviews, see, A. Leike, Phys. Reports **317** (1999) 143; T. G. Rizzo, in *Proc. of the 1996 DPF/DPB Summer Study on New Directions for High Energy Physics (Snowmass96)*, eConf **C960625**, NEW136 (1996) [arXiv:hep-ph/9612440]; S. Godfrey, Phys. Rev. **D51**, 1402 (1995) [arXiv:hep-ph/9411237]; M. Cvetič, S. Godfrey, in *"Electroweak Symmetry Breaking and Beyond the Standard Model"*, T. Barklow, S. Dawson, H. Haber, J. Siegrist (eds.), World Scientific 1995.
10. J. L. Hewett and T. G. Rizzo, Phys. Rept. **183**, 193 (1989).
11. L. S. Group *et al.*, arXiv:hep-ph/0410364.
12. T. G. Rizzo, arXiv:hep-ph/0303056.
13. A. Freitas, Phys. Rev. **D70**, 015008 (2004) [arXiv:hep-ph/0403288].
14. M. Dittmar, A. S. Nicollerat and A. Djouadi, Phys. Lett. **B583**, 111 (2004) [arXiv:hep-ph/0307020]; F. del Aguila, M. Cvetic and P. Langacker, Phys. Rev. **D48**, 969 (1993) [arXiv:hep-ph/9303299]; J. L. Hewett and T. G. Rizzo, Phys. Rev. **D45**, 161 (1992).
15. T. G. Rizzo, Phys. Rev. **D55**, 5483 (1997) [arXiv:hep-ph/9612304].
16. S. Godfrey, P. Kalyniak, B. Kamal, A. Leike, Phys. Rev. **D61**, 113009 (2000).
17. J. Hewett, in *Proc. of the 1996 DPF/DPB Summer Study on New Directions for High Energy Physics (Snowmass96)*, eConf **C960625**, NEW139 (1996) [arXiv:hep-ph/9704292].
18. M. Doncheski, S. Godfrey, P. Kalyniak, B. Kamal, A. Leike, Phys. Rev. **D63**, 053005 (2001).

19. J. Hewett and M. Spiropulu, Ann. Rev. Nucl. Part. Sci. **52**, 397 (2002) [arXiv:hep-ph/0205106].
20. N. Arkani-Hamed, S. Dimopoulos, and G. Dvali, Phys. Lett. **B429**, 263 (1998), and Phys. Rev. **D59**, 086004 (1999); I. Antoniadis, N. Arkani-Hamed, S. Dimopoulos, and G. Dvali, Phys. Lett. **B436**, 257 (1998).
21. C. D. Hoyle, D. J. Kapner, B. R. Heckel, E. G. Adelberger, J. H. Gundlach, U. Schmidt and H. E. Swanson, Phys. Rev. **D70**, 042004 (2004) [arXiv:hep-ph/0405262].
22. G.F. Giudice, R. Rattazzi, and J.D. Wells, Nucl. Phys. **B544**, 3 (1999).
23. T. Han, J. Lykken, and R.J. Zhang, Phys. Rev. **D59**, 105006 (1999).
24. E.A. Mirabelli, M. Perelstein, and M. Peskin, Phys. Rev. Lett. **82**, 2236 (1999); K. Cheung and W.Y. Keung, Phys. Rev. **D60**, 112003 (1999).
25. G.W. Wilson, LC-PHSM-2001-010; A. Vest, LC-TH-2000-058, http://www.desy.de/ lcnotes.
26. L. Vacavant and I. Hinchliffe, J. Phys. **G27** 1839 (2001).
27. H. Murayama and J. D. Wells, Phys. Rev. D **65**, 056011 (2002) [arXiv:hep-ph/0109004].
28. J.L. Hewett Phys. Rev. Lett. **82**, 4765 (1999); T.G. Rizzo, Phys. Rev. **D59**, 115010 (1999).
29. T. G. Rizzo, arXiv:hep-ph/0303056.
30. F. Gianotti *et al.*, arXiv:hep-ph/0204087; V. Kabachenko, A. Miagkov, and A. Zenin, ATLAS Physics Note, ATL-PHYS-2001-012 (2001).
31. T. G. Rizzo, JHEP **0302**, 008 (2003) [arXiv:hep-ph/0211374].
32. T. G. Rizzo, JHEP **0210**, 013 (2002) [arXiv:hep-ph/0208027].
33. J. L. Hewett and D. Sadri, Phys. Rev. **D69**, 015001 (2004) [arXiv:hep-ph/0204063].
34. I. Antoniadis, Phys. Lett. **B246**, 377 (1990); J. D. Lykken, Phys. Rev. **D54**, 3693 (1996) [arXiv:hep-th/9603133]; I. Antoniadis and M. Quiros, Phys. Lett. **B392**, 61 (1997) [arXiv:hep-th/9609209].
35. See, for example, T.G. Rizzo and J.D. Wells, Phys. Rev. **D61**, 016007 (2000); P. Nath and M. Yamaguchi, Phys. Rev. **D60**, 116006 (1999); M. Masip and A. Pomarol, Phys. Rev. **D60**, 096005 (1999); W.J. Marciano, Phys. Rev. **D60**, 093006 (1999); L. Hall and C. Kolda, Phys. Lett. **B459**, 213 (1999); R. Casalbuoni, S. DeCurtis and D. Dominici, Phys. Lett. **B460**, 135 (1999); R. Casalbuoni, S. DeCurtis, D. Dominici and R. Gatto, Phys. Lett. **B462**, 48 (1999); A. Strumia, Phys. Lett**B466**, 107 (1999); F. Cornet, M. Relano and J. Rico, Phys. Rev. **D61**, 037701 (2000); C.D. Carone, Phys. Rev. **D61**, 015008 (2000); A. Delgado, A. Pomarol and M. Quiros, JHEP **0001**, 030 (2000).
36. See, for example, I. Antoniadis, C. Munoz and M. Quiros, Nucl. Phys. **B397**, 515 (1993); I. Antoniadis and K. Benalki, Phys. Lett. **B326**, 69 (1994), and Int. J. Mod. Phys. **A15**, 4237 (2000); I. Antoniadis, K. Benalki and M. Quiros, Phys. Lett. **B331**, 313 (1994); K. Benalki, Phys. Lett. **B386**, 106 (1996); T.G. Rizzo, Phys. Rev. **D61**, 055005 (2000).
37. See, for example, the analysis by G. Azuelos and G. Polesello in, G. Azuelos *et al.*, "The beyond the standard model working group: Summary report,"

Proceedings of Workshop on Physics at TeV Colliders, Les Houches, France, 21 May–1 Jun 2001, arXiv:hep-ph/0204031; G. Azuelos and G. Polesello, EPJ Direct (2003), ATLAS note SN-ATLAS-2003-023.
38. B. C. Allanach *et al.* [Beyond the Standard Model Working Group Collaboration], arXiv:hep-ph/0402295.
39. T. G. Rizzo, JHEP **0306**, 021 (2003) [arXiv:hep-ph/0305077].
40. N. Arkani-Hamed and M. Schmaltz, Phys. Rev. **D61**, 033005 (2000);
41. E.A. Mirabelli and M. Schmaltz, Phys. Rev. **D61**, 113001 (2000); G. C. Branco, A. de Gouvea and M. N. Rebelo, Phys. Lett. **B506**, 115 (2001) [arXiv:hep-ph/0012289]; D. E. Kaplan and T. M. P. Tait, JHEP **0111**, 051 (2001) [arXiv:hep-ph/0110126]; Y. Grossman and G. Perez, Phys. Rev. **D67**, 015011 (2003) [arXiv:hep-ph/0210053]; W. F. Chang and J. N. Ng, JHEP **0212**, 077 (2002) [arXiv:hep-ph/0210414]; C. Biggio, F. Feruglio, I. Masina and M. Perez-Victoria, Nucl. Phys. **B677**, 451 (2004) [arXiv:hep-ph/0305129]; B. Lillie and J. L. Hewett, Phys. Rev. **D68**, 116002 (2003) [arXiv:hep-ph/0306193]; B. Lillie, JHEP **0312**, 030 (2003) [arXiv:hep-ph/0308091]; Y. Grossman, R. Harnik, G. Perez, M. D. Schwartz and Z. Surujon, Phys. Rev. **D71**, 056007 (2005) [arXiv:hep-ph/0407260]; G. Perez and T. Volansky, arXiv:hep-ph/0505222.
42. T. G. Rizzo, Phys. Rev. **D64**, 015003 (2001) [arXiv:hep-ph/0101278].
43. N. Arkani-Hamed, Y. Grossman, and M. Schmaltz, Phys. Rev. **D61**, 115004 (2000).
44. T. Appelquist, H. C. Cheng and B. A. Dobrescu, Phys. Rev. **D64**, 035002 (2001) [arXiv:hep-ph/0012100].
45. T. G. Rizzo, Phys. Rev. **D64**, 095010 (2001) [arXiv:hep-ph/0106336].
46. G. Servant and T. M. P. Tait, Nucl. Phys. **B650**, 391 (2003) [arXiv:hep-ph/0206071], New J. Phys. **4**, 99 (2002) [arXiv:hep-ph/0209262]; H. C. Cheng, J. L. Feng and K. T. Matchev, Phys. Rev. Lett. **89**, 211301 (2002) [arXiv:hep-ph/0207125].
47. H. Georgi, A. K. Grant and G. Hailu, Phys. Lett. **B506**, 207 (2001) [arXiv:hep-ph/0012379]; M. Carena, T. M. Tait and C. E. Wagner, Acta Phys. Polon. **B33**, 2355 (2002) [arXiv:hep-ph/0207056].
48. H. C. Cheng, K. T. Matchev and M. Schmaltz, Phys. Rev. **D66**, 036005 (2002) [arXiv:hep-ph/0204342];
49. H. C. Cheng, K. T. Matchev and M. Schmaltz, Phys. Rev. **D66**, 056006 (2002) [arXiv:hep-ph/0205314].
50. J.L. Hewett and T. Tait, contribution to G. Weiglein *et al.* [LHC/LC Study Group], arXiv:hep-ph/0410364.
51. M. Battaglia, A. Datta, A. De Roeck, K. Kong and K. T. Matchev, arXiv:hep-ph/0502041.
52. L. Randall and R. Sundrum, Phys. Rev. Lett. **83**, 3370 (1999), and *ibid.*, 4690, (1999).
53. W.D. Goldberger and M.B. Wise Phys. Rev. Lett. **83**, 4922 (1999); O. DeWolfe, D.Z. Freedman, S.S. Gubser, and A. Karch, Phys. Rev. **D62**, 046008 (2000); M.A. Luty and R. Sundrum, Phys. Rev. **D62**, 035008 (2000).
54. H. Davoudiasl, J.L. Hewett, and T.G. Rizzo, Phys. Rev. Lett. **84**, 2080 (2000).

55. H. Davoudiasl, T.G. Rizzo, hep-ph/0104199.
56. H. Davoudiasl, J.L. Hewett, and T.G. Rizzo, Phys. Rev. **D63**, 075004 (2001).
57. B.C. Allanach, K. Odagiri, M.A. Parker, and B.R. Webber, JHEP **0009**, 019 (2000).
58. H. Davoudiasl, J.L. Hewett, and T.G. Rizzo, Phys. Lett. **B473**, 43 (2000), Phys. Lett. **B493**, 135 (2001), and Phys. Rev. **D63**, 075004 (2001); A. Pomarol, Phys. Lett. **B486**, 153 (2000); Y. Grossman and M. Neubert, Nucl. Phys. **B474**, 361 (2000); R. Kitano, Phys. Lett. **B481**, 39 (2000); S. Chang *et al.*, Phys. Rev. **D62**, 084025 (2000); T. Ghergetta and A. Pomarol, Nucl. Phys. **B586**, 141 (2000).
59. J. L. Hewett, F. J. Petriello and T. G. Rizzo, JHEP **0209**, 030 (2002) [arXiv:hep-ph/0203091].
60. H. Davoudiasl, J. L. Hewett and T. G. Rizzo, JHEP **0304**, 001 (2003) [arXiv:hep-ph/0211377].
61. V. D. Barger, T. Han and J. Ohnemus, Phys. Rev. **D37**, 1174 (1988); I. Hinchliffe, Int. J. Mod. Phys. **A4**, 3867 (1989).
62. For a review, see, A. Djouadi, J. Ng, and T.G. Rizzo, in Electroweak Symmetry Breaking and New Physics at the TeV Scale, ed. T. Barklow *et al*, (World Scientific, Singapore), 416 (1996).
63. J. Maalampi, K. Mursula and M. Roos, Nucl. Phys. **B207**, 233 (1982).
64. P. Langacker and D. London, Phys. Rev. **D38**, 886 (1988).E. Nardi, E. Roulet and D. Tommasini, Nucl. Phys. **B386**, 239 (1992), Phys. Lett. **B327**, 319 (1994).
65. A. Djouadi, Z. Phys. **C63**, 317 (1994); G. Azuelos and A. Djouadi, Z. Phys. **C63**, 327 (1994).
66. F.M.L. Almeida Jr. *et al*, Phys. Rev. **D51**, 5990 (1995).
F.M.L. Almeida Jr. *et al*, Phys. Rev. **D62**, 075004 (2000).
67. F.M.L. Almeida Jr. *et al*, Phys. Rev. **D63**, 075005 (2001).
68. W. Buchmüller, R. Rückl and D. Wyler, Phys. Lett. **B191**, 442 (1987).
69. M. Heyssler, R. Rückl, and H. Spiesberger, in *International Workshop on Linear Colliders*, Sitges, Spain, 1999, hep-ph/9908319.
70. M. A. Doncheski and S. Godfrey, Phys. Rev. **D49**, 6220 (1994); Phys. Rev. **D51**, 1040 (1995); Phys. Lett. **B393**, 355 (1997); J.L. Hewett and S. Pakvasa, Phys. Lett. **B227**, 178 (1989).
71. J.L. Hewett and T.G. Rizzo, Phys. Rev. **D36**, 3367 (1987); J. Blümlein and R. Rückl, Phys. Lett. **B304**, 337 (1993); J. Hewett in *2nd International Workshop on Physics and Experimentation with* e^+e^- *Linear Colliders*, Waikoloa, HI, 1993, hep-ph/9308321.
72. S. Abdullin and F. Charles, Phys. Lett. **B464**, 223 (1999).
73. J.L. Hewett and T.G. Rizzo, Phys. Rev. **D56**, 5709 (1997).
74. S. Cullen, M. Perelstein, and M.E. Peskin Phys. Rev. **D62**, 055012 (2000).
75. E. Dudas, J. Mourad, Nucl. Phys. **B575**, 3 (2000); E. Accomando, I. Antoniadis, K. Benakli, Nucl. Phys. **B579**, 3 (2000);
76. G. Veneziano, Nuovo. Cim. **A57**, 190 (1968).
77. For a review of non-commutative theories, see, N. Seiberg, and E. Witten, J. High Energy Phys, **9909**, 032 (1999); M. R. Douglas and N. A. Nekrasov, Rev. Mod. Phys. **73**, 977 (2001) [arXiv:hep-th/0106048].

78. The Feynman rules for NCQED and more general non-commutative gauge theories can be found in: A. Armoni, Nucl. Phys. **B593**, 229 (2001) [arXiv:hep-th/0005208]; M. M. Sheikh-Jabbari, JHEP **9906**, 015 (1999) [arXiv:hep-th/9903107]; T. Krajewski and R. Wulkenhaar, Int. J. Mod. Phys. **A15**, 1011 (2000) [arXiv:hep-th/9903187]; I. Y. Aref'eva, D. M. Belov, A. S. Koshelev and O. A. Rytchkov, Nucl. Phys. Proc. Suppl. **102**, 11 (2001) [arXiv:hep-th/0003176].
79. J. L. Hewett, F. J. Petriello and T. G. Rizzo, Phys. Rev. **D64**, 075012 (2001) [arXiv:hep-ph/0010354].

CHAPTER 8

QCD

Philip N. Burrows
Department of Physics, Queen Mary, University of London
Mile End Rd., London E1 4NS, United Kingdom
E-mail: p.burrows@qmul.ac.uk

A programme of tests of Quantum Chromodynamics at the next-generation Linear Collider is described.

1. Introduction

1.1. *Hadronic Final States*

The production of hadronic final-states by a variety of interactions is illustrated in Fig. 1. In electron-positron annihilation hadronic activity is, by construction, limited to the final state, making the study of hadronic events cleaner and simpler relative to lepton-hadron and hadron-hadron collisions, from both the experimental and theoretical points-of-view. On the experimental side there are no remnants of the beam particles to add confusion to the interpretation of hadronic structures, and, apart from initial and final-state photon radiation effects, the hadronic centre-of-mass frame coincides with the laboratory frame. On the theoretical side the absence of hadrons in the incoming beams removes dependence on the limited knowledge of the parton density functions of hadrons, as well as rendering calculations at a given order of perturbation theory easier to perform because there are generally fewer strong-interaction Feynman diagrams to consider. Electron-positron annihilation thus provides an ideal environment for precise tests of Quantum Chromodynamics (QCD).

A large number of e^+e^- colliders has been constructed over the past 25 years. The range of CM energies Q extends from a few GeV up to over 200

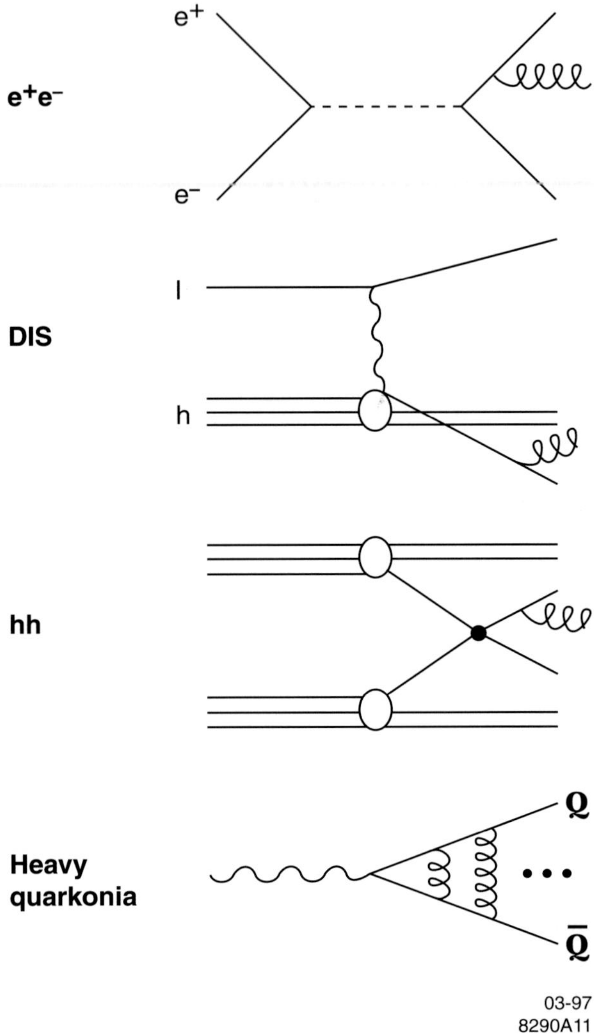

Fig. 1. Schematic of the production of hadronic final-states by different interactions: e^+e^- annihilation, deep inelastic lepton-hadron scattering, hadron-hadron collisions, and production and decay of heavy quarkonia.

GeV at the CERN LEP-II collider, and a wealth of tests of QCD has been performed; see *e.g.* [1]. Here I discuss the possibilities for QCD studies at a future high-energy linear e^+e^- collider.

1.2. *Quantum Chromodynamics*

QCD is our theory of the strong interaction between quarks and gluons [2]. It is a non-Abelian Yang-Mills gauge theory that describes the interactions of a triplet of spin-1/2 quarks possessing the colour quantum number ($c = r,\ b,\ g$) via an octet of vector gluons. The spinor quark fields $q_c(x)$ transform as the fundamental representation of the SU(3) group, whilst the gluon fields $A_\mu^a(x)$ ($a = 1,\ 2,\ \ldots,\ 8$) transform according to the adjoint representation. The SU(3) colour transformations are generated by the 3×3 matrices $T^a = \lambda^a/2$, where λ^a are the Gell-Mann matrices which obey the commutation relations:

$$[T^a, T^b] \quad = \quad i\, f^{abc}\, T^c \tag{1}$$

and f^{abc} are the structure constants of SU(3). The Lagrangian has the form:

$$\mathcal{L} \quad = \quad -\frac{1}{4} F^a_{\mu\nu} F^{\mu\nu\ a} \ + \ \overline{q}\ (i\gamma_\mu D^\mu \ - \ m)\ q \tag{2}$$

where $F_{\mu\nu}$ is the field strength tensor:

$$F^a_{\mu\nu} \quad = \quad \partial_\mu A^a_\nu - \partial_\nu A^a_\mu + g f^{abc} A^b_\mu A^c_\nu \tag{3}$$

and D_μ is the covariant derivative:

$$D_\mu \quad = \quad \partial_\mu - igT^a A^a_\mu(x), \tag{4}$$

g is the bare coupling of the theory, m the bare mass of the quark field and the gluons are massless.

Following Ref. [3], the "essential features" of QCD may be summarised as:

- quarks with spin 1/2 exist as colour triplets
- gluons with spin 1 exist as colour octets
- the coupling $q\overline{q}g$ exists
- the couplings ggg and $gggg$ exist
- the couplings are equal
- the coupling decreases as $1/\ln Q^2$

Over the past 30 years these essential features have been demonstrated, both qualitatively and quantitatively, through careful measurements of light-quark (u, d, s, c, b) systems at e^+e^-, hadron-hadron and lepton-hadron scattering experiments [1, 4]. It is now beyond reasonable doubt that QCD *is* the correct theory of strong interactions.

1.3. *QCD and the Linear Collider*

The Linear Collider (LC) will add incisive new information both via precision measurements and by exploring the qualitatively new top-quark system. The Collider offers the possibility of testing QCD at high Q^2 in the experimentally clean, theoretically tractable e^+e^- environment. In addition, $\gamma\gamma$ interactions will be delivered free by Nature through the interactions of virtual photons from the fields around the colliding electrons and positrons; a dedicated $\gamma\gamma$ collider is an additional option (see Chapter 9), and would allow detailed measurements of the relatively poorly understood photon structure. Some of the main motivations for LC QCD studies are:

- Tests of QCD at the highest energy scales available in $\mathrm{e}^+\mathrm{e}^-$ annihilation.
- Precise determination of the strong coupling, α_s.
- Measurement of the Q^2-dependence of α_s and constraints on the GUT scale.
- Studies of the $t\bar{t}(g)$ system.
- Symmetry tests in multi-jet final states using polarised electron (and possibly also positron) beams.

These and other topics will be reviewed in detail and the current state of studies for LC QCD analyses will be summarised.

2. Event Selection

2.1. *The Problem*

In order to perform the programme of QCD measurements just outlined it is of extreme importance to demonstrate that samples of $e^+e^- \to q\overline{q}$ ($q = u, d, s, c, b$) and $e^+e^- \to$ events can be selected with good efficiency, high purity and low bias. In QCD studies at and below the Z^0 resonance the primary background sources to $q\overline{q}$ events are $e^+e^- \to$ lepton-pairs, $e^+e^- \to e^+e^-\gamma^*\gamma^*$ with $\gamma^*\gamma^* \to$ hadrons, cosmic rays, and beam-related background events. Application of simple requirements on the number of charged tracks, visible energy, and energy balance of events has allowed selection of $q\overline{q}$ event samples of greater than 99% purity with over 90% efficiency [5].

At higher energies the situation is less straightforward, as other physics processes are present that have comparable or larger cross-sections than $q\overline{q}$ production, and which may also appear as hadronic final states. Perhaps the most serious case is $e^+e^- \to W^+W^-$, in which one or both Ws may decay into hadrons. From the viewpoint of QCD studies, therefore, the aim is to develop a set of cuts to isolate separately $q\overline{q}$ and events and to exclude primarily W^+W^- and Z^0Z^0 events.

2.2. *Initial State Radiation and Beamstrahlung*

Additional complications arise from the effects of initial state radiation and "beamstrahlung" (Chapter 1), which serve to reduce the effective CM energy available for the e^+e^- annihilation, and to cause events to be Lorentz-boosted along the beam axis. This was studied in [6], where it was shown that radiative effects are considerable and cause "event pileup" at the Z^0 resonance, where more events will be produced than at the nominal operating energy of the Collider. This situation is familiar from studies above the Z^0 resonance at LEP-II, and such radiative events can be removed by a cut on the effective hadronic CM energy $\sqrt{s\prime}$ [7].

2.3. *Selection Cuts*

A set of event selection cuts, based on kinematic quantities, was developed [6]. Events were required to satisfy:

- Charged particle multiplicity, $N_{charged} > 8$
- Thrust-axis polar angle, $|\cos\theta_T| < 0.8$
- Visible energy, $E_{vis}/Q > 0.5$
- Momentum balance, $(\Sigma|p_z|)/E_{vis} < 0.4$
- Minimum hemisphere mass, $M_{1,2} > 3$ GeV/c^2
- Hemisphere multiplicity, $N^{1,2}_{charged} \geq 4$
- Hemisphere mass, M_1/E_{vis} or $M_2/E_{vis} < 0.13$

The first two cuts are designed to reject lepton-pair, $\gamma^*\gamma^* \rightarrow$ hadrons and beam-related final states; the second cut also rejects some of the (t-channel) W^+W^- events; lower bounds on the hemisphere masses reject lepton pair events, whilst upper bounds reject , W^+W^- and Z^0Z^0 events. At $\sqrt{s} = 500$ GeV, assuming a detector with perfect efficiency within a polar-angle acceptance of $|\cos\theta| < 0.98$, these cuts select a sample comprising 83% $q\overline{q}$, 11% W^+W^-, 6% , and < 1% $\gamma\gamma$ events. Using this sample to form distributions of standard QCD 3-jet observables, it was found that binwise corrections at the level of 25% had to be applied in order to account for the bias to the distributions introduced by the kinematic cuts and residual background contamination. Clearly this situation needs to be improved in order to make precision measurements, for example a percent–level measurement of $\alpha_s(M_Z)$.

More recent studies have shown [8] that a large fraction of background events can be excluded, without bias, by eliminating the last kinematic cut and utilising a combination of a heavy-flavour veto and the beam polarisation. By selecting only right-polarised electron-beam produced events the (left-handed) W^+W^- contribution can be suppressed strongly (Fig. 2 [8]). Furthermore, the application of highly-efficient b-jet tagging can be used to reduce the contamination to the 1% level. For an electron polarisation of 80%, the selected sample is 82% pure in $q\overline{q}$ events, with backgrounds of 13% (W^+W^-), 4% (Z^0Z^0) and 1% (). For a higher beam polarisation of 90% the purity is improved to 87%, with an 8% W^+W^- background.

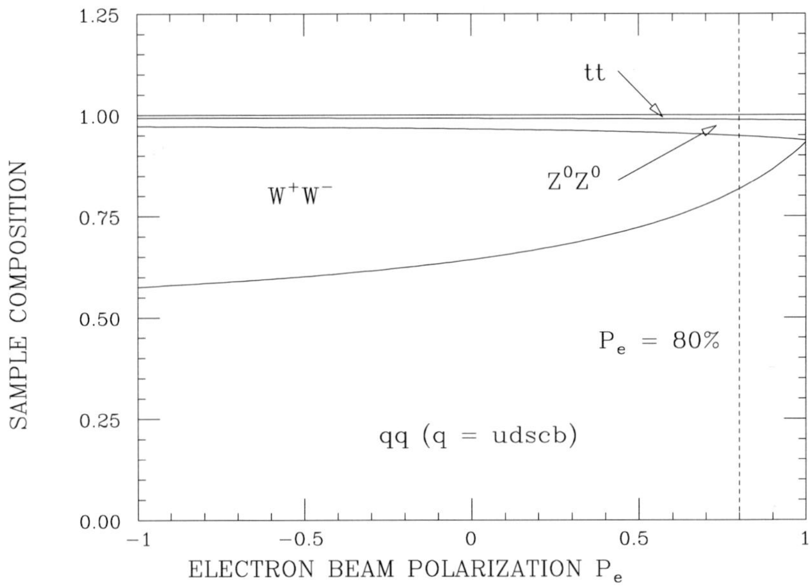

Fig. 2. Composition of event sample versus electron-beam polarisation.

3. Precise Measurement of α_s

3.1. *Introduction*

QCD contains in principle only one free parameter, the strong interaction scale Λ. Tests of QCD hence comprise comparisons of measurements of Λ in different processes and at different hard scales Q. If one knows Λ one may calculate the strong coupling $\alpha_s(Q)$ from the solution of the QCD renormalisation group equation [9].

Precise determination of the strong coupling α_s is important for a number of reasons. Since the uncertainty on α_s translates directly into an uncertainty on pQCD predictions, especially for high-order multijet processes which will form significant backgrounds to other physics studies at both the LC and the LHC, it is desirable to achieve as high a precision as possible. In addition, since the weak and electromagnetic couplings are known with much greater precision, the error on α_s represents the dominant uncertainty on our "prediction" of the scale for grand unification of the strong, weak and electromagnetic forces [10]. A more precise α_s measurement would also allow improved determination of the mass and width of the top quark from the threshold behaviour of the cross-section (Chapter 3).

Because of the large data samples taken in e^+e^- annihilationat the Z^0 resonance, it has become conventional to use as a yardstick $\alpha_s(M_Z)$, where M_Z is the mass of the Z^0 boson; $M_Z \approx 91.2$ GeV. Tests of QCD can therefore be quantified in terms of the consistency of the values of $\alpha_s(M_Z)$ measured in different experiments.

3.2. *Current Status*

Measurements of $\alpha_s(M_Z)$ have been performed in e^+e^- annihilation, hadron-hadron collisions, and in deep-inelastic lepton-hadron scattering experiments, covering a range of Q from roughly 1 to 200 GeV [11]. These measurements are consistent within the errors. An average yields $\alpha_s(M_Z) = 0.118 \pm 0.003$, implying that QCD has been tested to a precision of about 3%. This precision is however rather modest compared with the achievement of 0.1% level tests of the electroweak theory [12], and this is due primarily to the *theoretical uncertainties* that dominate most of the experimental measurements. These uncertainties are due to both the restriction of complete perturbative QCD calculations to low order, and non-perturbative ("hadronisation") effects that are presently incalculable in QCD.

3.3. *Measurement of α_s at the Linear Collider*

Several techniques for α_s determination will be available at the LC, and it is interesting to consider whether a substantially-improved measurement at the 1%–level of precision is possible.

3.3.1. *Event Shape Observables*

The determination of α_s from event "shape" observables that are sensitive to the 3-jet nature (see Fig. 3) of the particle flow has been pursued for 2 decades and is generally well understood [13]. In this method one usually forms a differential distribution, makes corrections for detector and hadronisation effects, and fits a pQCD prediction to the data, allowing α_s to vary. Examples of such observables are the thrust, jet masses and jet rates [5].

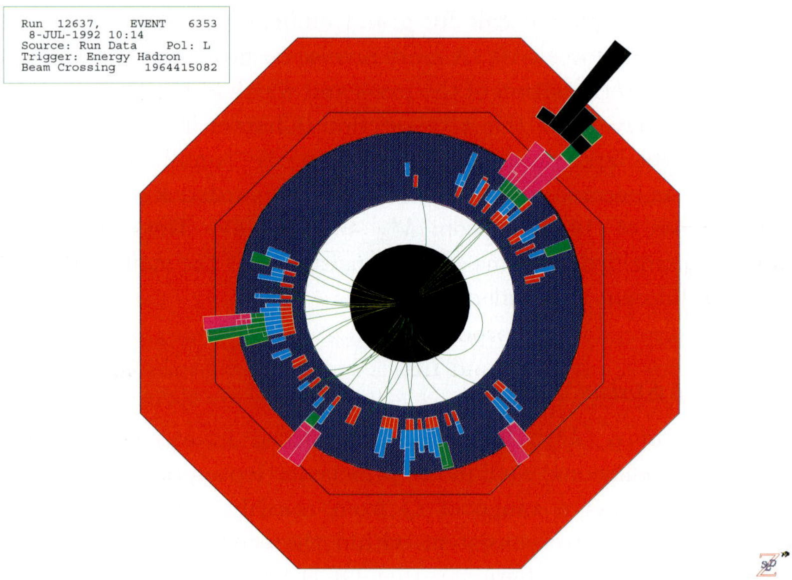

Fig. 3. A three-jet event recorded by SLD: the process $e^+e^- \rightarrow Z^0 \rightarrow q\bar{q}g$.

• **Statistical error:** The latest generation of such α_s measurements, from SLC and LEP, has shown that statistical errors below the 1% level can be obtained with samples of a few tens of thousands of hadronic events. With the current LC design luminosities of $2-3 \times 10^{34}/\text{cm}^2/\text{s}$ at $\sqrt{s} = 500$ GeV, hundreds of thousands of $e^+e^- \rightarrow q\bar{q}$ events would be produced each year, and a statistical error on $\alpha_s(M_Z)$ below the 0.5% level could be achieved.
• **Systematic error:** Detector systematic errors, which relate mainly to uncertainties on the corrections made for acceptance and resolution effects and are observable-dependent, are under control in today's detectors at the

$\Delta\alpha_s(M_Z) = 1 - 4\%$ level [5, 14]. If the LC detector(s) is (are) designed to be very hermetic, with good tracking resolution and efficiency, as well as good calorimetric jet energy resolution, all of which are required for the search for new physics processes, it seems reasonable to expect that the detector-related uncertainties can be beaten down to the $\Delta\alpha_s \simeq 1\%$ level or better.

• **Backgrounds:** As discussed in the previous section, $e^+e^- \rightarrow Z^0Z^0$, W^+W^-, or $t\bar{t}$ events will present significant backgrounds to $q\bar{q}$ events for QCD studies, and the selection of a highly pure $q\bar{q}$ event sample will not be as straightforward as at the Z^0 resonance. However, with application of the cuts discussed earlier, after statistical subtraction of the remaining backgrounds (the Z^0Z^0 and W^+W^- event properties have been measured accurately at SLC and LEPI/II), the residual bias on the event-shape distributions is expected to be under control at the 1% level on $\alpha_s(M_Z)$ [8].

• **Hadronisation uncertainties:** Since jets of final-state particles rather than partons (see Fig. 4) are observed in detectors it is necessary to correct hadronic distributions for the effects of the smearing of the particle momentum flow caused by hadronisation. These effects are usually estimated from Monte Carlo simulations incorporating hadronisation models. The models have been well tuned at SLC and LEP and are widely used for evaluating systematic effects. The size of the correction factor, and hence the uncertainty, is observable dependent, but the "best" observables measured at the Z^0 have corrections at the level of less than 10% [5] with corresponding uncertainties as low as $\Delta\alpha_s \simeq 1\%$.

It can be argued that non-perturbative corrections to jet final states in e^+e^- annihilationcan be parametrised in terms of inverse powers of the hard scale $\sqrt{s}$. For a generic observable X:

$$\delta X^{non-pert} \sim \Sigma_{n\geq 1}\frac{a_n}{Q^n}.$$

However, at leading order in perturbation theory:

$$X^{pert} \sim \frac{1}{\ln Q},$$

so that the ratio of non-perturbative to perturbative QCD contributions is dominated by a term of the form:

$$\frac{\delta X^{non-pert}}{X^{pert}} \sim \frac{\ln Q}{Q}.$$

Increasing $\sqrt{s}$ from 91 GeV to 500 GeV causes this ratio to decrease by a factor of 5, implying that hadronisation corrections should be 2% or less at

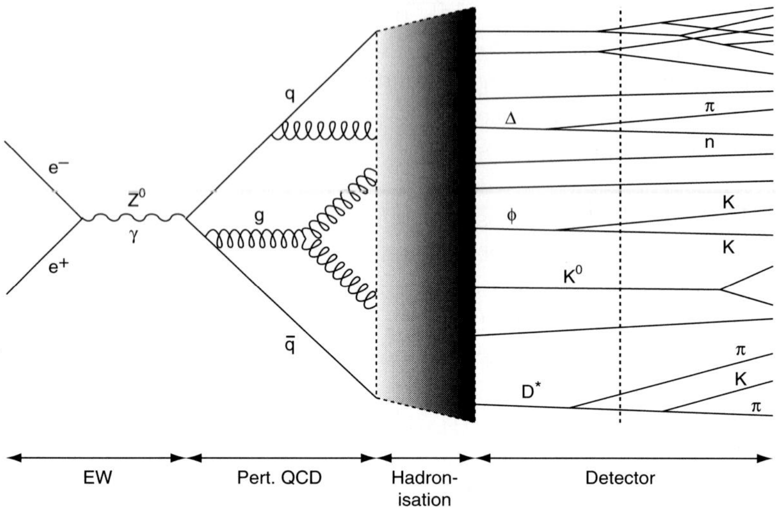

Fig. 4. Schematic of particle production in e^+e^- annihilation.

the LC. Assuming that these corrections can be estimated to much better than $\pm 50\%$, the hadronisation uncertainty should be substantially below 1% on $\alpha_s(M_Z)$.

A large amount of theoretical work [15] has added validity to the inverse power corrections approach, which for some observables appears to be in good agreement with current data. For example, the energy-dependence of the observable $1-T$ ($T=$ thrust [16]) is shown in Fig. 5 [15]; $\mathrm{O}(\alpha_s^2)$ pQCD combined with a non-perturbative term of the form $1/Q$ describes the data well. Explicit evaluation of hadronisation correction factors at 500 GeV has been performed using JETSET/PYTHIA [6, 14]. Results for $1-T$ are shown in Fig 6; away from the 2-jet region $T \to 1$, these confirm the expectations from the power corrections approach of hadronisation effects at the few per cent level.

• **Theoretical uncertainties:** Currently pQCD calculations of event shapes are available complete only up to $\mathrm{O}(\alpha_s^2)$ [17], although higher-order "resummed" calculations have been performed for some observables [18]. Since the data contain knowledge of all orders one must estimate the possible bias inherent in measuring $\alpha_s(M_Z)$ using the truncated QCD series. Though not universally accepted, it is customary to estimate this from the dependence of the fitted $\alpha_s(M_Z)$ value on the QCD renormalisation scale, yielding a large and dominant uncertainty of about $\Delta\alpha_s \simeq \pm 6\%$ [5]. Since

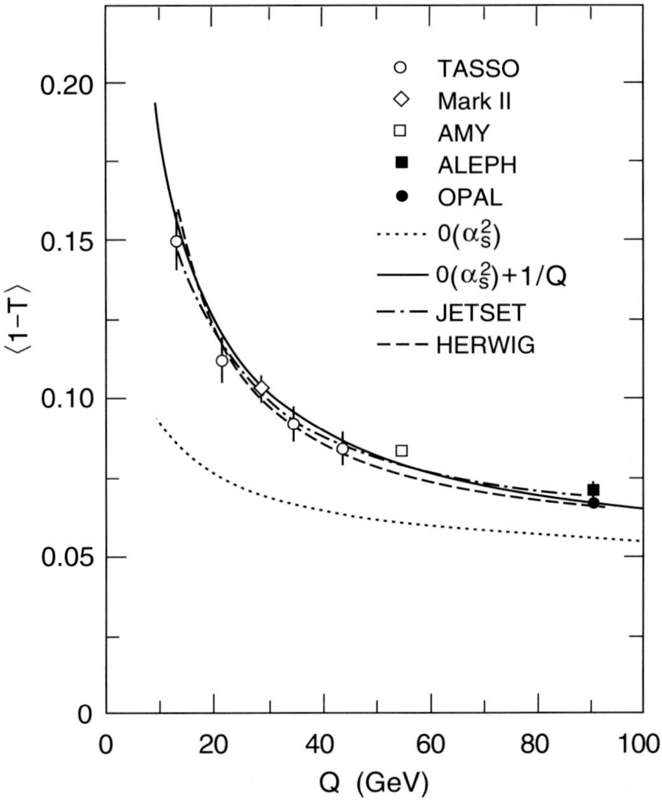

Fig. 5. Mean value of 1-Thrust vs. CM energy = Q.

the missing terms are $\mathrm{O}(\alpha_s^3)$, and since $\alpha_s(500\ \mathrm{GeV})$ is expected to be about 25% smaller than $\alpha_s(M_Z)$, one expects the uncalculated contributions to be almost a factor of two smaller at the higher energy, leading to an estimated uncertainty of $\pm 4\%$ on $\alpha_s(500\ \mathrm{GeV})$. However, translating to the yardstick $\alpha_s(M_Z)$ yields an uncertainty of $\pm 5\%$, only slightly smaller than currently. Therefore, an improved $\alpha_s(M_Z)$ measurement from event shapes will not be realised unless $\mathrm{O}(\alpha_s^3)$ contributions are calculated. There is reasonable expectation that this will be achieved within the next 5 years.

From this analysis it seems reasonable to conclude that achievement of the luminosity necessary for "discovery potential" at the LC will result in a $q\overline{q}$ event sample of sufficient size to measure $\alpha_s(M_Z)$ with a statistical uncertainty of better than 0.5%. Construction of detectors superior in per-

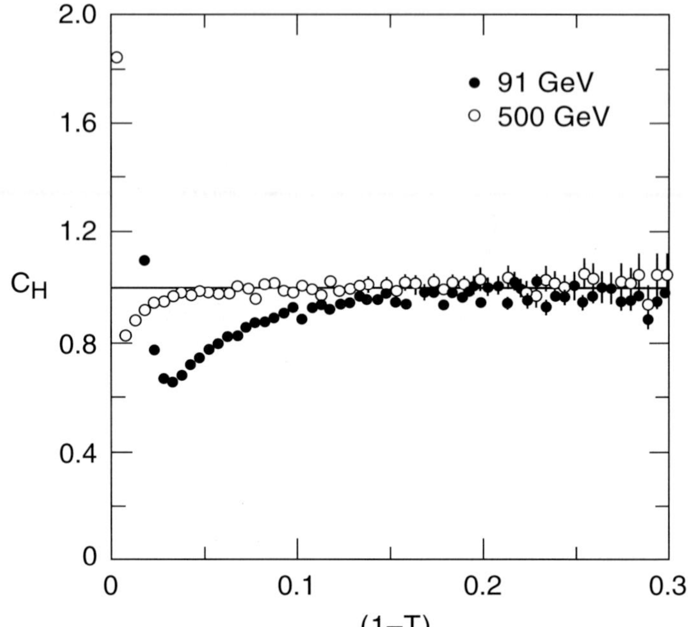

Fig. 6. Hadronisation correction factors calculated using JETSET.

formance to those in operation today at SLC and LEP will be necessary in order to reduce systematic errors to the 1% level. Hadronisation effects should be significantly smaller, implying a sub–1% uncertainty. However, unless $O(\alpha_s^3)$ contributions are calculated, $\alpha_s(M_Z)$ measurements at 500 GeV will be limited by theoretical uncertainties to a precision of $\pm 5\%$, only marginally better than that achieved at present.

3.3.2. *The $t\bar{t}$ System*

The value of α_s controls the strong potential that binds quarkonia resonances. In the case of production near threshold, the large top mass, m_t, and decay width, Γ_t, ensure that the top quarks decay in a time comparable with the classical rotation period of the bound system, washing out most of the resonant structure in the cross-section, σ. The shape of σ near threshold hence depends strongly on both m_t and α_s. Fits of next-to-leading-order (NLO) pQCD calculations to simulated measurements of σ showed [19] that the derived m_t is strongly correlated with α_s. Fixing α_s allowed the error on m_t to be reduced by a factor of 2. Since the main aim of such an exercise is to determine m_t as precisely as possible, the optimal strategy is to *input* α_s from elsewhere.

Moreover, recent NNLO calculations of σ near threshold [20] have caused consternation, in that the size of the NNLO contributions appears to be comparable with that of the NLO contributions, and the change in the threshold shape causes a shift of roughly 1 GeV in the value of the fitted m_t. This mass shift can be avoided by a judicious top-mass definition [20], which also reduces the m_t–α_s correlation [21]. However, the resulting cross-section theoretical normalisation uncertainty translates into an uncertainty of ±0.012 on $\alpha_s(M_Z)$, *i.e.*about 6 times larger than the estimated statistical error [21]. Although this may provide a useful "sanity check" of α_s in the system, it does not offer the prospect of a 1%-level measurement. Very recent progress indicates that the normalisation uncertainty can be reduced considerably by the inclusion of NNL logarithmic contributions in QCD perturbation theory [22].

A preliminary study has also been made [23] of the determination of α_s from $R_t \equiv \sigma/\sigma_{\mu^+\mu^-}$ *above* threshold. For $\sqrt{s} \sim 400$ GeV the theoretical uncertainty on R_t is roughly 3%; for $\sqrt{s} \geq 500$ GeV the uncertainty due to m_t is much less important and the uncertainty is smaller, around 0.5%. However, on the experimental side the limiting precision on R_t will be given by the uncertainty on the luminosity measurement. If this is as good as at LEPII, *i.e.*around 0.2%, then $\alpha_s(M_Z)$ could be determined with an experimental precision approaching 1%, which would be extremely valuable as a complementary precision measurement from the system.

There is also the possibility of determining α_s using $t\bar{t}g$ events, which have recently been calculated [24] at NLO. For reasonable values of the jet-resolution scale y_c the NLO contributions are substantial; for example, at $\sqrt{s} = 500$ GeV they are of the order of 40%, which is comparable with the situation for massless quarks. The discussion of unknown higher-order contributions above is hence also valid here, and $t\bar{t}g$ events will only be useful for determination of α_s once the NNLO contributions have been calculated. However, if the $t\bar{t}g$ event rate, R_3^t, could be measured precisely, and compared with the $q\bar{q}g$ event rate, R_3^q, to form $r^t = R_3^t/R_3^q$, the common systematic errors would cancel to a large degree. This would allow the ansatz of flavour-independence of strong interactions can be tested for the top quark, and a determination of the running of m_t in a similar manner to the running b-quark mass [25]. For example, a precision of 1% on r^t implies a measurement of $m_t(1\text{TeV})$ with an error of 5 GeV [24].

3.3.3. *A High-luminosity Run at the Z^0 Resonance*

There are strong physics motivations to run the Linear Collider at the Z^0 resonance, see *e.g.*Ref. [26], where a data sample of the order of 10^9 Z^0 events could in principle be collected within a few months of dedicated running at a luminosity of $5 \times 10^{33}/\mathrm{cm}^2/\mathrm{s}$. Such a "Giga Z^0" sample offers two additional options for α_s determination via measurements of the inclusive ratios $\Gamma_Z^{had}/\Gamma_Z^{lept}$ and $\Gamma_\tau^{had}/\Gamma_\tau^{lept}$. Both have been calculated to $\mathrm{O}(\alpha_s^3)$, so in terms of pQCD they are theoretically well understood. However, they are indirectly proportional to α_s, and require a very large event sample for a precise measurement. For example, the current LEP data sample of 16M Z^0 yields $\Delta\alpha_s(M_Z) = 2.5\%$ from $\Gamma_Z^{had}/\Gamma_Z^{lept}$. The statistical error could, naively, be pushed to below the $\Delta\alpha_s(M_Z) = 0.4\%$ level, but systematic errors arising from the hadronic and leptonic event selection will probably limit the precision to 0.8% [27]. This would be a very precise, reliable measurement. In the case of $\Gamma_\tau^{had}/\Gamma_\tau^{lept}$ the experimental precision from LEP and CLEO is already at the $\Delta\alpha_s(M_Z) = 1\%$ level. However, there has been considerable debate about the size of the theoretical uncertainties, with estimates as large as 5% [28]. If this situation is clarified, and the theoretical uncertainty is small, $\Gamma_\tau^{had}/\Gamma_\tau^{lept}$ may offer a further 1%-level $\alpha_s(M_Z)$ measurement.

4. Q^2 Evolution of α_s

In the preceeding sections I discussed the expected attainable precision on the yardstick $\alpha_s(M_Z)$. Translation of the measurements of $\alpha_s(Q)$ ($Q \neq M_Z$) to $\alpha_s(M_Z)$ requires the assumption that the "running" of the coupling is determined by the QCD β function. However, since the logarithmic decrease of α_s with Q is an essential component of QCD, reflecting the underlying non-Abelian dynamics, it is vital also to test this Q-dependence explicitly. Such a test would be particularly interesting if new coloured particles, such as squarks or gluinos (Chapter 4), were discovered, since they would increase the effective vacuum polarisation contributions to the QCD β-function, and cause a decrease in the running of α_s as a function of Q with respect to the standard QCD case. Furthermore, extrapolation of α_s to very high energies of the order of 10^{15} GeV can be combined with corresponding extrapolations of the dimensionless weak and electromagnetic couplings in order to constrain the coupling-unification, or GUT, scale [10].

Older data on hadronic final states data from e^+e^- annihilationat the PETRA, PEP, TRISTAN, SLC and LEP colliders span the range $14 \leq \sqrt{s} \leq 91$ GeV, although hadronisation uncertainties are large on the data below 25 GeV [29]. The highest-energy data, up to $\sqrt{s} \simeq 200$ GeV, are currently provided by LEPII. Simulated measurements of $\alpha_s(\sqrt{s})$ at $\sqrt{s}$ = 91, 500 and 800 GeV are shown in Fig. 7 [14], together with existing measurements, which span the range $20 \leq \sqrt{s} \leq 200$ GeV. The simulated point at $\sqrt{s} = 91$ GeV is based on the $\Gamma_Z^{had}/\Gamma_Z^{lept}$ technique, and those at 500 and 800 GeV are based on the event shapes technique. The last two include the current theoretical uncertainty, which yields a total error on each point equivalent to $\Delta\alpha_s(M_Z) = 4\%$. It is clear that the LC data would add significantly to the lever-arm, and would allow a substantially improved extrapolation to the GUT scale.

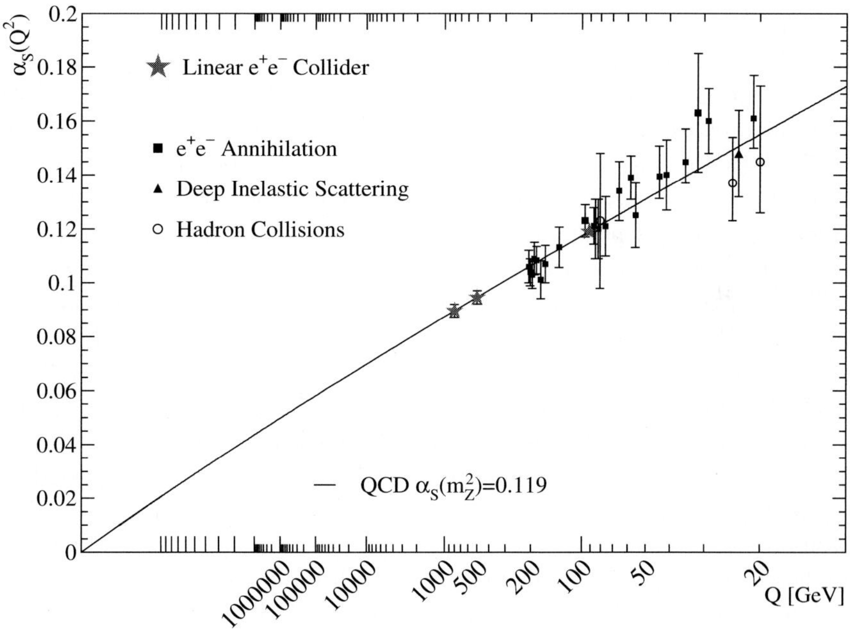

Fig. 7. α_s versus hard scale Q; for e^+e^- $Q = \sqrt{s}$. The line shows a NNLO QCD fit to the current data.

5. QCD Studies of $t\bar{t}g$ Events

5.1. *Soft Gluon Radiation in $t\bar{t}g$ Events*

The large mass and decay width of the top quark serve to make the study of gluon radiation in events a new arena for testing QCD. The large mass acts as a cutoff for collinear gluon radiation, and the large decay width acts as a cutoff for soft gluon radiation, allowing reliable perturbative QCD calculations to be performed; these effects are of course correlated. The latter case is particularly interesting. If the top width were infinite, top quarks would decay immediately to bottom quarks, and any gluons would be radiated from the secondary b's. If the top width were zero, top quarks would live forever and all radiation would be from the primary t's. In the case of a large but finite width, expected to be around 2 GeV for a top mass of 175 GeV/c^2, gluon radiation in events will be a coherent sum of contributions from these two limiting cases, with a degree of coherence regulated by the top width itself.

A previous theoretical study [30] of production above threshold, assuming $m_t = 140$ GeV/c^2, demonstrated interference effects in the angular distribution of soft gluons w.r.t. the top quark flight direction. This study has been updated for $m_t = 175$ GeV/c^2 at $\sqrt{s} = 1$ TeV, and the results are illustrated in Fig. 8 [31]. This shows the angular distribution of 5 GeV gluons w.r.t. the axis for the kinematic configuration in which the decay b-quark travels backwards w.r.t. the t flight direction. The dependence of the radiation pattern on the top decay width is strong. Similar effects are predicted in the spectrum of gluon radiation in events produced near their threshold [30]. Measurement of such effects would yield not only a dramatic demonstration of quantum interference in strong interactions, but might also provide an essential cross-check on the top-quark decay width; Γ_t may prove difficult to disentangle from measurements of the threshold cross-section and top momentum distributions, which also depend on α_s and m_t, as well as on the beam energy distribution (Chapter 1). This possibility is exciting, but will require a detailed hadron-level Monte Carlo simulation study, with the inclusion of a realistic detector, to demonstrate that the coherence effects can be measured with sufficient precision.

5.2. *Anomalous Chromomagnetic Top Quark Couplings*

The existence of anomalous couplings of quarks to gluons could manifest themselves via a modification of the rate and pattern of emitted gluon radiation, beyond effects such as those discussed in the last section. A

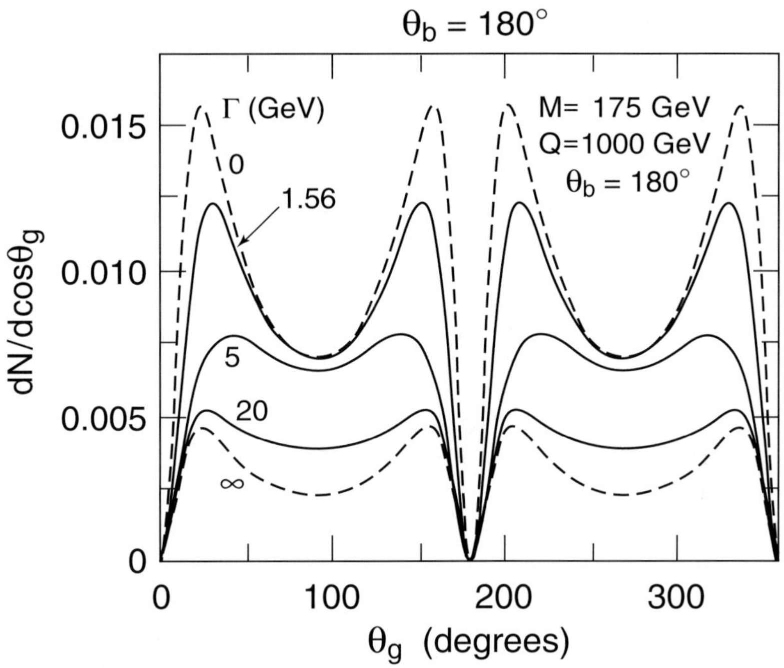

Fig. 8. Angular distribution of 5 GeV gluons w.r.t. axis, at $\sqrt{s} = 1$ TeV.

model-independent parametrisation of anomalous couplings in the strong-interaction Lagrangian may be written [32]:

$$\mathcal{L}^{q\bar{q}g} = g_s \bar{q} T_a \left(\gamma_\mu + \frac{i\sigma_{\mu\nu}k^\nu}{2m_t} (\kappa - i\tilde{\kappa}\gamma_5) \right) q G_a^\mu$$

where κ and $\tilde{\kappa}$ represent anomalous "chromomagnetic" and "chromoelectric" dipole moments, respectively. The chromoelectric moment gives rise to CP-violating effects. The phase space available for anomalous-moment effects increases with the CM energy.

In principle such moments may exist for quarks of any flavour, but new physics processes at high energy scales are more likely to couple to heavy quarks. A search for the anomalous b-quark chromomagnetic moment has been made using $Z^0 \to b\bar{b}g$ events [33], and motivates exploration in the $t\bar{t}g$ sector at the Linear Collider. This has been considered for $m_t = 175$ GeV/c^2 at $\sqrt{s} = 500$ GeV and 1 TeV [32]. For illustration the gluon energy spectrum in the latter case is shown in Fig. 9 for $\tilde{\kappa} = 0$ and large non-zero values of κ. A measurement of this spectrum for $\kappa = 0 = \tilde{\kappa}$, with a perfect detector and integrated luminosity corresponding to about 1 design year of running at 1 TeV, was simulated, and yielded 95% confidence-level limits of $|\kappa| \leq 0.03$ and $|\tilde{\kappa}| \leq 0.3$. The limits obtainable at 500 GeV are about a factor of two worse.

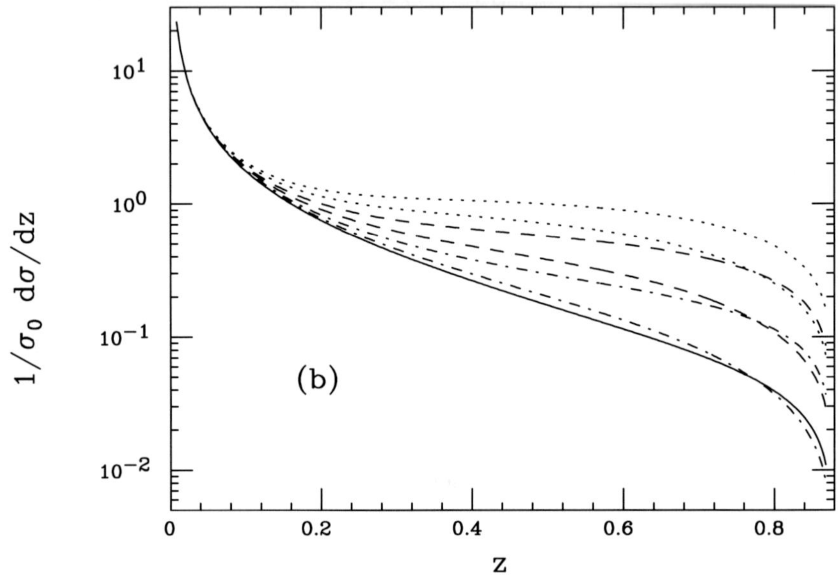

Fig. 9. Distribution of scaled gluon energy $z = E_{gluon}/E_{beam}$ at $\sqrt{s} = 1$ TeV. QCD (solid) $\kappa = \pm 1$ (dot-dash), ± 2 (dash), ± 3 (dots).

Limits of comparable statistical precision may be obtainable at the LHC via measurement of the inclusive top quark production cross-section, but this does not take into account theoretical uncertainties in the cross-section due to higher-order QCD effects and limited knowledge of structure functions.

6. Symmetry Tests Using Beam Polarisation

Production and transport of highly-polarised high-charge electron beams has been demonstrated at the SLAC Linear Collider (SLC), where stable operation at the Z^0 resonance with a beam polarisation of magnitude 0.77

has been achieved. The Z^0 bosons produced using longitudinally polarized electrons have polarization along the beam direction

$$A_Z = (P_{e^-} - A_e)/(1 - P_{e^-} \cdot A_e),$$

where P_{e^-} is the electron beam polarization, defined to be negative (positive) for a left-(right-) handed beam, and

$$A_e = 2v_e a_e/(v_e^2 + a_e^2)$$

with v_e and a_e the electroweak vector and axial vector coupling parameters of the electron, respectively. In order to reduce systematic effects the electron spin direction was reversed randomly pulse-by-pulse, thus achieving higher sensitivities to polarization-dependent asymmetries.

For polarized e^+e^- annihilation to three hadronic jets one can define the triple-product:

$$\vec{P_{e^-}} \cdot (\vec{k_1} \times \vec{k_2}),$$

which correlates the electron-beam polarization vector $\vec{P_{e^-}}$ with the normal to the three-jet plane defined by $\vec{k_1}$ and $\vec{k_2}$, the momenta of the highest- and the second-highest-energy jets respectively. The triple-product is even under reversal of CP, and odd under T_N, where T_N reverses momenta and spin-vectors without exchanging initial and final states. Since T_N is not a true time-reversal operation a non-zero value does not signal CPT violation and is possible in a theory that respects CPT invariance. If the quark (q) and antiquark ($\overline{q}$) jets can be explicitly identified, a corresponding triple-product can be defined:

$$\vec{P_{e^-}} \cdot (\vec{q} \times \vec{\overline{q}}),$$

in terms of the momentum vectors of the quark($\vec{q}$) and antiquark ($\vec{\overline{q}}$). This observable has a different symmetry property, being CP odd and T_N even.

The tree-level differential cross section for $e^+e^- \to q\overline{q}g$ for a longitudinally polarized electron beam and massless quarks may be written [34, 35]:

$$\frac{1}{\sigma}\frac{d\sigma}{d\cos\theta_N} = \frac{9}{16}[(1 - \frac{1}{3}\cos^2\theta_N) + \beta\, A_Z\, \cos\theta_N],$$

where θ_N is the polar angle of the vector normal to the event plane, $\vec{k_1} \times \vec{k_2}$, w.r.t. the electron beam direction. With $\beta|A_Z|$ representing the magnitude, the second term is proportional to the T$_N$-odd triple-product, and appears as a forward-backward asymmetry of the event-plane-normal relative to the electron polarization axis. It should be noted that the sign and magnitude of this term are different for the left and right beam polarisations.

Standard Model T_N-odd contributions of this form at the Z^0 resonance have been investigated [35]. The triple-product vanishes identically at tree level, but non-zero contributions arise from higher order processes involving: (a) QCD rescattering of massive quarks [34], (b) QCD triangle of massive quarks [36], and (c) electroweak rescattering via W and Z exchange loops. Due to various cancellations these contributions are found to be very small at the Z^0 resonance and yield values of the correlation parameter $|\beta| \leq 10^{-5}$ [35]. Because of this background-free situation, measurement of the cross section is sensitive to physics processes beyond the Standard Model that give $\beta \neq 0$. The first experimental study of this quantity, for inclusive-flavour Z^0 decay events, was made by the SLD Collaboration [37], yielding limits: $-0.022 < \beta < 0.039$. A more recent study, for $Z^0 \rightarrow b\bar{b}g$ events, yielded [38]: $-0.045 < \beta < 0.016$ (CP-even case) and $-0.082 < \beta < 0.012$ (CP-odd case).

The dominant Standard Model contributions to $< \cos\theta_N >$ above the Z^0 resonance arise from W– and Z^0–rescattering processes. The energy-dependence of these contributions is illustrated in Fig. 10. They reach their maximum magnitude between 300 and 500 GeV CM energy. For any energy they remain less than 2 parts in 10^5, which is immeasurably small. Therefore, at high energy, unless contributions from rescattering via pairs of vector bosons turn out to be large, θ_N is a potentially "background-free" observable for searches for deviations from the Standard Model, for example due to rescattering of new gauge bosons that couple only to baryon number [39]. At the LC, above threshold these observables will provide an additional search-ground for anomalous effects in the $t\bar{t}g$ system.

7. Further Important Topics

7.1. *Casimir Factor Analysis*

The colour SU(3) group can be characterised by constants known as Casimir factors that are defined by:

$$\Sigma_a (T^a T^a)_{ij} = \delta_{ij} C_F \tag{5}$$

$$\Sigma_{a,b} (f^{abc} f^{abd}) = \delta^{cd} N_C \tag{6}$$

$$Tr[T^a T^b] = \delta^{ab} T_F \tag{7}$$

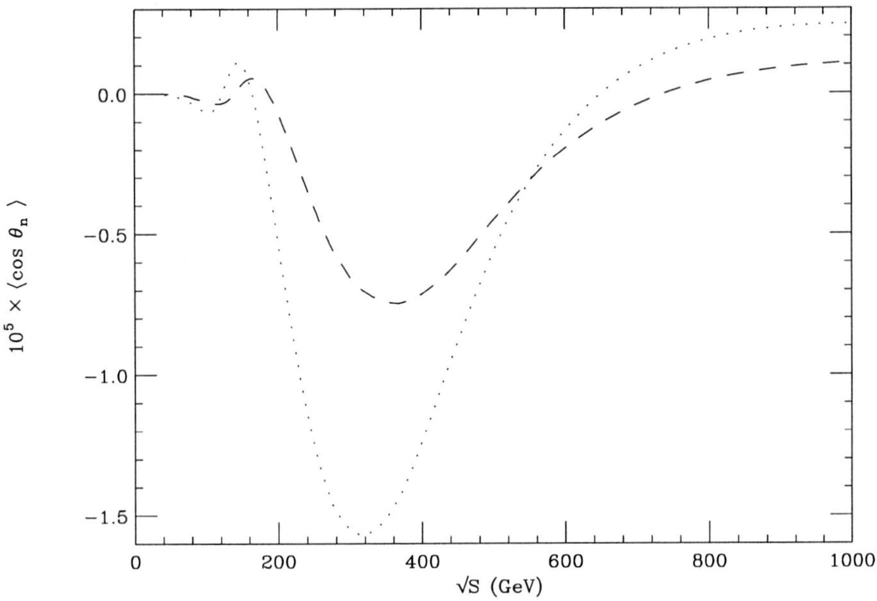

Fig. 10. Expectation value of $\cos\theta_N$ versus CM energy; contributions from W (long dashes) and Z^0 (short dashes) rescattering.

where N_C corresponds to the "number of colours". The tree-level couplings involved in 4-jet events may be classified in terms of these Casimir factors [1] (Fig. 11). The amplitude-squared corresponding to $q \to qg$ is proportional to C_F, that corresponding to $g \to q\bar{q}$ is proportional to T_F, and that corresponding to $g \to gg$ is proportional to N_C. The improved b-quark identification capabilities of the LC detectors (Chapter 1) should allow better separation of 4-jet events of the first category, and significantly improved determination of the Casimir factors compared with current results [40] (see Fig. 12).

7.2. *Charged Particle Multiplicity*

The difference between the particle multiplicity in heavy- (b, c) and light-quark events is predicted [41] to be independent of CM energy. Precise measurements have been made at the Z^0, but measurements at other energies are statistically limited in precision, rendering a limited test of this important prediction. High-precision measurements at the LC would add the lever-arm for a powerful test.

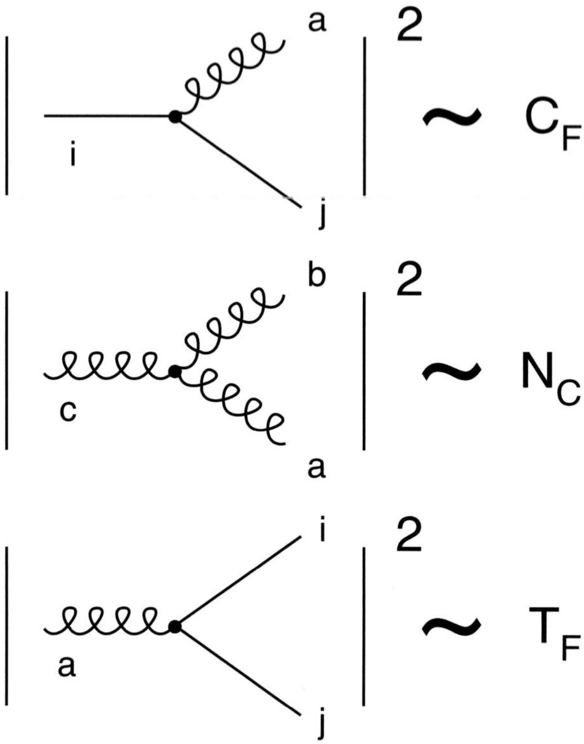

Fig. 11. Casimir classification of tree-level couplings.

7.3. *Colour Reconnection Effects*

Colour reconnection and Bose-Einstein correlations are important quantum mechanical effects in their own right; it is also important to study them precisely since they may affect the precision with which the masses of heavy particles, such as the $W^{\pm}$ and top-quark, can be reconstructed kinematically via their multijet decays [42].

7.4. *Hadronisation Studies and Renormalon Physics*

The understanding of hadronisation effects in terms of power corrections can be extended by measurements of event-shape observables over the wide range of $\sqrt{s}$ values accessible to the Linear Collider.

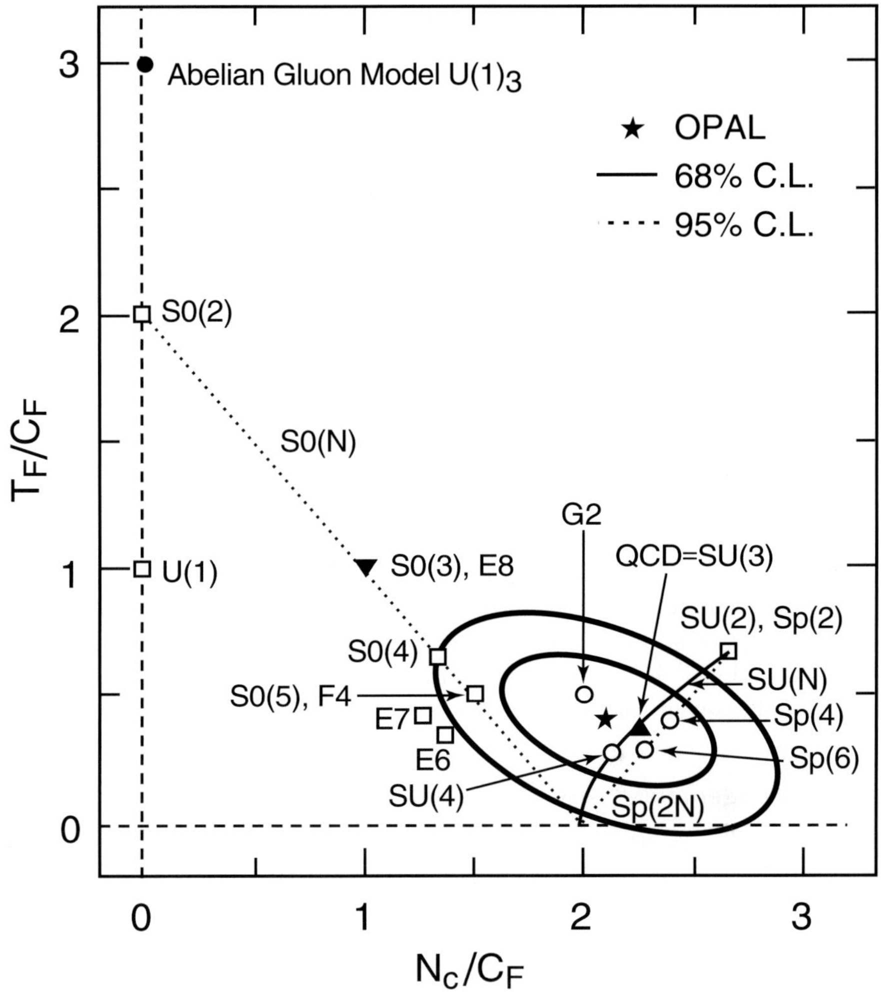

Fig. 12. The T_F/C_F versus N_C/C_F plane showing the expectations from various gauge groups; QCD is in good agreement with the data.

7.5. *Jet Properties and Detector Design*

Studies of the expected properties of jets at CM energies in the range [43] $0.5 \leq \sqrt{s} \leq 2$ TeV have been updated [44] in order to specify the requirements on the performance of detectors for the LC.

8. Summary and Conclusions

Tests of QCD will enrich the physics programme at future high energy linear e^+e^- colliders. Measurement of $\alpha_s(M_Z)$ at the 1% level of precision appears feasible experimentally via: (1) measurement of $\Gamma_Z^{had}/\Gamma_Z^{lept}$ in a high-statistics run at the Z^0 resonance, (2) precise measurement of the top-quark production cross-section well above threshold, or (3) measurements of hadronic event shape observables. The latter will require considerable theoretical effort to calculate $O(\alpha_s^3)$ contributions in QCD perturbation theory. The lever-arm provided by high-energy measurements of $\alpha_s(\sqrt{s})$ will allow improved constraints on the running of the coupling and more accurate extrapolation to the GUT scale. Longitudinal electron beam polarisation can be exploited to perform symmetry tests using multi-jet final states.

Quantum coherence is expected to give rise to interesting gluon radiation patterns in events, which could be used to constrain the top quark decay width. Measurement of the gluon radiation spectrum would also constrain anomalous top quark chromomagnetic couplings. Realistic hadron-level Monte Carlo simulations, including detector effects, need to be performed to evaluate these possibilities quantitatively. Efficient separation of pure samples of $q\overline{q}$ and events will be complicated by high backgrounds from hadronic W^+W^- and Z^0Z^0 events, but good progress has been made in developing event selection cuts.

Acknowledgements

I thank W. Bernreuther, O. Biebel, A. Brandenburg, L. Dixon, D. Miller, L. Orr, T. Rizzo and B. Schumm for helpful contributions.

References

1. P.N. Burrows, Proc. 24th SLAC Summer Institute on Particle Physics, SLAC-Report-508, 101 (1997).
2. H. Fritzsch, M. Gell-Mann, H. Leutwyler, Phys. Lett. **B47**, 365 (1973); D.J. Gross, F. Wilczek, Phys. Rev. Lett. **30**, 1343 (1973); H.D. Politzer, Phys. Rev. Lett. **30**, 1346 (1973).
3. G. Kramer, Springer Tracts in Modern Physics, Vol. 102 (1984).
4. R.K. Ellis, W.S.J. Stirling, B.R. Webber, 'QCD and Collider Physics', (Cambridge University press, 1996).
5. See *e.g.*SLD Collab., K. Abe *et al.*, Phys. Rev. **D51**, 962 (1995); OPAL Collab, P.D. Acton *et al.*, Z. Phys. **C59**, 1 (1993).
6. S. Bethke, Proc. Workshop on Physics and Experiments with Linear e^+e^- Colliders, September 1991, Saariselka, Finland (World Scientific), p. 575.

7. See *e.g.*, OPAL Collab., K. Ackerstaff *et al.*, Z. Phys. **C75**, 193 (1997).
8. B. Schumm, hep-ex/9612013 (1996).
9. Particle Data Group, D.E. Groom *et al.*, Eur. Phys. J. **C15**, 85 (2000).
10. U. Amaldi *et al.*, Phys. Lett. **B281**, 374 (1992).
11. See *e.g.*P.N. Burrows, Proc. International Symposium on Radiative Corrections, Cracow, Poland, August 1–5 1996; Acta Phys. Pol. **B28**, 467 (1997). S. Bethke, J. Phys. **G26**: R27 (2000).
12. A. Gurtu, Proc. XXX International Conference on High Energy Physics, Osaka, Japan, July 27–August 2, 2000, p. 107.
13. See *e.g.*P.N. Burrows, Proc. XXVIII International Conference on High Energy Physics, Warsaw, Poland, July 25–31 1996, Eds. Z. Adjuk, A.K. Wroblewski, (World Scientific, 1997), p. 797.
14. O. Biebel, LC-PHSM-2000-003 (2000); hep-ex/9912051.
15. B.R. Webber, Phys. Lett. **B339**, 148 (1994).
16. E. Farhi, Phys. Rev. Lett. **39**, 1587 (1977).
17. See *e.g.*, Z. Kunszt *et al.*, CERN 89–08 Vol I, p. 373 (1989).
18. S. Catani *et al.*, Nucl. Phys. **B407**, 3 (1993).
19. JLC Group, S. Matsumoto *et al.*, KEK Report 92-16 (1992) p. 53. P. Comas *et al.*, Proc. Workshop on Physics and Experiments with Linear Colliders, Sept. 8–12 1995, Morioka-Appi, Japan (World Scientific, 1996), Ed. A. Miyamoto *et al.*, p. 455.
20. A. Hoang, Proc. Worldwide Study on Physics and Experiments with Future Linear e^+e^- Colliders, Sitges, Barcelona, Spain, April 28–May 5 1999, Universitat Autonoma de Barcelona 2000, Ed. E. Fernandez, A. Pacheco, p. 69.
21. M. Martinez, Proc. Worldwide Study on Physics and Experiments with Future Linear e^+e^- Colliders, Sitges, Barcelona, Spain, April 28–May 5 1999, Universitat Autonoma de Barcelona 2000, Ed. E. Fernandez, A. Pacheco, p. 63.
22. A.H. Hoang, A.V. Manohar, I.W. Stewart, T. Teubner, Phys. Rev. **D65**, 014014 (2002).
23. W. Bernreuther, private communications.
24. A. Brandenburg, Proc. Worldwide Study on Physics and Experiments with Future Linear e^+e^- Colliders, Sitges, Barcelona, Spain, April 28–May 5 1999, Universitat Autonoma de Barcelona 2000, Ed. E. Fernandez, A. Pacheco, p. 222.
25. DELPHI Collab., P. Abreu *et al.*, Phys. Lett. **B418**, 430 (1998); A. Brandenburg *et al.*, Phys. Lett. **B468**, 168 (1999); ALEPH Collab., R. Barate *et al.*, Eur. Phys. J. **C18**, 1 (2000).
26. K. Mönig, Proc. Worldwide Study on Physics and Experiments with Future Linear e^+e^- Colliders, Sitges, Barcelona, Spain, April 28–May 5 1999, Universitat Autonoma de Barcelona 2000, Ed. E. Fernandez, A. Pacheco, p. 405.
27. M. Winter, private communications.
28. See *eg.* M. Neubert, Nucl. Phys. **B463**, 511 (1996); G. Altarelli, P. Nason, G. Ridolfi, Z. Phys. **C68**, 257 (1995).
29. TASSO Collab., W. Braunschweig *et al.*, Phys. Lett. **B214**, 286 (1988).

30. L. Orr, Proc. Workshop on Physics and Experiments with Linear e^+e^- Colliders, 26–30 April 1993, Waikoloa, Hawaii; World Scientific, Eds. F.A. Harris *et al.*, Vol. II p. 670.
31. L. Orr, private communications.
32. T. Rizzo, Phys. Rev. **D50**, 4478 (1994); SLAC-PUB-7317 (1996), hep-ph-9610373.
33. SLD Collab., K. Abe *et al.*, Phys. Rev. **D60**, 092002 (1999).
34. K. Fabricius *et al.*, Phys. Rev. Lett. **45**, 867 (1980); J. G. Körner *et al.*, Phys. Lett. **94B**, 207 (1980).
35. A. Brandenburg, L. Dixon, Y. Shadmi, Phys. Rev. **D53**, 1264 (1996).
36. K. Hagiwara *et al.*, Nucl. Phys. **B358**, 80 (1991).
37. SLD Collaboration, K. Abe *et al.*, Phys. Rev. Lett. **74**, 4173 (1995).
38. SLD Collab., K. Abe *et al.*, Phys. Rev. Lett. **86**, 962 (2001).
39. C. Carone, H. Murayama, Phys. Rev. Lett. **74**, 3122 (1995) ; D. Bailey, S. Davidson, Phys. Lett. **B348**, 185 (1995).
40. OPAL Collab., R. Akers *et al.*, Z. Phys. **C65**, 367 (1997); DELPHI Collab., P. Abreu *et al.*, Phys. Lett. **B414**, 401 (1997).
41. Y. Dokshitzer *et al.*, Phys. Rev. Lett. **69**, 3025 (1992).
42. T. Sjöstrand, V.A. Khoze, Proc. Worldwide Study on Physics and Experiments with Future Linear e^+e^- Colliders, Sitges, Barcelona, Spain, April 28–May 5 1999, Universitat Autonoma de Barcelona 2000, Ed. E. Fernandez, A. Pacheco, p. 217.
43. P.N. Burrows, G. Ingelman, Z. Phys. **C34**, 91 (1987).
44. G.R. Bower, Proc. Worldwide Study on Physics and Experiments with Future Linear e^+e^- Colliders, Sitges, Barcelona, Spain, April 28–May 5 1999, Universitat Autonoma de Barcelona 2000, Ed. E. Fernandez, A. Pacheco, p. 1049.

CHAPTER 9

Gamma-Gamma and Other Options

Tohru Takahashi
Graduate School of Advenced Sciences of Matter,
Hiroshima University
Higashi-Hiroshima, 739-8530, Japan
E-mail: tohrut@hiroshima-u.c.jp

In this chapter, we describe overview of possible options of the Linear Collider such as, e^-e^-, $e\gamma$, and $\gamma\gamma$ collider.

1. Introduction

The linear accelerator for the e^+e^- collider will provide us the highest energy and the most intense electron and positron beam. So that it is natural to come up with the idea of using the electron beam as a source of secondary beams rather than using it as the primary beam of interactions. The simplest case is to replace the positron beam of the e^+e^- collider with the electron beam. So the facility is now a high energy e^-e^- collider. The other possibility is to convert the electron beam to the photon beam by the backward Compton scattering. In this case, it is possible to build a $\gamma\gamma$ collider or an $e\gamma$ collider.

The feasibility of these optional interaction has been studied for years [1–4], however, its usefulness for physics study and feasibility of the construction of these facilities have not been explored as performed in the e^+e^- collider. This is mainly because most of the linear collider communities have concentrated on the topics with e^+e^- collision and have not paid much attention to opportunities with the options. However, recent studies revealed that there are interesting and important subjects to be explored with them. In addition, for the $\gamma\gamma$ and $e\gamma$ collider, rapid development of the laser technology makes the construction of the high average power laser within the scope of near future. Thus, physics with other than the

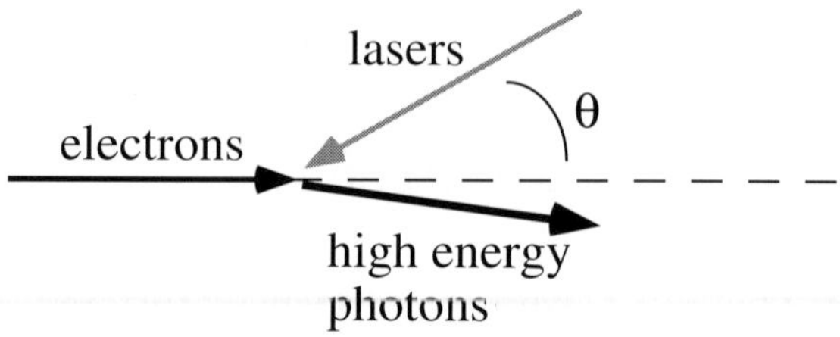

Fig. 1. Kinematics of backward Compton scattering.

e^+e^- interaction is getting more and more attention by the linear collider communities.

The construction of these options, however, need modification to the accelerator complex as well as facilities such as the laser system, an additional electron beam source and a second detector. So that these options have to be considered within the road map toward the linear collider project from the very beginning.

In this chapter, to give a basic information for the consideration above, we will try to give general idea of how these options look like in the sense of the beam provided by these new facilities, *i.e.*, energy, luminosity and polarization distribution, and then discuss some examples of the topics with them.

2. Luminosity

2.1. $\gamma\gamma$ *Collider*

In $\gamma\gamma$ colliders, photon beams are generated by the backward Compton scattering of electron- and laser-beams just before the interaction point. The maximum energy of the generated photon is expressed as

$$E_\gamma\Big|^{max} = \frac{x}{x+1}E_e. \tag{1}$$

Here, x is a kinematics parameter of the Compton scattering, defined as

$$x = \frac{4E_e\omega_L}{m_e^2}\cos^2(\theta/2), \tag{2}$$

where E_e, ω_L and θ are the electron energy, laser photon energy and angle between the electron beam and the laser beam (see Fig. 1). A calculated photon energy spectrum generated by the backward Compton scattering is shown in Fig. 2. As seen from the figure, the energy distribution of the photons depends on the polarization of the electrons (P_e) and the laser

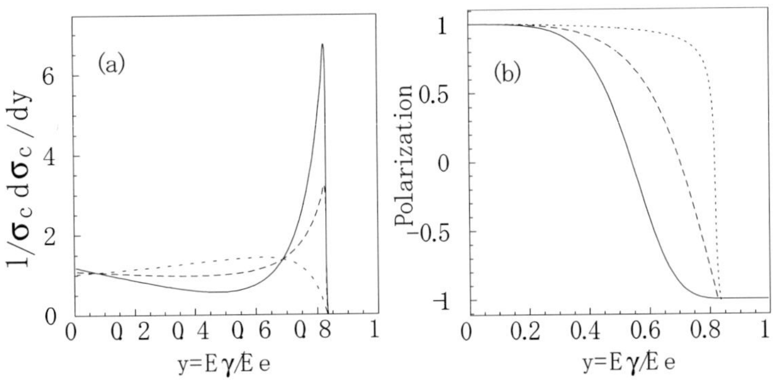

Fig. 2. Calculated energy(a) and polarization(b) of back-scattered photons by laser-Compton scattering. Three lines are for different combination of the election and the laser polarization, *i.e.*, $P_eP_L = -1$(solid), $P_eP_L = 0$(dash) and $P_eP_L = +1$(dots), respectively.

(P_L) beam, *i.e.*, one can obtain a broad energy distribution ($P_eP_L = +1$) or peaked distribution ($P_eP_L = -1$) by controlling the polarization.

According to equation (1), the maximum photon energy increases as x becomes higher. However, when x exceeds $2(1+\sqrt{2}) \approx 4.83$, the energy of a Compton-scattered photon and a laser photon system exceeds the threshold of e^+e^- pair creation. This pair-creation process could waste generated high-energy photons and could be an additional source of detector background so that, at least as the first assumption, x parameter is restricted to be smaller than 4.83, corresponding to the maximum photon energy of about 80% of the electron energy. In the case that the electron beam energy of 250 GeV, laser wave length of $1\mu m$, which matches wave length of typical solid state lasers, makes $x \approx 4.7$.

Since the precise estimation of the luminosity of $\gamma\gamma$ colliders requires a detail consideration of the laser-Compton scattering and geometry of the interaction region, which need a simulation of the laser-Compton scattering as well as the beam-beam interaction, we first estimate the luminosity of $\gamma\gamma$ colliders by a simple analysis. With the typical parameters of the photon collider, the Compton-conversion probability for an electron, *i.e.*, mean number of Compton interactions of an electron in a laser pulse, is assumed to be 1. With this assumption, the conversion efficiency k, is calculated as;

$$k = \sum_{n=1}^{\infty} P_n^1 = 1 - e^{-1} \approx 0.63 \qquad (3)$$

where P_n^1 is a probability that an electron encounters n laser photons when

Table 1. Major parameters of $\gamma\gamma$ collisions at JLC-I.

Parameter set			(a)	(b)	(c)
e^-					
Beam energy	E_e	GeV	250	103	80
Particles/bunch	N		0.95@10	←	←
Repetition rate	f_{rep}	Hz	150	←	←
bunches/pulse	n_b		95	←	←
Bunch length	σ_z	μm	120	←	←
Bunch sizes at IP	σ_x^*/σ_y^*	nm	680/12.8	77/12.2	88/10.7
Beta func. at IP	β_x^*/β_y^*	mm	0.5/0.8	0.3/0.5	0.3/0.3
Norm. emittance	$\varepsilon_{xn}/\varepsilon_{yn}$	nm·r	4500/100	4000/60	←
Vertical offset		σ_y^*	1	←	←
CP-IP distance	d	mm	7.0	3.3	2.8
Geom. luminosity	$L_{geom.}^{ee}$	cm^{-2}s^{-1}	11.8@33	6.78@33	6.79@33
Laser					
Wave length	λ_L	μm	1.053	←	0.37
Pulse energy	E_L	J	1.0	←	←
Pulse length	σ_z^L	μm	230	←	120
r.m.s spot size	σ_0^L	μm	3.17	←	1.53
x parameter	$4\omega_L E_e/m_e^2$		4.51	1.86	4.11

its average is one. Thus the number of scattered photons (N_γ) is $0.63N$, where N is the number of electrons in a bunch. Usually, we are interested in the high energy part of the spectrum and if we take high energy part ($E_\gamma > 65\%$) of the spectrum, the effective conversion efficiency k' can be regarded as $k' \approx 0.3$. Therefore, the luminosity of the $\gamma\gamma$ collider is expressed approximately as;

$$L_{\gamma\gamma} = k'^2 L_{ee} \approx 0.3^2 L_{ee} \approx 0.1 L_{ee}, \tag{4}$$

where L_{ee} is the geometric luminosity of e^-e^- collisions defined as;

$$L_{ee} = \frac{N^2 f}{4\pi\sigma_x^e\sigma_y^e} \tag{5}$$

with N, f, σ_x^e and σ_y^e being number of electrons in a bunch, its repetition rate, horizontal and vertical bunch size at the interaction point, respectively. It should be noted that the geometric luminosity is not necessary to be the same as the e^+e^- collider. Since the effect of the beam-beam interaction is less serious than e^+e^- interaction, the geometric luminosity is possibly higher than the e^+e^- collider.

More precise estimation of the luminosity spectrum can be performed using numerical simulation with specific parameters of the electron and the laser beam. As an example, the parameters for the JLC $\gamma\gamma$ collider are shown in Table 1 [5]. The parameter sets are one for the high energy operation with 250 GeV electron beam and two for low energy operation. Difference of two parameter sets for low energy operation is the wave length of the laser, *i.e.*, one with $1\mu m$ and one with $0.37\mu m$. For both cases, the maximum photon energy is set at 60 GeV with 120 GeV Higgs boson in mind. From the physics point of view, the laser wave length of $0.37\mu m$ is preferable because x parameter is higher (4.11). However development of high power laser for $0.37\mu m$ seems very challenging. Thus, we have to consider parameters with a conservative laser parameter.

The luminosity distributions simulated using CAIN [6], a simulation program for the laser-electron and the beam-beam interaction, for the high energy operation is shown in Fig. 3. We have $\gamma\gamma$, $e\gamma$ and e^-e^- interaction

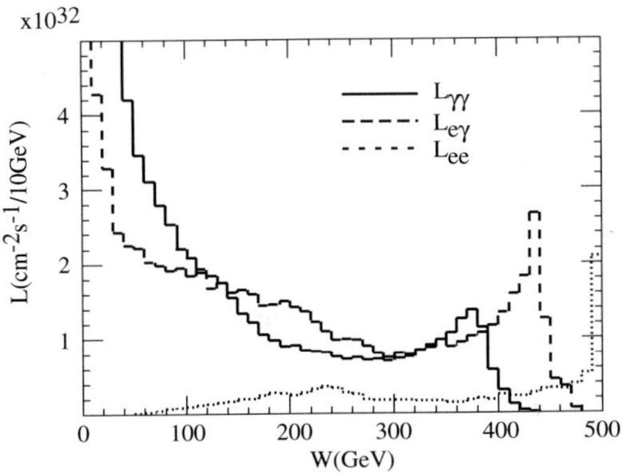

Fig. 3. Luminosity distributions simulated by CAIN[6] for collisions, the solid, dashed and dotted lines are luminosities of $\gamma\gamma$, $e\gamma$ and e^-e^- collisions, respectively, and $\sqrt{s}$ is their center-of-mass energies.

simultaneously as well as low energy interaction caused by beamstrahlung photons.

An important feature that should be mentioned is that the generated photon beam is polarized as shown in Fig. 2 and the polarization is almost 100% at the high energy edge. Using this feature, for example, the Higgs bosons are produced with the $J_z = 0$ combination of the initial photons while the background process as $\gamma\gamma \rightarrow f\bar{f}$ are suppressed with this helicity configuration. It is also very important and unique for the $\gamma\gamma$ collider that the photon beam can be linearly polarized. It is easily realized by flashing the linearly polarized laser beam on to the electron beam which is not necessary to be polarized in this case. The linear polarization transfer from

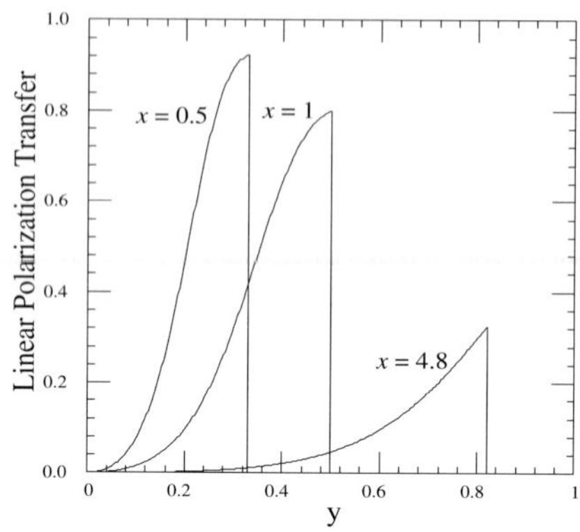

Fig. 4. Linear-polarization transfers for $x = 0.5$, 1 and 4.8, as a function of the scaled energy, $y \equiv E_\gamma / E_e$.

the laser beam to the Compton photon is shown in Fig. 4. The polarization transfer is a function of x parameter and is higher for smaller x, *i.e.*, lower photon energy. It is more than 90 % when $x = 0.5$ but is about 30 % at $x = 4.8$ though it is a preferable value for high energy photon generation. This has to be taken into account to design experiment with the linearly polarized photon beam.

Figure 5 shows luminosity distribution (CAIN simulation) when parameters are set for linearly polarized photons(see also 3.3). Plotted in the upper figure are absolute luminosities for two combination of colliding photons when the parameters are set to maximize parallel combination (denoted as $L^{\parallel}_{\parallel}$ and $L^{\parallel}_{\perp}$). The real polarization is not 100 % and it can be expressed in term of the analyzing power as

$$A_{lum} = \frac{L^{\parallel}_{\parallel} - L^{\parallel}_{\perp}}{L^{\parallel}_{\parallel} + L^{\parallel}_{\perp}} \tag{6}$$

which is plotted in the lower part of Fig. 5.

2.2. *eγ Collider*

The $e\gamma$ collider mode can be obtained by turning off the laser for one of the two electron beam at the $\gamma\gamma$ collider operation. By the same discussion on the luminosity of the $\gamma\gamma$ collider, the luminosity of the high energy part of the $e\gamma$ luminosity distribution can be estimated as

$$L_{e\gamma} = k' L_{ee} \approx 0.3 L_{ee}. \tag{7}$$

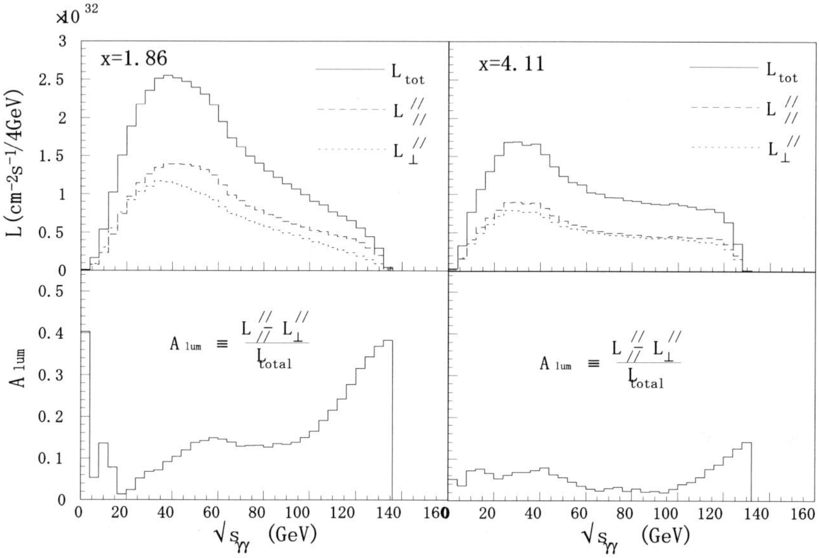

Fig. 5. Linearly polarized luminosities and the asymmetries at 120 GeV $\gamma\gamma$ collisions, which are simulated by CAIN [6], where $L_{\parallel}^{\parallel}$(dashed), $L_{\perp}^{\parallel}$ (dotted) and $L_{\text{total}} \equiv L_{\parallel}^{\parallel} + L_{\perp}^{\parallel}$(solid). The left and right figures correspond to the parameter sets of (b) and (c) in table 1, respectively.

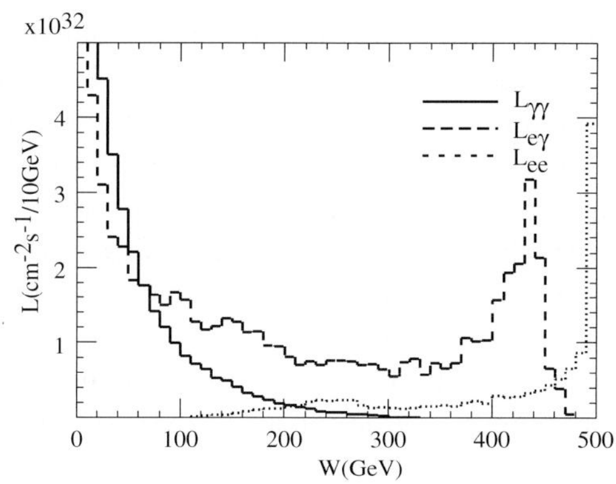

Fig. 6. Luminosity distributions for $e\gamma$ collider mode simulated by CAIN [6], where the solid, dashed and dotted lines are luminosities of $\gamma\gamma$, $e\gamma$ and e^-e^- collisions, respectively, and $\sqrt{s}$ is their center-of-mass energies.

The result of the of numerical simulation is shown in Fig. 6. Comparing the luminosity for the $e\gamma$ collisions in Figs. 3 and 6, it is found that both

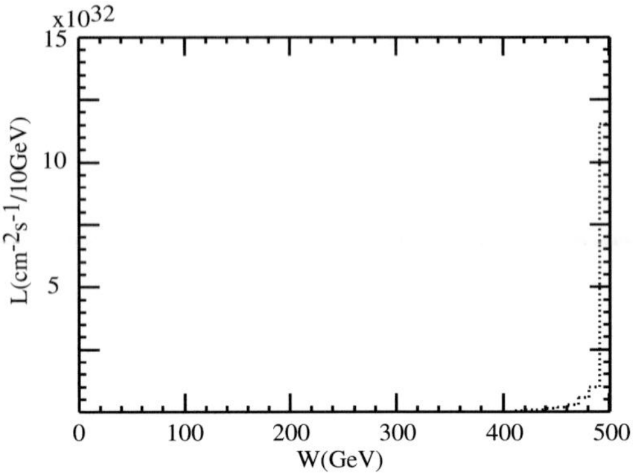

Fig. 7. Luminosity distributions for e^-e^- collider mode simulated by CAIN [6].

luminosities are almost the same amount though the later is set for the $e\gamma$ collision mode. This is because, in $\gamma\gamma$ mode, only one third of the electron is available for the high energy collision, however, the $e\gamma$ collision has two combination. These factors cancel each other out, resulting in almost same amount of luminosity for $e\gamma$ and $\gamma\gamma$ mode.

2.3. e^-e^- *Collider*

The collision of the same sign beam suffers from the anti-pinch effect due to the repulsive force between them. It requires a careful tuning of the beam parameters because small beam size just increases the beamstrahlung effect but may not gain the luminosity. As an example of the parameter, a parameter set for JLC/NLC (ISG-B) [8] e^+e^- collider were used by replacing a positron beam with the electron beam. In this case, the luminosity was about 20% of the e^+e^- luminosity with the same parameters. A simulated luminosity distribution is show in Fig. 7.

An optimization of the luminosity of the e^-e^- collider was discussed by Thompson [7]. In the reference we found an analytical estimation of the anti-pinch factor for the e^-e^- collision. The e^-e^- luminosity can be increased by increasing number of electrons in a bunch and/or decrease length of the bunch. However, it is, of course, necessary to tune these parameter in actual machine parameter optimization. According to the discussion, it may be possible to optimize the luminosity of the e^-e^- collider up to 30% of the e^+e^- collider.

Table 2. Proposed electron beam parameters of the $\gamma\gamma$ collider optoins for the JLC, NLC and TESLA.

Parameter set			JLC	NLC	TESLA
e^-					
Beam energy	E_e	GeV	250	267.5	250
Particles/bunch	N		0.95@10	1.5@10	2.0@10
Repetition rate	f_{rep}	Hz	150	120	5
bunches/pulse	n_b		95	95	2.5k
bunch spacing	t	ns	2.8	2.8	$\approx$ 300
Bunch length	σ_z	μm	120	156	300
Bunch sizes at IP	σ_x^*/σ_y^*	nm	67.8/12.8	166/3.0	88/4.3
Beta func. at IP	β_x^*/β_y^*	mm	0.5/0.8	4.0/0.065	1.5/0.3
Norm. emittance	$\varepsilon_{xn}/\varepsilon_{yn}$	nm·r	4500/100	3600/71	2500/30
CP-IP distance	d	mm	7.0	1.0	2.1
Geom. luminosity	L_{ee}	$\mathrm{cm}^{-2}\mathrm{s}^{-1}$	11.8@33	41.0@33	118@33

2.4. *Summary of the Luminosities*

Roughly speaking, typical luminosity of the options are:

- $L_{\gamma\gamma}^{eff} \approx 0.1 L_{ee}$
- $L_{e\gamma}^{eff} \approx 0.3 L_{ee}$
- $L_{e\gamma}^{eff} \approx 0.3 L^{e^+e^-}$

A question for the $\gamma\gamma$ and the $e\gamma$ collider is that how large the geometric luminosity can be. The effort on the optimization is going on for each linear collider project, *i.e.*, JLC, NLC, and TESLA. All these efforts have the constraint that the parameters of the linac are not too different from those of the e^+e^- collider. However, the current optimization results still have some differences depending on how much differences from e^+e^- machines are acceptable. In Table 2, we show main parameters for the electron beam for the JLC, the NLC [14] and the TESLA [3]. It appears that TESLA assumed the highest geometric luminosity of $118 \times 10^{33}\mathrm{cm}^{-2}\mathrm{s}^{-1}$ while the JLC is the most conservative one($11.8 \times 10^{33}\mathrm{cm}^{-2}\mathrm{s}^{-1}$).

3. Higgs Boson in $\gamma\gamma$ Interaction

The study of Higgs boson, particularly measurement of two photon decay width, primarily motivates the $\gamma\gamma$ collider. The schematic diagram of the coupling of the Higgs boson to two photons is shown in Fig. 8. As discussed in the previous section, the luminosity distribution of the $\gamma\gamma$ collider is not monochromatic, so that convolution of the spectrum with the production cross section is necessary to estimate number of events expected. For this

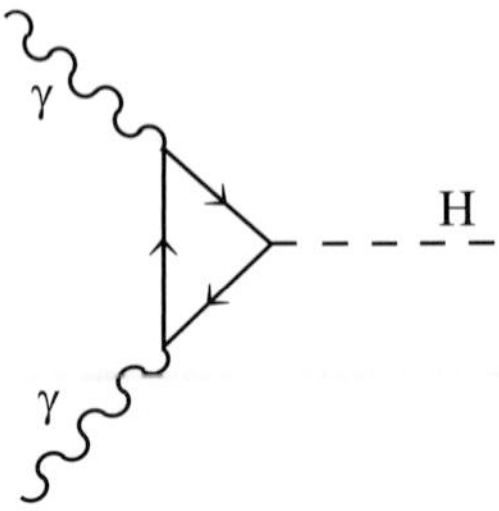

Fig. 8. Couplings of the Higgs boson to two photons.

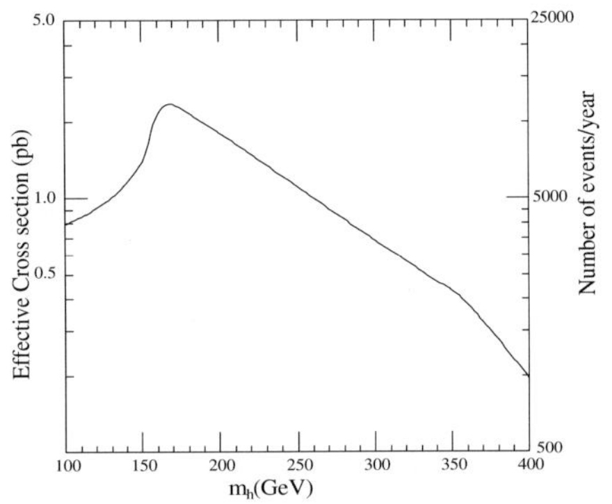

Fig. 9. The effective production section and expected number of events/year for the Higgs boson production in a $\gamma\gamma$ collider.

purpose, an effective cross section is defined as:

$$\hat{\sigma} \equiv \frac{1}{L_{\gamma\gamma}^{tot}} \sum_{J=0,2} \int \sigma_{\gamma\gamma \to h}^{J}(s_{\gamma\gamma}) \frac{dL_{\gamma\gamma}}{d\sqrt{s_{\gamma\gamma}}} d\sqrt{s_{\gamma\gamma}} \tag{8}$$

The number of events is estimated as:

$$N = \int Ldt \cdot \hat{\sigma}. \tag{9}$$

The effective cross section defined in Eq. (8) and expected number of events per year is plotted in Fig. 9 as function of Higgs boson mass with the luminosity distribution expected with the JLC $\gamma\gamma$ collider.

It is found that order of 5,000 events can be expected with typical parameter as an option of the JLC.

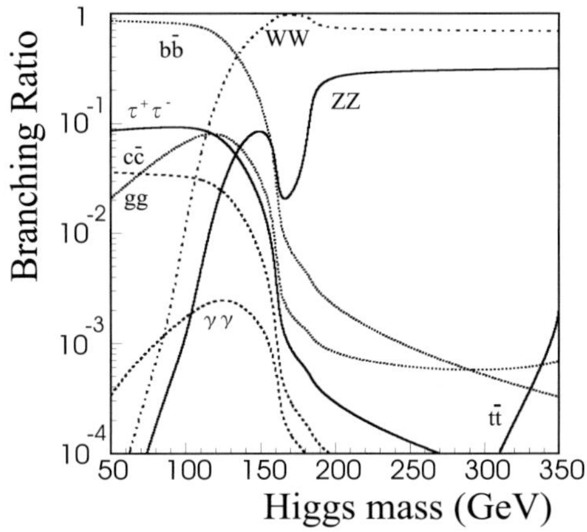

Fig. 10. Branching ratio for the standard model Higgs boson for a 176 GeV top-quark mass.

3.1. *Two Photon Decay Width of the Light Higgs Bosons*

As two photons couple to the Higgs boson only through a loop diagram of massive charged particles, any kind of massive charged particle contributes to the two-photon decay width of the Higgs boson. Therefore, the measurement of the two photon decay width of the Higgs boson provides comprehensive information of the particles which couple to the Higgs boson. Regarding physics beyond the standard model, a deviation of the two-photon width from the standard model prediction indicates an additional contribution from unknown particles, and is a signature of physics beyond the standard model. For example, the supersymmetric extension of the standard model (MSSM) has additional charged particles, such as scalar fermions, charged Higgs bosons and charginos. Since the masses of these particles partly originate from the Higgs mechanism of the electroweak symmetry breaking, the presence of theses particles results in a shift of the two-photon decay amplitude of the Higgs boson from its value of the SM. In fact, a minimal extension of the standard model(MSSM) predicts the ratio of two-photon decay widths $(\Gamma(h^0 \to \gamma\gamma, \text{MSSM})/\Gamma(H \to \gamma\gamma, \text{SM}))$ to be as much as 1.2 for the lightest Higgs boson having a mass of 120 GeV [10].

The light Higgs boson in the SM mainly decays into a $b\bar{b}$ pair, as is shown in Fig. 10, and b hadrons can be tagged effectively by current b-tagging technology [11]. Therefore, $b\bar{b}$ events are the best signal of the light Higgs bosons. The main backgrounds are the continuum $\gamma\gamma \to q\bar{q}$ processes and are dominantly produced by the initial photon collision in the $J_z = 2$

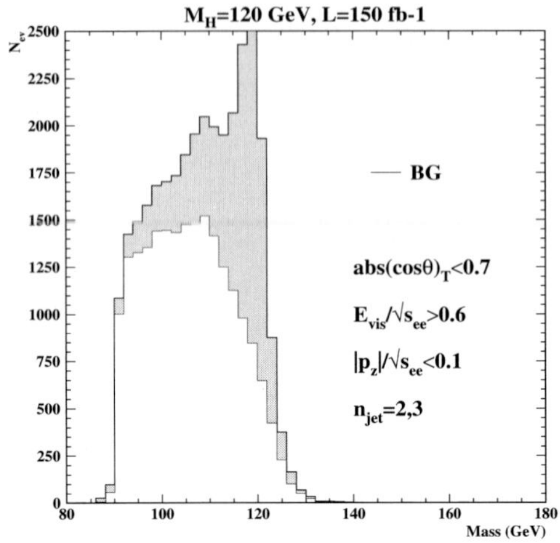

Fig. 11. Mass distributions for Higgs signal and heavy quark background for $m_h = 120$ GeV [13].

angular momentum state, while the Higgs boson signals are only accessible by the $J_z = 0$ collisions. This feature allows good background suppression by controlling beam polarization. It was reported, however, that the effect of QCD corrections to $\gamma\gamma \to q\bar{q}$ may be large, since the helicity suppression in the $J_z = 0$ channel does not work due to gluon emission. It could be a serious source of background in the light Higgs boson detection. Previous studies [12] took this effect into account with an approximation. The complete estimate has been reported by G. Jikia and Soldner-Rembold [13]. The mass plot of the signal and background is shown in Fig. 11 [13]. It has been shown by all of recent works [12-14], statistical accuracy on measurement of $\Gamma(h \to \gamma\gamma)Br(h \to b\bar{b})$ is estimated to be about 2 %. The $Br(h \to b\bar{b})$ can be measured by the e^+e^- collision with the accuracy of 2 % for 120 GeV Higgs boson [16] so that $\Gamma(h \to \gamma\gamma)$ is obtained with 3 % accuracy.

3.2. *Heavy Higgs Bosons*

A hunt for the heavy Higgs Boson is recognized to be an important issue for the photon collider. As it is produced singly through the s-channel process, the mass reach of the photon collider is larger than the e^+e^- mode. Particularly, recent study for the Higgs boson search in the MSSM by the ATLAS group put the spotlight on it [15]. Figure 12 is the discovery plot for the Higgs boson in the parameter space of the MSSM. As seen from the picture, in the region of moderate $tan\beta$ and $m_A > 250$ GeV, only h_0 is ob-

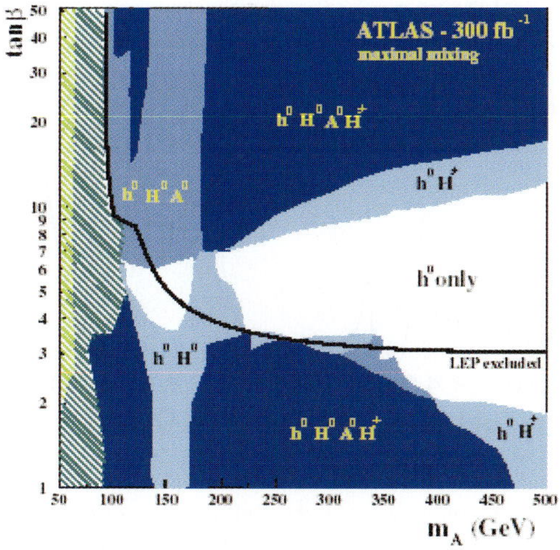

Fig. 12. Number of Higgs Boson observed with 300 fb^{-1} with ALTAS at LHC[15].

served by the LHC. In addition, if m_A is heavier than about 300 GeV, the process $e^+e^- \to HA$ is not possible by the e^+e^- collider of $\sqrt{s} = 500$ GeV. In this area, referred as the wedge, the photon collider is the only way to produce the heavy Higgs bosons. Asner, *et al.* performed a systematic study on the Higgs search in the wedge by the photon collider [14]. They pointed out that the photon collider is able to find the heavy Higgs boson up to its maximum energy.

3.3. *Measurement of CP Nature of the Higgs Boson*

Using the linear polarization of the photon beam, it is possible to measure asymmetry of the Higgs boson production as;

$$A_{CP} \equiv \frac{\sigma_{\parallel} - \sigma_{\perp}}{\sigma_{\parallel} + \sigma_{\perp}} \tag{10}$$

which directly reflects CP nature of the produced Higgs boson, where $\sigma_{\parallel}(\sigma_{\perp})$ is the production cross section of the Higgs boson when the polarization of the initial photons are set in parallel (perpendicular) to each other. In the real situation, the photon beam is not 100% polarized as shown in Fig. 5 and an observed asymmetry is smeared as;

$$A_{CP}^{ops} \equiv A_{lum} A_{CP} \tag{11}$$

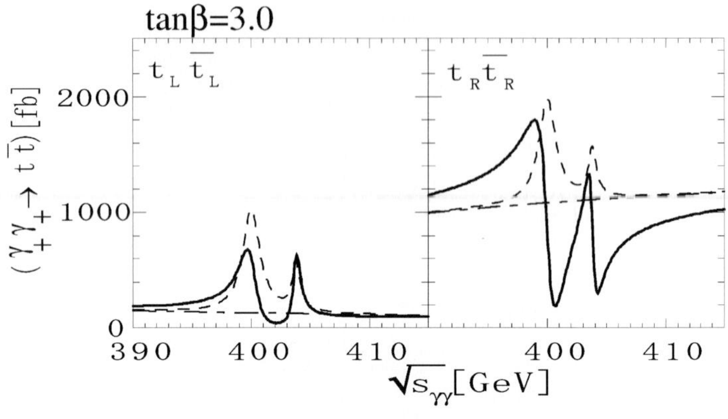

Fig. 13. The cross sections of $\gamma_+\gamma_+ \to t_L\bar{t}_L$ (left), $t_R\bar{t}_R$ for $tan\beta = 3$. The solid curves show the cross sections with Higgs resonances, the dashed curves is the ones obtained by neglecting the interference, the dot-dashed lines are the continuum contribution, respectively [17].

where A_{lum} is an analyzing power defined in Eq. (6). This process may be the best way to observe CP nature of the Higgs boson with the linear collider including e^+e^- mode. A feasibility of this measurement was studied by Asner *et al.* [14]. They claimed that for the light Higgs boson of $m_h =$ 120 GeV of CP = 1, it is possible to determine the CP of the Higgs boson with δCP/CP ≈ 0.23.

The MSSM predicts two relatively heavy Higgs bosons of CP even and odd state and the masses of these two can be very close in some parameter space of the MSSM. The recent study shows that if this is the case, it might be possible to observe an interference of CP even (H) and CP odd(A) Higgs bosons via their top quark decay (See Fig.13) [17].

4. Supersymmetry

4.1. *Sfermion and Chargino Production by $\gamma\gamma$ Collisions*

Production of sfermions by $\gamma\gamma$ interaction, $\gamma\gamma \to \tilde{f}\bar{\tilde{f}}$ where $\tilde{f} = (\tilde{e}_{L,R}, \tilde{\mu}_{L,R}, \tilde{\tau}_{1,2}; \tilde{u}_{L,R}, \tilde{d}_{L,R}, \cdots, \tilde{t}_{1,2})$, is a simple SUSY QED.

The production cross section is shown in Fig. 14 as a function of its mass [1, 18]. Except for the selectron production, the production cross section in $\gamma\gamma$ collision is much larger than that of e^+e^- collision in most of mass range. This is because the s-wave production in $\gamma\gamma$ interaction while it has the p-wave suppression in e^+e^- interactions.

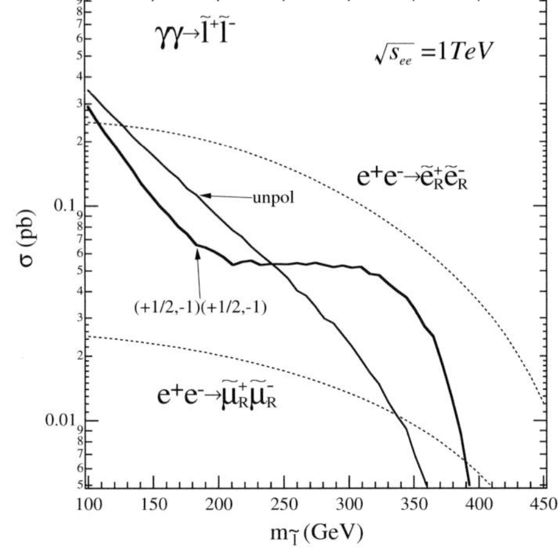

Fig. 14. Slepton mass dependence of the total cross section for each initial beam polarization (λ_1^e, P_1^c) and (λ_2, P_2^c). $\sqrt{s_{ee}}$ is taken at 1 TeV [18].

The chargino can also be produced in $\gamma\gamma$ interaction. The chargino mass dependence of the cross section is shown in Fig. 15, where $\sqrt{s_{ee}} = 1$ TeV is assumed [1, 18]. For comparison, the total cross sections for $e^+e^- \to \tilde{\chi}_1^+\tilde{\chi}_1^-$ is plotted. The $\gamma\gamma$ cross section is larger than that of e^+e^- if $\sqrt{s}$ is enough above the mass threshold. It is also seen that the beam polarization significantly enhances the cross section. In the real study, however, background from the $\gamma\gamma \to W^+W^-$ could be very serious and its contamination has to be studied.

In addition, even if we could get enough number of events with reasonable signal to background ratio, it is still not clear that the $\gamma\gamma$ interaction could give additional information to e^+e^- interaction. It should be noted that, in e^+e^- interaction, beautiful methods for the systematic study of the SUSY parameter and for testing SUSY GUT scenario has been demonstrated (see Chapter 4). The role of the $\gamma\gamma$ option to the supersymmetry is yet to be studied.

4.2. *Selectron in the $e\gamma$ Collisions*

In $e\gamma$ interaction, the single production of selectron, $e\gamma \to \tilde{e}\tilde{\chi}^0$ is advantageous over e^+e^- interactions. Assuming $\tilde{\chi}^0$ as the LSP, we find that the process has a lower mass threshold of $m_{\tilde{e}_R} + m_{\tilde{\chi}_1^0}$ than that of $2m_{\tilde{e}_R}$ for the selectron pair production at e^+e^- colliders. Figure 16 shows the selectron

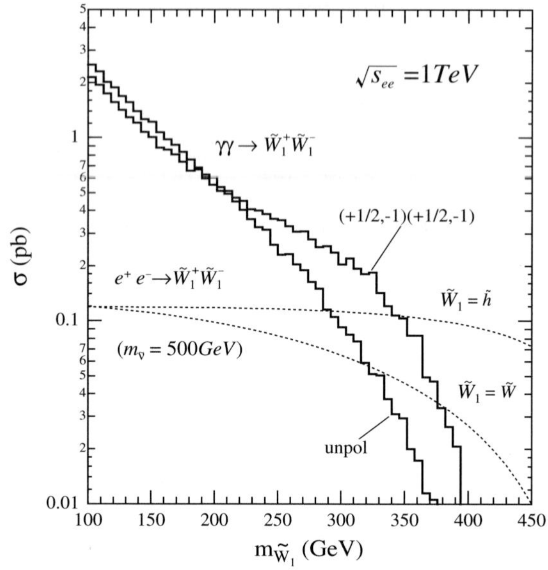

Fig. 15. Chargino mass dependence of the total cross sections for each initial beam polarization (λ_1, P_1^c) and (λ_2, P_2^c). $\sqrt{s_{ee}}$ is at 1 TeV [18].

mass dependence of the total cross section [18]. We find a large cross section for $e\gamma \rightarrow \tilde{e}_R\tilde{\chi}_1^0$, even if the selectron is rather heavy, $m_{\tilde{e}_R} \geq \sqrt{s}/2$. It is also seen that the cross section depends on the beam polarization and is effective to enhance selectron signals.

4.3. Selectron in e^-e^- Collisions

The e^-e^- interaction has a unique feature for threshold behavior of the production of the selectrons. In the case of the same helicity selectron production in the e^+e^- collision such as, $e^+e^- \rightarrow \tilde{e}_R^+\tilde{e}_R^-$, it suffers from β^3 behavior at the threshold. On the contrary, in the e^-e^- collision, the process, $e^-e^- \rightarrow \tilde{e}_R^-\tilde{e}_R^-$, has β shape at the threshold. It is because total angular momentum $J = 0$ in the e^-e^- initial state so that $\tilde{e}_R^-\tilde{e}_R^-$ is produced in s-state. The production cross section of the right handed selectron near the threshold is shown in Fig. 17 for both e^+e^- and e^-e^- collisions [20]. It is clear that the e^-e^- collision has advantage over the e^+e^- collision at the threshold behavior of the production cross section which allows a precise determination of the mass of the $\tilde{e}_R$ by the energy scan [20]. A precise estimate of the mass determination, including the beam-beam effect at the

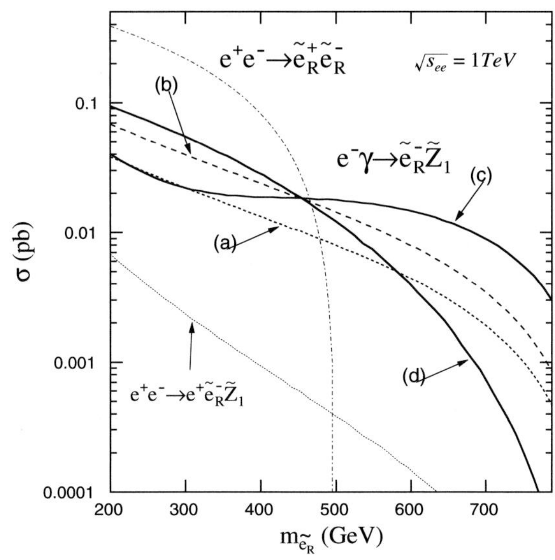

Fig. 16. Selectron mass dependence of the total cross sections for each initial beam polarization $(\lambda_1, \lambda_2, P_2^c) = (0, 0, 0)$ (a), $(+1/2, 0, 0)$ (b), $(+1/2, +1/2, -1)$ (c) and $(+1/2, -1/2, +1)$ (d). We take $\sqrt{s_{ee}} = 1$ TeV and $m_{\tilde{\chi}_1^0} = m_{\tilde{B}} = 100$ GeV [18].

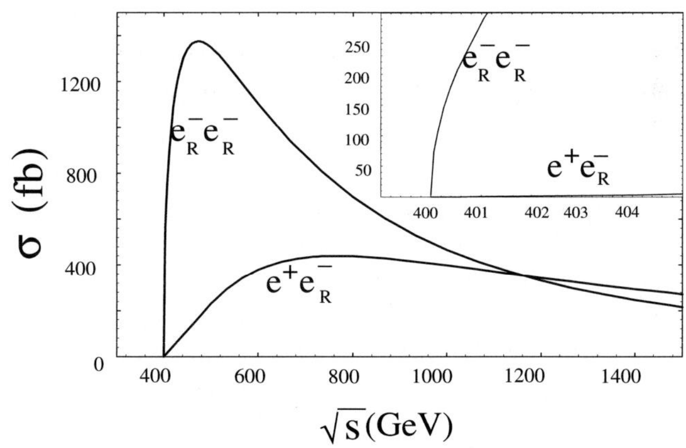

Fig. 17. Threshold behavior of the production cross section of right handed selectron pairs in e^+e^- and e^-e^- collision for $m_{\tilde{e}_R} = 200$ GeV [20].

interaction point, has been reported by Feng and Peskin [21]. It has been pointed out that it may be possible to determine selectron mass with accuracy of 100 MeV with the integrated luminosity of 1 fb^{-1} while more that 100 fb^{-1} is needed for e^+e^- interaction.

5. Precise Measurement of Standard Model Particles

So far we discussed the physics assuming discovery of new particles, either it is within the framework of the standard model or beyond the standard model. As it has been shown historically, the indirect measurement of physics via precise measurements sometimes shows us the direction to the next experiments. In this sense the importance of the precise measurement is more significant in the case that no new particles are found by the e^+e^- linear collider or the LHC. In this section, we discuss possibility of precise measurement of the standard model particles by $\gamma\gamma$, $e\gamma$ and e^-e^- interaction and their complementarity to the e^+e^- mode.

5.1. *Anomalous Coupling of the W Bosons*

The sensitivity for the anomalous couplings of W bosons, $\delta\kappa_\gamma, \lambda_\gamma$, by high energy $\gamma\gamma$ and $e\gamma$ interaction has been discussed by several authors [22–24]. All these studies pointed out that, both $\gamma\gamma$ and $e\gamma$ interaction play a complementary role to e^+e^- interaction for this measurement.

The cross section of W pair production in the $\gamma\gamma$ collision is about 90 pb for $\sqrt{s_{\gamma\gamma}} > 200$ GeV and is almost independent of the $\gamma\gamma$ center-of-mass energy. This cross section is O$(10-100)$ higher than the fermion pair production, and is also O$(10-100)$ higher than the W pair production in e^+e^- collisions. However, it has to be mentioned that even if the total cross section of the W boson pair production is large in high energy $\gamma\gamma$ interaction, the amplitude for W_TW_T production is more and more dominant in higher energy. Since the information of the anomalous coupling is included in the longitudinal W boson amplitude, going to higher energy may not improve the sensitivity to the anomalous coupling [22]. In this sense, e^+e^- mode is advantageous in higher energy, however, at $\sqrt{s} \approx 500\ GeV$, $\gamma\gamma$ mode is still useful to explore the anomalous coupling of W bosons.

For e^+e^- interaction, the feasibility of the measurement has been investigated by Monte Carlo simulation including detector effect [25]. Since effect of the beam (and luminosity) energy distribution and polarization is more important for $\gamma\gamma$ and $e\gamma$ options, it has to be taken into account to estimate sensitivity for the precise measurement. The feasibility of a measurement of the W pair production cross section in the $\gamma\gamma$ collision and its

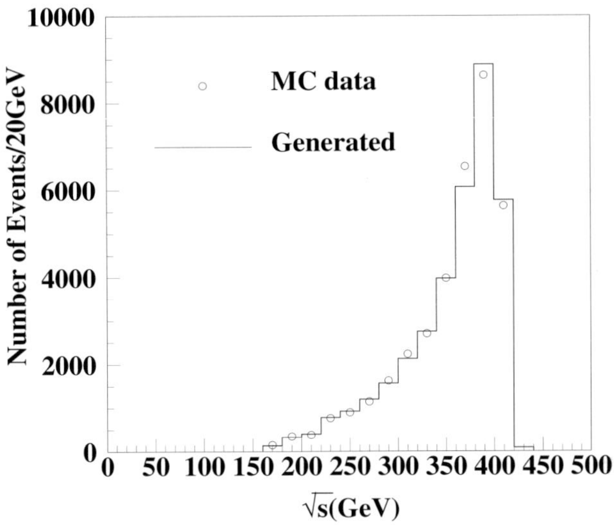

Fig. 18. Number of W pair events as a function of the center-of-mass energy. The solid line is the number of generated and circles are the efficiency-corrected number of detected events. The normalization corresponds to a total integrated luminosity of 10 fb^{-1}.

sensitivity to the anomalous coupling of W bosons to photons have been investigated by the Monte Carlo simulation [26]. The W pair events were selected by 4-jet events, *i.e.*, for the case that both the W bosons decayed into quark pairs. The production cross section for W pairs weighted by luminosity,

$$\widehat{\sigma}_{WW} \equiv \sum_{J=1,2} \int \frac{dL^J_{\gamma\gamma}}{d\sqrt{s}} \sigma^J_{WW} d\sqrt{s}$$

was approximately 50 pb. Thus, the expected number of events with an integrated luminosity of 10 fb^{-1} is about 75000. The detection efficiency was found out to be 15 % so that we could expect about 11,000 events/10 fb^{-1}. The reconstructed energy of WW system was plotted in Fig. 18.

For the obtained number of events, the sensitivity to the anomalous coupling, $\delta\kappa_\gamma$ and λ_γ, were examined as shown in Fig. 19. As can be seen from the figure, the feasibility for a measurement of the anomalous coupling of W bosons by $\gamma\gamma$ colliders is comparable to that from e^+e^- colliders and is complementary.

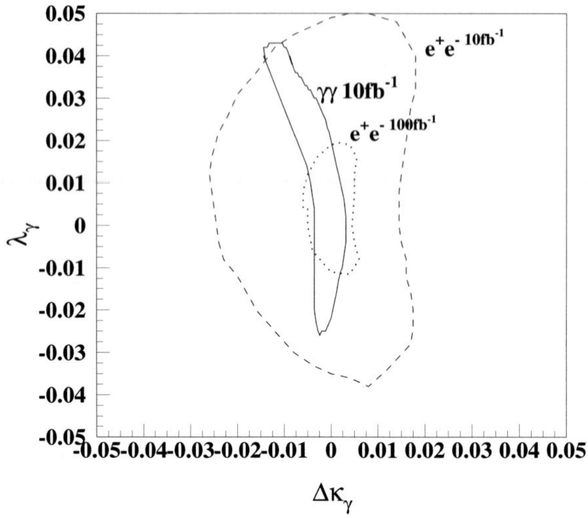

Fig. 19. Expected 90% limit in W's anomalous coupling parameter $\delta\kappa_\gamma$, λ_γ. An estimation with an e^+e^- collision is also shown by the dotted and dashed lines. The total integrated luminosity for $\gamma\gamma$ is assumed to be 10 fb^{-1}.

This feasibility is derived only from measurement of the total cross section, and improvements are expected by a comprehensive angular analysis using the decay property of W bosons. [a]

5.2. *Top Quarks in the $\gamma\gamma$ Interaction*

As discussed in the previous section, the top quark pairs are important as a final state of the heavy Higgs bosons. In addition, there are several interesting and important top quark topics in the $\gamma\gamma$ and the $e\gamma$ interactions such as;

- study of CP violating coupling.
- measurement of the anomalous $\gamma\gamma t$ coupling [27],
- search for EDM using the linear polarization [28] or using the lepton from the top quark decay [29].
- measurement of the Wtb coupling by the single top production in the $e\gamma$ interaction [30].

[a] In the Linear Collider Workshop 2002, a new analysis which taken into account angular distribution as well as total cross section was reported. It probably will appear in the proceedings of the workshop.

The detail of the discussion can be found in reference [31] and CP violation in top physics is extensively dealt by Atwood *et al.* in reference [32] and also in chapter 10 in this book.

The experimental aspect of these studies are, however, not investigated to date except for the estimate of number of detected events and initial try of the top quark reconstruction [33]. In the study number of top quark events are estimated for both six-jet and lepton+four-jet mode, assuming the JLC detector and the JLC $\gamma\gamma$ collider. When taking the convolution of the production cross section and the luminosity spectrum of the JLC $\gamma\gamma$ collider, about 4,000 events of top quark pairs are produced in a year. The most serious background in this case is the W boson pairs. Even if the event topology is different from the W boson pairs, it is still main source of the background due to large production cross section; *i.e.*, $\sigma_{ww} \approx 300\sigma_{t\bar{t}}$. Since the top quark decays as $t \to Wb$, good b-quark tagging efficiency is essential to eliminate W boson contamination. In the study, the detection efficiency was found to be 9 % for lepton+four-jet and 15.6 % for six-jet mode, corresponding to 360 and 620 events per year with the signal to background ratio of 10.

Toward the full reconstruction of the top quark events for further analysis, reduction of wrong assignment of the W boson and the b quark jet is to be an issue. In top quark decay, there will always be two b quark jets in an event and one has to choose the right partner of the W boson to reconstruct the top quark. The initial study shows that probability of the miss-assignment is not negligible (order of 10%) and further study is necessary for this issue.

5.3. *Precise Determination of Weinberg Angle by e^-e^- Interaction*

The use of polarized beam for the precise measurement has been well verified by the SLD experiment [34]. The precise measurement of the Møller scattering at the e^-e^- interaction has the advantage for this purpose [35]. Since both beam can be polarized in the e^-e^- mode, we can expect higher effective polarization as;

$$P_{eff} = \frac{2P_e}{1 + P_e^2} \tag{12}$$

here the same polarization for both beams are assumed. With currently achieved 77% electron beam polarization [36], it is already reaching 97% and if we assume 90% polarization for the electron beam in the future, it

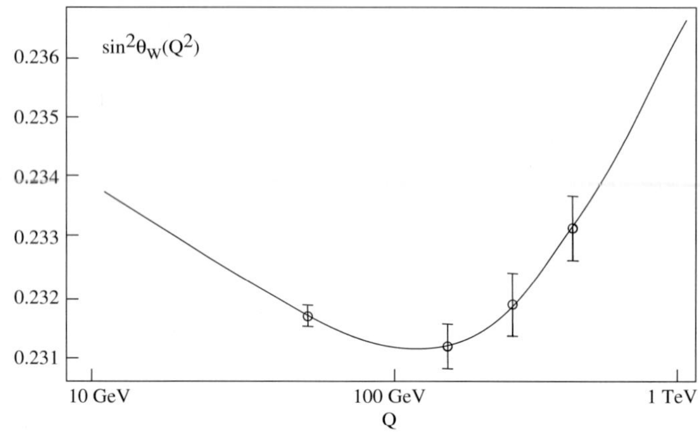

Fig. 20. Running of $sin^2\theta_W(Q^2)$ compared with potential future e^-e^- collider measurements at $\sqrt{s} = 1$ TeV [35].

goes up to 99%. This allows e^-e^- interaction to measure A_{LR} with double beam polarization as;

$$A_{LR}{\prime}^{(2)} = P_{eff} A_{LR}^{(2)} = P_{eff} \frac{\sigma_{LL} - \sigma_{RR}}{\sigma_{LL} + \sigma_{RR}} \tag{13}$$

where $\sigma_{LL}(\sigma_{RR})$ is the cross section of Møller scattering with two left-handed(right-handed) beams. Because Møller is a t-channel process, the Q^2 depends on the scattering angle as;

$$Q^2 = ys \tag{14}$$

where y is a kinematical parameter related with the scattering angle with respect to the beam axis as;

$$y = \frac{1 - \cos\theta}{2}. \tag{15}$$

Figure 20 shows a potential of e^-e^- interaction for the measurement of the Weinberg angle at $\sqrt{s} = 1$ TeV [35]. It is seen that single measurement gives the $sin^2\theta_W$ at several Q^2 and its precision is better for lower Q^2 corresponding to the small angle scattering.

5.4. *Search for the Extra-Dimensions*

The scenario of the gravity in extra-dimension proposed by Arkani-Hamed, Dimopoulos and Dvali is of great interest for the linear collider physics

[37]. The detail of the scenario and its feasibilities in the linear colliders are discussed in Chapter 7. Here, we summarize possibilities to explore them using the photon collider options.

In the scenario, gravity lives in 3+n dimensional space while the standard model particle lies in the 3 dimensional wall of the 3+n dimension. The Plank scale, M_{pl} in the three dimension space is then related with the gravitational energy scale in the 3+n dimension, M_s, as

$$M_{pl}^2 = R^n M_s^{n+2} \tag{16}$$

where R is the size of the n-dimensional extra-space. Note that the gravitational potential obey $1/r^{2+n}$ law with in the distance below R. It has been pointed out that, assuming $M_s \approx 1$ TeV, R is order of 0.1 mm which is not excluded by current experiment on the gravity measurement and this energy range (≈ 1 TeV) can be accessible by the linear collider experiment.

Rizzo discussed the sensitivity of various processes in the linear collider experiment [38]. It was pointed out that using, $\gamma\gamma \to W^+W^-$, it is possible to explore M_s up to $11\sqrt{s}$ while it is about $(4-5)\sqrt{s}$ in other channel like $e^+e^- \to f\overline{f}$.

6. Other Experimental Issues

6.1. *Hadronic Cross Section*

The hadronic cross section is an important issue for the $\gamma\gamma$ collider. In the physics aspect, it gives us information on the structure function of the high energy (and polarized) photon. On the other hand, the cross section of the hadronic cross section is expected to be large and getting larger with the energy. So that the hadron events may potentially be a serious background to the physics processes.

The hadronic cross section has been measured in e^+e^- and γp interactions for years and data for the structure function has been accumulated however projection to the high energy $\gamma\gamma$ collision still has large ambiguity. The current situation on the hadronic cross section is discussed by Godbole and her colleagues [39] as illustrated in Fig. 21. Even each model is consistent with the experimental measurement at low energy region, there is large model dependence on the prediction of the cross section in high energy $\gamma\gamma$ collision. From the QCD point of view, on the other hand, the $\gamma\gamma$ collider could have a capability to distinguish these models and also provide information on the structure function of polarized photon which could never be done in previous experiments.

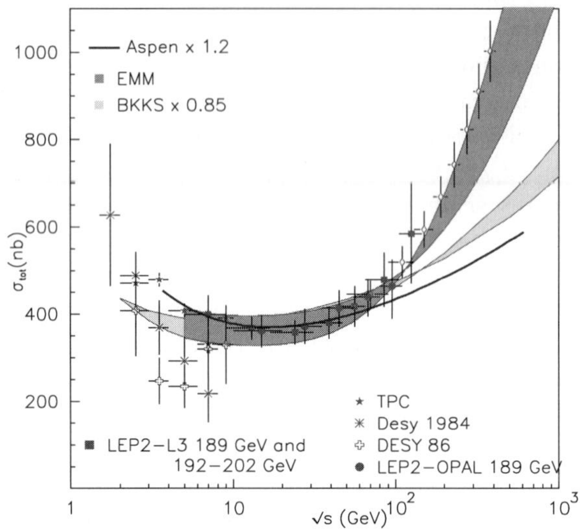

Fig. 21. The total $\gamma\gamma$ cross-section as a function of the energy with various model calculations and data obtained at low energy experiments [1, 39].

6.2. *Luminosity Measurement*

At the e^+e^- collider, the Bhabha scattering has a large cross section and is a good channel to monitor the luminosity [43, 44].

In the $\gamma\gamma$ collider, there is no such process having a large cross section as Bhabha scattering in the e^+e^- interaction. In addition, the luminosity spectrum of the $\gamma\gamma$ collision is far from monochromatic and information of the polarization is essential for physics study. Therefore, to pursue good physics at the $\gamma\gamma$ collider, measurement of differential luminosity as:

$$\frac{dL_{\gamma\gamma}}{d\sqrt{s_{\gamma\gamma}}d\eta_{\gamma\gamma}}. \tag{17}$$

is essential. Here, $\eta_{\gamma\gamma}$ is the rapidity of the $\gamma\gamma$ system defined as

$$\eta_{\gamma\gamma} = \log\sqrt{w_1/w_2},$$

with w_1 and w_2 being the energies of the two colliding photons.

For a good measurement of the luminosity, physics processes must have large cross sections and can be selected with high efficiency and small backgrounds. The possible candidate processes are; $\gamma\gamma \to e^+e^-$, $\gamma\gamma \to \mu^+\mu^-$, $\gamma\gamma \to W^+W^-$. If we assume that W is the standard model gauge boson,

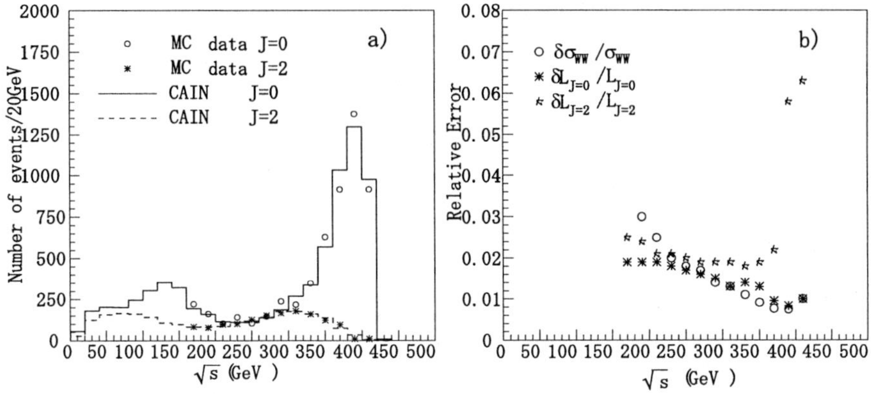

Fig. 22. a) Reconstructed events in the process $\gamma\gamma \to e^+e^-$, where the circles and asterisks are the measured luminosities with J=0 and J=2, while the solid and dashed lines are the generated luminosities with J=0 and J=2, respectively. b) Estimated statistical error for the luminosity measurement, assuming the integrated luminosity of 10 fb^{-1}.

the W pair production is a good candidate for a luminosity measurement at a center-of-mass energy greater than 200 GeV. The lepton pairs are good processes, however, it is only sensitive to the $J = 2$ component of the luminosity so that it is necessary to measure total luminosity by other processes or to measure $J = 0$ component by other methods.

Telnov proposed a possible way to measure polarized luminosity using lepton pairs [45]. When the $\gamma\gamma$ collider is running with $P_L P_e = -1$, the helicities of two photon beams are usually chosen to have the luminosity distribution of either the $J = 0$ or $J = 2$ dominant distribution. These two modes can be easily exchanged if both P_L and P_e are flipped simultaneously for one of the photon beams. This operation keeps $P_L P_e = -1$ but changes the sign of the helicity of the generated photons, making it possible to exchange between $J = 0$ and $J = 2$ mode without, in ideal condition, changing luminosity distributions. Thus, with this helicity-flipping method, the $J = 0$ luminosity can be directly given by the measured $J = 2$ luminosity.

A feasibility study for a luminosity measurement was performed for the process $\gamma\gamma \to e^+e^-$ and for $\gamma\gamma \to W^+W^-$ [26, 1]. The reconstructed luminosity distribution and the statistical errors on the luminosity measurements are shown in Fig. 22.

As shown in this figure, for both electron and W pairs, the statistical errors are expected to be $1 \sim 2\%$ for each energy bin.

Table 3. Possible topics studied by the linear collider.

Scenario	e^+e^-	$\gamma\gamma/e\gamma$	e^-e^-
Light Higgs			
	Discovry	$\Gamma(h \to \gamma\gamma)$	
	Branching Ratio		
	Coupling	CP	
		by linear polarization	
	$\Gamma(h \to all)$		
Heavy Higgs	up to $\approx$ 300 GeV	Could cover wedges	
		CP interference	
SUSY			
	Discovery	Selectron with $e\gamma$	Selectron
	SUSY spectroscopy		
No new particles			
	Anomalous coupling	Anomalous coupling	
	WW scattering		WW scattering
	Contact int.		Contact int.
	Weinberg angle (Z pole?)		Weinberg angle
	Top threshold physics	pol. top factory	
	TOP CP		
Extra-dimension	$M_s \leq 6\sqrt{s}$	$M_s \leq 11\sqrt{s}$	$M_s \leq 6\sqrt{s}$
		by $\gamma\gamma \to WW$	

7. Summary

By expanding its initial particles, from the e^+e^- to e^-e^-, $e\gamma$ and $\gamma\gamma$, we will be able to expand physics opportunities of the linear collider experiments to where it is hardly accessible by or complementary to the e^+e^- interactions. An overview of the physics feasibility shows that there are many topics which are interesting as well as important as shown in Table 3.

At the same time, it has to be noted that the feasibility of physics with these options is not investigated as well as has been demonstrated in the e^+e^- interaction. In addition, particularly for the $\gamma\gamma$ and the $e\gamma$ operation of the linear collider, technical issues such as development of the high peak and average power laser or the optics layout around the interaction region are yet to be solved.

To clarify these issues and to see how it is possible to accommodate options is the work that has to be done early in new millennium.

The author of this chapter would like to thank members of the photon collider sub-group in the ACFA LC working group. Contents of this chapter are based on collaboration between them. He also would like to thank his colleagues of photon collider groups in North America and European LC working group for fruitful discussions and excellent works in this field which he quoted in this article.

References

1. I. Watanabe *et al.*, KEK Report 97-17.
2. *Zeroth Order Design Report for the Next Linear Collier* SLAC-474 (1996); *Physics and Technology of the Next Linear Collider*, SLAC Report 485 (1996).
3. TESLA Technical Design Report, Part 6, Chapter 1, hep-ex/0108012.
4. See for example, C. A. Heusch ed. *Proceeding of the third electron-electron linear collider workshop*, Int. J. of Mod. Phys. **A15**, (16) (2000).
5. K. Yokoya, private communication.
6. K. Yokoya, http://www-acc-theory.kek.jp/members/cain/default.html.
7. K. A. Thompson, Int. J. of Mod. Phys. **A15**, (16) 2485 (2000).
8. KEK-Report 2000-7 (2000).
9. See for exmaple, Proc. International Worksop on High Energy Photon Colliders. Nucl. Instru. and Meth. **A472**, (2001).
10. D.L. Borden, D.A. Bauer, and D.O. Caldwell, Phys. Rev. **D48**, 4018 (1993).
11. T. Abe, *et al.*, Nucl. Instr. Meth. **A447**, 90 (2000).
12. T. Ohgaki, T. Takahashi and I. Watanabe, Phys. Rev. **D56**, 1723 (1997).
13. G. Jikia, S. Soldner-Rembold Nucl. Phys. Proc. Suppl. **82**, 373 (2000).
14. D. Asner, J.Gronberg, and J.Gunion, hep-ph/0110320 (2001).
15. T. Tregzger, SNOWMASS-2001-P111
16. S. Yamashita, talk given in LCWS99, Sitges, Spain, May, 1999, To be published in Nucl. Phys. Proc. Suppl.
17. E. Asakawa *et al.*, Eur. Phys. J. **C14**, 335 (2000).
18. A. Goto and T. Kon, Europhys. Lett. **19**, 575 (1992): T. Kon and A. Goto, Phys. Lett. **B295**, 324 (1992).
19. T. Tsukamoto, *Proc. of the Workshop on Physics and Experiments at Linear e^+e^- Colliders*, eds. F. A. Harris, S. L. Olsen, S. Pakvasa and X. Tata, (World Scientific, Singapore, 1993), Vol. II, p. 814
20. J. Feng, Int. J. of Mod. Phys. **A15**, num. 16, 2355 (2000).
21. J. Feng and M. Peskin, Phys. Rev. **D64**, 115002 (2001).
22. M. Baillargeon, G. Belanger, F. Boudjema, ENSLAPP-A-473/94
23. S. Y. Choi and F. Schremmpp, Phys. Lett. **B272**, 149 (1991).
24. E. Yehudai, Phys. Rev. **D44**, 3434 (1991).
25. A. Miyamoto, KEK-preprint 93-116.
26. T. Takahashi, *Proc. of Workshop on Physics and Experiments with Linear colliders* World Scientific (1996) 681.
27. A. Djouadi, J. Ng, T. Rizzo *et al.*, hep-ph/9504210.
28. S. Y. Choi and K. Hagiwara, Phys. Lett. **B359**, 360 (1995).
29. S. Poulise and S. D. Rindani, Phys. Rev. **D57**, 5444 (1998); **61**, 119902 (2000)(E).
30. see for example, E. Boos, L. Duduko, T.Ohl, Eur. Phys. J. **C11**, 473 (1999).
31. J. Hewett, Int. J. Mod. Phys. **13**, 2389 (1998).
32. D. Atwood, S. Bar-Shalom, G. Eilam, A. Soni, Phys. Rep. **347**, 1 (2001).
33. T. Takahashi and K. Ikematsu, Int. J. Mod. Phys. **A15**, No.6, 2599 (2000).
34. K. Abe *et al.*, Phys. Rev. Lett. **84**, 5945 (2000).

35. A. Czarnecki and W. J. Marciano, Int. J. of Mod. Phys. **A15**, num. 16, 2365 (2000).
36. M. Wood, Proceeding of SPIN96 (ed. C. W. Jager *et al.*), 621 (1997).
37. N. Arkano-Hamed. S. Dmopoulos, G. Dvali, Phys. Letts. **B429**, 263 (1998).
38. T. Rizzo SLAC-PUB-8238, 1999.
39. R. M. Godbole, G. Pancheri, and A. De Roech LC-TH-2001-030.
40. M. Gluk, E.Reya and A. Vogt, Z. Phys. **C67**, 433 (1994).
41. G. Schuler and T. Sjostrand, Phys. Lett. **B376**, 193 (1996).
42. JLC-I, KEK-Report 92-16
43. M. N. Frary and D. J. Miller, DESY Report 92-123A (1992) 379.
44. N. Toomi et.al, Phys. Lett. **B429**, 162 (1998).
45. V. Telnov, Nucl. Instru. Meth. **A355**, 3 (1995).

CHAPTER 10

CP Violation at the Linear Collider

David Atwood[a] and Amarjit Soni[b]

[a]*Department of Physics and Astronomy, Iowa State University Ames, IA 50011, U.S.A.*

[b]*High Energy Theory Group, Brookhaven National Laboratory Upton, NY 11973, U.S.A.*

We briefly review here the key physics opportunities that the linear collider offers for exploring CP-violation phenomena. In particular, production and decays of the top quark provide an important avenue for this study. The CP-phase in the CKM-paradigm causes completely negligible effects here and any observed CP asymmetry must originate from beyond the SM effects. The fact that the top decays provide information on its spin orientation is a powerful tool that can be used in this context especially to study dipole moment (or form factor) effects. Reactions such as $e^+e^- \to t\bar{t}$, $e^+e^- \to Z + h$, $t\bar{t}+$ h, Z, g and $\nu_e\overline{\nu}_e$ are also discussed.

1. Introduction

An important class of physics that can be studied at the e^+e^- linear collider is that related to the CP violation phenomena; in particular involving the top quark. Just as the bottom quark is extremely suitable for studying CP-violation due to the Standard Model's CKM-paradigm [1], the top quark is highly suitable for studying non-standard sources of CP-violation. There are several reason for this. For one thing, the top quark is exceedingly insensitive to the CP-odd phase in the CKM matrix; that phase can only cause completely negligible asymmetries. Recall that this is in sharp contrast to the bottom quark wherein the SM causes large CP-asymmetries. Thus any observation of CP violation in top production or decay is extremely likely to be due to non-standard source(s) of CP violation. Furthermore, in many extensions of the SM, sizeable CP violation effects can occur involving the top quark. Due to the very large mass of the top quark it has an extremely short lifetime and it is unable to bind itself into a hadron before it undergoes weak decay. As a result its decay products contain information regarding its CP interaction. In this context the spin of the top quark becomes a very important handle. The cleanliness of an e^+e^- machine becomes valuable in studying the detail dynamics of production and decays of the top quark to unravel the CP violation effects. In addition to $e^+e^- \to t\bar{t}$, $e^+e^- \to hZ$, $t\bar{t}$ $h(Z, g)$ are also promising for searching CP violating effects due to beyond the SM physics.

2. Single Top Polarimetry

At the LC, processes which involve the top quark are likely to be important probes for CP violation. One feature of top decays that is key to such experimental tests of CP violation is the fact that they are self-analyzing their polarization.

In particular, consider the decay $t \to bW^+ \to be^+\nu$. Because of the chiral nature of the W coupling, in the rest frame of the top, it will be fully polarized in the direction of the positron momentum. This follows from the fact that the amplitude for the top-quark decay is proportional to [10]:

$$[\overline{u}_b \gamma_\mu (1-\gamma_5) u_t]\,[\overline{u}_\nu \gamma^\mu (1-\gamma_5) v_e] \ . \tag{1}$$

On Fierz transformation this becomes

$$\frac{1}{2}\,[\overline{u}_b (1+\gamma_5) u_\nu^c]\,[\overline{v}_e^c (1-\gamma_5) u_t] \ , \tag{2}$$

where $u^c = C\overline{u}^T$, C being the charge conjugation matrix. Therefore in the rest frame of the top quark, the top spin is in the direction of the positron three-momentum. Of course a similar statement follows in the case of $t \to bW^+ \to b\bar{d}u$ where the top is fully polarized in the direction of the $\bar{d}$ but as a practical matter one cannot tell which jet is the $\bar{d}$.

Even though the polarization cannot be recovered in fully hadronic decays, it is important to obtain at least partial information in such decays.

To do this we would like to use the momenta of the decay particles to construct a "handle", a vector in the top rest frame which approximates the top decay. If $\vec{P}_h$ is the direction of the handle, then the polarization density matrix of the top is:

$$\rho = \frac{1}{2}(1 + \epsilon\vec{\sigma}\cdot\vec{P}_h) \tag{3}$$

where ϵ is the analyzing power of the particular handle used. For instance, in the case of semileptonic top decay we use the positron momentum as the handle and in this case $\epsilon = 1$.

More generally, if we use the momentum of some particle A in the final state which has an analyzing power ϵ^A then if the top is initially spin up along the z-axis, the angular distribution of A is given by:

$$\frac{d\Gamma}{d\cos\theta_A} = a\left(1 + \epsilon^A \cos\theta_A\right) . \tag{4}$$

Thus we can define ϵ^A in terms of the expectation value of $\cos\theta_A$

$$\epsilon^A = 3\langle\cos\theta_A\rangle , \tag{5}$$

where θ_A is the angle between $\vec{p}_A$ and the spin of the top (*i.e.* z-axis).

If we wish to measure polarization of the top along the z-axis, the optimal way to determine the polarization of the top is to experimentally measure $\langle\cos\theta_A\rangle$. In this way, the number of top decays that needs to be observed to determine the top polarization to within 1-σ is given by

$$N_t^{(1\sigma)} = \frac{3}{\left[\epsilon_t^A \Pi_t\right]^2} , \tag{6}$$

where Π_t is the polarization of the top quark,

$$\Pi_t \equiv \frac{N(\uparrow) - N(\downarrow)}{N(\uparrow) + N(\downarrow)} , \tag{7}$$

$N(\uparrow)$ being the number of top quarks with spins along the positive z axis.

Let us consider three concrete handles on the top spin:

(1) Semi-leptonic decays ($t \to b\ell^+\nu_\ell$) have an analyzing power of $\epsilon^\ell = 1$ if you take the lepton momentum as the handle. Likewise in the unrealistic case of hadronic decays where you can identify the $\overline{d}$ jet $\epsilon^d = 1$.

(2) For hadronic decays of the W, if the momentum of the W can be determined, then it is easy to see that

$$\epsilon^W = \frac{m_t^2 - 2m_W^2}{m_t^2 + 2m_W^2} \simeq 0.4 . \tag{8}$$

This reflects the fact that the W-momentum is an imprecise indicator of the $\overline{d}$ type jet in the decay $W^+ \to \overline{d}u$. Here the $\overline{d}$ jet plays the same role as the e^+ in the previous case.

(3) From the above two points, it should be clear that we can increase ϵ from 0.4 to 1 if we can identify the $\overline{d}$-type jet, which plays the role of the e^+. Since this jet is one of the two that reconstructs to the W in method 2 we are approximating it with the W-momentum. In the limit that m_t is much greater than m_W, the two quark jets would coincide

with the W-momentum and $\epsilon_t^W \to 1$; however, with the given top mass, the result is far from this limit.

We have not, however, exploited the full information available in the kinematics of the decay. It was suggested in [51] that the energy distribution of the two jets could give additional information concerning which is the desired $\bar{d}$-jet. In particular, in $W^+ \to \bar{d}u$, the $\bar{d}$-jet is on the average less[a] energetic than the u-jet: and so, if we take the polarization to be in the direction of the less energetic of these two jets, one would expect some improvement. Indeed, in this case we obtain

$$\epsilon_t^{\text{low energy jet}} \simeq 0.5 \,. \tag{9}$$

This is close to the theoretical limit, $\epsilon_t^{\text{opt}} \simeq 0.63$, which can be obtained using the method [12] of optimized observables discussed above for determining the polarization where one restricts the experimentally available information to quantities which are symmetric under the interchange of the two jets.

3. Top Pair Polarimetry

As discussed in the previous section, the decay of the top quark is "self-analyzing" since from the decay products one can determine a spin vector S_t for a given top quark from its decay products. In this section, we will discuss the correlation of top and anti-top states in a $t\bar{t}$ system, for instance produced via $e^+e^- \to t\bar{t}$, $\gamma\gamma \to t\bar{t}$ or $e^+e^- \to t\bar{t}H$. This will give us a simple way of identifying observables which indicate the presence of correlations which can only come from new physics, in particular due to CP violation. We will apply this to the case $e^+e^- \to t\bar{t}$ where the CP violation appears through electric dipole moment of the top-quark.

In the typical decay chain $t \to bW$ with subsequent two body decay of the W, if one can determine which of the fermions is the lower member of a doublet (*i.e.* the charge lepton or the down-type quark) then the decay is self polarizing and one determines the spin vector S_t where $S_t^2 = -1$. In a more general situation, one obtains a spin vector $0 \leq -S_t^2 \leq 1$. In this section we will first frame the discussion assuming that the top decay is self polarizing and then consider how to extend the methods to the more general case.

---

[a] In passing we wish to mention that while we find the $\bar{d}$-jet to be on average less energetic then the u-jet, Ref. [51] found it to be the other way around.

Let us now work in the rest frame of the $t\bar{t}$ pair where we take the $+z$ axis to be in the direction of the top momentum. It is convenient to decompose a general $t\bar{t}$ state in terms of helicity states which are CP eigenstates. There are four helicity states which we will denote as follows:

$$\begin{aligned} V_R &= t_R\bar{t}_L \\ V_L &= t_L\bar{t}_R \\ V_0 &= \frac{1}{\sqrt{2}}\left(t_R\bar{t}_R + t_L\bar{t}_L\right) \\ P &= \frac{1}{\sqrt{2}}\left(t_R\bar{t}_R - t_L\bar{t}_L\right) \end{aligned} \tag{10}$$

where t_R is spin up along the $+z$ axis, t_L is spin down along the $+z$ axis and $\bar{t}_{L,R} = CP\ (t_{R,L})$.

Of course the spins of the top quarks are not seen directly but rather the spin vectors can be determined for each decay. Let us denote by θ_1 and ϕ_1 the polar angles of the spin vector of the top in its boosted frame where θ_1 is the angle between the boosted $+z$ axis and S_t with ϕ being the azimuthal angle. Likewise let θ_2 and ϕ_2 be the polar angles of the anti-top spin in its boosted frame. Each of the polarization states in the above will have a decay amplitude which is a function of $\{\theta_1,\ \phi_1,\ \theta_2,\ \phi_2\}$. These amplitudes will in turn be multiplied by coefficients which depend on couplings and the other kinematic variables of the event. Thus, we can write the overall amplitude A for the interaction as:

$$A = \sum_i c_i(\omega)a_i(\theta_1,\phi_1,\theta_2,\phi_2) \tag{11}$$

where i ranges over the four helicity states, a_i is the decay amplitude for each helicity state and c_i is the amplitude for producing each of the helicity states which depends on ω, the other kinematic variables of the problem.

The decay distribution depends on $|A|^2$ and it is useful to write this in the following way:

$$|A|^2 = \sum_i |c_i|^2|a_i|^2 + 2\sum_{i<j} Re(c_ic_j^*\ a_ia_j^*) \tag{12}$$

If a manifestly CP violating effect is to be observed, it will be due to one of the interference terms in the second summation above. This can come about as follows:

(1) c_i and c_j could have the same CP properties while a_i and a_j have opposite CP.

(2) c_i and c_j have opposite CP properties while a_i and a_j have the same CP.
(3) One or more of the c_i functions does not have a well defined CP symmetry.

In case (1), the interference would be between the V states which are CP $= +1$ and the P state which is CP $= -1$. If this is the only manifestation of CP violation, top polarimetry is essential to observe a CP violating signal. This case would apply, for instance if the $t\bar{t}$ pair were produced by a gauge boson which had an electric dipole-like coupling or by a scalar with mixed scalar and pseudoscalar couplings. In case (2) and (3) it is often the case that the CP violation can be discovered without the need to consider top polarization specifically. For instance, in the reaction $e^+e^- \to \gamma^* \to t\bar{t}$ if the electron had an electric dipole moment but the top did not, then case (2) would apply but such an effect could only be discovered by considering different combinations of electron polarization.

We would now like to develop formulas for the amplitudes a_i and their interferences. To do this we need to specify a convention for the normalization and phase of these amplitudes. We will use the Dirac spinor basis of [28] normalized such that $\overline{u}u = 2m_t$. In these terms, the amplitudes are given by:

$$
\begin{aligned}
a_{V_R} &= \overline{u}(S_1)\gamma_R v(S_2) \\
a_{V_L} &= \overline{u}(S_1)\gamma_L v(S_2) \\
a_{V_0} &= \overline{u}(S_1)\gamma_z v(S_2) \\
a_P &= \overline{u}(S_1)\gamma_5 v(S_2)
\end{aligned}
\tag{13}
$$

where $\gamma_{L,R} = (\gamma_x \pm i\gamma_y)/\sqrt{2}$.

If we define $\chi_{1(2)}$ to be the 2 component spinor along the direction of $S_{1(2)}$ in the $t(\bar{t})$ frame, the above reduce to

$$
\begin{aligned}
a_{V_R} &= -\sqrt{s}\chi_1^\dagger \sigma_R \tilde{\chi}_2 \\
a_{V_L} &= -\sqrt{s}\chi_1^\dagger \sigma_L \tilde{\chi}_2 \\
a_{V_0} &= -2m_t \chi_1^\dagger \sigma_z \tilde{\chi}_2 \\
a_P &= \sqrt{s}\chi_1^\dagger \tilde{\chi}_2
\end{aligned}
\tag{14}
$$

where s is the invariant mass of the $t\bar{t}$ pair and $\tilde{\chi}_2 = \epsilon\chi_2^*$ where:

$$
\epsilon = \begin{pmatrix} 0 & 1 \\ -1 & 0 \end{pmatrix}
$$

The magnitude of these amplitudes are:

$$
\begin{aligned}
|a_{V_R}| &= \sqrt{2}\sqrt{s}\cos\frac{\theta_1}{2}\cos\frac{\theta_2}{2} \\
|a_{V_L}| &= \sqrt{2}\sqrt{s}\sin\frac{\theta_1}{2}\sin\frac{\theta_2}{2} \\
|a_{V_0}| &= 2m_t\sin\frac{\hat{\Theta}_{12}}{2} \\
|a_P| &= \sqrt{s}\sin\frac{\Theta_{12}}{2}
\end{aligned}
\tag{15}
$$

where Θ_{12} is the angle between S_1 and S_2 and $\hat{\Theta}_{12}$ is the angle between S_1 rotated by π about the z-axis and S_2.

The phases of the amplitudes are convention dependent. If we define the following convention for spinors:

$$
\chi_{1,2} = \begin{pmatrix} e^{-i\phi/2}\cos\frac{\theta}{2} \\ e^{+i\phi/2}\sin\frac{\theta}{2} \end{pmatrix}
\tag{16}
$$

Then, in this convention the amplitudes including the phases are:

$$
\begin{aligned}
a_{V_R} &= -\sqrt{2}\sqrt{s}\cos\frac{\theta_1}{2}\cos\frac{\theta_2}{2}e^{+i\frac{\phi_1+\phi_2}{2}} \\
a_{V_L} &= \sqrt{2}\sqrt{s}\sin\frac{\theta_1}{2}\sin\frac{\theta_2}{2}e^{-i\frac{\phi_1+\phi_2}{2}} \\
a_{V_0} &= -2m_t\left(\sin\frac{\theta_2+\theta_1}{2}\cos\frac{\phi_1-\phi_2}{2} + i\sin\frac{\theta_2-\theta_1}{2}\sin\frac{\phi_1-\phi_2}{2}\right) \\
a_P &= \sqrt{s}\left(\sin\frac{\theta_2-\theta_1}{2}\cos\frac{\phi_1-\phi_2}{2} + i\sin\frac{\theta_2+\theta_1}{2}\sin\frac{\phi_1-\phi_2}{2}\right)
\end{aligned}
\tag{17}
$$

In the cross terms of the amplitudes, the phase convention for the spinors cancels and the expressions can be simplified considerably:

$$
\begin{aligned}
a_{V_0}a_P^* &= -m_t\sqrt{s}\ \ e_z\cdot[i(S_1+S_2) - S_1\times S_2] \\
a_{V_R}a_P^* &= (s/2)(e_z+S_1)\times(e_z+S_2)\cdot e_R \\
a_{V_L}a_P^* &= (s/2)(e_z-S_1)\times(e_z-S_2)\cdot e_L \\
a_{V_R}a_{V_0}^* &= (s/2)(e_z+\hat{S}_1)\times(e_z+\hat{S}_2)\cdot e_R \\
a_{V_L}a_{V_0}^* &= -i(s/2)(e_z-\hat{S}_1)\times(e_z-\hat{S}_2)\cdot e_L \\
a_{V_R}a_{V_L}^* &= -s(\hat{S}_1\cdot e_R)(\hat{S}_2\cdot e_R)
\end{aligned}
\tag{18}
$$

where $e_{x,y,z}$ are the unit vectors in the x, y, and z directions and $e_{R,L} = (e_x \pm ie_y)/\sqrt{2}$.

Let us now consider the application to the process $e^+e^- \to t\bar{t}$ where there might be a CP violating electric dipole moment coupling present. In this case the coupling of a photon to $t\bar{t}$ is:

$$\bar{t}\left(\frac{2}{3}e\gamma^\mu + ed_t^\gamma q^\nu \sigma^{\mu\nu}\right)t \tag{19}$$

For simplicity let us assume first that only the photon exchange is present and that the electrons are polarized with right handed chirality. If we take the event plane to be the $x-z$ plane and let θ_e be the angle of the electron momentum, one can easily calculate the c_i coefficients:

$$\begin{aligned} c_{V_R} &= q_t\sqrt{2}\frac{e^2}{s}(\cos\theta_e + 1) \\ c_{V_L} &= q_t\sqrt{2}\frac{e^2}{s}(\cos\theta_e - 1) \\ c_{V_0} &= -2q_t\frac{e^2}{s}\sin\theta_e \\ c_P &= id_t^\gamma\frac{2e^2}{s}\sqrt{s-4m_t^2}\sin\theta_e \end{aligned} \tag{20}$$

where $q_t = 2/3$.

This case is clearly (1) above and so CP violation will result from the interference of a_P and the a_V states; top polarization is required to observe such a CP violation.

Let us now consider the experimental observables which we can use to extract the CP violating asymmetry which arises from dipole moment effects. Qualitatively, eqn. (18) gives us some insight into the form which such observables take.

Consider first the case of a real dipole moment. The term $a_{V_0}a_P^*$ is proportional to the azimuthal difference of the S vectors about the z-axis. This term is proportional to m_t and thus suppressed at high energy. The term $a_{V_R}a_P^*$ is a triple product which has the same sign as the azimuthal angle difference about the x-axis measured from the negative z-axis while $a_{V_L}a_P^*$ is proportional to a similar angle difference measured from the positive z-axis. These two terms will be dominant at high energies.

Let us next consider an imaginary dipole moment. Here the term $a_{V_0}a_P^*$ corresponds to an excess of $t_R\bar{t}_R$ over $t_L\bar{t}_L$. At high energies the terms $a_{V_{RL}}a_P^*$ again dominate. The geometric picture is similar to the real case except here the azimuthal angle is about the y-axis. This asymmetry can thus be generated when the spins lie in the decay plane while for the real dipole moment the asymmetries rely on spin components perpendicular to the decay plane.

In reality, of course, the reaction proceeds through both Z and γ exchange and indeed the top quark may have electric dipole like couplings both to the gamma (EDM) and Z-boson (ZDM). If we define the total CP violating coupling as:

$$\bar{t}\left(ed_t^\gamma q^\nu \sigma^{\mu\nu} E_\nu + ed_t^Z q^\nu \sigma^{\mu\nu} Z_\nu\right) t \tag{21}$$

then the amplitudes become:

$$\begin{aligned} c_{V_R} &= A\sqrt{2}\frac{e^2}{s}(\cos\theta_e + 1) \\ c_{V_L} &= (A - B\beta)\sqrt{2}\frac{e^2}{s}(\cos\theta_e - 1) \\ (A + B\beta)c_{V_0} &= -2q_t\frac{e^2}{s}\sin\theta_e \\ c_P &= id_{tot}\frac{2e^2}{s}\sqrt{s - 4m_t^2}\sin\theta_e \end{aligned} \tag{22}$$

where $A = 1 - \frac{(3-8x_w)}{12(1-x_w)(1-x_Z)}$; $B = \frac{1}{4(1-x_w)(1-x_Z)}$ and $d_{tot} = d_t^\gamma - \sqrt{\frac{x_w}{(1-x_w)(1-x_Z)^2}}d_t^Z$ in which $x_w = \sin^2\theta_w$ and $x_Z = \frac{m_z^2}{s}$.

Let us consider now the effective asymmetries given by various operators. Since the dependence of the interference terms on angle is fixed by the above expressions, we can consider operators specifically tailored to be sensitive to the possible dipole moments we have considered. In particular if we write the moments as:

$$\begin{aligned} d_t^\gamma &= (D_R^\gamma + iD_I^\gamma)/m_t \\ d_t^Z &= (D_R^Z + iD_I^Z)/m_t \end{aligned} \tag{23}$$

for each of the four dimensionless quantities $\{D_R^\gamma, D_I^\gamma, D_R^Z, D_R^Z\}$, we denote the contribution to the angular distribution linear in these as $\mathcal{A}_{R,I}^{\gamma,Z} D_{R,I}^{\gamma,Z}$ while the distribution in the limit $d \to 0$ is $\mathcal{S}$. We can thus define in each case the operator $O_1 = sign(\mathcal{A})$ which is ± 1 at each point in phase space and $O_2 = \mathcal{A}/\mathcal{S}$ which is the optimal observable. The effective asymmetry $\alpha = |<O>|/\sqrt{<O^2>}$ is plotted in Fig. 1 as a function of $\sqrt{(s)}$. In this graph, the solid line is $\alpha(D_R^\gamma, O_1)/D_R^\gamma$, the dashed line the optimal observable $\alpha(D_R^\gamma, O_2)/D_R^\gamma$ which is somewhat larger as expected. Likewise for the imaginary parts $\alpha(D_I^\gamma, O_1)/D_I^\gamma$ is the dotted line and $\alpha(D_I^\gamma, O_2)/D_I^\gamma$ is the dot-dashed line. In the case of the ZDM we have plotted the optimal observables $\alpha(D_R^Z, O_2)/D_R^\gamma$ with the long dashed line and $\alpha(D_I^Z, O_2)/D_R^\gamma$

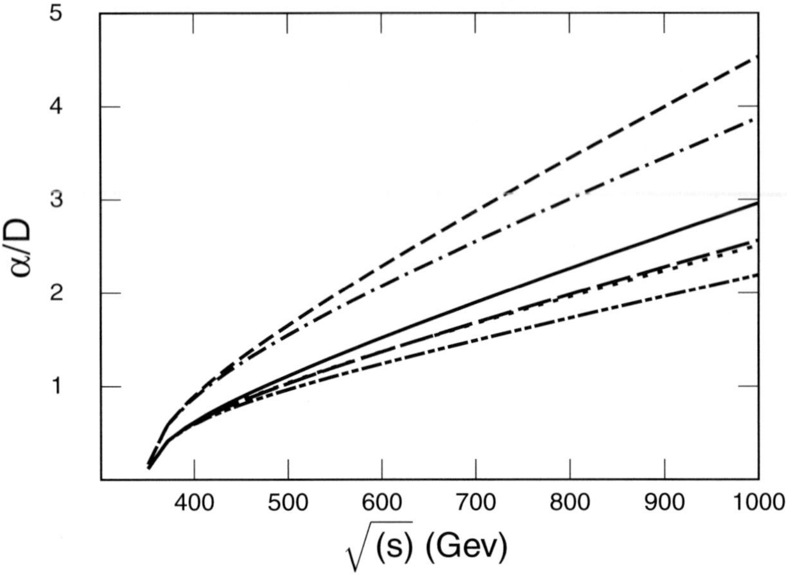

Fig. 1. This figure shows various asymmetries in $e^+e^- \to t\bar{t}$ resulting from real and imaginary EDM and ZDM for the top quark as a function of $\sqrt{s}$. For a Real EDM, the solid line shows $\alpha(D_R^\gamma, O_1)/D_R^\gamma$ and the dashed line shows $\alpha(D_R^\gamma, O_2)/D_R^\gamma$. For an imaginary EDM, the dotted line shows $\alpha(D_I^\gamma, O_1)/D_I^\gamma$ and the dash-dot line shows $\alpha(D_I^\gamma, O_2)/D_I^\gamma$. For a real ZDM, the long dashed line shows $\alpha(D_R^Z, O_2)/D_R^Z$, and for an imaginary ZDM, dash-dot-dot line shows $\alpha(D_I^Z, O_2)/D_I^Z$. We assume the electrons have right helicity and the positrons have left helicity.

with the dot-dot-dash line. In the case of an accelerator working at a specific $\sqrt{s}$, it is clearly best to consider observables tailored to the combination d_{tot}.

3.1. *Constraints on Top Dipole Moments*

Using this method, one can thus constrain the real and imaginary parts of the EDM and ZDM of the top at an e^+e^- collider operating above threshold. Assuming that the value of the anomalous moment remains constant, generally the highest energies possible will give the best bounds. Of course this will only be true until the energy of the collider becomes comparable to whatever degrees of freedom are generating the effective coupling.

It is well known that such moments cannot be generated at measurable levels in the Standard Model; thus the goal of probing such moments is to place limits on extensions to the standard model many of which can contribute at one loop level. The order of magnitude of such effects in SUSY [49, 23, 48, 18, 19] or extra Higgs [71, 26, 42, 73, 4, 70, 11] extensions to the Standard Model is typically in the range $10^{-18} - 10^{-20}$ e-cm.

Various studies have considered operators which are sensitive to the top pair correlations resulting from such moments [12, 52, 53, 30, 67, 6, 25].

In Table 1 from [25] we show the 3-σ bounds which can be placed on the four kinds of moments using optimized observables at a collider with $\sqrt{s} =$ 500 GeV or 800 GeV and an integrated luminosity of 200 fb^{-1} with various beam polarizations. As can be seen, the results in this case are $O(10^{-18})$ hence it can be expected that such studies at a LC will begin to put useful bounds on these CP violating moments.

Table 1. Attainable 3-σ sensitivities with integrated luminosity of 200 fb^{-1} to the CP-violating dipole moment form factors in units of 10^{-18} e-cm, with ($P_e = \pm 1$) and without ($P_e = 0$) beam polarization.

	20 fb^{-1}, $\sqrt{s} = 500$ GeV			50 fb^{-1}, $\sqrt{s} = 800$ GeV		
	$P_e = 0$	$P_e = +1$	$P_e = -1$	$P_e = 0$	$P_e = +1$	$P_e = -1$
$\delta(\Re e d_t^\gamma)$	4.6	0.86	0.55	1.7	0.35	0.23
$\delta(\Re e d_t^Z)$	1.6	1.6	1.0	0.91	0.85	0.55
$\delta(\Im m d_t^\gamma)$	1.3	1.0	0.65	0.57	0.49	0.32
$\delta(\Im m d_t^Z)$	7.3	2.0	1.3	4.0	0.89	0.58

4. Three Particle Final States

4.1. *Tree Level $e^+e^- \to t\bar{t}H$ and $e^+e^- \to t\bar{t}Z$*

Let us now turn our attention to the reactions

$$e^+(p_+) + e^-(p_-) \to t(p_t) + \bar{t}(p_{\bar{t}}) + h(p_h) \;, \tag{24}$$

$$e^+(p_+) + e^-(p_-) \to t(p_t) + \bar{t}(p_{\bar{t}}) + Z(p_Z) \;, \tag{25}$$

which may exhibit large CP violation asymmetries in a 2HDM [9, 15, 14, 54]. Note that for reasons discussed below, for the CP-violating effects in $e^+e^- \to t\bar{t}h$ as well as in $e^+e^- \to t\bar{t}Z$, only the two scalars out of the three neutral Higgs of the 2HDM are relevant since the pseudo-scalar has no tree level coupling to two gauge bosons. We denote these two neutral Higgs particles by h and H corresponding to the lighter and heavier states respectively. We will denote a generic neutral Higgs by $\mathcal{H}$, then $\mathcal{H} = h$ or H in the 2HDM. In more general multi-higgs doublet models the situation would be qualitatively the same except that there would be additional scalar states which would participate. Here we will discuss this signature in the context of the 2HD model.

A novel feature of these reactions is that the effect arises at the tree-graph level. As a consequence, one can construct new type of asymmetries which are $\propto$ (tree$\times$tree)/(tree$\times$tree) and are therefore *a-priori* of $\mathcal{O}(1)$. This stands in contrast to loop induced CP-violating effects such as the top EDM which are likely to be suppressed by a small coupling being $\propto$ (tree$\times$loop)/(tree$\times$tree). As we shall see asymmetries of the order of 10% may be possible. Note that since there is no rescattering phase any signal must be T_N-odd, *i.e.* involving triple products. The simplest case is where we take observables proportional to the center of mass frame triple product of the electron momentum with the momenta of the two top quarks.

In both cases, $t\bar{t}h$ and $t\bar{t}Z$, Higgs(Z) emission off the t and $\bar{t}$ interferes with the Higgs(Z) emission off the s-channel Z-boson (see Fig. 2) [9, 15, 14]. Thus, the processes $e^+e^- \to t\bar{t}h$ and $e^+e^- \to t\bar{t}Z$ provide two independent, but analogous, promising probes to search for the signatures of the same CP-odd phase, residing in the $t\bar{t}\mathcal{H}$ coupling, if the value of $\tan\beta$ (the ratio between the two VEV's in a 2HDM) is in the vicinity of 1.

Although these reactions are not meant (necessarily) to lead to the discovery of a neutral Higgs, they will, no doubt, be intensely studied at the LC since they stand out as very important channels for a variety of reasons. In particular, they could perhaps provide a unique opportunity to observe the top-Higgs Yukawa couplings directly [56, 39, 40, 74, 55, 57, 43]. In [56] and [55], using a generalization of the optimal observables technique, Gunion *et al.* have extended the initial work [9] on CP violation in $e^+e^- \to t\bar{t}h$ to include a detailed cross-section analysis such that all Higgs Yukawa couplings combinations can be extracted. A similar analysis which also uses the optimized observable technique for $e^+e^- \to t\bar{t}Z$ is given in Ref/ [54]. Likewise, a detailed cross-section analysis of the reaction $e^+e^- \to t\bar{t}Z$ in the SM was performed by Hagiwara *et al.* [57, 43]. There, it was found that the Higgs exchange contribution of diagram (b) on the right hand side of Fig. 2 will be almost invisible in a TeV e^+e^- collider for neutral Higgs masses in the range $m_h < 2m_t$.

Thus, a "simple" cross-section study for these reactions may also come in handy for searching for new physics. However, one should keep in mind that, from the experimental point of view, asymmetries, *i.e.*, ratios of cross-section, are easier to handle and, in particular, CP-violating signals are likely to be definitive signals for new physics.

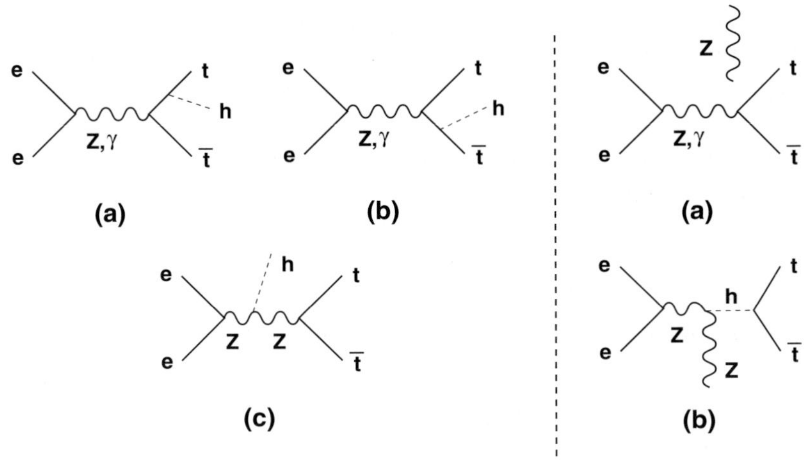

Fig. 2. Tree-level Feynman diagrams contributing to $e^+e^- \to t\bar{t}h$ (left hand side) and $e^+e^- \to t\bar{t}Z$ (right hand side) in the unitary gauge, in a 2HDM. For $e^+e^- \to t\bar{t}Z$, diagram (a) on the right hand side represents 8 diagrams in which either Z or γ are exchanged in the s-channel and the outgoing Z is emitted from e^+, e^-, t or $\bar{t}$.

The unitary gauge Feynman diagrams for the reactions of Eqs. (24) and (25) are shown in Fig. 2. The only place where CP violation can arise in the standard model is at the $t\bar{t}h$ coupling. Hence CP violating observables must arise through the interference of graphs where the higgs couples with the top quark with the other graphs.

The relevant pieces of the interaction Lagrangian involving the $t\bar{t}\mathcal{H}^k$ Yukawa and the $ZZ\mathcal{H}^k$ couplings can be written in the following general form:

$$\mathcal{L}_{\mathcal{H}^k tt} = -\frac{g_W}{\sqrt{2}}\frac{m_t}{m_W}\mathcal{H}^k \bar{t}\left(a_f^k + i b_f^k \gamma_5\right) t \ , \tag{26}$$

$$\mathcal{L}_{\mathcal{H}^k ZZ} = g_W m_W C_Z c^k \mathcal{H}^k Z^\mu Z_\nu \ , \tag{27}$$

where C_Z, a_t^k and b_t^k are model dependent parameters. For the standard model higgs, $C_Z = m_Z^2/m_W^2$, $a_t = 1/\sqrt{2}$ and $b_t = 0$. CP violation is present only if $b_t \neq 0$ and C_Z and a_t do not both vanish.

In 2HDM, only two out of the three neutral Higgs can simultaneously have a coupling to vector-bosons and a pseudoscalar coupling to fermions which we will denote h for the lighter and H the heavier higgs boson. In such models, CP violation is proportional to the mass splitting so these two particles can not be degenerate. Here we will assume that the heavier has a relatively large mass, $750 - 1000\ GeV$ for the $t\bar{t}Z$ final state. Of course if one of the higgs is openly produced in $t\bar{t}h$, the cancellation between the two higgs states is not relevant and then a large splitting is not required.

For both the $t\bar{t}h$ and $t\bar{t}Z$ final states processes, we denote the tree-level polarized differential cross section by $\Sigma_{(j)f}$, where $f = t\bar{t}h$ or $f = t\bar{t}Z$ corresponding to the $t\bar{t}h$ or $t\bar{t}Z$ final states, respectively, and $j = 1(-1)$ for the left(right) polarized incoming electron beam. $\Sigma_{(j)f}$ can be subdivided into its CP-even ($\Sigma_{+(j)f}$) and CP-odd ($\Sigma_{-(j)f}$) parts which are T_N-even and T_N-odd respectively. One can also consider T_N-even CP violation where the rescattering phase appears due to the imaginary part of the higgs propagator but these effects tend to be smaller; we will not discuss them further here.

$$\Sigma_{(j)f} = \Sigma_{+(j)f} + \Sigma_{-(j)f} \ . \tag{28}$$

These can be written in the form

$$\begin{aligned} \Sigma_{+(j)f} &= \sum_i g^i_{+f} F^i_{+(j)f} \ , \quad \mathrm{CP-even} \ , \\ \Sigma_{-(j)f} &= \sum_i g^i_{-f} F^i_{-(j)f} \ , \quad \mathrm{CP-odd} \ , \end{aligned} \tag{29}$$

where g^i_{+f}, g^i_{-f}, are different combinations of the Higgs couplings $a^{\mathcal{H}}_t, b^{\mathcal{H}}_t, c^{\mathcal{H}}$ and $F^i_{+(j)f}, F^i_{-(j)f}$ are kinematic functions of phase space.

For the CP-even part, the relevant terms in the expansion for the case of $e^+e^- \to t\bar{t}h$ are [56, 9, 14]:

$$g^1_{+t\bar{t}h} = (a^h_t)^2 \ , \ g^2_{+t\bar{t}h} = (b^h_t)^2 \ , \ g^3_{+t\bar{t}h} = (c^h)^2 \ , \ g^4_{+t\bar{t}h} = a^h_t c^h \ . \tag{30}$$

In the case of $e^+e^- \to t\bar{t}Z$, in addition to the SM diagrams not involving Higgs propagators, there are the following terms [15, 14]:

$$\begin{aligned} g^1_{+t\bar{t}Z} &= (a^{\mathcal{H}}_t c^{\mathcal{H}}) \Re\mathrm{e}(\Pi_{\mathcal{H}}) \ , \\ g^3_{+t\bar{t}Z} &= (a^{\mathcal{H}}_t c^{\mathcal{H}})^2 \Re\mathrm{e}(\Pi_{\mathcal{H}}) \ , \\ g^5_{+t\bar{t}Z} &= (b^{\mathcal{H}}_t c^{\mathcal{H}})^2 \Re\mathrm{e}(\Pi_{\mathcal{H}}) \ , \end{aligned} \tag{31}$$

where

$$\Pi_{\mathcal{H}} \equiv \left(s + m_Z^2 - m_{\mathcal{H}}^2 - 2p \cdot p_Z + i m_{\mathcal{H}} \Gamma_{\mathcal{H}}\right)^{-1} \ , \tag{32}$$

$p \equiv p_- + p_+$ and $\Gamma_{\mathcal{H}}$ is the width of $\mathcal{H} = h$ or H.

The CP-odd coupling combinations are

$$g^1_{-t\bar{t}h} = b^h_t c^h \ , \tag{33}$$

for the $t\bar{t}h$ final state and

$$g^1_{-t\bar{t}Z} = b_t^{\mathcal{H}} c^{\mathcal{H}} \Re\text{e}(\Pi_{\mathcal{H}}) \; , \tag{34}$$

for $t\bar{t}Z$ final state.

Of course the total cross section arises from the integral over phase space of the CP-even component, $\Sigma_{+(j)f}$

$$\sigma_{(j)f} = \int \Sigma_{+(j)f}(\Phi) d\Phi \; , \tag{35}$$

where Φ stands for the phase-space variables. These cross sections $\sigma_{t\bar{t}h}$ and $\sigma_{t\bar{t}Z}$ are shown in Fig. 3(a) and 3(b) (taken from [14]) as a function of m_h and $\sqrt{s}$, for Model II with $m_H = 750$ GeV and the set of values $\{\alpha_1, \alpha_2, \alpha_3\} = \{\pi/2, \pi/4, 0\}$ which we denote as set II. Set II is also adopted later when discussing the CP-violating effect. Furthermore, for the $t\bar{t}h$ final state we choose $\tan\beta = 0.5$ while for $t\bar{t}Z$ we choose $\tan\beta = 0.3$.

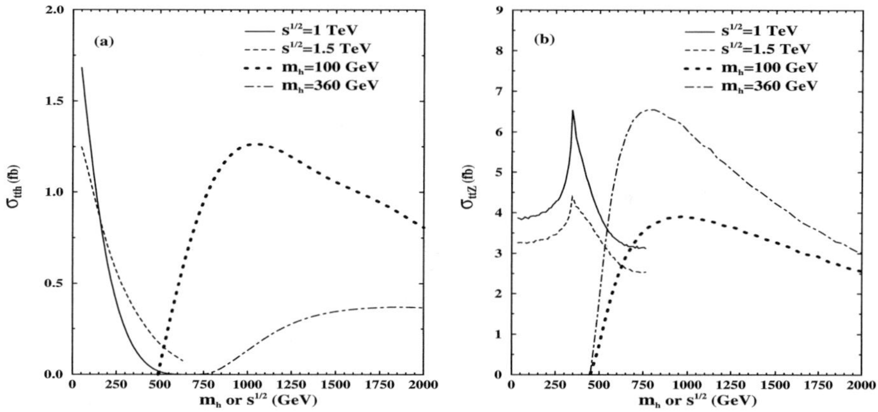

Fig. 3. The cross-sections (in fb) for: (a) the reaction $e^+e^- \to t\bar{t}h$ with $\tan\beta = 0.5$ and (b) the reaction $e^+e^- \to t\bar{t}Z$ with $\tan\beta = 0.3$, assuming unpolarized electron and positron beams, for Model II with set II and as a function of m_h (solid and dashed lines) and $\sqrt{s}$ (dotted and dotted-dashed lines). Set II means $\{\alpha_1, \alpha_2, \alpha_3\} \equiv \{\pi/2, \pi/4, 0\}$.

Let us now consider the CP violating (T_N-odd) asymmetries which can arise in these reactions. Such effects are proportional to the couplings $b_t^h c^h$ (in the $t\bar{t}Z$ case there is an additional similar piece corresponding to the heavy Higgs H). The corresponding CP-odd kinematic functions, $F^1_{-(j)t\bar{t}h}$, $F^1_{-(j)t\bar{t}Z}$, are T_N-odd. If the spins of the final state particles are not observed, these must be proportional to the only non-vanishing Levi-Civita tensor present, $\epsilon(p_-, p_+, p_t, p_{\bar{t}})$. The explicit expressions for these are: (recall that $j = 1(-1)$ for left(right) polarized incoming electron beam)

$$F^1_{-(j)t\bar{t}h} = -\frac{1}{\sqrt{2}}\left(\frac{g_W^3}{c_W^3}\right)^2 \frac{m_t^2}{m_Z^2}\Pi_{Zh}\Pi_Z T_t^3 c_j^Z \epsilon(p_-, p_+, p_t, p_{\bar{t}}) \times$$
$$\{j(\Pi_t^h + \Pi_{\bar{t}}^h)\left[(s - s_t - m_h^2)(3w_j^- - w_j^+) + m_Z^2(w_j^- - w_j^+)\right] +$$
$$T_t^3 c_j^Z \Pi_Z(\Pi_t^h - \Pi_{\bar{t}}^h) f\} \ , \quad (36)$$

$$F^1_{-(j)t\bar{t}Z} = -\sqrt{2}\left(\frac{2g_W^3}{c_W^3}\right)^2 \frac{m_t^2}{m_Z^2}\Pi_Z T_t^3 c_j^Z \epsilon(p_-, p_+, p_t, p_{\bar{t}}) \times$$
$$\{j(\Pi_t^Z + \Pi_{\bar{t}}^Z)\left[m_Z^2 w_j^- + (s_t - s) w_j^+\right] +$$
$$T_t^3 c_j^Z \Pi_Z(\Pi_t^Z - \Pi_{\bar{t}}^Z) f\} \ , \quad (37)$$

where $s \equiv 2p_- \cdot p_+$ is the CM energy of the colliding electrons, $s_t \equiv (p_t + p_{\bar{t}})^2$ and $f \equiv (p_- - p_+) \cdot (p_t + p_{\bar{t}})$. Also

$$\Pi^h_{t(\bar{t})} \equiv \left(2p_{t(\bar{t})} \cdot p_h + m_h^2\right)^{-1} \ , \ \Pi^Z_{t(\bar{t})} \equiv \left(2p_{t(\bar{t})} \cdot p_Z + m_Z^2\right)^{-1} \ , \quad (38)$$

$$\Pi_Z \equiv \left(s - m_Z^2\right)^{-1} \ , \ \Pi_\gamma \equiv s^{-1} \ , \ \Pi_{Zh} \equiv \left((p - p_h)^2 - m_Z^2\right)^{-1} \ . \quad (39)$$

and

$$w_j^{\pm} \equiv \left(s_W^2 Q_t - \frac{1}{2}T_t^3\right) c_j^Z \Pi_Z \pm Q_t s_W^2 c_W^2 \Pi_\gamma \ , \quad (40)$$

where $s_W(c_W)$ is the sin(cos) of the weak mixing angle θ_W, Q_f and T_f^3 are the charge and the z-component of the weak isospin of a fermion, respectively, and $c^Z_{-1} = 1/2 - s_W^2$, $c_1^Z = -s_W^2$.

Since at tree-level there cannot be any absorptive phases, CP-violating asymmetries only of the T_N-odd type are expected to occur in $\Sigma_{-(j)f}$. Note that in the $t\bar{t}Z$ case there is also a CP-odd, T_N-even piece, $b_t^{\mathcal{H}} c^{\mathcal{H}} \mathrm{Im}(\Pi_{\mathcal{H}}) \times F^{2(+)}_{-(j)t\bar{t}Z}$ [see Eq. (34)], in the differential cross section. However, being proportional to the absorptive part coming from the Higgs propagator, it is not a pure tree-level quantity.

It is possible to use a number of simple observables which are sensitive to the CP-odd triple product correlation in both of these cases as considered in [9] as well as the observable optimized to the expected CP-odd distribution:

$$O = \frac{\vec{p}_- \cdot (\vec{p}_t \times \vec{p}_{\bar{t}})}{s^{3/2}} \ , \ O_{\rm opt}(t\bar{t}h) = \frac{\Sigma_{-(j)t\bar{t}h}}{\Sigma_{+(j)t\bar{t}h}} \ ; \ O_{\rm opt}(t\bar{t}Z) = \frac{\Sigma_{-(j)t\bar{t}Z}}{\Sigma_{+(j)t\bar{t}Z}} \ . \quad (41)$$

Both observables are proportional to $\epsilon(p_-, p_+, p_t, p_{\bar{t}})$ but multiplied by different CP-even and T_N-even functions. Furthermore, $O_{\rm opt}(t\bar{t}h; t\bar{t}Z)$ are related to O through a multiplication by a CP-even function. It turns out that in this case the two observables give similar effective asymmetries (the optimal being slightly better as it must) we will mainly quote results for the optimal case.

Let us denote by N_{SD} the statistical significance with which an asymmetry can be measured in an ideal experiment. Thus $N_{SD} = A\sqrt{L}\sqrt{\sigma}$ (for each of the cross sections $\sigma = \sigma_{t\bar{t}h}, \sigma_{t\bar{t}Z}$) where L is the effective luminosity for fully reconstructed $t\bar{t}h$ or $t\bar{t}Z$ events. In particular, we take $L = \epsilon\mathcal{L}$, where $\mathcal{L}$ is the total yearly integrated luminosity and ϵ is the overall efficiency for reconstruction of the $t\bar{t}h$ or $t\bar{t}Z$ final states.

To see what kind of asymmetries might be possible, we show here sample calculation of [14] where we use set II defined before for the angles $\alpha_{1,2,3}$, *i.e.*, $\{\alpha_1, \alpha_2, \alpha_3\} = \{\pi/2, \pi/4, 0\}$[b]. In this case, Figs. 4(a) and 4(b) show the expected asymmetry and statistical significance in the unpolarized case, corresponding to $O_{\rm opt}$ in Model II for the $t\bar{t}h$ and $t\bar{t}Z$ final states, respectively. The asymmetry is plotted as a function of the mass of the light Higgs (m_h) where again, $m_H = 750$ GeV in the $t\bar{t}Z$ case. We plot $N_{SD}/\sqrt{L}$, thus scaling out the luminosity factor from the theoretical prediction.[c]

We remark that set II corresponds to the largest CP effect, though not unique since we are dealing with angles, *i.e.*, $\alpha_{1,2,3}$, which may be rotated by π or $\pi/2$ leaving the relevant combinations of angles with the same value (*e.g.*, $b_t^h \propto \sin\alpha_1 \sin\alpha_2$). In the $t\bar{t}h$ case $\tan\beta = 0.5$ is favored, however, the effect mildly depends on $\tan\beta$ in the range, $0.3 \lesssim \tan\beta \lesssim 1$ [9, 14]. In the $t\bar{t}Z$ case, the effect is practically insensitive to α_3 and is roughly proportional to $1/\tan\beta$, it therefore drops as $\tan\beta$ is increased. Nonetheless, we find that $N_{SD}/\sqrt{L} > 0.1$, even in the unpolarized case for $\tan\beta \lesssim 0.6$ [15, 14]; note that $N_{SD}/\sqrt{L}$ here is dimensionless if L is in fb^{-1}.

From Fig. 4(a) we see that, in the $t\bar{t}h$ case, as m_h grows the asymmetry increases while the statistical significance drops, in part because of the decrease in the cross-section. Evidently, the asymmetry can become quite large; it ranges from $\sim 15\%$, for $m_h \lesssim 100$ GeV, to $\sim 35\%$ for $m_h \sim 600$

[b]Recall that for the $t\bar{t}h$ final state we choose $\tan\beta = 0.5$ while for the $t\bar{t}Z$ final state we take $\tan\beta = 0.3$.

[c]As a reference value, we note that for $L = 100$ fb^{-1}, $N_{SD}/\sqrt{L} = 0.1$ will correspond to a 1-σ effect.

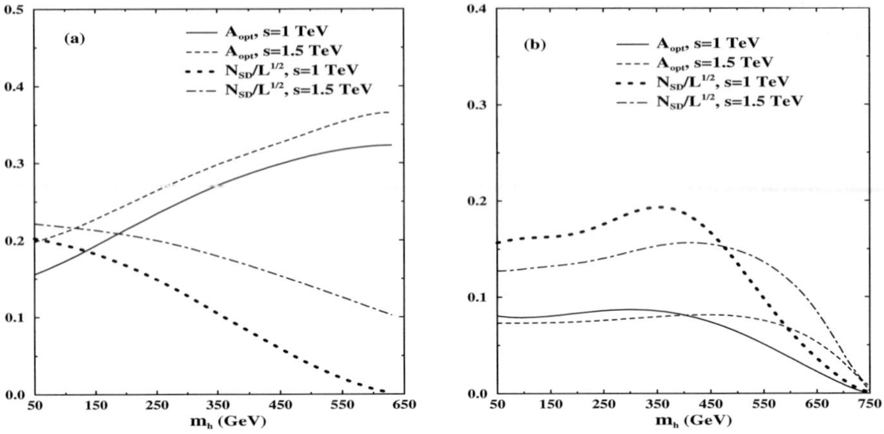

Fig. 4. The asymmetry, $A_{\rm opt}$, and scaled statistical significance, $N_{SD}/\sqrt{L}$, for the optimal observable $O_{\rm opt}$ for: (a) the reaction $e^+e^- \to t\bar{t}h$ with $\tan\beta = 0.5$ and (b) the reaction $e^+e^- \to t\bar{t}Z$ with $\tan\beta = 0.3$, as a function of the light Higgs mass m_h, for $\sqrt{s} = 1$ TeV and 1.5 TeV. All graphs are with set II of the parameters, as in Fig. 3.

GeV. However, the CP-effect is more significant for smaller masses of h, wherein $A_{\rm opt}$ is smaller. In contrast, Fig. 4(b) shows that, in the $t\bar{t}Z$ case, $A_{\rm opt}$ stays roughly fixed at around $7-8\%$ for $m_h \lesssim 2m_t$, and then drops till it totally vanishes at $m_h = m_H = 750$ GeV, due to the GIM-like mechanism. The scaled statistical significance $N_{SD}/\sqrt{L}$ behaves roughly as $A_{\rm opt}$. That is, $N_{SD}/\sqrt{L} \approx 0.1$–$0.2$ in the mass range 50 GeV $\lesssim m_h \lesssim$ 350 GeV, for both $\sqrt{s} = 1$ and 1.5 TeV.

The variation of the asymmetry with the energy is likewise shown in Figs. 5(a) and 5(b) We see that, in the case of $t\bar{t}h$, the CP-effect peaks at $\sqrt{s} \approx 1.1(1.5)$ TeV for $m_h = 100(360)$ GeV and stays roughly the same as $\sqrt{s}$ is further increased to 2 TeV. In the case of $t\bar{t}Z$, the statistical significance is maximal at around $\sqrt{s} \approx 1$ TeV and then decreases slowly as $\sqrt{s}$ grows for both $m_h = 100$ and 360 GeV. Contrary to the $t\bar{t}h$ case, where a light h is favored, in the $t\bar{t}Z$ case, the effect is best for $m_h \gtrsim 2m_t$. In that range, on-shell Z and h are produced followed by the h decay $h \to t\bar{t}$, thus, the Higgs exchange diagram becomes more dominant.

In summary, these two channels may turn out to be the most important probes at the LC for CP violation that may arise with extra higgs doublets. This is largely the result of the fact that because two tree graphs are interfering, if the CP phase is large, O(10%) asymmetries can naturaly arise since there is no other suppression factor. Indeed, if large tree-tree interference effects are indeed present, the detailed study of both the CP-odd and CP-even components will provide information about the nature of the Higgs couplings. References [56] and [54] give a detailed description of the extraction of each possible product of couplings from the phase space distribution.

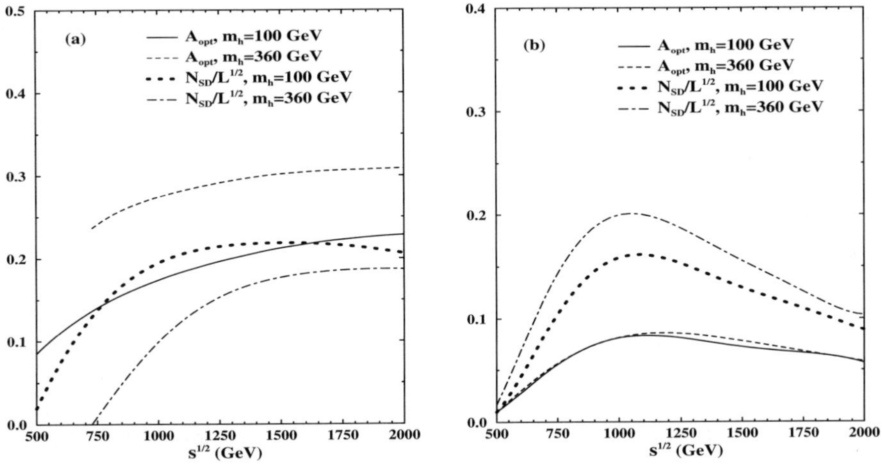

Fig. 5. The asymmetry, $A_{\rm opt}$, and scaled statistical significance, $N_{SD}/\sqrt{L}$, for the optimal observable $O_{\rm opt}$ for: (a) the reaction $e^+e^- \to t\bar{t}h$ with $\tan\beta = 0.5$ and (b) the reaction $e^+e^- \to t\bar{t}Z$ with $\tan\beta = 0.3$, as a function of the CM energy $\sqrt{s}$, for $m_h = 100$ GeV and $m_h = 360$ GeV. All graphs are with set II of the parameters, as in Fig. 3.

4.2. *CP Asymmetries in $e^+e^- \to Zh$ with Subsequent Higgs Decay $h \to t\bar{t}$*

Let us now consider the same $Zt\bar{t}$ final state in the scenario where $m_h > 2m_t$ and the energy of the LC is large enough to produce an on-shell Zh. In this case a significant contribution to the $Zt\bar{t}$ final state will be from Zh with subsequent $h \to t\bar{t}$ and if one segregates events with $t\bar{t}$ invariant mass near m_h, this process should dominate the amplitude.

Because there is only one tree level graph involved with an intermediate scalar, the $t\bar{t}$ pair produced in this way looses all information about the geometry of the rest of the event. Thus, there will be no triple product correlations of the momenta of the final state particles; so, CP violation here can only be manifested in the polarizations of the $t\bar{t}$ pair.

Indeed, as with the EDM, we are interfering P and V, $t\bar{t}$ helicity states that arise from scalar and pseudoscalar couplings to the scalar. To this end, general methods for tracing CP-odd and CP-even $t, \bar{t}$ spin-correlations, in $h \to t\bar{t}$, was introduced in [24]. There, it was assumed that an on-shell Higgs, with $m_h > 2m_t$, is produced through, for example, $e^+e^- \to Zh, \ell^+\ell^- h$ or even $\mu^+\mu^- \to h$, and that its rest system can be reconstructed. In [49], a helicity asymmetry in $h \to t\bar{t}$ was suggested, where $e^+e^- \to Zh$ was explicitly assumed as the Higgs production mechanism. We will describe below these two works. Other related works can be found in [51, 21, 33, 22, 61, 60, 63, 7].

In [24] the asymmetries in terms of the top spins

$$\mathcal{S}_1 = \hat{\mathbf{k_t}} \cdot (\mathbf{s}_t - \mathbf{s}_{\bar{t}}) \ , \tag{42}$$

$$\mathcal{S}_2 = \hat{\mathbf{k_t}} \cdot (\mathbf{s}_t \times \mathbf{s}_{\bar{t}}) \ , \tag{43}$$

$$\mathcal{S}_3 = \mathbf{s}_t \cdot \mathbf{s}_{\bar{t}} \ , \tag{44}$$

were considered where $\mathbf{s}_t(\mathbf{s}_{\bar{t}})$ is the spin operator of $t(\bar{t})$ and $\mathbf{k_t}$ is the top quark 3-momentum in the $t\bar{t}$ CM frame. $\mathcal{S}_1$ and $\mathcal{S}_2$ are CP-odd, where $\mathcal{S}_1$ is T_N-even resulting from the interference of P with V_0 and $\mathcal{S}_2$ is T_N-odd resulting from the interference of P with $V_+ - V_-$. Thus, a non-zero expectation value of $\mathcal{S}_1$ will also require absorptive parts, while $< \mathcal{S}_2 > \neq 0$ can be generated already at the tree-level. $\mathcal{S}_3$ is CP-even and is also generated at the lowest order (*i.e.*, tree-level). For a general $t\bar{t}h$ Yukawa interaction Lagrangian as in Eq. (26), it was found that the spin-asymmetries, $\mathcal{S}_{1,2,3}$ in Eqs. (42)–(44), depend only on one combination of the couplings a_t^h and b_t^h

$$r_t = \frac{b_t^h}{a_t^h + b_t^h} \ , \tag{45}$$

where a value of r_t between 0 and 1 indicates some mixture of scalar and pseudo-scalar couplings.

These spin-asymmetries, $\mathcal{S}_{1,2,3}$ can be translated to correlations between momenta of the t and $\bar{t}$ decay products. One very efficient way to do that would be to use momenta of an hadronic and a semi-leptonic decay of the $t\bar{t}$ pair [24]. Fig. 6 shows the number of events needed for a 3-σ observation of the spin-correlations $\mathcal{S}_{1,2,3}$ as a function of the parameter r_t, assuming $m_h = 400$ GeV and the product Br for $t\bar{t}$ pair of $\frac{4}{27}$. We see from the figure that with a data sample of about 1500 ($h \to t\bar{t}$) events and with values of $r_t \approx 0.18 - 0.52$ there could be a 3-σ effect in the $\mathcal{S}_2$ asymmetry. Furthermore, with a simultaneous study of $\mathcal{S}_2$ and $\mathcal{S}_3$ with these ≈ 1500

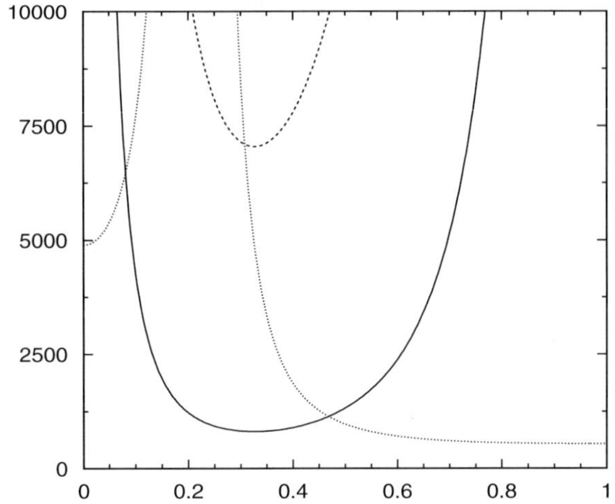

Fig. 6. The cross-sections (in fb) for: (a) the reaction $e^+e^- \to t\bar{t}h$ with $\tan\beta = 0.5$ and (b) the reaction $e^+e^- \to t\bar{t}Z$ with $\tan\beta = 0.3$, assuming unpolarized electron and positron beams, for Model II with set II and as a function of m_h (solid and dashed lines) and $\sqrt{s}$ (dotted and dotted-dashed lines). Set II means $\{\alpha_1, \alpha_2, \alpha_3\} \equiv \{\pi/2, \pi/4, 0\}$.

events, there could be a 3-σ sensitivity to r_t from about 0.18 to its maximum value of 1.

It seems that production of a few thousand of the $h \to t\bar{t}$ events through the Bjorken mechanism, $e^+e^- \to hZ$, is feasible. As an illustration if we consider a CP-mixed neutral Higgs of mass around 400 GeV and with non-vanishing scalar and pseudoscalar couplings to fermions and with a ZZh coupling of strength similar to the SM, then the cross-section for $e^+e^- \to hZ$ is about a few fb's for e^+e^- CM energies in the range of 500–1000 GeV [75]. Thus with a yearly integrated luminosity of around 100 fb^{-1}, hundreds of hZ pairs may be produced.

Another interesting CP violating asymmetry, studied in [49], is:

$$\mathcal{O}^h_{tt} = \frac{\Gamma(++) - \Gamma(--)}{\Gamma(++) + \Gamma(++)}, \tag{46}$$

where $\Gamma(++)$ and $\Gamma(--)$ are the decay widths of the lightest Higgs-boson h, into a $t\bar{t}$ pair of the indicated helicities.

Since $\mathcal{O}^h_{tt}$ is CP-odd and T_N-even, it requires a CP-odd as well as a CP-even absorptive phase. To generate the CP-odd phase, simultaneous presence of the scalar and pseudoscalar $t\bar{t}h$ couplings are needed. For $m_h > 2m_t$, an absorptive phase will occur at one-loop level from the diagrams in

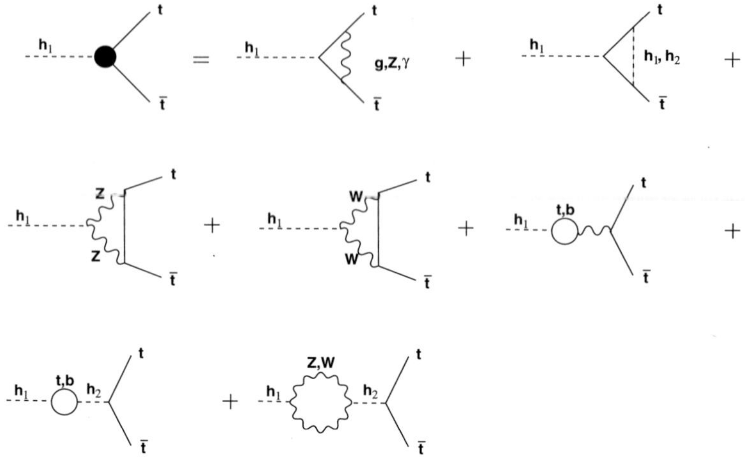

Fig. 7. The 1-loop diagrams contributing to the asymmetry $\mathcal{O}^h_{tt}$ in a 2HDM.

Fig.(7). In [49] it was found that asymmetries of several tens of percents may be generated by this mechanism. Also with an integrated luminosity $\approx 100\ fb^{-1}$, at least for values of $\beta \approx 0.5$, statistically significant signals may result.

5. Some Other Topics, in Brief

5.1. *Threshold Studies of* $e^+e^- \to t\bar{t}$

This is interesting and can be very useful in constraining top-dipole moments. Chapter on the *Top Quark Physics* has a section discussing this capability.

5.2. $e^+e^- \to t\bar{t}g$

This reaction has been extensively studied [75], with and without beam polarization, especially for its capability to set limits on top-dipole moments. It seems to be somewhat less good than $e^+e^- \to t\bar{t}$ by roughly an order of magnitude.

5.3. $e^+e^- \to t\bar{t}\nu_e\bar{\nu}_e$

This reaction is interesting [76, 75] as it can give rise to largish CP-asymmetries originating from a CP-odd phase in neutral Higgs sector appearing in some extensions of the SM. However, the energy carried by neutrinos in the final state presents a serious experimental challenge in reconstructing the $t, \bar{t}$ rest frames as is required to measure the polarization asymmetries when the top quark aniquark decay semileptonically. This problem is not there when they decay hadronically but then it becomes difficult to distinguish decays of t from those of $\bar{t}$.

Acknowledgements

This research was supported in part by Contract Nos. DE-FG02-94ER40817(IOWA) and DE-AC02-98CH10886 (BNL).

References

1. N. Cabibbo, Phys. Rev. Lett. **10**, 531 (1963); M. Kobayashi and T. Maskawa, Prog. Th. Phys. **49**, 652 (1973).
2. E. Accomando *et al.*, (ECFA/DESY LC Physics Working Group), Phys. Rep. **299**, 1 (1998).
3. K. Ackerstaff *et al.*, (OPAL Collaboration), Zeit. f. Phys. **C74**, 403 (1997).
4. C.H. Albright, J. Smith and S.-H.H. Tye, Phys. Rev. **D21**, 711 (1980).
5. H. Anlauf, W. Bernreuther and A. Brandenburg, Phys. Rev. **D52**, 3803 (1995); Erratum-*ibid.* **D53**, 1725 (1996).
6. T. Arens and L.M. Sehgal, Phys. Rev. **D50**, 4372 (1994).
7. T. Arens, U. Gieseler, and L.M. Sehgal, Phys. Lett. **B339**, 127 (1994).
8. D. Atwood and A. Soni, Phys. Rev. Lett. **74**, 220 (1995).
9. D. Atwood *et al.*, Phys. Rev. **D53**, 1162 (1996).
10. D. Atwood, A. Aeppli and A. Soni, Phys. Rev. Lett. **69**, 2754 (1992).
11. D. Atwood, S. Bar-Shalom and A. Soni, Phys. Rev. **D51**, 1034 (1995).
12. D. Atwood and A. Soni, Phys. Rev. **D45** 2405 (1992).
13. M.S. Baek, S.Y. Choi and C.S. Kim, Phys. Rev. **D56**, 6835 (1997).
14. S. Bar-Shalom, `hep-ph/9710355`.
15. S. Bar-Shalom, D. Atwood and A. Soni, Phys. Lett **B419**, 340 (1998).
16. S. Bar-Shalom, D. Atwood, G. Eilam and A. Soni, Z. Phys. **C72** 79 (1996).
17. V. D. Barger and R. J. N. Phillips, *Collider Physics*, (Addison-Wesley, New York, 1987).
18. A. Bartl, E. Christova and W. Majerotto, Nucl. Phys. **B460**, 235 (1996); Erratum-*ibid.* Nucl. Phys. **B465**, 365 (1996).
19. A. Bartl, E. Christova, T. Gajdosik and W. Majerotto, Nucl. Phys. **B507**, 35 (1997); Erratum-*ibid.*, **B531**, 653 (1998).
20. G. Baur, K. Hencken and Trautmann, J. Phys. **G24**, 1657 (1998).
21. W. Bernreuther and A. Brandenburg, Phys. Lett. **B314**, 104 (1993).
22. W. Bernreuther and A. Brandenburg, Phys. Rev. **D49**, 4481 (1994).
23. W. Bernreuther and P. Overmann, Z. Phys. **C61**, 599 (1994).
24. W. Bernreuther, A. Brandenburg and M. Flesch, Phys. Rev. **D56**, 90 (1997).

25. W. Bernreuther, A. Brandenburg and P. Overmann, hep-ph/9602273, in e^+e^- *Linear Collisions: The Physics Potential*, p. 49, ed. P. M. Zerwas (1995).
26. W. Bernreuther, T. Schröder and T.N. Pham, Phys. Lett. **B279**, 389 (1992).
27. I. I. Bigi *et. al.*, Phys. Lett. **B181** 157 (1986).
28. J. D. Bjorken and S. D. Drell, *Relativistic Quantum Mechanics*, McGraw-Hill (1964).
29. R. Brinkman *et al.*, Nucl. Instrum. Meth. **A406**, 13 (1998).
30. L. Brzezinski, B. Grządkowski and Z. Hioki, Int. J. Mod. Phys. **A14**, 1261 (1999).
31. For a review see A.J. Buras, hep-ph/9806471, in *Probing the Standard Model of Particle Interactions*, eds. F. David and R. Gupta (Elsevier Science, 1998).
32. C. Caso *et al.*, *Review of Particle Physics*, Eur. Phys. J. **C3**, 1 (1998).
33. D. Chang, W.-Y. Keung, and I. Phillips, Phys. Rev. **D48**, 3225 (1993).
34. S.Y. Choi and K. Hagiwara, Phys. Lett. **B359**, 369 (1995).
35. E. Christova and M. Fabbrichesi, Phys. Lett. **B315**, 338 (1993).
36. A. Czarnecki and B. Krause, Phys. Rev. Lett. **78**, 4339 (1997).
37. M. Davier *et. al.*, Phys. Lett. **B306** 411 (1993).
38. M. Diehl and O. Nachtmann, Zeit. f. Phys. **C62**, 397 (1994).
39. A. Djouadi, J. Kalinowski and P.M. Zerwas, Mod. Phys. Lett. **A7**, 1765 (1992).
40. A. Djouadi, J. Kalinowski and P.M. Zerwas, Z. Phys. **C54**, 255 (1992).
41. J. F. Donoghue, Phys. Rev. **D18**, 1632 (1978).
42. C.D. Froggatt, R.G. Moorhouse and I.G. Knowles, Nucl. Phys. **B386**, 63 (1992).
43. K. Fujii, in proceedings of the *4th KEK Topical Conference on Flavor Physics*, (1996) and references therein.
44. I. Ginzburg *et al.*, Pizma ZhETF **34**, 514 (1981).
45. I. Ginzburg *et al.*, JETP Lett. **34**, 491 (1982).
46. I. Ginzburg *et al.*, Nucl. Instr. & Meth. **205**, 47 (1983).
47. I. Ginzburg *et al.*, Nucl. Instr. & Meth. **219**, 5 (1984).
48. B. Grządkowski, Phys. Lett. **B305**, 384 (1993).
49. B. Grządkowski, Phys. Lett. **B338**, 71 (1994).
50. B. Grządkowski and J.F. Gunion, Phys. Lett. **B294**, 361(1992).
51. B. Grządkowski and J.F. Gunion, Phys. Lett. **B350**, 218 (1995).
52. B. Grządkowski and Z. Hioki, Nucl. Phys. **B484**, 17 (1997).
53. B. Grządkowski and Z. Hioki, Phys. Lett **B391**, 172 (1997).
54. B. Grzadkowski and J. Pliszka, Phys. Rev. **D60** 115018 (1999).
55. J.F. Gunion and X.-G. He, hep-ph/9609453, in proceedings of the *1996 DPF/DPB Summer Study for New Directions in High Energy Physics*, Snowmass, Colorado, June (1996).
56. J.F. Gunion, B. Grządkowski and X.-G. He, Phys. Rev. Lett. **77**, 5172 (1996).
57. K. Hagiwara, H. Murayama and I. Watanabe, Nucl. Phys. **B367** 257 (1991).
58. T. Han, T. Huang, Z. H. Lin, J. X. Wang and X. Zhang, Phys. Rev. **D61** 015006 (2000).

59. F. Harris ed., Proceedings of the *Workshop on Physics and Experiments with Linear e^+e^- Colliders* (World Scientific, Singapore, 1993).
60. X.-G. He, J.P. Ma, and B.H.J. McKellar, Mod. Phys. Lett. **A9**, 205 (1994).
61. X.-G. He, J.P. Ma, and B.H.J. McKellar, Phys. Rev. **D49**, 4548 (1994).
62. [J. L. Hewett (1993)] hepph9406302 For a review see J. L. Hewett, `hep-ph/9406302`, in proceedings of the *SLAC Summer Inst. on Particle Physics: Spin Structure in High Energy Processes*, Stanford, CA, (1993).
63. M. Krämer, J. Kuhn, M. L. Strong and P. M. Zerwas, Z. Phys. **C64**, 21 (1994).
64. S. Kuhlman *et al.*, (NLC ZDR Design Group and NLC Physics Working Group), `hep-ex/9605011`.
65. A. Miyamoto and Y. Fujii eds., Proceedings of the *Workshop on Physics and Experiments with Linear* e^+e^- *Colliders*, (World Scientific, Singapore, 1996).
66. H. Murayama and M. E. Peskin, Ann. Rev. Nucl. Part. Sci. **49**, 513 (1996).
67. P. Poulose and S.D. Rindani, Phys. Rev. **D57**, 5444 (1998).
68. T. Rizzo, `hep-ph/9610373`, talk given at the *1996 DPF/DPB Summer Study on New Directions for High-energy Physics*, Snowmass, CO (1996).
69. E. P. Shabalin, Sov. J. Nucl. Phys. **28**, 75 (1978).
70. K. Shizuya and S.-H.H. Tye, Phys. Rev. **D23**, 1613 (1981).
71. A. Soni and R.M. Xu, Phys. Rev. Lett. **69**, 33 (1992).
72. V. Telnov, Turk. J. Phys. **22**, 541 (1998).
73. S. Weinberg, Phys. Rev. Lett. **37**, 657 (1976).
74. X. Zhang, S. K. Lee, K. Whisnant and B. L. Young, Phys. Rev. **D50**, 7042 (1994).
75. See. p. 133 in D. Atwood *et al*, Physics Reports, V. 347(2001).
76. D. Atwood and A. Soni, hep-ph/9607481.

CHAPTER 11

Overall Perspective

Keisuke Fujii[a] and Michael E. Peskin[b]
[a]*KEK, Japan*
E-mail: keisuke.fujii@kek.jp
[b]*Stanford Linear Accelerator Center, 2575 Sand Hill Rd*
Menlo Park, CA 94025, U.S.A.
E-mail: mpeskin@slac.stanford.edu

1. Grand View of Particle Physics

The progress of elementary particle physics in the latter half of the twentieth century was guided by great questions about the properties of atomic nuclei: What were the forces that held nuclei together? What determined the structure of protons and neutrons? What was the mechanism of radioactive decay? As the century ended, elementary particle physicists could feel confident that the answers to these questions were in hand. Experiments using high-energy collisions had exposed the basic constituents of nucleons and had demonstrated their essential similarity to leptons and photons. A unified picture of the strong and weak nuclear forces had been created, with common ingredients to the theory of electromagnetism. In particular, the experimental study of the Z^0 boson, one of the basic quanta of weak interactions, had verified aspects of this theory to high precision [1, 2, 3]. This theory, buttressed by its experimental successes, is now known as the 'Standard Model'.

These successes bring forward new questions and define a new territory to be explored in the twenty-first century. The new description of the strong, weak, and electromagnetic interactions relies on the properties of the basic particles of the model—quarks, leptons, photons, gluons, and weak-interaction bosons. But it also depends on the properties of the vacuum state, the quantum state of empty space in which these particles move. Most strikingly, all particle masses in the model arise because the vacuum state is asymmetric with respect to fundamental symmetries of the equations of the weak interactions. This asymmetry must be a phenomenon similar to the spontaneous ordering of condensed-matter systems in their ground states, such as one sees in magnets or binary alloys. In those systems, the ordering is preferred energetically due to well-understood interatomic forces. For the space-time vacuum, the energy that stabilizes the asymmetric ground state can only come from new particles and forces that are still undiscovered.

At the same time, new developments in cosmology made clear that there are new elementary particles not contained in the Standard Model. About 20% of the energy content of the universe is now known to be accounted for by 'dark matter', a species of neutral, weakly interacting matter that is neither quark, nor lepton, nor neutrino. About 70% is contained in the gravitational energy of the vacuum, a quantity that cannot even be computed, much less explained, within our current theory [4].

The search for the new particles required for these phenomena and the understanding of their role in Nature will be one of the major themes of physics in the new century. The scale of energy of the required new interactions is of the order of hundreds of GeV. This value comes independently from two separate arguments, one based on the energy scale of the symmetry-breaking in weak interactions, the other based on the cosmic density of dark matter. This means that these interactions should make themselves apparent in the experiments at the CERN Large Hadron Collider (LHC) that will begin in just a few years. Though it is possible that these experiments will turn up only one new particle, the long-sought Higgs boson, it is more likely that these experiments will open the door to a whole new sector of elementary particles.

It will be a challenge to characterize these particles, to determine their quantum numbers and interactions, and to discover the organizing structures of the new physical interactions. In the study of heavy quarks and leptons and of the W and Z bosons, the crucial experiments that determined the quantum numbers and couplings of the new particles were carried out by production of the particles in electron-positron (e^+e^-) collisions. The reaction of e^+e^- annihilation has a number of special features that allow experimenters to obtain detailed information about produced particles. Production cross sections typically depend only on the spins and electroweak quantum numbers of the produced particles. This implies that new particles are produced at substantial rates, comparable to the production rates of familiar particles. It also means that the quantum numbers of new particles can be determined straightforwardly by measurement of the production cross sections. Because new particle production is a large fraction of the whole event sample, special triggers are unnecessary and so all decay modes of new particles can be recorded and measured. In addition, the kinematics of e^+e^- annihilation is simple, allowing complete spin analyses of particle production from which one can extract spins and coupling constants.

In the remainder of this chapter, we will review a number of proposals for new physics at the energy scale of a few hundred GeV. For each model, we will survey the ways in which the special features of the experimental environment of e^+e^- collisions can be used to bring out important insights about the physics. The various chapters of this book expand on this discussion with many more specific examples. Even more detail can be found in the documents on the future e^+e^- experimental program that have been prepared by the European, American, and Asian study groups [5, 6, 7].

As we are completing this overview, the International Committee on Future Accelerators (ICFA), the organization of the world's major high-energy physics laboratories, has announced a global initiative to construct a 500 GeV e^+e^- collider based on the technology of superconducting accelerating cavities—the International Linear Collider (ILC) [8]. The initiative would realize ideas that were put forward often in the previous century, that the resources of the whole world community should be pooled in the pursuit of the deepest questions about the structure of Nature. In this volume, we present, for the new century, the physics that justifies such a project of global scale.

2. Highlights of the LC Physics Programs

In this section, we review some of the major issues that will be studied at the ILC. Each of these topics has been presented as the subject of a chapter of this book.

2.1. *The Higgs Boson*

The minimal description of the asymmetric vacuum of the electroweak theory is a model with one new scalar field and one new physical particle. This particle is called the 'Higgs boson'. Its field is assigned a potential energy that stabilizes the asymmetric state. In this model, the single Higgs boson and Higgs field is directly responsible for the masses of the W and Z bosons and the masses of the quarks and leptons.

We believe it is unlikely that the minimal model could be correct. The model is more a phenomenological description than a theory; it does not explain the electroweak vacuum asymmetry but only allows an asymmetric vacuum state for a certain choice of parameters. However, deeper theories that do explain why an asymmetric vacuum is energetically favored often contain a particle similar to the Higgs boson and a field similar to the Higgs field that obtains a nonzero vacuum expectation value. Thus, however one feels about the minimal model as a fundamental theory, it is crucial to discover this particle and to study it in great detail. In the minimal model—in fact, in almost every model consistent with the precision measurements of electroweak interactions—the mass of the Higgs boson is less than 250 GeV. It is very likely that this particle will be discovered at the LHC, but the LHC will provide only the first-level tests of the model.

If the Higgs boson is solely responsible for generating the masses of elementary particles, the Higgs couplings will be exactly proportional to mass. To the extent that we can test this prediction for a variety of massive species, we can observe the connection between symmetry-breaking and mass generation characteristic of the minimal electroweak model. To the extent that this prediction is violated, we find clues to the additional particles responsible for mass generation in more complex models. Thus it is important to test the Higgs boson couplings as accurately as possible for as many species as possible.

As is explained in Chapter 2 of this book, the ILC offers an excellent setting to observe the whole profile of the Higgs boson. The Higgs boson h^0 is produced in a number of reactions. Of these, the simplest is $e^+e^- \to Z^0h^0$. In this reaction, the Higgs boson is tagged by its recoil against a Z^0 boson. All decay modes of the Higgs boson can be recorded, even a decay to two invisible particles, and the branching fractions can be measured with very small systematic errors. The total cross section for this reaction measures the Higgs coupling to the Z^0, that is, the fraction of the Z^0 mass for which the h^0 is responsible.

The power of this reaction is supplemented by a number of other processes. The reaction $e^+e^- \to \nu\bar{\nu}h^0$ has a cross section that is large and increasing with energy. This reaction allows one to make a high precision measurement of the Higgs coupling to the W, and also to measure Higgs decay modes with small branching fractions, to 10^{-3} and below. At high center of mass energies of about 1 TeV, the reactions $e^+e^- \to t\bar{t}h^0$, $e^+e^- \to \nu\bar{\nu}h^0h^0$, $e^+e^- \to Z^0h^0h^0$ allow one to measure the Higgs coupling to the top quark and the Higgs boson self-coupling.

2.2. *Supersymmetry*

A model that can explain the electroweak vacuum asymmetry must contain many new ingredients beyond the simple Higgs field. The best developed such model is the supersymmetric extension of the Standard Model [9]. In this model, each quark, lepton, and gauge boson is accompanied by a partner of opposite statistics. For quarks and leptons, one adds scalars with the same Standard Model gauge quantum numbers; for gauge bosons, one adds new fermions. It can be seen that the model requires a minimum of two Higgs fields, plus associated fermions.

The supersymmetric Standard Model has a number of appealing features. The model contains the Higgs scalar field in a natural way, as one of many scalar fields in the model. However, it also contains interactions, related by symmetry to the Higgs field coupling to the top quark, that generate an unstable potential for this field that energetically favors an asymmetric vacuum state. The renormalization group equations for the model are consistent with grand unification of the strong, weak, and electromagnetic interactions with the value of $\sin^2\theta_w$ measured at LEP. The model naturally has a light Higgs boson and generates only small corrections to the electroweak predictions of the Standard Model. The model contains a stable neutral fermion that is an excellent candidate for the particle that comprises the cosmic dark matter.

Supersymmetry might not be the correct model of the physics of electroweak symmetry breaking, but we expect that the model that is correct will be at least as complex in order to share these attractive properties. For this reason, supersymmetry provides an interesting test case for proposed experiments or for a new accelerator facility. We would like to know how detailed a picture of new interactions we can assemble from experimental data.

To verify that new particles at the energy scale of a few hundred GeV have supersymmetry, we need to verify the predictions of the theory at many levels. First, the qualitative properties of the new particles should be observed; new particles should be found with the gauge quantum numbers of quarks, leptons, and gauge bosons. The masses of these particles should be measured as accurately as possible. Second, the spins of the particles should be determined, to check that the new particles have statistics opposite to the particles of the Standard Model. Third, the couplings of the new particles should be measured, to see that these have the correct symmetry relation to the Standard Model couplings. Fourth, the interactions that are responsible for the Higgs field potential should be measured, to verify that the model does correctly provide the energetics of the asymmetric vacuum state. Fifth, the interactions of the dark matter particle should be measured, to check whether the correct number density of this particle is generated in the early universe.

As is explained in Chapter 4 of this book, the whole program that we have described can be carried out at an e^+e^- collider. For most particles, the cross sections for pair production from e^+e^- depend in a characteristic way on the gauge quantum numbers and spins of the produced particles. Thus, the observation of pair-production not only allows high-precision mass measurements and studies of the decay modes, but it also pinpoints the identities of the new particles. The exceptions to this rule are the supersymmetry partners of the electrons and the W bosons. For these particles, characteristic features of the cross sections are sensitive to the couplings and, in fact, accurately test the symmetry predictions of supersymmetry. Measurements of the properties of the supersymmetry partners of top quarks and Higgs bosons provide very clean determinations of the parameters that determine the Higgs field potential. The various supersymmetry partners decay to final states that include the neutral fermion responsible for dark matter. By making precision measurements of these decays, it is possible to determine the properties of the (unobserved) dark matter particle and to measure the couplings that determine its dynamics in the early universe.

To the extent that supersymmetry is a representative model of the new set of interactions responsible for the asymmetric vacuum state of the electroweak interactions, we confidently expect that we will be able to determine the nature of these new interactions at the ILC.

2.3. *The Top Quark and the W Boson*

Whatever the new physics that we find at the hundred GeV mass scale, it will be interesting to determine its relation to the heaviest particles of the Standard Model, the top quark and the Z and W bosons.

For the top quark, the ILC provides first of all a very clean method for determining the mass and width. As is explained in Chapter 3 of this book, it is possible in e^+e^- collisions to measure the mass of the top quark to better than 100 MeV. This essentially eliminates the error in the top quark mass as a source of uncertainty in models of new physics. We should note that the ILC also allows a precision determination of α_s to better than 1%, addressing another ambiguity in the Standard and non-standard model predictions.

The study of top quark pair-production and decay allows the couplings of the top quark to the photon, Z, W, and gluon to be determined to percent accuracy. In models such as supersymmetry in which the couplings of the Higgs particles are essentially weak, only small corrections to the Standard Model predictions are expected. However, in models in which the Higgs boson is composite or strongly coupled, these experiments would reveal significant anomalies.

The process $e^+e^- \to W^+W^-$ provides an even more sensitive way to probe for strong coupling in the Higgs field interactions. The mechanism by which the W boson obtains mass from the Higgs field causes the W to absorb a degree of freedom from the Higgs field, which becomes the longitudinal polarization state of the W. The process of W boson pair production can be studied in great detail at an e^+e^- collider, with enough observables to make a complete polarization analysis of the final states. This allows one to determine the couplings of individual polarization states of the W and thus to measure properties of the absorbed Higgs field state.

2.4. *Z^0's and Other 4-fermion Interactions*

Another type of new dynamics that might be associated with new physics at the energy scale of hundreds of GeV is the appearance of new resonances that couple to fermion-antifermion pairs. Models of new physics with extra Z bosons, with composite quarks and leptons, and with additional space dimensions all contain such resonances, with properties that, in each case, are characteristic of the broader conceptual scheme. We have noted above that there are some models that violate the general connection between small corrections to precision electroweak predictions and the presence of a light Higgs boson. These models typically work by modifying the couplings of the Z to the Standard Model fermions, compensating these changes by the effects of additional resonances. If, despite expectations, a light Higgs boson does not appear at the ILC, it will be very important to look for these modifications of fermion pair couplings.

As is explained in Chapter 7 of this book, pair production of fermions from e^+e^- can be measured to fractions of a percent, giving a striking sensitivity to this new physics. Because it is possible to determine the pair-production cross sections for all species of quarks and leptons, identifying specifically the heavy quarks c and b and the two separate polarization states of the τ, the pattern of deviations can be matched to specific models of new resonances. Because the couplings of new Z bosons depend on electron polarization, the use of polarized beams doubles the number of independent and useful observables. The process of annihilation through a virtual photon or Z has a characteristic angular distribution. Observation of a more complex angular dependence would signal the influence of higher-spin resonances. For example, models with extra dimensions often have large fermion-antifermion couplings to spin 2 resonances associated with the gravitational field; the presence of such resonances would be seen unambiguously.

In this section, then, we have illustrated many different ways in which the precision and the level of detail with which reactions can be studied in e^+e^- annihilation can give information about new physics.

3. Relation of the ILC to the LHC

As we have pointed out above, the new physics responsible for the asymmetric electroweak vacuum should be observed at the CERN Large Hadron Collider. Once it has become clear that we have reached the energy of new fundamental interactions, it will become compelling to attack the mysteries of these new interactions with a variety of experimental techniques. From our viewpoint today, it is somewhat more difficult to see what essential information the ILC will bring to the picture that will come from the LHC data. However, we only need to look again at the physics menu presented in the previous section to see that the ILC will have a important role. We will review this briefly here. A very detailed examination of the relation of the ILC and the LHC has been presented in [10].

As a facility for discovering new physics, the LHC has great strengths. The energy of the LHC allows the discovery of color triplet and color octet particles up to masses of about 2 TeV. The dominant backgrounds to processes with large momentum transfer are QCD 2- and 3-jet reactions. These reactions are well understood theoretically and should be well modeled by simulation programs tuned to data from the Fermilab Tevatron. New physics reactions typically involve multi-jet signatures together with leptons or missing transverse momentum. These events should be apparent as an anomalous contribution above the levels predicted for the Standard Model. The LHC detectors have excellent capabilities to measure jets, leptons, and photons, allowing some well-designed measurements to give precision information.

However, the LHC has well-known weaknesses in analyzing these events. In a large momentum transfer reaction at the LHC, most of the energy and momentum goes forward in the direction of one beam or the other and is not measured. If the event contains one missing neutral particle, it is still possible to do some kinematic reconstruction; if two neutral particles are produced, one can only measure invariant masses and decay patterns within some model of the underlying physics. Many decay modes of new particles, especially decays to quarks, cannot be recognized above strong interaction backgrounds. So the picture that we will obtain from the LHC is a partial view in several respects.

With this in mind, we can look back at the physics topics discussed in the previous section. Consider first the Higgs boson. A light Higgs boson should be discovered at the LHC in its rare decay mode $h^0 \to \gamma\gamma$. The LHC should also be able to detect the $\tau^+\tau^-$ and Z^0Z^{0*} decays of the Higgs boson in the reaction of W^+W^- fusion tagged by forward quark jets. The decay to $b\bar{b}$ is extremely difficult to observe. Because of the QCD uncertainties in the Higgs production cross sections, the true precision measurements involve only ratios of branching ratios among these modes. At best, we could learn that the Higgs boson agrees with the predictions of the minimal model to 10–15% accuracy in one or two such ratios, a result that could well hide discrepancies coming from the correct model of the electroweak asymmetry.

If supersymmetry is present at the few hundred GeV mass scale, the LHC should see many indications of the production of supersymmetric particles. The LHC experiments should observe a level of multijet plus missing energy events at more than 10 times the Standard Model level in events with TeV momentum transfer, and significant anomalies in multileptons, Z plus jet production, and other modes. Studies of kinematic endpoints in dilepton or jet plus lepton distributions can give some mass splittings in the associated spectroscopy to percent-level precision. However, these measurements bring us only part way through the first level of the questions about supersymmetry spectroscopy that we have listed above. In particular, it is very difficult to measure the spins and electroweak quantum numbers of particles produced in hadronic collisions. Because some decay modes cannot be observed, it is difficult to determine coupling strengths in production or decay processes. So it is very difficult at the LHC to observe the characteristic features of supersymmetry.

The LHC will produce enormous samples of top quarks and W and Z bosons. However, it is very difficult at the LHC to determine the specific couplings of the top quark, especially the top quark coupling to the Z^0, which is most likely to be modified by coupling to the symmetry-breaking sector. Some anomalies in W couplings might be observed at the LHC in

$q\bar{q}$ annihilation to $W\gamma$, but the detailed polarization analysis that we have described in the previous section requires complete knowledge of the event kinematics, a property of e^+e^- reactions that is not available at the LHC.

Because of the high energy of the LHC, it will be possible to detect narrow resonances that couple to $q\bar{q}$ and decay to $\ell^+\ell^-$. However, for resonances at 2 TeV and above, we will likely learn from the LHC only the mass and the $\ell^+\ell^-$ production cross section. There are no observables that demonstrate the species- and spin-dependence of the couplings that we would expect to obtain from measurements in e^+e^- annihilation. In this case, there is an obvious synergy of the LHC and ILC measurements; knowing the mass of the resonance, we can convert deviations observed in e^+e^- to absolute coupling strengths for the resonance.

A pattern is evident here, one especially apparent in the case of supersymmetry but present in all of the other examples as well. Because the LHC gives only partial information about new physics, the LHC data must be interpreted in the context of a model. If we knew the correct model to use, we could extract precision information. But the true nature of the physics of the electroweak asymmetry is a mystery. At best, the LHC will allow us to propose models of this physics. In the LHC era, these will stand as hypotheses that will need to be checked and differentiated by further experiment. The ILC, with its ability to look globally at new physics events, will be just the tool that we will need to test and validate or exclude these hypotheses.

The accelerator signatures of cosmic dark matter provide an especially clear illustration of this pattern. As we have noted, the observed density of non-baryonic matter in the universe leads to a estimate for the mass of the dark matter particle in the range of 100 GeV. If this idea is correct, it is likely that most new physics reactions at the LHC will produce the dark matter particle. This particle must be stable, so it should carry a conserved quantum number. Then LHC reactions will actually produce these particles in pairs. These reactions will have missing transverse momentum and multijets. We have already noted that this is an ideal signature for the LHC.

Models for the dark matter particle include ones based on supersymmetry, in which the particle could be the supersymmetric partner of the photon, the Higgs boson, the right-handed neutrino, or the graviton, and models based on extra dimensions, in which this particle could be an excited state of the photon, the neutrino, or the graviton. The various possibilities are distinguished by the spin of the dark matter particle and its detailed

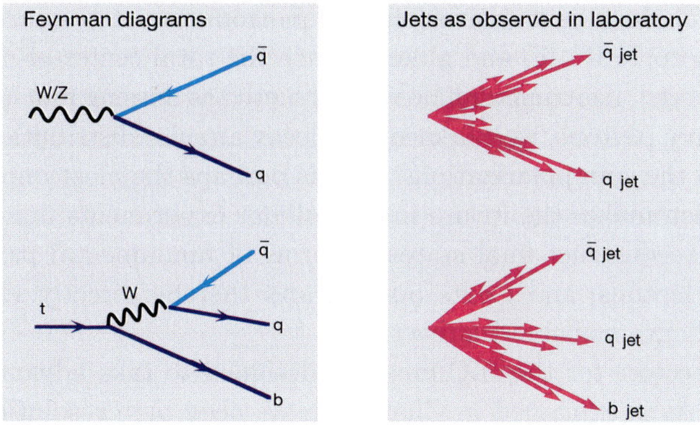

Fig. 1. Schematics illustrating the correspondence between underlying decay Feynman diagrams and resultant jets as observed in a laboratory for heavy 'partons' like W/Z and t.

scheme of couplings. The LHC will tell us that the dark matter particle is being produced, but it will not let us learn what particle this is. Many hypotheses will be consistent with the LHC data. To sort out which is the correct one, we will need the precision and clarity of e^+e^- experiments.

4. Experimentation at the ILC

The future linear collider experiments will inherit all virtues of the e^+e^- colliders in the past, including well-defined initial states consisting of structureless particles, an electron and a positron, with a controlled center of mass energy, and final states with their production cross sections essentially determined by spins and electroweak quantum numbers of the particles involved in the reactions. As stressed already, this implies that once a production threshold is crossed, one can obtain, without any special triggers, a clean sample of new particles with a signal to background ratio of $O(1)$ before any selection cuts. This allows detailed studies of various decay modes.

These advantages of e^+e^- colliders will be further enhanced by new features of the future linear collider that are even more powerful. First of all, jets become jettier and calorimetric resolutions improve with energy. So at the linear collider energies we will be able to identify heavy 'partons' such as W and Z bosons and t quarks by reconstructing their hadronic products as jets, as shown in Fig. 1. Full reconstruction of final states in this way is possible only in the clean environment of an e^+e^- collider. Jet reconstruction will allow us to measure the four-momenta of all final-state partons, including light quarks, charm, bottom, and top quarks, charged leptons, photons, W, Z, and gluons. Since the total center-of-mass energy can be observed, neutrinos can be reconstructed as missing four-momentum. For the heavy partons, we can measure decay angular distributions that are sensitive to the spin polarizations. This is perhaps the most important new feature which makes the future linear collider experiments unique: we will be able to reconstruct final states in terms of fundamental particles such as quarks, leptons, and gauge bosons, and thereby directly visualize the Feynman diagrams behind a reaction.

The detectors for the ILC must be designed to take advantage of this capability. As is explained in Chapter 1, we need high resolution tracking and calorimetry for optimized energy flow measurements that ensure the required jet-invariant-mass resolution for the heavy parton identification. We need high resolution vertexing for heavy-flavor tagging, and hermeticity for neutrino detection as missing momenta. The tiny beam spot size of the linear collider at the interaction point can be used to great advantage to precisely define the position of primary vertices of any reactions of interest in the plane perpendicular to the beam axis. This will enable us to identify most of the secondary and even some of the tertiary vertices of events with heavy flavors.

Another remarkable advantage of the future linear collider experiments is the availability of a highly polarized electron beam, possibly supplemented by a polarized positron beam. As an example, consider the reaction e^+e^- to W^+W^-, which is interesting both as a signal of strongly interacting Higgs boson physics and as a major background for many new particle searches. The largest contribution comes from transversely-polarized W bosons states. In the electroweak symmetry limit where the gauge boson masses are negligible, we can analyze this process in terms of weak eigenstates; only exchanged particles with nonzero $SU(2)$ quantum numbers contribute. This process then involves only two diagrams: an s-channel diagram with a triple gauge boson coupling and a t-channel diagram with a neutrino exchange, as shown in Fig. 2. The right-handed electron has no coupling to weak $SU(2)$, and so both diagrams disappear for a right-handed polarized electron beam. Because of this, we can use the electron beam polarization to suppress this process to high degree when it is a background, or to enhance it relative to other processes when it is a signal. Thus, at the linear collider, Feynman diagrams are not only visible, but also selectable. In this way, the linear collider experiments measure S-matrix elements for processes in which four-momenta and spins of external partons are specified

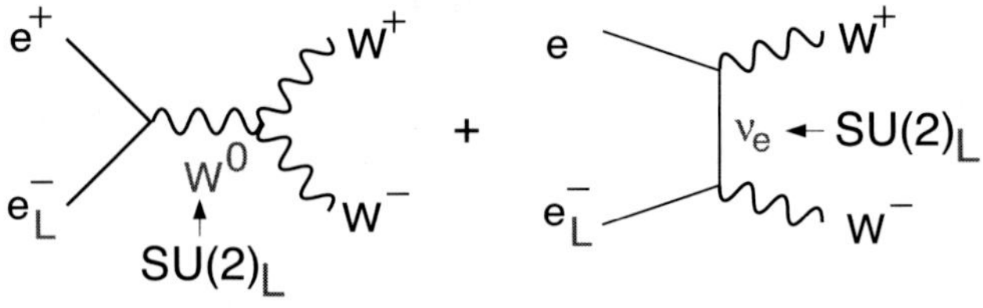

Fig. 2. Feynman diagrams for the $e^+e^- \to W^+W^-$ process in the electroweak symmetry limit.

experimentally. This comes close to the dream measurements of elementary scattering processes.

Together with these opportunities, though, the linear collider environment brings some new difficulties. As explained in Chapter 1, since the collider operates in a single pass mode, one has to squeeze the transverse size of its beam bunches to nanometer size to achieve the required luminosity. Such high density beam bunches produce an electromagnetic field which is strong enough to significantly bend the particles in the opposing bunches, generating bremsstrahlung photons. Because of this phenomenon, called 'beamstrahlung', electrons or positrons in the beam bunches lose part of their energies before collision. When plotted as a function of the effective center of mass energy, the differential luminosity distribution shows a long tail towards the low energy region due to this effect. The distribution still peaks sharply at the nominal center of mass energy, corresponding to collisions essentially without beamstrahlung. The existence of this sharp peak implies that for most physics programs we can benefit from the well-defined initial state energy that has been the traditional benefit of e^+e^- colliders. For precision work, it is necessary to measure and account for the distribution in beam energies; methods for measuring this effect are described in Chapter 1. Some processes acquire new backgrounds from the beamstrahlung tail. Also, for some precision measurements, the natural beam energy spread from the main linacs can be a source of uncertainty. Top pair production at threshold is a example of a process for which this effect is important (see Chapter 3).

The intense beam-beam interactions also induce other new kinds of backgrounds which are potentially serious problems for the linear collider experiments. The photons from direct radiation and beamstrahlung can create low-energy e^+e^- pairs, other lepton pairs, and also low-energy hadron events ($\pi^+\pi^-$, $\pi\rho$, etc.). It should be possible to overcome these problems by proper design of the detector and the beam delivery system, with masking at small angles and time-stamping of particles from different colliding bunches. This issue is discussed in Chapter 1.

5. Benchmark Processes for Evaluating the ILC Detectors

Any detector design should take into account the new features of e^+e^- annihilation at linear colliders described above. The detector models discussed in Chapter 1 are indeed based on the ambitious concept of reconstructing final states in terms of known fundamental particles as viewing Feynman diagrams. Any of these general-purpose 4π detector models is equipped with a state-of-the-art vertex detector. In all cases, the central tracking detector and the calorimeter is placed in a solenoidal magnetic field of up to several Tesla to achieve good resolution and hermeticity. The designs also require that final focus quadrupole magnets and a background mask system be integrated into the detector system. The final focus magnets and the mask system should be considered as part of the detector system. It is noteworthy that this situation has triggered active interactions between machine designers and particle physicists.

The detector system should then meet the following performance goals:

- Efficient and high purity $b/c/(u, d, s, g)$ tagging, for studying top, Higgs, and other new particles decaying into them.
- Recoil mass resolution limited by natural beam energy spread and not by trackers, to perform precision measurement of the Higgs mass and to confirm the narrowness of the Higgs width.
- 2-jet invariant mass resolution comparable with the natural widths of W and Z, to efficiently separate these particles in hadronic final states.
- Hermeticity, to indirectly detect invisible particles such as neutrinos and dark matter.
- A well designed masking system and time stamping capability for beam-beam backgrounds.

The first of these requirements sets a performance goal for a vertex detector, while the second imposes the most stringent constraint on the tracking system. It should be emphasized that calorimetry at this level will be based on the energy-flow strategy of separately measuring the charged and neutral components of a jet. Thus, the third point requires not only a good calorimeter, but also a good track-cluster matching capability. The requirement for time stamping capability has been somewhat relaxed by the choice of the cold RF technology for the ILC.

In any case, the detector parameters should be optimized so as to make maximal use of the potential of the collider. We will now discuss a set of reference reactions that we feel would be useful for evaluating the performance of linear collider detectors against the requires that we have just reviewed. Other lists of reference reactions are also available; for example, the list of the World-Wide Study, which more elaborately covers the physics menu of the linear collider, can be found at [11].

The reactions we will discuss are:

- $e^+e^- \rightarrow Zh$
- $e^+e^- \rightarrow Zhh$
- $e^+e^- \rightarrow t\bar{t}h$
- $e^+e^- \rightarrow t\bar{t}$
- $e^+e^- \rightarrow \tilde{e}_R^+\tilde{e}_R^-,\ \tilde{\mu}_R^+\tilde{\mu}_R^-$
- $e^+e^- \rightarrow \tilde{\chi}_1^+\tilde{\chi}_1^-$
- $e^+e^- \rightarrow \tilde{\tau}_1^+\tilde{\tau}_1^+$
- $e^+e^- \rightarrow \tilde{\chi}_1^0\tilde{\chi}_1^0(GMSB)$

For each reaction, we will point out the important measurements that should be considered in designing and comparing linear collider detectors.

5.1. $e^+e^- \rightarrow Zh$

As is explained in Chapter 2, the reaction $e^+e^- \rightarrow Z^0h^0$ is the most important channel to study properties of the light Higgs boson expected from the electroweak data. Depending on the decay modes of the Z and the h bosons in the final state, there will be a variety of final-state configurations.

Let us consider first the case in which the Z boson decays into $\nu\bar{\nu}$, which has a branching fraction of about 20%. The signature of this channel is the final state with an invariant mass consistent with that of the Higgs boson and the missing mass consistent with that of the Z boson. From the experimental point of view, this channel will provide us with a very clean way to measure the Higgs boson mass with the jet-invariant-mass method, since the visible final-state particles must all come from the Higgs boson and hence there will be no jet clustering ambiguity. This process is hence one of the best to benchmark the energy flow measurement performance of the detector system.

When the Z boson decays into a lepton (e/μ) pair, which comprises a branching fraction of about 7%, we can detect the Higgs boson production even if it decays invisibly, by measuring the recoil mass against the lepton pair that has an invariant mass consistent with that of the Z boson. This is a great advantage of the e^+e^- collider, allowing an absolute measurement of the total cross section for Higgs production through $e^+e^- \to Zh$ and providing absolute normalization for various branching fraction measurements. The recoil mass resolution depends on the beam energy spread and the momentum resolution for the lepton pair from the Z decay. The requirement that the recoil mass resolution be limited not by the detector resolution but by the beam energy spread will put the most stringent constraint on the desirable momentum resolution of the central tracking device, $\sigma_{p_t}/p_t \lesssim 10^{-4} p_t[\text{GeV}]$, which is typically one order-of-magnitude better than those at the LEP and SLC experiments. With such an excellent momentum resolution, one can hope to detect a rare decay mode such as $h^0 \to \mu^+\mu^-$.

Finally, when the Z boson decays hadronically, which occurs with a branching fraction of about 70%, the final state will be occupied by multiple jets. The advantage of the e^+e^- collider is that we can exploit such dominant hadronic decay modes by the jet invariant mass method: when the Higgs boson decays into $b\bar{b}$, the final state will consist of four jets, two of them coming from the Z decay and hence having an invariant mass consistent with m_Z and the other two coming from the Higgs boson decay and hence having an invariant mass consistent with m_h. This is in contrast with the case of hadron colliders, in which one has to often give up using the dominant decay mode because of the huge QCD background. Nevertheless, in order to take full advantage of this, we need a state-of-the-art energy flow measurement system and well designed jet clustering algorithm, combining all the available information from various detector components including the vertex detector, and particle ID devices, if any, not to mention the central tracker and the calorimeter.

Once the Higgs boson production is detected and classified according to the Z boson decay modes as described above, we can then study in detail properties of the Higgs boson in any of the above three modes. As already stated in the previous section, we can measure basic quantum numbers of the Higgs boson such as its mass, width, and spin to a good accuracy, a necessary step to confirm that what we will see is indeed a scalar particle with a narrow width. For this purpose, we should be able to fully reconstruct the final state and carry out angular analysis for the production and decay.

We can then measure various branching fractions, which will be crucial to understand the mass generation mechanism. In particular we need to be able to distinguish the expected two-jet decays: $h^0 \to b\bar{b}$, $c\bar{c}$, gg, which will require a precision vertex detector that can resolve the secondary and the tertiary vertices if any from the primary vertex with high efficiency and purity. Notice that such measurements will greatly benefit from the linear collider's nanometer level beam spot that will precisely define the primary vertex position and the small beam pipe that will allow us to place the first layer of the vertex detector as close as 1.5 cm to the beam axis.

5.2. $e^+e^- \to Zhh$

At higher energies, we will have double Higgs production processes, such as $e^+e^- \to Z^0h^0h^0$. As emphasized in Chapter 2, this process has fundamental importance, since it provides us with a handle to measure the Higgs self-coupling and hence to touch upon the force that makes the Higgs field condense in the vacuum. The production cross section of this process is tiny, of the order of 1 fb, and its measurement is an experimental challenge. If the $h^0 \to b\bar{b}$ decay dominates, the final state will contain four b-jets that can be a clean signature of the Zhh production, again provided that we will have a precision vertex detector that can identify b-jets with high efficiency and purity. Notice also that, since the production cross section is small, we need to rely on the dominant decay mode of the final-state Z boson. Unambiguous reconstruction of such a 6-jet final state will be an experimental challenge and will benchmark the energy flow measurements and b-tagging.

5.3. $e^+e^- \to t\bar{t}h$

The $e^+e^- \to t\bar{t}h^0$ process is another good example for benchmarking the energy flow measurements and b-tagging. The final state involves eight jets, of which four are b-flavored if the Higgs boson decays into $b\bar{b}$. If the Higgs decays into WW^* the final state will be even more complicated and will contain ten jets. Reconstruction of such a complicated final state will really be an experimental challenge. This process is, however, of fundamental importance, since it will provide a unique opportunity to measure the top Yukawa coupling, other than the possible indirect measurement at the $t\bar{t}$ threshold described in Chapter 3.

As we have emphasized earlier, the branching fraction measurements for the Higgs boson will give some of the most crucial data to understand

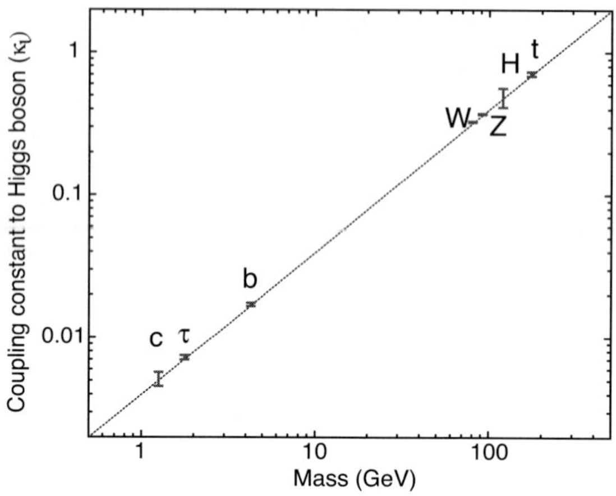

Fig. 3. The relation between the mass of a particle and its coupling to the Higgs boson in the case of the Standard Model with $m_h = 120$ GeV.

the mass generation mechanism in the Standard Model and most of its extensions. The three reactions we have just discussed provide a set of expectations of the measurements of the various Higgs boson couplings, which we show in Fig. 3 for the case of a 120 GeV Higgs boson [12]. These measurements indeed provide a stringent test of the proportionality of the Higgs boson couplings to the particle mass. If the Higgs sector is non-minimal, we expect to see a deviation from the simple linear expectation, with the pattern of deviations indicating the nature of the Higgs spectrum.

5.4. $e^+e^- \to t\bar{t}$

As detailed in Chapter 3, the $e^+e^- \to t\bar{t}$ process has many new features due to the heaviness of the top quark. For the top quark, the direct decay to bW is kinematically allowed and proceeds, in the Standard Model, with a branching fraction of essentially 100%. The decay occurs before the top quark enters the non-perturbative QCD regime, thereby allowing the calculation of threshold cross sections using perturbative QCD. We will be able to extract various information from the threshold shape such as the mass and width of the top quark, the strong coupling constant, and even the top Yukawa coupling if the Higgs boson is not too heavy. Such a measurement requires, however, good control of the beam energy spread and beamstrahlung together with their accurate monitoring (see Chapter 1), and hence comprises one of the best cases to benchmark the adequacy of the luminosity spectrum (a machine property) and its monitoring system (a part of the detector).

Since the toponium system decays not via $t\bar{t}$ annihilation but via independent single decays, the $t\bar{t}$ threshold will provide us for the first time with the opportunity to measure the momentum wave function of a quarkonium system through the momentum measurement of their decay products. This means that the process is also very useful to benchmark the energy flow measurement.

It should also be noted that the $e^+e^- \to t\bar{t}$ process also sets a typical benchmark to exercise the full parton-level reconstruction of the final state, making full use of the information on the energy flow, the secondary and/or tertiary vertices.

The full reconstruction is expected to proceed as follows: Given a good vertex detector, we can identify two b-jets as b and $\bar{b}$ quarks from the t and $\bar{t}$ decays, respectively. The remaining partons in the final state should then be coming from the two Ws from the t and $\bar{t}$ decays. If one of the Ws decays leptonically and the other hadronically, the remaining two jets can be uniquely assigned to one of these two W bosons. By combining this W and one of the two b-jets previously identified with the vertex detector, we can reconstruct a t or $\bar{t}$ as a 3-jet system that has an invariant mass consistent with the top mass. By identifying the charge of the isolated energetic lepton with that of the leptonically decayed W, we can tell whether the reconstructed 3-jet system is actually a t or $\bar{t}$. It should be noted that if one of the two jets from the hadronically decayed W is c-tagged by the vertex detector, we can even tell which of the two jets is up-type and which is down-type. The leptonically decayed W can also be reconstructed by combining the isolated lepton with the missing momentum. It has been shown that such a full parton-level reconstruction is indeed possible with good precision, even in the presence of initial state radiation and beamstrahlung, by means of properly designed kinematic fitting [13]. Kinematic fitting can also be done in the threshold region, though here one should take care that, because of QCD binding, the t and $\bar{t}$ cannot simultaneously be on mass-shell.

5.5. $e^+e^- \to \tilde{e}_R^+\tilde{e}_R^-$, $\tilde{\mu}_R^+\tilde{\mu}_R^-$

Let us now turn our attention to physics beyond the Standard Model. We start with the simplest reactions involving supersymmetry, the pair production of the right-handed selectron ($\tilde{e}_R$) or smuon ($\tilde{\mu}_R$), followed by their two body decays to a lepton plus a Lightest SUSY Particle (LSP), $\tilde{\chi}_1^0$. As described in Chapter 4, these processes have a very simple final state, an acoplanar lepton pair. They provide us with a model-independent way to measure the slepton and the LSP masses through the end-point measurement. The end-point measurement is not only one of the most important experimental techniques but also is a typical place where detector resolution might play some important role. These processes also provide one of the best examples to demonstrate power of the beam polarization.

Since the power of the beam polarization has already been detailed in Chapter 4, let us concentrate on the experimental aspect of the end-point measurement in what follows. It is first of all noteworthy that initial-state radiation (ISR) and beamstrahlung will not change the end-point locations, since the reduction of the effective beam energy, if sizable, makes the higher end-point (E_+) lower and the lower end-point (E_-) higher. Fig. 4 clearly demonstrates this fact.

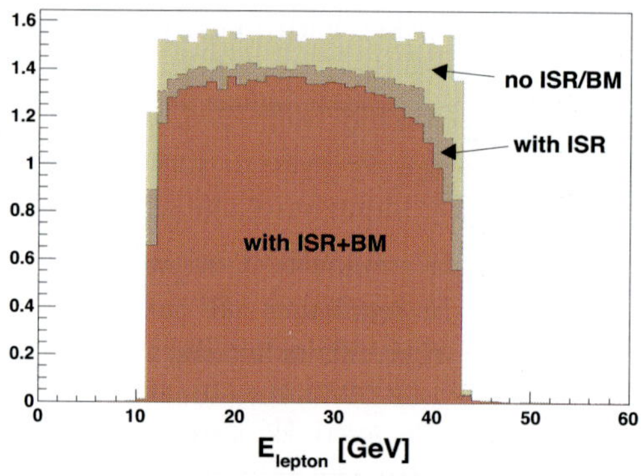

Fig. 4. A typical muon energy distributions for $e^+e^- \to \tilde{\mu}_R^+\tilde{\mu}_R^-$ followed by $\tilde{\mu}_R^\pm \to \mu^\pm\tilde{\chi}_1^0$ (dotted) with neither ISR nor beam effects (BM), (dot-dashed) with ISR only, and (solid) with both ISR and beam effects: $(m_{\tilde{\mu}}, m_{LSP}) = (142, 118)$GeV at $\sqrt{s} = 350$GeV.

Denoting the lower and higher end points by E_- and E_+, we can write the measurement error (σ_M) on the parent slepton mass (M) in the following form:

$$\sigma_M = \left(\frac{M}{2}\right)\left(\frac{E_+ - E_-}{E_+ + E_-}\right)\sqrt{\left(\frac{\sigma_{E_+}}{E_+}\right)^2 + \left(\frac{\sigma_{E_-}}{E_-}\right)^2}.$$

The lepton energy will be measured by the central tracker with an energy (momentum) resolution proportional to the energy; parametrize this as $\sigma_E/E = aE$. Then the above equation reduces to

$$\sigma_M = \left(\frac{M}{2}\right)\left(\frac{E_+ - E_-}{E_+ + E_-}\right) a\sqrt{E_+^2 + E_-^2}.$$

If we substitute a typical momentum resolution, $a \simeq 5 \times 10^{-5}\ \mathrm{GeV}^{-1}$, assumed for the linear collider central tracker, we obtain σ_M of about 0.1 GeV for the example shown in Fig. 4. Since the expected statistical error for the parent mass from the end point measurement is $\sigma_M \simeq 0.7$ GeV for 20 fb^{-1}, the detector resolution will become comparable with the statistical error only when the integrated luminosity reaches 1 ab^{-1}, provided that the statistical error scales as $1/\sqrt{N}$. This means that the tracker resolution described in Section 5.1 is good enough for the measurement[a].

5.6. $e^+e^- \to \tilde{\chi}_1^+\tilde{\chi}_1^-$

The end-point measurement is also applicable to the chargino mass determination for $e^+e^- \to \tilde{\chi}_1^+\tilde{\chi}_1^-$ followed by $\tilde{\chi}_1^\pm \to W^\pm\tilde{\chi}_1^0$. By measuring the end points of the jet pairs corresponding to the final-state Ws, we can in principle determine the masses of the chargino and the neutralino (the LSP) as with the sleptons discussed in the previous subsection. In this case we need to retain the W mass in the end-point formula. The resolution formula then becomes

$$\sigma_M^2 = \left(\frac{\partial M}{\partial E_+}\right)^2 \sigma_{E_+}^2 + \left(\frac{\partial M}{\partial E_-}\right)^2 \sigma_{E_-}^2.$$

with

$$\left(\frac{\partial M}{\partial E_\pm}\right) = \left(\frac{M}{E_+ + E_-}\right)\left[\left(\frac{E}{M}\right)^2 \frac{E_\mp + E_\pm\sqrt{\frac{E_\mp^2 - m_W^2}{E_\pm^2 - m_W^2}}}{E_+ + E_-} - 1\right].$$

[a]Note also that if we reach this level of precision, we may start worrying about the effect of the finite width of the parent particle. As a matter of fact, the width of the right-handed smuon in our example is about 0.7 GeV, which is indeed comparable with the ultimately expected precision.

Notice that, since we will rely on the energy flow measurement for jets from the W decays, we need to use a calorimetric energy resolution, parametrized as

$$\sigma_{E_\pm}/E_\pm = b/\sqrt{E_\pm}$$

here. The formula gives $\sigma_M \simeq 8$ GeV for $(m_{\tilde{\chi}_1^\pm}, m_{\tilde{\chi}_1^0}) = (220, 118)$GeV at $\sqrt{s} = 500$GeV, which is already comparable with the expected statistical error for an integrated luminosity of 50 fb^{-1}. This means that we need to understand the resolution function for the energy flow measurements to the accuracy that stands up to the expected statistical error, a challenge for experimenters.

There is yet another experimental challenge. Since the final state contains two LSPs, which are invisible, the use of leptonic decay of the final-state Ws means more than two missing momenta, which make the reconstruction of the production angle impossible. We have to rely on the 4-jet mode where both of the final-state Ws decay hadronically, in order to measure the differential cross section for the $e^+e^- \to \tilde{\chi}_1^+\tilde{\chi}_1^-$ process. This requires the determination of the charge of at least one of the two jet pairs reconstructed as Ws so as to tell which is W^- and which is W^+. One way to do this on an event-by-event basis is to identify a c-jet from a W and then decide whether the charmed hadron in the c-jet contains a c or $\bar{c}$ quark by measuring the vertex charge in combination with information from particle identification devices for K/π separation. This is one of the most important cases to benchmark the charm ID capability of the vertex detector and the necessity of particle ID devices.

5.7. $e^+e^- \to \tilde{\tau}_1^+\tilde{\tau}_1^+$

Let us now consider the hermeticity issue of the detector system. As stressed at the beginning of this section, the hermeticity is a must to indirectly detect invisible particle such as neutrinos or the LSPs. This is particularly important for the SUSY process: $e^+e^- \to \tilde{\tau}_1^+\tilde{\tau}_1^+$, where the missing P_T signal will be weakened by the final-state τ decays. In order to detect the missing P_T signal, it is essential to be able to veto low angle e^+/e^-s to suppress two-photon backgrounds. Fig. 5 shows the P_T distributions of the signal ($e^+e^- \to \tilde{\tau}_1^+\tilde{\tau}_1^-$) and the background from the two-photon process ($e^+e^- \to e^+e^-\tau^+\tau^-$), for $\theta_{acop} > 30°$ and $|\cos\theta_{jet}| < 0.8$. The figure clearly demonstrates the importance of the smallest possible minimum veto angle, without which the significant part of the signal would be swamped

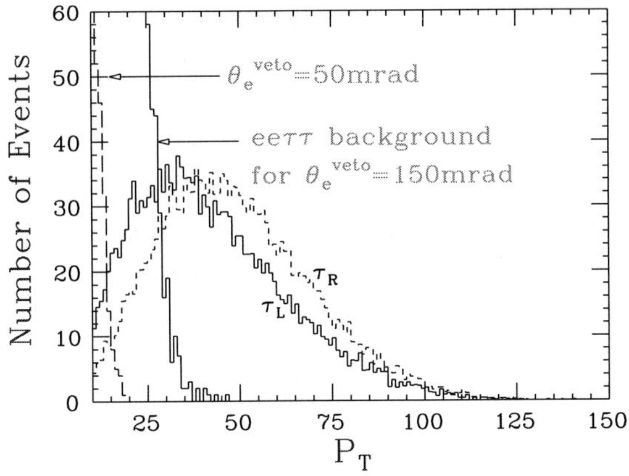

Fig. 5. P_T distributions for the signal and the background $e^+e^-\tau^+\tau^-$ events at $\sqrt{s} = 500$ GeV after the cuts: $|\cos\theta_{jet}| < 0.8$ and $\theta_{acop} > 30°$, for different minimum veto angles. 10^4 signal events were generated with $m_{\tilde{\tau}} = 150$ GeV and $m_{\tilde{\chi}_1^0} = 100$ GeV, corresponding to $\int Ldt = 100\ fb^{-1}$.

by the huge two-photon background, as is seen for the case of $\theta_e^{veto} = 150$ mrad.

5.8. $e^+e^- \to \tilde{\chi}_1^0\tilde{\chi}_1^0$ *(GMSB)*

Finally, we consider the Gauge Mediated SUSY Breaking (GMSB) scenario discussed in Chapter 4. In the case of GMSB, the LSP is the gravitino ($\tilde{G}$) and the LSP ($\tilde{\chi}_1^0$) assumed in the previous subsections will be a candidate for the Next to Lightest SUSY Particle (NLSP) and will decay into $\gamma + \tilde{G}$. Since the neutralino then becomes visible if it decays within the detector, we will be able to directly detect the $e^+e^- \to \tilde{\chi}_1^0\tilde{\chi}_1^0$ process in this scenario. Since the NLSP may have a long life time, it would decay after flying over a macroscopic distance. The detector should hence be able to detect such visibly displaced decay vertices of the NLSP. This requires the electromagnetic calorimeter to have fine granularity to determine the electromagnetic shower axis with good enough accuracy to point back to these displaced decay vertices. This is an experimental challenge and hence one of the most stringent requirements to benchmark the calorimeter. In the GMSB scenario, the stau makes another candidate for the NLSP. If its life time is long enough, we would see a pair of heavily ionizing slow moving particles from $e + e^- \to \tilde{\tau}^+\tilde{\tau}^-$. In such a case, dE/dx measurement in the central tracker would become important together with a proper data acquisition system to record slowly moving stable particles.

6. Choice of the LC Energy and Luminosity

The physics program of the ILC that we have sketched above has implications for the choice of energy and luminosity for the ILC. The program of the ILC contains three components—the study of the Standard Model particles W and t, the precision study of the Higgs boson, and the study of new particles associated with electroweak symmetry breaking. All three components recommend an the initial ILC program at 500 GeV in the center of mass with an integrated luminosity of about 500 fb^{-1}. The ILC is planned to be expandable in energy to about 1 TeV, with an integrated luminosity of 1000 fb^{-1} at its second stage. Beyond the general benefits of higher energy for exploration, we will see that some specific reactions in the mainstream program benefit from this extension.

Consider first the study of t and W. Here the masses and Standard Model cross sections are known. For the top quark, the pair-production cross section is about 500 fb, depending on the electron beam polarization. A program of 100 fb^{-1} at the pair-production threshold at 350 GeV is needed for the precision determination of the top quark mass. The determination of top quark form factors to percent accuracy requires a luminosity sample several times larger; this might be taken either at the threshold or in the continuum. For W bosons, the LEP experiments have already reached the pair-production threshold at 160 GeV. However, there are advantages in going to higher energy, since the effects of anomalous couplings indicative of W mixing with the Higgs sector are multiplied by factors s/m_W^2. A luminosity sample of 300 fb^{-1} at 500 GeV determines these couplings to better than part-per-mil accuracy [6].

Next, we discuss experiments on the Higgs boson. The first requirement for such experiments is that the reaction $e^+e^- \to Z^0h^0$ be kinematically accessible. Within the Standard Model, the mass of the Higgs boson is severely constrained by the agreement of precision electroweak experiments with the predictions of the Standard Model. A heavy Higgs boson gives a quantum correction proportional to $\log(m_h^2/m_W^2)$ that is not observed in the data. The absence of this correction restricts the mass of the Higgs boson to be less than 260 GeV as a 95% confidence limit [14]. Then this reaction can be studied at the 500 GeV ILC. The cross section for $e^+e^- \to Z^0h^0$ is 50-100 fb at high energy, higher near threshold for a light Higgs boson. Then a luminosity sample of 500 fb^{-1} offers a sample of 25-50 $\times 10^3$ Higgs bosons. This is adequate to determine the branching ratios even for rare modes of Higgs boson decay to the percent level of accuracy.

The experiments on the Higgs coupling to top and on the Higgs self-coupling that we have described in Section 3.1 and 5.1 require higher energy to achieve a small statistical uncertainty. The Higgs self-coupling measurement requires 1000 fb^{-1} at 1 TeV to achieve a 10% uncertainty. If the minimal Higgs theory turned out to be correct, this experiment, in the second stage of the ILC program, would be the definitive test.

Now we turn to physics beyond the Standard Model. Since we do not know the exact nature of the new physics at the next energy scale, it is difficult to make precise statements about the requirements for the collider energy. However, any particle with nonzero electroweak quantum numbers is necessarily pair-produced in e^+e^- annihilation with a cross section of the order of the QED annihilation cross section of 300 fb at 500 GeV. An event sample of 500 fb^{-1} will give of the order of 10^5 pairs of each type of new particle for detailed study. Thus, if new particles should appear in the energy range of the ILC, the luminosity already required for Higgs boson studies will be quite sufficient for precision studies of the new physics.

We can make more definite statements about the energy required for finding new physics within the context of specific models. The best-studied example is supersymmetry. Supersymmetry is attractive as a basis for models of electroweak symmetry breaking because it forces radiative corrections to the Higgs boson mass to cancel. That cancellation, however, is effective only if the supersymmetric particles are sufficiently light. Theoretical estimates for the masses of supersymmetric particles based on this idea, reviewed in [15], give as a consensus that the superpartner of the W boson should be lighter than 250 GeV. By this criterion, the lightest supersymmetric particles should appear at a 500 GeV e^+e^- collider, with the possibility of extension to 1 TeV giving a reasonable margin of safety. As we have discussed in Section 3, the ILC can make important contributions beyond the LHC even if it produces only the lightest supersymmetric particles, since only at the ILC can one make the definitive identifications of spin and quantum numbers needed to confirm the symmetry predictions of supersymmetry. If many new particles appear at 500 GeV, one might question the adequacy of the luminosity sample. However, in [16], run scenarios are worked out that cover Standard Model and Higgs physics as well as measurements of new particles and even precision mass determinations by threshold scans, within a luminosity budget of 500 fb^{-1}.

Another criterion for the ILC energy comes from models with a heavy Higgs boson. We have already noted that, within the Standard Model, there is an upper bound on the Higgs boson mass. In fact, the quantum correction that is responsible for this bound appears in all models that generalize the minimal Higgs theory. Thus, it is only possible for the Higgs boson to be heavier than 300 GeV consistently with the precision electroweak experiments if some effect of new physics provides a quantum correction that cancels the correction due to the heavy Higgs. Most models with new physics, for example, supersymmetric models, respect the above-mentioned bound of 260 GeV on the Higgs mass. However, a few scenarios are known in which this cancellation can occur by tuning of parameters. These scenarios are reviewed in [17]. Typically, the new physics required in these models introduces new signatures that can be studied at a 500 GeV e^+e^- collider.

The most robust way to raise the Higgs boson mass is to enlarge the weak interaction gauge group; in such models, new Z and W resonances at about 2 TeV compensate the precision electroweak effects. Similar effects of resonances in the 2 TeV region appear in models with extra dimensions. With an event sample of 500 fb^{-1}, a 500 GeV e^+e^- collider is already sensitive to such resonances up to masses of 4–6 TeV, from the fact that s-channel resonances perturb the precise electroweak predictions for $e^+e^- \to f\overline{f}$. The region of sensitivity is comparable to but somewhat greater than the range of the LHC. We have already noted in Section 3 that, if a resonance is seen at the LHC and its mass is determined, the detailed pattern of couplings of the resonance can be measured through its effects on $e^+e^- \to f\overline{f}$. Thus, even in the most pessimistic case in which the Higgs boson does not appear at the ILC energy of 500 GeV, experiments at this energy with an event sample of 500 fb^{-1} will necessarily provide important information about the new physics responsible for electroweak symmetry breaking and the next energy scale of particle interactions.

7. Conclusions

We have thus reviewed the motivation and the experimental program for the next-generation e^+e^- collider, the ILC. The experiments of the past decade in high-energy physics have confirmed the Standard Model of strong, weak, and electromagnetic interactions to a remarkable level. Yet, in doing so, they have focused attention on new questions that lie just beyond the domain of this model. Two of those questions—the origin of electroweak symmetry breaking and the origin of cosmic dark matter—signal new physics at the energy scale of order 100 GeV. It is very likely that any further progress in understanding the fundamental interactions will need to take this new physics into account. Thus, we need to go to the 100 GeV energy scale and study that physics directly. We will need experimental techniques for that study that will give us definitive information on the nature of these new forces.

The tools for the direct study of the energy scale of electroweak symmetry breaking will be those of the hadron collider LHC and the e^+e^- collider ILC. The LHC will give us our first glimpse of the new particles and interactions that come into play. We have emphasized, however, that the e^+e^- collider ILC adds to the LHC an exceptionally incisive set of tools. We have explained how these tools can be used to visualize individual Feynman diagrams that contribute to the underlying physics processes. With this capability, we can dissect the new couplings that appear and measure particle masses, quantum numbers, and spins in a model-independent way.

We do not know today where this knowledge will take us. It is possible that the new century will see a succession of new energy scales and interactions. It is equally possible that the discovery of supersymmetry or some other compelling paradigm will point us directly to a picture of unification that encompasses all of the interactions of Nature. For the moment, we know only that the knowledge of the next set of fundamental interactions is within our grasp. If we can build the colliders and detectors that we now envision—with the next e^+e^- linear collider at the center—we can take that step toward the understanding of the deep structure of the universe.

Acknowledgments

The work of MEP is supported by the US Department of Energy under grant DE–AC02–76SF00515, while that of KF is partily supported by Japan-Europe (UK) Research Cooperation Program and JSPS-CAS Scientific Cooperation Program under the Core University System.

References

1. M. Martinez, R. Miquel, L. Rolandi and R. Tenchini, Rev. Mod. Phys. **71**, 575 (1999).
2. P. B. Renton, Rept. Prog. Phys. **65**, 1271 (2002) [arXiv:hep-ph/0206231].
3. J. Erler and P. Langacker, in S. Eidelman *et al.* [Particle Data Group Collaboration], Phys. Lett. **B592**, 1 (2004).
4. O. Lahav and A. Liddle, in S. Eidelman *et al.* [Particle Data Group Collaboration], Phys. Lett. **B592**, 1 (2004).
5. J. A. Aguilar-Saavedra *et al.*, 'TESLA Technical Design Report, Part III: Physics at a Linear Collider', DESY-2001-011 (2001), arXiv:hep-ph/0106315.

6. T. Abe, *et al.*, 'Linear Collider Physics Resource Book for Snowmass 2001', in *Proc. of the APS/DPF/DPB Summer Study on the Future of Particle Physics*, N. Graf, ed., arXiv:hep-ex/0106055-58.
7. K. Abe, *et al.*, 'Particle Physics Experiments at JLC', KEK report 2001-11, arXiv:hep-ph/0109166.
8. Internatinal Committee on Future Accelerators, `http://www.fnal.gov/directorate/icfa/Summary_ICFA-ILCSCmtg_Beijing.html`.
9. S. P. Martin, in *Perspectives in Supersymmetry*, G. Kane, ed. (World Scientific, 1997), arXiv:hep-ph/9709356.
10. G. Weiglein, *et al.*, arXiv:hep-ph/0410364.
11. The WWS group, "Reference Reactions for the World-wide LC Study", `http://blueox.uoregon.edu/~lc/wwstudy/reference.html`.
12. "GLC Project", KEK Report 2003-7.
13. K.Ikematsu, *et al*, Eur. Phys. J. **C29**, 1 (2003) [arXiv:hep-ph/0302214].
14. LEP Electroweak Working Group, `http://lepewwg.web.cern.ch/LEPEWWG/`.
15. J. Bagger *et al.*, arXiv:hep-ex/0007022.
16. M. Battaglia, *et al.*, in *Proc. of the APS/DPF/DPB Summer Study on the Future of Particle Physics*, N. Graf, ed., arXiv:hep-ph/0201177.
17. M. E. Peskin and J. D. Wells, Phys. Rev. **D64**, 093003 (2001) [arXiv:hep-ph/0101342].cd .

Index